Lecture Notes on Data Engineering and Communications Technologies

Volume 55

Series Editor

Fatos Xhafa, Technical University of Catalonia, Barcelona, Spain

The aim of the book series is to present cutting edge engineering approaches to data technologies and communications. It will publish latest advances on the engineering task of building and deploying distributed, scalable and reliable data infrastructures and communication systems.

The series will have a prominent applied focus on data technologies and communications with aim to promote the bridging from fundamental research on data science and networking to data engineering and communications that lead to industry products, business knowledge and standardisation.

Indexed by SCOPUS, INSPEC.

All books published in the series are submitted for consideration in Web of Science.

More information about this series at http://www.springer.com/series/15362

P. Karuppusamy · Isidoros Perikos ·
Fuqian Shi · Tu N. Nguyen
Editors

Sustainable Communication Networks and Application

Proceedings of ICSCN 2020

Springer

Editors
P. Karuppusamy
Department of EEE
Shree Venkateshwara Hi-Tech Engineering
Erode, Tamil Nadu, India

Isidoros Perikos
Department of Computer Engineering and Informatics
University of Patras
Patras, Greece

Fuqian Shi
College of Information and Engineering
Wenzhou Medical University
Wenzhou, China

Tu N. Nguyen
Department of Computer Science
Purdue University Fort Wayne
Fort Wayne, IN, USA

ISSN 2367-4512 ISSN 2367-4520 (electronic)
Lecture Notes on Data Engineering and Communications Technologies
ISBN 978-981-15-8676-7 ISBN 978-981-15-8677-4 (eBook)
https://doi.org/10.1007/978-981-15-8677-4

This Springer imprint is published by the registered company Springer Nature Singapore Pte Ltd.
The registered company address is: 152 Beach Road, #21-01/04 Gateway East, Singapore 189721, Singapore

This book is gratefully dedicated to all the researchers and editors in the field of Sustainable Communication Networks and Applications.

Foreword

It is with deep satisfaction that I write this Foreword to the proceedings of the ICSCN 2020 held at Surya Engineering College (SEC), Erode, India, during 06–07 August 2020.

This conference brought together researchers, academics and professionals from all over the world, experts in sustainable networking technology, sustainable applications and sustainable computing and communication technologies.

This conference particularly encouraged the interaction of research students and developing academics with the more established academic community in an informal setting to present and to discuss new and current work. The papers contributed the most recent scientific knowledge known in the field of ultra-low-power sustainable system, sustainable vehicular ad hoc networks, Internet-enabled infrastructures for sustainability and sustainable mobility and vehicle management. Their contributions helped in making the conference as outstanding as it has been. The local organizing committee members and volunteers have put much effort ensuring the success of the day-to-day operation of the meeting.

We hope that this program will further stimulate research in sustainable big data frameworks, energy and power-constrained devices, low-power communication technologies, sustainable vehicular ad hoc networks, smart transport systems and smart data analytics techniques.

We thank all authors and participants for their contributions.

Dr. E. Baraneetharan
Conference Chair, ICSCN 2020
Associate Professor & Head
Department of EEE, Surya
Engineering College
Erode, India

Preface

This conference proceedings volume contains the written versions of most of the contributions presented during the conference of ICSCN 2020. The conference provided a setting for discussing recent developments in a wide variety of topics including communications, networks and sustainable applications. The conference has been a good opportunity for participants coming from various destinations to present and discuss topics in their respective research areas.

This conference tends to collect the latest research results and applications on intelligent data communication technologies and networks. It includes a selection of 53 papers from 293 papers submitted to the conference from universities and industries all over the world. All of the accepted papers were subjected to strict peer-reviewing by 2–4 expert referees. The papers have been selected for this volume because of quality and the relevance to the conference.

We would like to express our sincere appreciation to all authors for their contributions to this book. We would like to extend our thanks to the keynote speakers, all the referees for their constructive comments on all papers. Especially, we would like to thank to the organizing committee for their hard work. Finally, we would like to thank Springer publications for producing this volume.

Dr. P. Karuppusamy
Shree Venkateshwara Hi-Tech Engineering College
Erode, India

Dr. Fuqian Shi
Rutgers Cancer Institute of New Jersey
New Jersey, USA

Dr. Isidoros Perikos
Professor
Department of Computer Engineering
and Informatics
University of Patras
Patras, Greece

Dr. Tu N. Nguyen
Professor
Director of Network Science
Department of Computer Science
Purdue University Fort Wayne
Fort Wayne, USA

Acknowledgements

ICSCN 2020 would like to acknowledge the excellent work of our conference organizing committee, keynote speakers for their presentation during 06–07 August 2020. The organizers also wish to acknowledge publicly the valuable services provided by the reviewers.

On behalf of the editors, organizers, authors and readers of this conference, we wish to thank the keynote speakers and the reviewers for their time, hard work and dedication to this conference. The organizers wish to acknowledge Thiru. Andavar. A. Ramasamy, Ln. K. Kalaiyarasan, Dr. S. Vijayan, Prof. E. Baraneetharan for the discussion, suggestion and cooperation to organize the keynote speakers of this conference. The organizers wish to acknowledge publicly the valuable services provided by the reviewers. Many thanks to all persons who helped and supported this conference. We would like to acknowledge the contribution made to the organization by its many volunteers. We would like to like to acknowledge the contribution made to the organization by its many volunteers and members contribute their time, energy and knowledge at a local, regional and international level.

We also thank all the chairpersons and conference committee members for their support.

Contents

About the Editors

Dr. P. Karuppusamy is working as a Professor and Head in the Department of Electrical and Electronics Engineering at Shree Venkateshwara Hi-Tech Engineering College, Erode, India. In 2017, he had completed doctorate in Anna University, Chennai, and in 2007, he had completed his postgraduate Power Electronics and Drives in Government College of Technology, Coimbatore, India. He has more than 12 years of teaching experience. He has published more than 60 papers in national and international journals and conferences. He has acted as Conference Chair in IEEE and Springer international conferences and Guest Editor in reputed journals. His research area includes modeling of PV arrays and adaptive neuro-fuzzy model for grid connected photovoltaic system with multilevel inverter.

Dr. Isidoros Perikos received the Diploma of Computer Engineer in 2008, the M.Sc. and the Ph.D. on Computer Science from the Department of Computer Engineering and Informatics, University of Patras, in 2010 and 2016, respectively. He is currently an Assistant Professor (adjust) at the Computer Engineering and Informatics Department. His main research interests include artificial intelligence, machine learning, data mining and knowledge extraction, human–computer interaction, natural language processing, and affective computing. He has published over 80 papers in international conferences, journals, and workshops. He is a member of IEEE, ACM, and the Artificial Intelligence in Education Society.

Dr. Fuqian Shi is currently working as a Professor at Wenzhou Medical University, College of Information and Engineering, Prague. He had completed his Ph.D. in Computer Science and Application at Zhejiang University, P.R. China, and completed his M.S. in Control Theory and Engineering at Zhejiang University of Technology, P.R. China. He had published more than 100 papers in national and international journals. He is the reviewer and editorial board member in many reputed journals. His research interest includes computer networks, computer programming, computer graphics, image processing, data structure, operating system, and medical informatics.

Dr. Tu N. Nguyen (Senior Member, IEEE) received the Ph.D. degree in Electronic Engineering from the National Kaohsiung University of Science and Technology (formerly, National Kaohsiung University of Applied Sciences) in 2016. He was a Postdoctoral Associate with the Department of Computer Science & Engineering, University of Minnesota—Twin Cities in 2017. In 2016, he joined the Missouri University of Science and Technology as a Postdoctoral Researcher with the Intelligent Systems Center. He is currently an Assistant Professor with the Department of Computer Science, Purdue University Fort Wayne. His research interests include design and analysis of algorithms, network science, cyber-physical systems, and cybersecurity. He has served as the TPC Chair for the NICS 2019, SoftCOM (25th), and ICCASA 2017, the Publicity Chair for iCAST 2017 and BigDataSecurity 2017, and the Track Chair for ACT 2017. He has also served as a technical program committee member for more than 70 premium conferences in the areas of network and communication such as INFOCOM, Globecom, ICC, and RFID. He has been serving as an Associate Editor for the EURASIP Journal on Wireless Communications and Networking since 2017 and IEEE ACCESS since 2019. He has also been the Editorial Board of Cybersecurity journal, Internet Technology Letters since 2017, the International Journal of Vehicle Information and Communication Systems since 2017, the International Journal of Intelligent Systems Design and Computing since 2017, and IET Wireless Sensor Systems since 2017.

A Long Short-Term Memory (LSTM) Model for Business Sentiment Analysis Based on Recurrent Neural Network

Md. Jahidul Islam Razin, Md. Abdul Karim, M. F. Mridha, S. M. Rafiuddin Rifat, and Tahira Alam

Abstract Business sentiment analysis (BSA) is one of the significant and popular topics of natural language processing. It is one kind of sentiment analysis techniques for business purpose. Different categories of sentiment analysis techniques like lexicon-based techniques and different types of machine learning algorithms are applied for sentiment analysis on different languages like English, Hindi, Spanish, etc. In this paper, long short-term memory (LSTM) is applied for business sentiment analysis, where recurrent neural network is used. LSTM model is used in a modified approach to prevent the vanishing gradient problem rather than applying the conventional recurrent neural network (RNN). To apply the modified RNN model, product review dataset is used. In this experiment, 70% of the data is trained for the LSTM and the rest 30% of the data is used for testing. The result of this modified RNN model is compared with other conventional RNN models and a comparison is made among the results. It is noted that the proposed model performs better than the other conventional RNN models. Here, the proposed model, i.e., modified RNN model approach has achieved around 91.33% of accuracy. By applying this model, any business company or e-commerce business site can identify the feedback from their customers about different types of product that customers like or dislike. Based on the customer reviews, a business company or e-commerce platform can evaluate its marketing strategy.

Md. Jahidul Islam Razin · Md. Abdul Karim (✉) · S. M. Rafiuddin Rifat · T. Alam
Department of Computer Science and Engineering, University of Asia Pacific, 74/A Green Road, Dhaka, Bangladesh
e-mail: karim.cse007@gmail.com

Md. Jahidul Islam Razin
e-mail: razin.cse@gmail.com

S. M. Rafiuddin Rifat
e-mail: rifat.cse@uap-bd.edu

T. Alam
e-mail: tahira.cse@uap-bd.edu

M. F. Mridha
Bangladesh University of Business and Technology, Dhaka, Bangladesh
e-mail: firoz@bubt.edu.bd

P. Karuppusamy et al. (eds.), *Sustainable Communication Networks and Application*, Lecture Notes on Data Engineering and Communications Technologies 55,
https://doi.org/10.1007/978-981-15-8677-4_1

Keywords Business sentiment analysis · Product reviews · Recurrent neural network · LSTM

1 Introduction

Nowadays, the Internet is a fundamental part of our daily life. All fields of information are rising exponentially every moment, where it is easier to share our opinions on e-commerce websites, forums and media for various types of products and services. It provides essential information about some different number of domains and some social applications. It is difficult to handle these huge amounts of data manually. For this reason, business sentiment analysis (BSA) is very feasible, which also provides an idea on the requirements of people. It has become a popular research topic in natural language processing domain. Analysis of reviews can quickly extract information from a text and can also define the target and opinion polarity. Various types of social applications and websites can use BSA to forecast consumer patterns, economic policies and stock market forecasting.

Researchers perform business sentiment analysis using various machine learning (ML) techniques. ML's support learning models include support vector machines (SVM) [1], logistic regression [2], naive Bayes (NB) [3], random forests [4] and so on. Artificial neural network (ANN) is one of the areas of ML [5]. ANN also has various forms like recurrent neural network (RNN) [6] and convolution neural network (CNN) [7]. Artificial neural networks are mainly constructed using three layers. These are the input layer, hidden layer and output layer. This concept is extended in deep learning. Deep learning is constructed using more than two hidden layers. How much deeper the network is defined by the layers used in the hidden layer. Deep learning is a part of artificial neural network. Deep learning gives much better services than other ML techniques like SVM, NB, etc. These techniques are used in all natural language processing fields like all types of recognition (speech, entity, pattern) [8] and computer vision techniques. For its accuracy level, it has become very popular nowadays. Good accuracy can be achieved if the representation of data is perfect. Conventional ML algorithms are depending on handcrafted characteristics. But deep learning needs high computation ability and storage to increment the number of hidden layers as comparison to conventional ML algorithms and getting better performance. Deep learning technique can adapt more swiftly as it will get more and more training data. As business sentiment analysis needs to process a lot of data for the prediction, and this motivated us to apply deep learning technique for business sentiment analysis [9].

The importance of consumer views and the vast amounts of centric data are available to the public and the unpredictability of the market climate has led the organization to introduce monitoring and tracking measures such as sentiment analysis, which is mainly performed here as business sentiment analysis.

The principal contribution of the proposed research work can be summed up as follows. This paper presents the study of the business sentiment function by a list of product feedback. Using the proposed model, the reviews are divided into three categories, which are positive, neutral and negative. A well-defined text dataset is used to apply long short-term memory (LSTM) as a part of modified RNN model that gives better accuracy result in comparison to conventional RNN model. This is a different approach when compared to traditional feed-forward networks [10].

A feed-forward network takes a limited text for predicting the next word. Here, RNN can easily use previous words for prediction. An RNN model sees the text as a signal made up of terms, where it is presumed that recursive weights reflect short-term memory. LSTM is different, which is a part of the RNN structure in the ANN network. It is allowed for both long and short patterns and it sorts out the problem of vanishing gradient. So, LSTM is now approved in several numbers of applications and it is a promising practice for business sentiment analysis.

This article is structured as follows. Section 2 explains the related works and the motivation of this research. Section 3 discusses about the methodology. Section 4 discusses about implemented tools. Section 5 explains result and analysis. Section 6 concludes the paper.

2 Related Work

Several numbers of researches have been performed on sentiment analysis for different languages using different types of neural networks. Some of them are summarized as follows.

Xu et al. [11] applied supervised learning algorithms like the perceptron algorithm, naive Bayes and support vector machine to predict the reviewer's ratings. They used 70% data for training and 30% data for testing. They also determine the precision and recall values for the different classifiers.

Deep learning is very popular in this area of sentiment analysis. Sujata Rani, Parteek Kumar, applied convolution neural network for sentiment analysis [12]. After cross-validation, they have used 50% data as train data and 50% for test data.

Mikolov et al. [13] proposed a model applying recurrent neural network (RNN); it is used for processing sequential text data. An RNN model is depending on the input layer, hidden layers and output layer. When an RNN model is performed, the input layer is added with the hidden layer, which is arranged for a new input layer to calculate the hidden layers. This is working as a loop and it has been repeated after a limit of time '*t*.' All the information from the previous layer is successfully reversal using it. For this, the performance is increased and also helping to squint all words in sequential order. So, this model is beneficial for looping and sequencing facilities.

However, the RNN model covers the orderly structure of the text and it can fix a short-term dependence problem. But it is not possible to fix the long-term dependence problem using RNN, because it cannot gain any knowledge properly from the long term. If there is an interval between two relative texts and the current location is larger, then there will always be a problem in the RNN model. As there are many layers in the back-propagation through time optimization algorithm (BPTT), it is indicating the loss of information while training. So, long short-term memory (LSTM) is used for our business sentimental analysis, where it provides better results than the traditional RNN model.

3 Methodology

This section will describe our method for business sentiment analysis from reviewer's text data. For business sentiment analysis, long short-term memory (LSTM) is used. The LSTM model is designed for avoiding long-term dependency. Remembering that long-term dependency is the default behavior of LSTM. This model is most popular with researchers. Within this model, the RNN node is replaced by the LSTM cell, which can easily store previous knowledge. For this reason, business sentiment analysis using LSTM is better than other methods.

Our proposed method is divided into four phases:

3.1 Discussion about RNN and LSTM architecture.
3.2 Data processing.
3.3 Training the model.
3.4 Testing new data.

3.1 Discussion About RNN and LSTM Architecture

LSTM is a subsection of RNN and there are many similarities between them. This work has proposed some structural diagram and brief discussion about LSTM and RNN.

Figure 1 [14] provides an illustration of a traditional RNN model where $X(t)$ is input, $h(t)$ is output and A is the neural network that can gain knowledge from the previous. One output goes to another and passed the information easily.

Figure 2 [14] provides a clear knowledge of a single LSTM cell. LSTM model is kind of similar like standard RNN, except that the memory block is just replaced into the hidden layer, as in Fig. 4 [14]. The symbols in Fig. 3 [14] have the following meanings. X is using for scaling of information, σ means Sigmoid layer, 'tan$h$' means hyperbolic Tangent layer, $h(t-1)$ means sigmoid layer, 'tan$h$' means hyperbolic tangent layer, $h(t-1)$ means output of last LSTM unit, $c(t-1)$ means memory from last LSTM unit, $X(t)$ means current input, $c(t)$ means new updated memory,

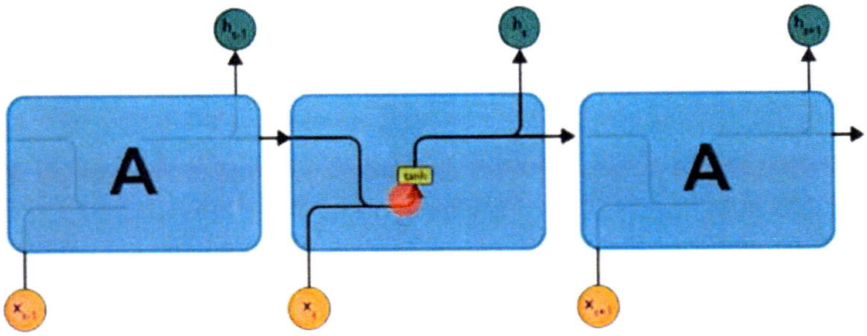

Fig. 1 The structural diagram of RNN model

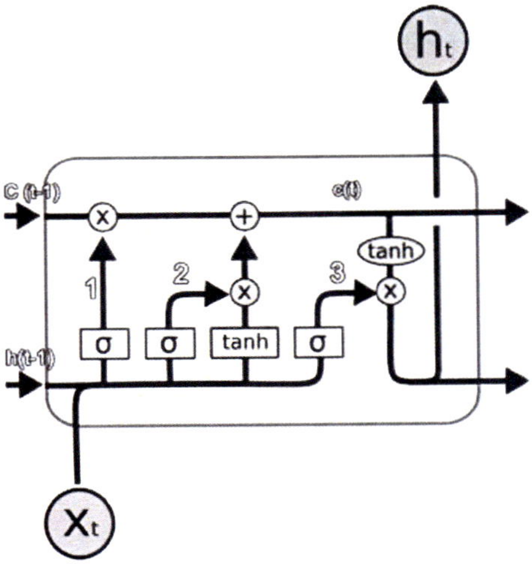

Fig. 2 Structural diagram of LSTM cell (a)

$h(t)$ means current output. In RNN, only the hyperbolic tangent layer is using to overcome the vanishing gradient problem [15]. The second derivative of the hyperbolic tangent function can solve the problem before going to zero. But it cannot forget or remember anything. That is why the sigmoid layer is also used in LSTM model with the hyperbolic tangent layer. Sigmoid can output 0 or 1, which is used to forget or remember the information.

The first step in the LSTM is to find the useless text which is not needed and also thrown away from the cell state. It is achieved by a sigmoid layer, called a forget layer. New pieces of information will be processed in the cell in the second level, which will be decided by a sigmoid layer called the gate layer of the data. In the third stage, the old cell state updates into the new cell satellite with the above-mentioned input gates and forgets gates information. The output gate essentially defines the output value, which is dependent on the cell state.

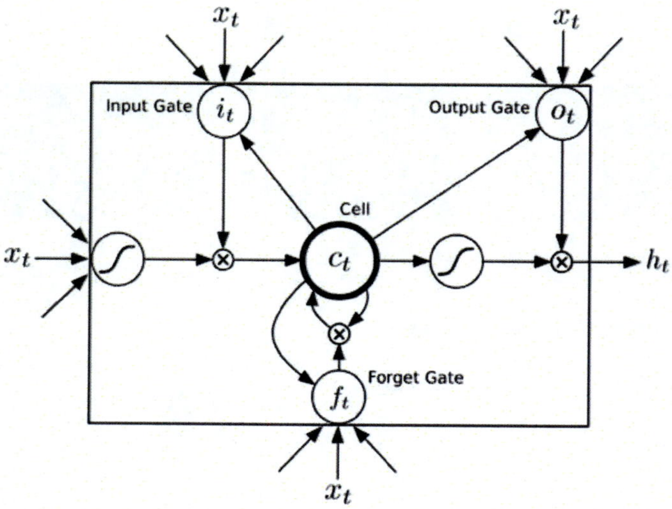

Fig. 3 Structural diagram of LSTM cell (b)

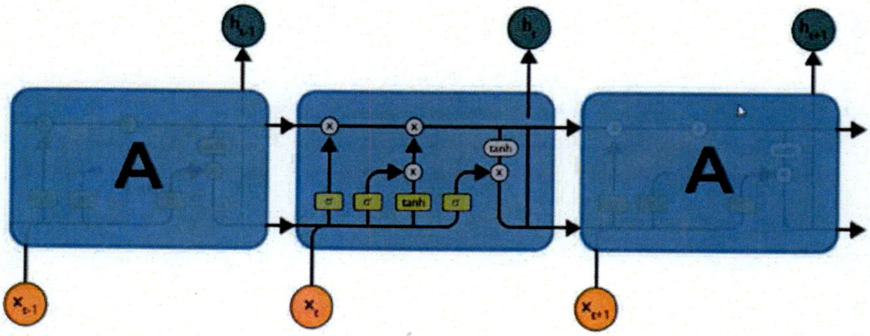

Fig. 4 LSTM architecture

Mainly, four steps of calculation process in LSTM.

- Initially calculates the forget gate values and input gate values.
- Update the cell stage of LSTM.
- Calculate output gate's value.
- Lastly update the cell state [16–19].

All equations are shown in below [14].
Input Gates:

$$i_t = \sigma(x_t U^i + h_{t-1} W^i) \tag{1}$$

```
Layer (type)                 Output Shape              Param #
=================================================================
embedding_1 (Embedding)      (None, None, 32)          640000
_________________________________________________________________
lstm_1 (LSTM)                (None, 100)               53200
_________________________________________________________________
dense_1 (Dense)              (None, 1)                 101
=================================================================
Total params: 693,301
Trainable params: 693,301
Non-trainable params: 0
_________________________________________________________________
None
```

Fig. 5 LSTM network

Forget gates:

$$f_t = \sigma(x_t U^f + h_{t-1} W^f) \tag{2}$$

Cells:

$$\underline{C_t} = \tanh(x_t U^g + h_{t-1} W^g) \tag{3}$$

$$C_t = \sigma(f_t \times C_{t-1} + i_t \times \underline{C_t}) \tag{4}$$

Output gates:

$$O_t = \sigma(x_t U^o + h_{t-1} W^o) \tag{5}$$

Cell output:

$$h_t = \tanh(C_t) \times O_t \tag{6}$$

RNN and LSTM consider the problem of the vanishing gradient via a sequenced sentence of text. RNN model will display lower error rate. But when there is a long text argument, LSTM is more effective at overcoming the vanishing gradient problem. In our business sentiment analysis, the modified RNN-based LSTM model is applied. The proposed model is shown in Figs. 5 and 6.

3.2 Data Processing

To train the proposed model, a business review dataset is collected from amazon.com on product analysis. The dataset was created by Web scraping or APIs. Researchers

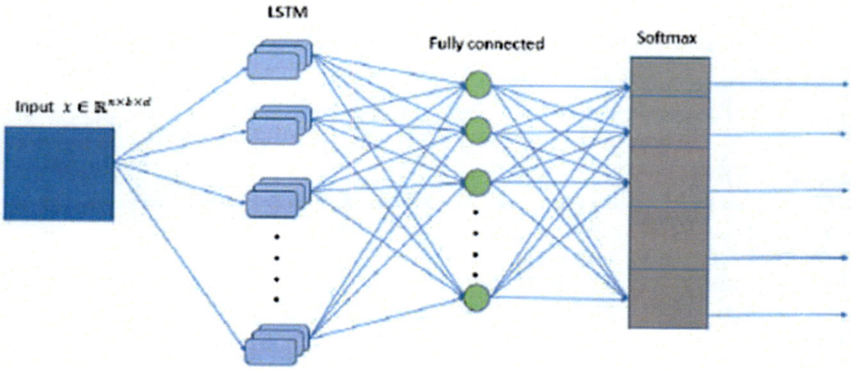

Fig. 6 LSTM architecture

compiled the information from Amazon Review Information (ARD) when it comes to Amazon datasets [20]. This dataset contained punctuation and HTML tags. In order to vanish these, a function is used for taking a text string as a parameter and then preprocess the string to remove punctuation and HTML tags from the given string. These punctuation and HTML tags are replaced with an empty space. All the single characters and multiple spaces are removed. At last for each term, the analysis text is naturally divided into emotions of the business class. Then, a tool is used for vector representation to transform every term into a vector using the vector equation:

$$v \in R_1 \times d \tag{7}$$

which is known as the word embedding. Here, word2vector tools are used. Then the cycle was followed by the natural order; the sentence of participles was traversed from left to right, which is the forward calculation of LSTM. The results of the production depended on the probabilities of the word at 't' time and also the sequence of vocabulary giving before 't' time. Finally, the error was determined by the likelihood of a common distribution of all the words in the sentences.

3.3 *Training the Model*

After preprocessing the data and splitting the dataset, LSTM model is used to train our model. LSTM layer is created with 100 neurons and also added a dense layer with sigmoid activation function.

The process is introduced as follows:

- Due to their emotional marks, all training data is divided into three groups, has positive, negative and neutral. The LSTM models are then trained in each data

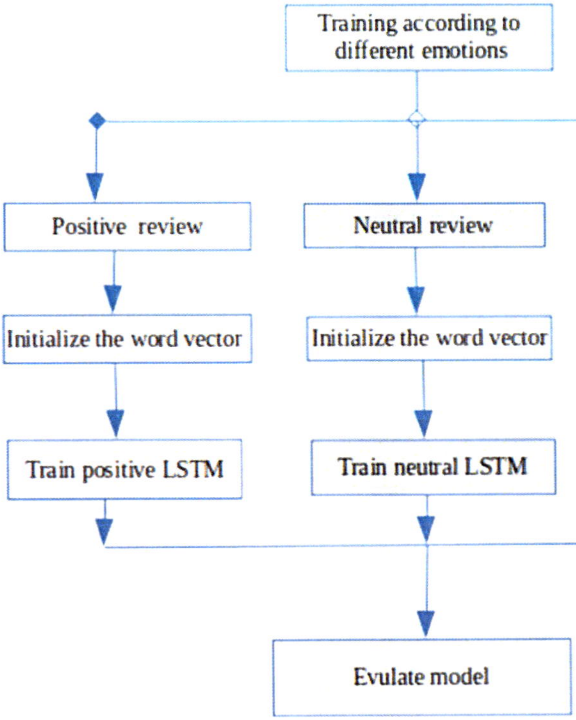

Fig. 7 Business sentimental emotion classification

category and with multiple LSTM models resulting in them as well. This is done for equal ratings.

- To get a new input review, the LSTM models are available in the training phase evaluated on the new input review. The model which is giving the smallest error value is assigned to the new input review.

Figure 7 is showing the structure of the processing of the training phase: This model could overcome the vanishing gradient problem completely than the conventional RNN model. It also better performs in many experiments, like as structure with conjunctions, such as, 'not only…but also…,' 'However,' 'in addition,' etc.

3.4 Testing the Model

After training the model, new text data is used for testing our model. When a new product review is coming, this model has classified the new review as negative, positive or neutral.

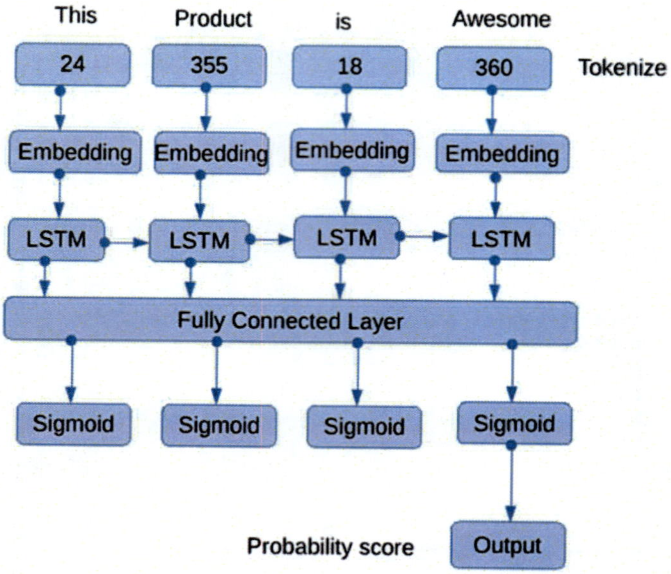

Fig. 8 RNN-LSTM working process in business sentiment analysis

Figure 8 described the working process of our RNN-LSTM model in business sentiment analysis.

4 Implementation Tools

The proposed work has used the Jupyter Notebook [21], which is an open-source Web platform. It has helped to develop an environment to perform our experiments and also creating and sharing all types of documents like code, text, equations and visualizations. The LSTM-RNN model is developed using TFLearn (deep learning library) Python packages, which are installed at the top of the TensorFlow [22]. The proposed work also uses Keras (NNs API) [23], which is written with Python and runs in TensorFlow.

5 Result and Analysis

The analysis of the outcome along with the result for the entire system is discussed in this section. The outcome is splitted into three parts. This is an experimental dataset (Sect. 5.1), Experiment Result Analysis (Sect.5.2) and Comparison (Sect.6).

Table 1 Accuracy of LSTM model

Epoch	Training accuracy	Testing accuracy
10	0.9504	0.8954
30	0.9551	0.9092
50	0.9623	0.9193

5.1 Experimental Dataset

Amazon Review Information Dataset (ARD) [20] contains 142.8 million ratings and a variety of metadata. This work has taken a selected amount of data from this dataset. After taking the data from dataset, all the data has been categorized into positive, neutral and negative. Then Word2Vec tools are used to create word embedding for our LSTM model. After that, it has been gone through the LSTM layer. There is a Dense() layer, which is the final layer of the model. It can crunch all the output that is coming from the LSTM layer and convert it as a single numeric value of 0.0 and 1.0.

Total of 25,000 product reviews, which are divided into a 70% items for the training set and a 30% items for the testing set. Then train and test have been performed by the proposed model. It achieved 95.04% (from Table 1) accuracy on the training data and 89.54% accuracy on the test data for 10 epochs. Then increasing the epoch size into 20, our accuracy for training data 95.51% and accuracy for test data 90.92%. Again, the epoch size is increased from 20 to 50, then the accuracy on training data will be 96.23% and accuracy on testing the data will be 91.33%. If epoch size is increased more than 50 iteration, the deference between training and test result is increased a lot and overfitted. That is why the epoch size is not increased more than 50. Figure 9 is showing the training and testing loss; after 20 iteration, here blue line are showing training loss and orange line are showing testing loss.

Figure 10 is showing the training and testing accuracy; here, blue line is indicating training accuracy and orange line is indicating testing accuracy. Our model test data accuracy is 91.33%.

5.2 Experiment Results Analysis

After training and testing this model, it can easily classify a new product review, which is previously unseen. Previously unseen product review of, ' The product was not so great.' Here, the prediction probability value is around 0.1368. From Table 2: When the value is less than 0.30, the model predicts it as very bad. If the value is greater than 0.30, then it is predicted as bad. When the value is greater than 0.50, then it is positive. When the value is greater than 0.60, then it is predicted as good. When it is greater than 0.80, then it is predicted as excellent. For multi-valued analysis, it is required to encode as positive (0, 0, 1), neutral (0, 1, 0) and negative (1, 0, 0).

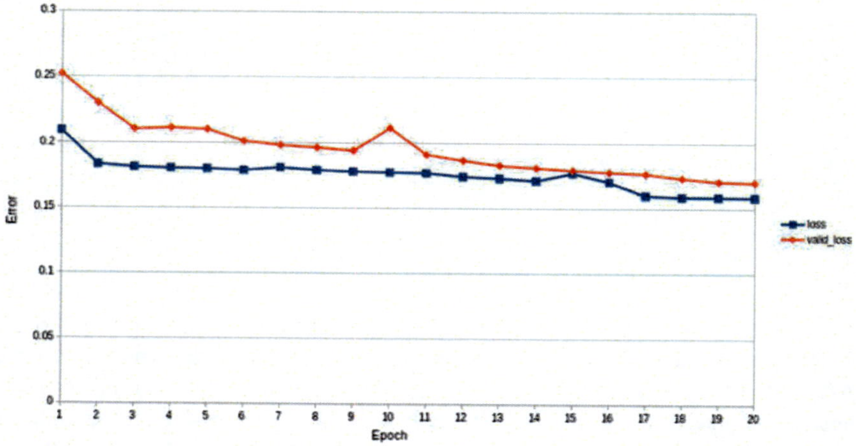

Fig. 9 Loss versus epoch graph

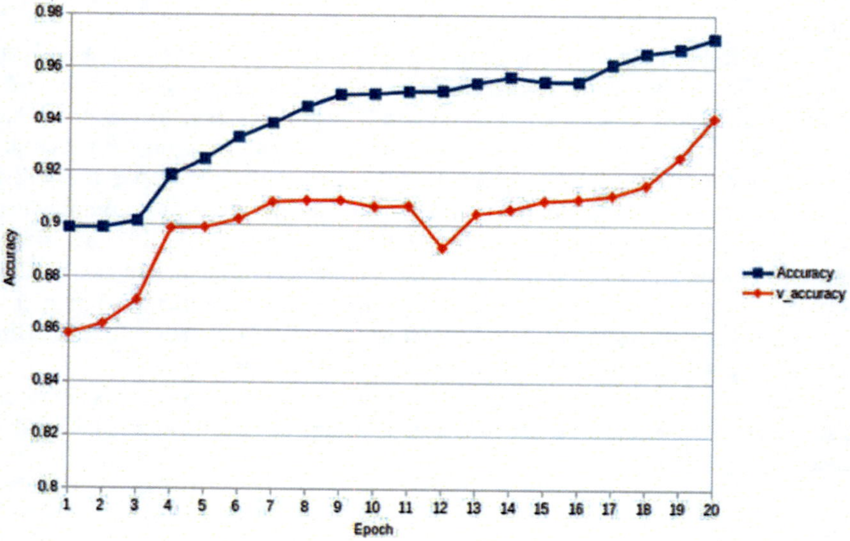

Fig. 10 Accuracy versus epoch graph

Table 2 Review type

Prediction	Result
Greater than 0. 80	Excellent
Greater than 0.60	Better
Greater than 0.50	Good
Less than 30	Very bad
Greater than 0.30	Bad

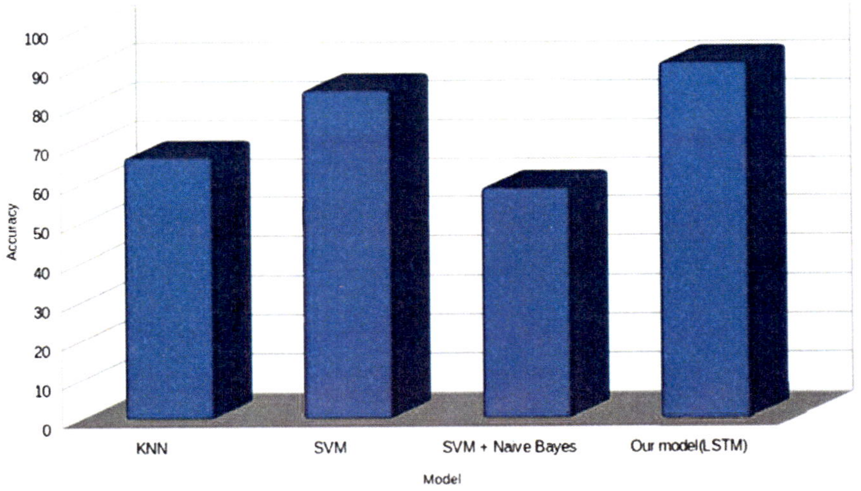

Fig. 11 Comparison with other models (Accuracy)

6 Comparison

Mohammad Rezwanul Huq and Ahmad Ali proposed a two model for sentiment classification [24]. Their model is based on k-nearest neighbor (KNN) and another model is based on a support vector machine (SVM). 84.32% accuracy is achieved based on KNN model and 67.03% accuracy is achieved based on SVM model. Xu et al. [11] applied SVM and naive Bayes model to predict the reviewer's rating. 88.8% accuracy is achieved on training data by this model and 59.1% accuracy is achieved on testing data. Figure 11 is showing a comparison graph between different models and their accuracy. Here, our model has been achieved 91.33% accuracy, which is better and maximum than other proposed models.

7 Conclusion

In this paper, an LSTM model based on RNN is used to evaluate the business sentiment. It covers all the sequences and performance of business sentiment analysis. This model is used for multi-classification in product reviews. This model can be used for any kind of product review datasets for different languages. In the future, it will be attempted to improve our algorithm for obtaining a better accuracy rate by preprocessing the dataset before feed into the model and by proper feature engineering and tuning. The dataset is required to adjust depending on our local market, which is more balanced than this one. Further, a Web tool will be developed for our local market in order to help the local businesses.

References

1. Vishwanathan, S.V.M., Murty, M.N.: SSVM: A simple SVM algorithm. In: Proceedings of the 2002 International Joint Conference on Neural Networks. IJCNN'02 (Cat. No. 02CH37290) vol. 3, pp. 2393–2398. IEEE (2002)
2. Wright, R.E.: Logistic regression (1995)
3. Zhang, H.: The optimality of Naive Bayes. AA, **1**(2), 3 (2004)
4. Breiman, L.: Random Forests. Mach. Learn. **45**(1), 5–32 (2001)
5. Agatonovic-Kustrin, S., Beresford, R.: Basic concepts of artificial neural network (ANN) modeling and its application in pharmaceutical research. J. Pharm. Biomed. Anal. **22**(5), 717–727 (2000)
6. Mikolov, T., Karafiát, M., Burget, L., Černocký, J, Khudanpur, S.: Recurrent neural network based language model. In: Eleventh Annual Conference of the International Speech Communication Association (2010)
7. Huang, W., Qiao, Y., Tang, X.: Robust scene text detection with convolution neural network induced mser trees. In: European Conference on Computer Vision, pp. 497–511. Springer, Cham (2014)
8. Liang, J., Koperski, K., Dhillon, N.S., Tusk, C., Bhatti, S., Evri Inc.: NLP-based entity recognition and disambiguation. U.S. Patent 8,594,996 (2013)
9. Bhatia, V., Rani, R.: Dfuzzy: a deep learning-based fuzzy clustering model for large graphs. Knowl. Inf. Syst. **57**(1), 159–181 (2018)
10. Schmidt, W.F., Kraaijveld, M.A., Duin, R.P.: Feed forward neural networks with random weights. In: International Conference on Pattern Recognition p. 1. IEEE Computer Society Press (1992)
11. Xu, Y., Wu, X., Wang, Q.: Sentiment analysis of yelp's ratings based on text reviews (2014)
12. Rani, S., Kumar, P.: Deep learning based sentiment analysis using convolution neural network. Arabian J. Sci. Eng. **44**(4), 3305–3314 (2019)
13. Mikolov, T., Yih, W.T., Zweig, G.: Linguistic regularities in continuous space word representations. In: Proceedings of the 2013 Conference of the North American Chapter of the Association for Computational Linguistics: Human Language Technologies, pp. 746–751 (2013)
14. Graves, A.: Long short-term memory. In: Supervised Sequence Labelling with Recurrent Neural Networks, pp. 37–45. Springer, Berlin, (2012)
15. Zhang, X., Lu, L., Lapata, M.: Top-down tree long short-term memory networks. arXiv preprint, arXiv:1511.00060 (2015)
16. Graves, A.: Supervised sequence labelling. In: Supervised Sequence Labelling with Recurrent Neural Networks, pp. 5–13. Springer, Berlin, (2012)
17. Soutner, D., Müller, L.: Application of LSTM neural networks in language modelling. In International Conference on Text, Speech and Dialogue, pp. 105–112. Springer, Berlin, (2013)
18. Sundermeyer, M., Tüske, Z., Schlüter, R., Ney, H.: Lattice decoding and rescoring with long-span neural network language models. In: Fifteenth Annual Conference of the International Speech Communication Association (2014)
19. Sundermeyer, M., Ney, H., Schlüter, R.: From feedforward to recurrent LSTM neural networks for language modeling. IEEE/ACM Trans. Audio, Speech. Lang. Process. **23**(3), 517–529 (2015)
20. He, R., McAuley, J.: Modeling the visual evolution of fashion trends with one-class collaborative filtering. WWW (2016)
21. Perkel, J.M.: Why Jupyter is data scientists' computational notebook of choice. Nature **563**(7732), 145–147 (2018)

22. Abadi, M., Barham, P., Chen, J., Chen, Z., Davis, A., Dean, J., Devin, M., Ghemawat, S., Irving, G., Isard, M., Kudlur, M.: Tensorflow: a system for large-scale machine learning. In: 12th USENIX Symposium on Operating Systems Design and Implementation (OSDI 16), 265–283 (2016)
23. Gulli, A., Pal, S.: Deep Learning with Keras. Packt Publishing Ltd (2017)
24. Huq, M.R., Ali, A., Rahman, A.: Sentiment analysis on Twitter data using KNN and SVM. IJACSA. Int. J. Adv. Comput. Sci. Appl. **8**(6), 19–25 (2017)

An Automatic Violence Detection Technique Using 3D Convolutional Neural Network

Md. Abdul Karim, Md. Jahidul Islam Razin, Nahid Uddin Ahmed, Md Shopon, and Tahira Alam

Abstract In the field of machine learning, the deep learning technique plays a very important role as it is useful in various real-life domains. As various crimes and misdeeds are occurring in various public places because of lack of proper monitoring, a number of methods have been proposed for detecting violence from videos. Automatic violence detection has gained increased research importance in case of video surveillance. However, they suffer from various limitations and most of the times it depends on special criteria. In this perspective, this paper proposes an effective violence detection method from videos using 3D convolutional neural network. The proposed methodology uses machine learning and deep learning techniques for improving the accuracy. Comprehensive performance analyses have proven that the proposed method achieves high performance in case of detecting violence from videos. The experimental results also prove that the proposed technique outperforms various other existing methods for detecting violence from videos.

Keywords Machine learning · 3D convolutional neural network · Computer vision · Violence detection · Deep learning

Md. Abdul Karim · Md. Jahidul Islam Razin (✉) · N. U. Ahmed · M. Shopon · T. Alam
Department of Computer Science and Engineering, University of Asia Pacific, 74/A Green Road, Dhaka, Bangladesh
e-mail: razin.cse@gmail.com

Md. Abdul Karim
e-mail: karim.cse007@gmail.com

N. U. Ahmed
e-mail: nahiduddin.cse@gmail.com

M. Shopon
e-mail: shopon.cse@uap-bd.edu

T. Alam
e-mail: tahira.cse@uap-bd.edu

P. Karuppusamy et al. (eds.), *Sustainable Communication Networks and Application*, Lecture Notes on Data Engineering and Communications Technologies 55,
https://doi.org/10.1007/978-981-15-8677-4_2

1 Introduction

Deep learning is considered as one of the most powerful domains of machine learning. It builds models that assist computers to do things that humans can do naturally. In deep learning, the model learns to perform classification tasks directly from image, text, sound, or video. These models can acquire higher accuracy, sometimes even outperforming human-level performance. Models are trained by using the large labeled dataset and neural network architectures [1] having a number of layers.

In deep learning, a convolutional neural network [2] is considered as a class of deep neural networks that most commonly applied for analyzing the visual imagery. Three different types of convolution layers are present in CNN architecture, namely 1D, 2D, and 3D convolutional layer. 1D convolutional layer can perform in one direction. It is used for text and sound data recognition. 2D convolutional layer is basically used for image data for sliding a kernel in two directions with 3 dimensions (height, width, and color of the image). 3D convolutional layer is very popular for 3D image classifications and recognition. In 3D convolutional layer, a kernel can easily move in 3 directions with 4 dimensions where the last dimension represents the color. 3D convolutional layer is widely used for recognizing different types of diseases from different medical data.

Convolutional neural networks have become widely popular as they can be used for solving various real-life problems. For example, this paper is focused on detecting violence from video data. The use of CCTV is becoming popular day by day for ensuring security. Both government and non-government organizations are spending a lot of money in this sector. However, CCTV is not properly used for the sake of our safety because of various reasons, such as not having sufficient manpower for monitoring the CCTV footages, not having modern control room with modern equipments, etc. Therefore, for solving these problems many methods and techniques [3] have been introduced for automatically detecting violence from video data [4]. They use artificial neural network (ANN), support vector machine (SVM), 2D convolutional neural network, etc. for building their models. However, most of them depend on some special criteria and their performance decrease if the video quality is not good. All these factors motivated us to propose an effective model for detecting violence from video data, which does not depend on any criteria and performs well even if the video quality is not good.

There are many reasons behind selecting 3D convolutional neural network for building our model. Other machine learning algorithms need to create some features from the image and then feed those features to those algorithms. This feature extraction is a time-consuming work. Some algorithms also use image-level pixel values as a vector of features. For example, for training a 28×28 image, SVM uses each of the pixel values as a feature and thus gets 784 features. On the other hand, 3D convolutional neural network automatically extracts features from images. A pixel-vector algorithm loses a lot of spatial interaction between pixels, but on the other hand, a 3DCNN effectively uses adjacent pixel information. It effectively downsamples the image and then uses a prediction layer at the end. For these reasons, 3D convolutional neural network is more effective compared to other machine learning methods.

An effective model is proposed for detecting violence using 3D convolution neural network (3DCNN). For training our model, the dataset is collected from Kaggle. The name of the dataset is 'Real-Life Violence Situation Dataset,' and it is proposed by Soliman and Kamal [5]. This dataset contains both violent and non-violent video data. Even though the quality of the videos in this dataset is poor, our model can easily detect violence automatically. As the model is automated, there is no need of any human interaction for detecting violence. Only a signal needs to be sent to the nearest security control room.

The main contributions of our paper are given as follows:

- Proposed an effective model for detecting violence automatically from video data.
- The proposed model can detect violence regardless of the quality of the videos and is not dependent on any specific criteria.
- As 3DCNN is used, it automatically extracts features from the data.
- Effective data processing technique is used for the model.
- Real-life applications of this technique show the suitability of our research work.
- Comprehensive experimental analyses establish the supremacy of our proposed method.

The remaining part of our paper is arranged as follows:

- Section 2 presents some related research works to get insight of our proposed research work.
- The proposed method is explained in Sect. 3.
- Section 4 has described the proposed experimental results and also a comparison with other methods.
- Finally, Sect. 5 concludes the remarks and discussions on our future plans.

2 Related Works

The first method for violence recognition from videos was proposed in 1998 by Giannakopulo [6]. It identified violence by detecting blood and flame from videos. It also used the characteristics of motion and sound for detecting violence. On the other hand, Clarin [7] detected violence from movie sequences. The sequences that involve actions and produces blood from extreme punching, kicking, etc. are identified as violent.

Gao [8] used two different types of datasets for detecting violence. In their model, they used different types of classifiers and methods. When they used SVM classifier [9–11] with ViF method, they got 88.01% accuracy. On the other hand, when they used the same classifier with ViF and OViF method they got 91.93% accuracy. Again, when they used these same methods with SVM and AdaBoost [12] classifier, then they got 92.81% accuracy.

Nievas proposed a technique [13] for recognizing fight from video. For training their model, they used the KTH dataset which has videos of various activities including fights. They divided the videos into fights and non-fights and labeled them accordingly. For violence detection, they used MoSIFT [14] and Space-Time Interest Points (STIP) [15, 16] method. Kumar [17] proposed a method for HDR video encoding that is capable for capturing a high dynamic video range than traditional 8-f-stops method.

However, all these proposed methods depend on some specific features of the videos. But in real-life scenarios, those features may not be available in the videos. For example, for crime detection from video footages, many methods depend on the audio also. On the other hand, many methods depend on the quality of the videos and some methods require gray-scale videos. These features are not always available in all CCTV cameras, and even if they are, then it becomes very costly. However, our model does not depend on any special feature or criteria. It can detect violence even if the quality of the videos is poor. While dealing with video data, the optimization and prepossessing of data are very difficult. These problems are omitted by tuning various features of our 3DCNN model and also by using effective techniques in our data preprocessing step. For these reasons, our method achieved better performance than the other methods proposed for detecting violence from video data.

3 Proposed Model

This section will describe the method proposed for detecting violence from videos. For violence detection, convolutional neural network is used [18]. As it is mainly dealt with video dataset, 3D convolutional neural network is used. The advantages of our proposed method over other methods can be described in two folds. First, like other methods, it does not depend on some specific features or the quality of the videos. Other algorithms also need to create some features from the data and then feed those features to the algorithms. But as 3D convolutional neural network is used, it automatically extracts features from the data. Second, by using effective techniques in our data processing step and by tuning various features of our 3DCNN model, the problems regarding video data optimization is omitted. Most of the methods will suffer from the complexities of video data optimization and preprocessing, but this part is handled effectively. For these reasons, our method performs better than the other methods.

Our proposed method is divided into 3 phases:

1. Data Processing
2. Training the Model
3. Testing New Video Data

They are described below:

3.1 Processing

OpenCV is an open-source python library that is used for solving complex computer vision tasks. OpenCV is used to extract the frames from the videos. After extracting the frames, all the frames were resized into 50 × 50 dimensions. After that, the 2D frames are converted into 3D frames using openCV as 3DCNN is used. The frame depth was 15. The frames were transformed into grayscale after resizing. In grayscale images, very less information is required to distinguish a particular image from other images. After that, the preprocessed frames were converted into Numpy Array [19] and were labeled according to their class (violent or non-violent) using class vector. For feeding our model, it is converted into binary class matrix. They are used as input for the proposed model.

3.2 Training the Model

Here, 3D convolution neural network (CNN) model is used for training our model. The design of the layers of CNN is different than other neural networks as shown in Fig. 1. CNN has various layers like the input layer, convolution layer, hidden layer, polling layer, and output layer. Convolution layer is the key feature of CNN. This layer is used for extracting edges, corners, colors, etc. And it also helps in case of identifying different shapes, digits, etc. The 3D convolution layer is used with a filter size 32 and 3 × 3 pooling layer.

Various deep learning techniques are also used for improving the accuracy of the model. The ReLU activation layer is used [20] for amplifying the positive results and down sampling the negative results. Softmax activation function [21] is used so that the classifying model produces only two probabilistic values (0 or 1). Max-pooling [22] layer is used for reducing the number of parameters from image matrices and also for selecting the largest parameter from the matrices. A 3 * 3 max-pooling is

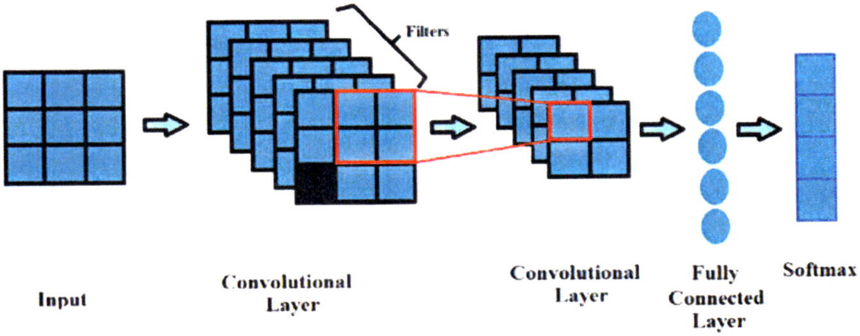

Fig. 1 Convolutional neural network architecture

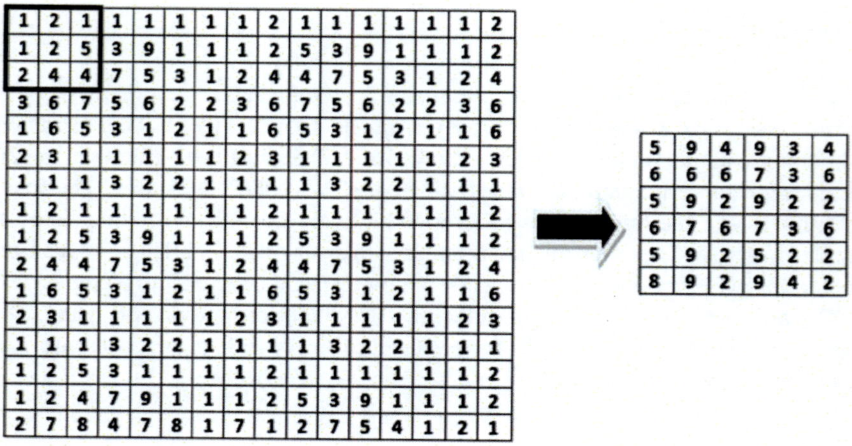

Fig. 2 Max pooling

shown in Fig. 2. The dropout is also used for regularizing the proposed model. At last, the Nadam optimizer [23] is used for updating the networking width in training data.

3.3 Testing New Video Data

After training the model, the unlabeled new video data is used for testing the proposed model. The data frames are randomly selected from the dataset, but this time it is not labeled as violent or non-violent. The proposed model has classified them into violent and non-violent.

4 Result and Analysis

This section has described the implementation tools, experimental dataset, experimental results, and performance comparison with other methods.

4.1 Implementation Tools

Jupyter Notebook [24], which is an open-source web application, is used here. The 3DCNN model is developed by using python package TFLearn (deep learning library) [25], which is built on top of TensorFlow [26]. Keras (NNs API) [27] which

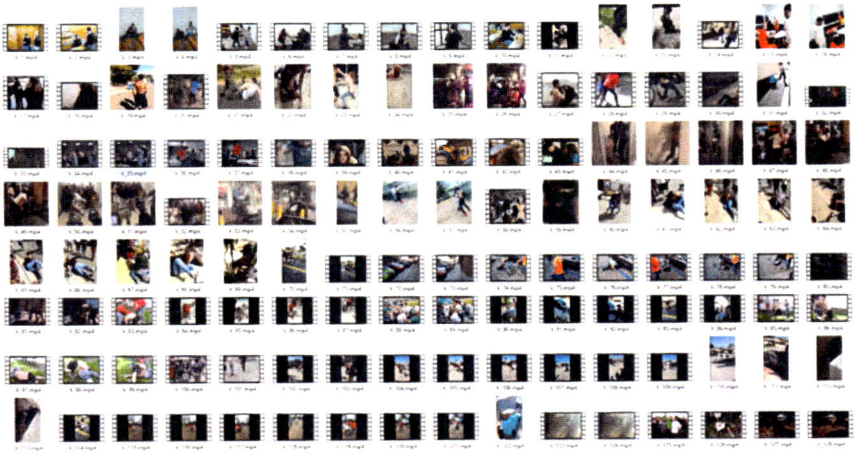

Fig. 3 Violent dataset sample

is written with python is used for further process. The proposed method has also used OpenCV, which is an open-source computer vision library for processing the video data.

4.2 Experimental Data Set

For training our model, a dataset named 'Real-Life Violence Situation Dataset' is collected from Kaggle that is basically proposed by Solaiman and Kamal [5]. It contains a total of 2000 videos. Each video is categorized as either violent or non-violent. Half of the videos are violent and the other half are non-violent. Each video has a length of five seconds. Figure 3 is showing some of the violent videos, and Fig. 4 is showing some of the non-violence videos.

It has used 80% data for training the model and 20% data for testing the model. That means 1400 out of the 2000 videos were used for training our model and the rest of the 600 videos were used for testing our model. Our model has achieved 96.17% accuracy in case of identifying violence from the test data outperforming many other methods.

4.3 Performance Analysis

For evaluating the performance of our method, experiments are performed on the mentioned dataset and our model has achieved satisfactory accuracy in case of identifying violence. The proposed dataset is trained for various epoch sizes. It helped

Fig. 4 Non-violent dataset sample

Table 1 Accuracy of the model (varying epoch size)

Epoch	Accuracy (train) (%)	Accuracy (test) (%)
10	97.21	95.33
30	97.43	95.33
50	97.85	96.00
100	97.93	96.17

our model to increase the performance. However, if the epoch size is increased to more than 100, then it causes overfitting. Table 1 shows the accuracy for different epoch sizes by considering both training and testing data.

Figure 5 depicts the training and testing loss graph after 50 iterations. Here, the Blue line is indicating the training loss and the Green line is indicating the testing loss according to the number of iterations. On the other hand, Fig. 6 represents training and testing accuracy graph after 50 iterations. Here, the Blue line is indicating the

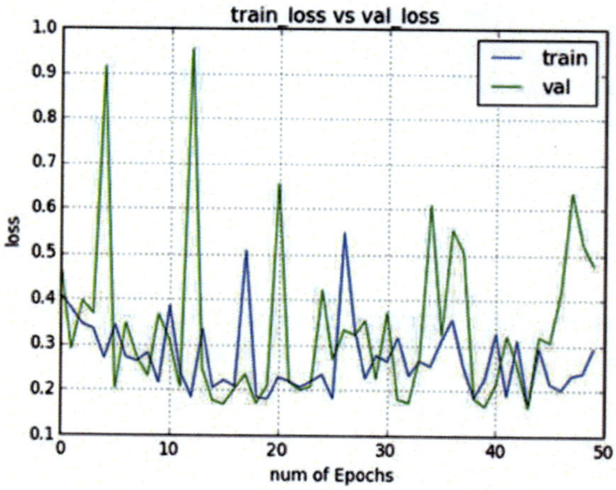

Fig. 5 Loss graph (50 iterations)

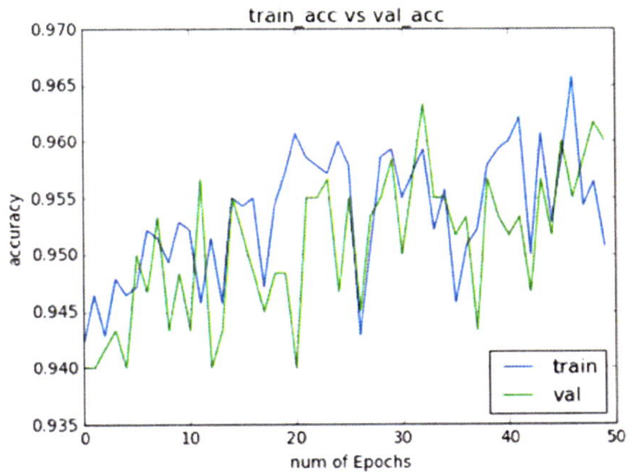

Fig. 6 Accuracy graph (50 iterations)

training accuracy and Green line is indicating the testing accuracy according to the number of iterations. Here, the performance is better compared to the performances for 10 and 30 iterations. Figure 7 depicts the training and testing accuracy graph for 100 iterations. Here, the Blue line is indicating the training loss and the Green line is indicating the testing loss according to the number of iterations. This graph shows that the training and testing accuracy is increasing with the increasing number of iterations. Here, the performance is better compared to the performance for 50 iterations. On the other hand, Fig. 8 represents the training and testing loss graph after 100 iterations. Here, the Blue line is indicating the training loss and Green line is indicating the testing loss according to the number of iterations. This graph shows that the training and testing loss is decreasing with the increasing number of iterations.

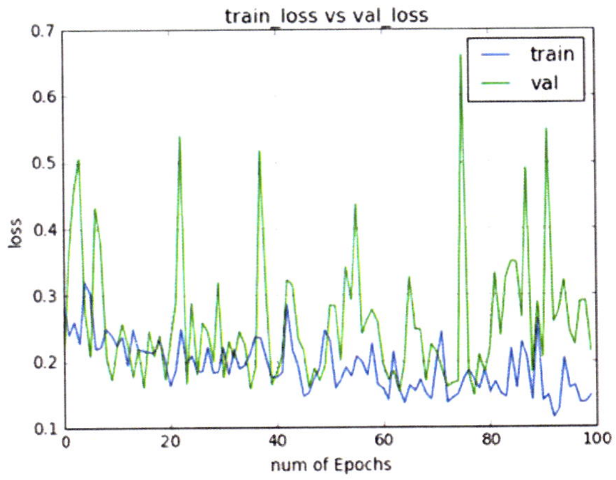

Fig. 7 Accuracy graph (100 iterations)

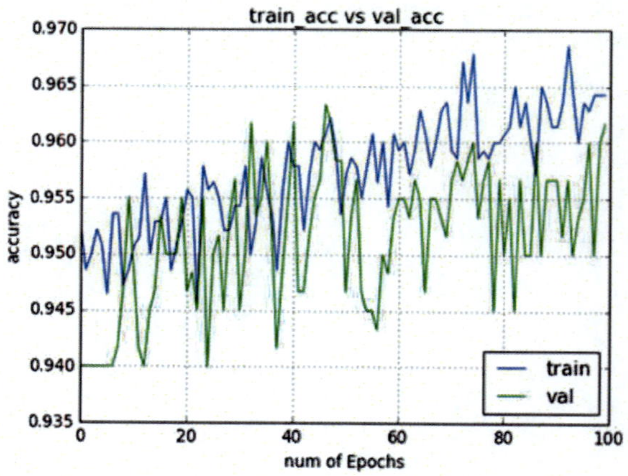

Fig. 8 Loss graph (100 iterations)

Table 2 Accuracy comparison of different models

Model name	Accuracy (%)
Fused detection	58.5
ViF with SVM	88.01
ViF + OviF with SVM	91.93
ViF + OviF with SVM + Adaboost	92.81
STIP and MoSIFT	90
Our model	**96.16**

5 Comparison

For establishing the supremacy of our method, the performance of our model is compared for detecting the violence from video data with other methods [6, 8, 13]. It is observed that our method outperforms these other methods for achieving higher accuracy.

Table 2 is showing different model name and their accuracy, and Fig. 9 is showing a comparison graph of different models in case of their accuracy. It clearly shows that our method performs better in case of identifying violence from video dataset.

6 Conclusion

Recognition of aggressive and vicious activities from videos is an important application area. Many methods have been proposed for detecting violence from videos. Such methods can be useful in video surveillance in public areas, prisons, psychiatric centers, etc. The primary aim of this work is to propose a method for detecting violence from video feeds. 3D convolutional neural network is used for classifying

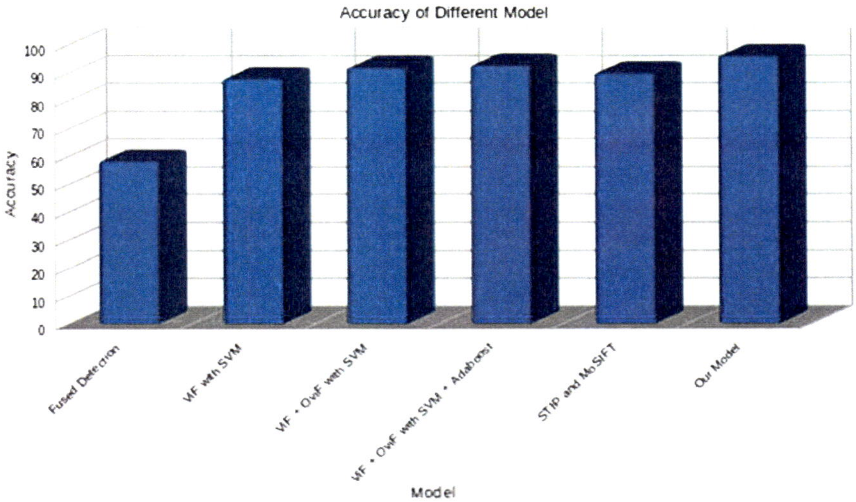

Fig. 9 Comparison with other models

the videos. Like other methods, our method does not depend on any specific criteria of the videos and is capable of identifying violence even if the video quality is poor. The proposed work has achieved 96.17% accuracy by using this architecture. The future research work includes developing a dataset that focuses on the context of our country.

References

1. Agatonovic-Kustrin, S., Beresford, R.: Basic concepts of artificial neural network (ANN) modeling and its application in pharmaceutical research. J. Pharm. Biomed. Anal. **22**(5), 717–727 (2000)
2. Guo, X., Chen, L., Shen, C.: Hierarchical adaptive deep convolution neural network and its application to bearing fault diagnosis. Measurement **93**, 490–502 (2016)
3. Ji, S., Xu, W., Yang, M., Yu, K.: 3D convolutional neural networks for human action recognition. IEEE Trans. Pattern Anal. Mach. Intell. **35**(1), 221–231 (2012)
4. Vashistha, P., Singh, J.P., Khan, M.A.: A comparative analysis of different violence detection algorithms from videos. In Advances in Data and Information Sciences, pp. 577–589. Springer, Singapore (2020)
5. Fouad, A.E., Ali, S., Soliman, M., Osman, A.: Domestic violence inducing females' gynecological and urological problems: the forensic and social perspectives. Ain Shams J. Forensic Med. Clin. Toxicol. **32**(1), 20–30 (2019)
6. Nam, J., Alghoniemy, M., Tewfik, A.H.: Audio-visual content-based violent scene characterization. In: Proceedings 1998 International Conference on Image Processing. ICIP 1998 Cat. No. 98CB36269, vol. 1, pp. 353–357. IEEE, October 1998
7. Clarin, C., Dionisio, J., Echavez, M., Naval, P.: DOVE: detection of movie violence using motion intensity analysis on skin and blood. PCSC **6**, 150–156 (2005)

8. Gao, Y., Liu, H., Sun, X., Wang, C., Liu, Y.: Violence detection using oriented violent flows. Image Vis. Comput. **48**, 37–41 (2016)
9. Graf, H.P., Cosatto, E., Bottou, L., Dourdanovic, I., Vapnik, V.: Parallel support vector machines: the cascade SVM. In: Advances in Neural Information Processing Systems, pp. 521–528 (2005)
10. Mavroforakis, M.E., Theodoridis, S.: A geometric approach to support vector machine (SVM) classification. IEEE Trans. Neural Netw. **17**(3), 671–682 (2006)
11. Suykens, J.A., Vandewalle, J.: Least squares support vector machine classifiers. Neural Process. Lett. **9**(3), 293–300 (1999)
12. Hastie, T., Rosset, S., Zhu, J., Zou, H.: Multi-class adaboost. Stat. Interface **2**(3), 349–360 (2009)
13. Nievas, E.B., Suarez, O.D., García, G.B. Sukthankar, R.: Violence detection in video using computer vision techniques. In: International Conference on Computer Analysis of Images and Patterns, pp. 332–339. Springer, Heidelberg, August 2011
14. Chen, M.Y. Hauptmann, A.: MoSIFT: recognizing human actions in surveillance videos (2009)
15. Willems, G., Tuytelaars, T, Van Gool, L.: An efficient dense and scale-invariant spatio-temporal interest point detector. In European Conference on Computer Vision, pp. 650–663. Springer, Heidelberg, October 2008
16. Yuan, Y., Zheng, H., Li, Z., Zhang, D.: Video action recognition with spatio-temporal graph embedding and spline modeling. In: 2010 IEEE International Conference on Acoustics, Speech and Signal Processing, pp. 2422–2425. IEEE, March 2010
17. Kumar, T.S.: A novel method for HDR video encoding, compression and quality evaluation. J. Innov. Image Process. (JIIP) **1**(02), 71–80 (2019)
18. Bashar, A.: Survey on evolving deep learning neural network architectures. J. Artif. Intell. **1**(02), 73–82 (2019)
19. van der Walt, S., Colbert, S.C., Varoquaux, G.: The NumPy array: a structure for efficient numerical computation. Comput. Sci. Eng. **13**(2), 22–30 (2011)
20. Nair, V., Hinton, G.E.: Rectified linear units improve restricted boltzmann machines. In: Proceedings of the 27th International Conference on Machine Learning, ICML 2010, pp. 807–814 (2010)
21. Dunne, R.A., Campbell, N.A.: On the pairing of the softmax activation and cross-entropy penalty functions and the derivation of the softmax activation function. In: Proceedings of 8th Australian Conference on the Neural Networks, Melbourne, vol. 181, p. 185, June 1997
22. Wang, C., Pelillo, M., Siddiqi, K.: Dominant set clustering and pooling for multi-view 3d object recognition. arXiv preprint arXiv:1906.01592 (2019)
23. Kim, J., Kim, H.: An effective intrusion detection classifier using long short-term memory with gradient descent optimization. In: 2017 International Conference on Platform Technology and Service (PlatCon), pp. 1–6. IEEE (2017)
24. Perkel, J.M.: Why Jupyter is data scientists' computational notebook of choice. Nature **563**(7732), 145–147 (2018)
25. Tang, Y.: TF. Learn: TensorFlow's high-level module for distributed machine learning. arXiv preprint arXiv:1612.04251 (2016)
26. Abadi, M., et al.: TensorFlow: a system for large-scale machine learning. In: 12th USENIX Symposium on Operating Systems Design and Implementation (OSDI 2016), pp. 265–283 (2016)
27. Antonio, G., Pal, S.T.: Deep learning with Keras pipeline. In: Proceedings 1998 International Conference on Image. Packt Publishing Ltd. (2017)

An Online E-Cash Scheme with Digital Signature Authentication Cryptosystem

Md. Ashiqul Islam, Md. Sagar Hossen, Mosharof Hossain, Jannati Nime, Shahed Hossain, and Mithun Dutta

Abstract This paper is intended to enlighten the curious minds on how to use cryptocurrency easily in our day-to-day life. What bitcoin really is? The relation between bank and the user, who has bitcoin, describes potential system design, basic payment method using cryptocurrency, the payment gateway and explains it in the simplest way, and finally the conclusion. Cryptocurrency is one type of digital virtual currencies that have not physically existed. This proposed research work is mainly focused on bitcoin, which is the decentralized digital currency, and it conducts peer-to-peer connection, to make it safe and secure than other digital currency types or hand cash. This paper has proposed an online E-cash scheme with a digital signature authentication cryptosystem that has the tendency to replace the traditional fiat currency, and bitcoin is used instead of the conventional currency and payment system. We can exchange bitcoin to E-cash and E-cash to bitcoin also. This system will find a new way to protect the users from unauthorized transactions in online and offline E-cash system.

Keywords Bitcoin · Cryptocurrency · E-cash · Peer-to-peer connection · Digital signature

1 Introduction

Bitcoin and E-cash are the most common and well-known crypto-currencies in the world. By using bitcoin, users are able to do transactions with each other and users can exchange values digitally with the bank [1]. In this model, users are able to do transactions. Bitcoin or E-cash is used in various applications like bank, online super shop/market, and other such online application [2]. According to this situation,

Md. Ashiqul Islam (✉) · Md. Sagar Hossen · M. Hossain · J. Nime · S. Hossain
Daffodil International University, Dhaka, Bangladesh
e-mail: ashiqul15-951@diu.edu.bd

M. Dutta
Rangamati Science and Technology University, Rangamati, Bangladesh
e-mail: mithundutta92@gmail.com

P. Karuppusamy et al. (eds.), *Sustainable Communication Networks and Application*, Lecture Notes on Data Engineering and Communications Technologies 55,
https://doi.org/10.1007/978-981-15-8677-4_3

bitcoin can be changed into E-cash and E-cash can be changed into bitcoin. E-cash will be available in two types, they are

1. Network type
2. Smart Card type.

Users can complete the transaction by picking up one suitable way. If the user has bitcoin, then they will either exchange or withdraw E-cash from the bank as well. User can also make their payment in the super shop. Users can easily pay their bill by using bitcoin if the super shop has a bitcoin wallet. Otherwise, super shop will be unable to receive bitcoin because their receive system is only E-cash, then the customer has to convert their bitcoin into E-cash [3]. When the customers pay the bill, then the super-shop system sends a notification message to customers [4]. If the user gives the permission to the system to receive bill, an acknowledgment is generated automatically. Super shop can also deposit their bitcoin or E-cash in the bank [5]. Super shop can also exchange bitcoin into E-cash from the bank.

In this system, users order goods from the super shop and the user can make payment by bitcoin or E-cash. We think this method will be one of the easiest ways to online transactions all over the world [6]. The proposed approach components, incorporating the applicable criteria that follows.

2 Related Work

We know that there are many types of digital currencies in the world. And I got ideas about market capitalization [7] from different websites. Among them bitcoin, Litecoin, Ethereum, are notable. Peter mentions in his paper about its future uses [1]. We know about the close relationship between bitcoin and blockchain [8]. Bitcoin is just a fancy method of blockchain known as cryptocurrency. William mentions in his writing that bitcoin is like a memory [9]. In 2015, Raymaekers wrote in his paper about the challenges and future of using the cryptocurrency bitcoin [3]. Also, the use of bitcoin in e-commerce has led to a technological revolution [6]. Buying and selling products directly in e-commerce using bitcoin or digital currency has increased [10]. Blockchain technology and digital signature are used to find out the way of the secure transactions [11–13]. The hash function makes the digest to encrypt and decrypt the data with an asymmetric algorithm [14]. Above all, the massive growth of bitcoin has been discussed in the paper [15]. In [16] paper, author shows the group signature method that can use the banker and bank control transaction by group signature this system customer depends on the bank. Sattar J. Aboud writes his paper method of offline E-cash system, and we got information about his analysis account that can be hacked by insiders of the untrusted bank or malicious [17]. İn [18] paper, we can see the scheme of the E-cash system and its security level. In our papers, we make sure that our security system is definitely better than [18] other security systems.

3 Basic Model

We are living in the digital era of new technology where everything is digitalized. In future, there will be no hand cash, day-by-day the technology is improved and the human has increased their lifestyle at the next level. So, people are interested in E-cash more than hand cash. E-cash is easy to use and people are purchasing their goods easily by using E-cash system. So, we proposed an online E-cash scheme with a digital signature authentication cryptosystem to make their lives faster and easier. We want to use bitcoin in our proposed system to explore the new trend. Bitcoin is similar to cryptocurrency, and it is a decentralized peer-to-peer electronic digital currency in the world [19]. It does not have a central issuing authority or regulatory body that's why its process is faster and transactions are accurate [9]. In this paper, we make a relationship between the user, bank, and super shop/application. A user can withdraw E-cash online/offline both methods from the bank, and users can also convert their bitcoin into E-cash from the bank as well. A user can buy goods from the super shop/application, if the super shop/application has a bitcoin wallet, then they can receive the bill in bitcoin, otherwise, users have to pay the bill as E-cash. Super shop/application can deposit their E-cash in the bank and they can also convert their bitcoin into E-cash from the bank [1]. This simple online E-cash scheme is more secure, and this cryptocurrency method is more reliable and faster than any other currency systems (Fig. 1).

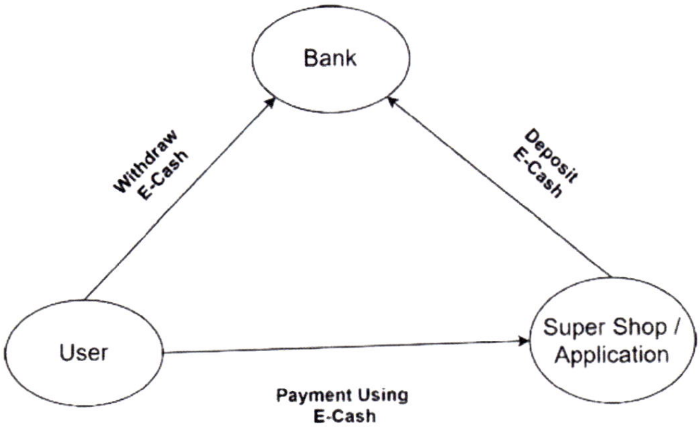

Fig. 1 Simple basic model

4 System Operation

4.1 System Design

There are three actors in the E-cash model. The first one is user, the second one is bank and another one is super shop/application. Users can withdraw E-cash from the bank. If the users have bitcoin, they can convert it into E-cash by the bank transaction and then they can withdraw E-cash as the same process easily. A user can pay any type of bill in the super shop by E-cash. If the super shop does not have any system to receive bitcoin, then the super shop can take their bill from the customer on E-cash method by their own super-shop application. With this application, customers will pay their bills easily. Super shop can deposit money to the bank on E-cash (Fig. 2).

In this system, we are showing the easiest way to the user to use bitcoin in real life. The system also represents how to make payment by using bitcoin. A user can order necessary goods from the super shop by using the super-shop apps. After ordering, user can complete the user's payment in two ways.

First one, a user can make payment by bitcoin. If the user wants to pay through bitcoin, the system will convert it into E-cash or money, and then the super shop will receive the money. After receiving money super shop will prove the payment [14].

The second one is, a user can pay on E-cash. If the system will convert E-cash to money than the system can receive payment from the customer, otherwise transaction will be cancel.

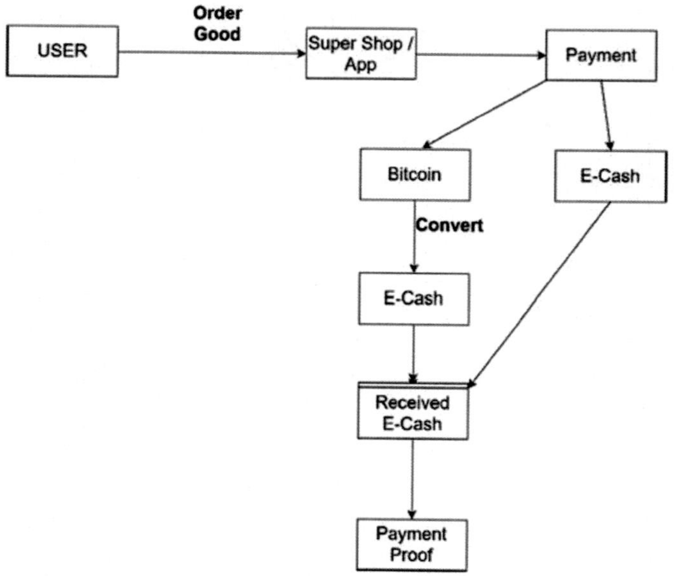

Fig. 2 Proposed system

4.2 Hash Function

Hashing algorithm is a mathematical procedure that takes the information contained in files and scrambles it to create a fixed-length string of numbers and characters [7]. In this paper, we approach SHA-256 to digest the information and encrypt it using the user's private key and users can share their public key to decrypt it [11, 12], and get the payment to the client. A strong cryptographic and blockchain technology process the whole system easily [14]. So that no one can freeze the account and the transaction is safe.

$$P(H(X) = i) \sim 1/N \tag{1}$$

Here,

- $[0, N]$ is a small fixed-length output.
- $i =$ infinite and it is a large input domain.
- Regular hash function $h = H(X)$ is using for the hash table.

We are using strong collision resistance to perform the hash function, and it can computationally infeasible to find any pair (X, Y) such that $H(X) = H(Y)$

4.3 Payment Gateway

In the payment gateway, the system represents a user on how to make payment. A user can make a payment on bitcoin then it will be converted into money by super-shop application. This application sends a notification with a request for verification to the customer. We will observe two options YES or No. If the customer presses YES then he/she gives access to generate OTP. Then a secret code will be sent through OTP to super-shop application [4]. The secret code will authenticate by the system. After authenticating, the system verify and confirm certificate acknowledgement, will be sent to the customer. Customers receive a short message of successful payment [14]. If the customer presses No, then the transaction will be canceled (Fig. 3).

Our approach proposed system is functionally related to credit or debit card system, but our approach system is more secure than any other system. The debit card or credit card system produce hand cash, and our approach system produce safe E-cash transaction.

4.4 Limitation of Some Research Works

(Table 1).

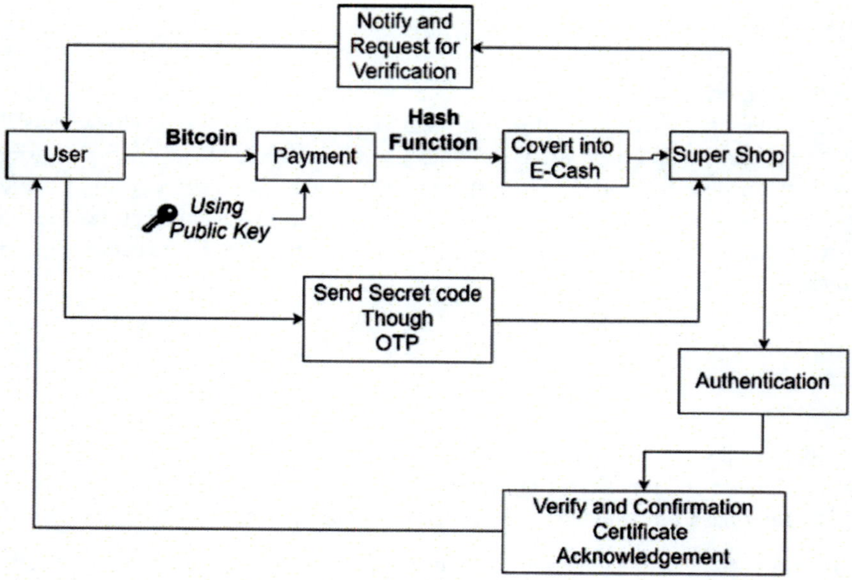

Fig. 3 Payment gateway

4.5 Proof of Work

Hashcash is a method that is generated proof of work to verify the data in an easy way [20]. Proof of work is a random process with low probability. The trial and error are required to generate valid proof [21]. Hashcash proof of work is using to generate a new block in the bitcoin.

$$\text{Hash}(B) \leq \text{balance}(U) * M/D \tag{2}$$

Here,

- Hash (B) is the hashing function.
- B is the block header of the current block.
- Balance (U) is a function that returns the balance of address U.
- M is the maximum value of the hashing function.
- D is the target difficulty.

4.6 System Algorithm

Step 1: Start the procedure.
Step 2: Input all of the goods (products) items.
Step 3: Get an online memo to purchase.

Table 1 Limitation of some research work

Author information	Limitation	Our approach system
Mauro Conti, Senior Member, IEEE, Sandeep Kumar E, Member, IEEE, Chhagan Lal, Member, IEEE, Sushmita Ruj, Senior Member, IEEE, "A Survey on Security and Privacy Issues of Bitcoin" arXiv:1706.00916v3 [cs.CR] 25 Dec 2017	Anonymity level depends on the number of participants, vulnerable to DoS (by stalling joint transactions), Sybil, and intersection attacks, prevents plausible deniability	Our approach system security is better than others. So, our system will be safe from any other attack as like as DoS
Aboud, Sattar. (2014). Analysis of Offline E-cash Schemes International Journal of Advanced Research	They cannot fulfill about crypto security	Blockchain technology, hash function are mainly focus on crypto base security to protect the system from unauthorized access
U. Mukhopadhyay, A. Skjellum, O. Hambolu, J. Oakley, L. Yu and R. Brooks, "A brief survey of Cryptocurrency systems," 2016 14th Annual Conference on Privacy, Security and Trust (PST), Auckland, 2016, pp. 745-752, https://doi.org/10.1109/pst.2016.7906988	Their model cannot ensure proper security-based work	We ensure high security with digital signature authentication [13]
H. Tewari and E. O. Nuallain, "Netcoin: A Traceable P2P Electronic Cash System," 2015 IEEE International Conference on Web Services, New York, NY, 2015, pp. 472-478, doi:10.1109/ICWS.2015.69	This system showed peer-to-peer transaction system but they did not connect to the bank this is an important thing about economic	Our approach system is decentralized and peers to peer connection in transaction [19]. We can establish a better relationship between the bank, user, and super shop/application

Step 4: Give the payment using bitcoin; if Yes, then convert the bitcoin into E-cash, otherwise E-cash has to be paid.
Step 5: Verify the payment; if Yes, then receive the money and get notification to the customer with payment confirmation and acknowledgment, otherwise stop the procedure.
Step 6: End the procedure.

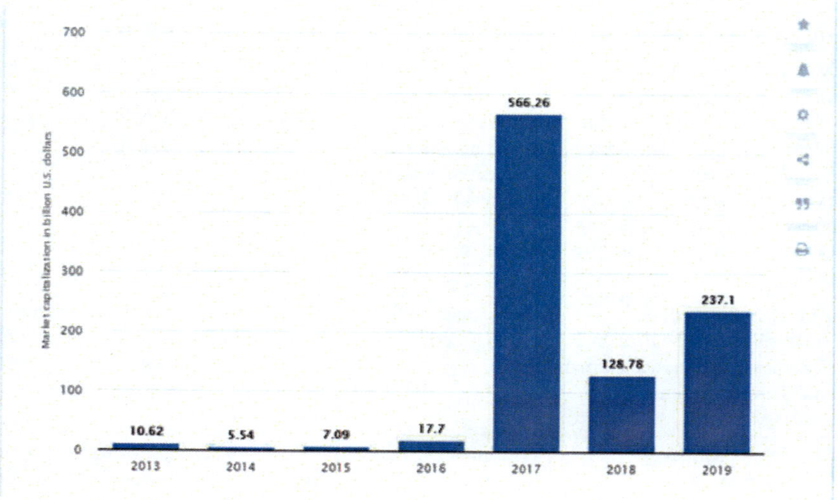

Fig. 4 Market capitalization of bitcoin from 2013 to 2019

5 Cryptocurrency Market Analysis

Based on the 2019 market analysis, bitcoin was the most valuable cryptocurrency in the world. From the world market analysis, bitcoin's fastest-growing market is in North America and the largest market in Europe, the Middle East, and Africa. The CAGR of bitcoin is 62.2% [7] (Fig. 4).

From the market capitalization of bitcoin 2013 to 2019 analysis report, we can see that bitcoin is the fastest-growing cryptocurrency in the world and day-by-day its value is increasing rapidly [7]. In 2017 bitcoin market capitalization was $566.26 bn USD, it was the highest market value of bitcoin and 2019 market capitalization was $237.1 bn USD. Today's market analysis bitcoin price is 9164.54 USD, and bitcoin cash price is 221.8 USD, where the Etherium price is 225.71 USD. In Europe/Latin America, 85.6% of people are using cryptocurrency [7]. It is the most valuable currency in the future and no one could use the hand cash. So, we are approaching this system based on market capitalization analysis of cryptocurrency.

6 Observation

Our approach proposed system efficiency is good, and it can be performed very well. The blockchain technology and cryptographic algorithm increase the system accuracy and make it more powerful. A comparative study between existing system and proposed system [22] is written below (Table 2).

Table 2 A comparative study of existing and proposed system

S. No.	Existing system	Proposed system
1	The existing system can produce physical hand cash	The proposed system is an approach to use digital currency
2	The physical currency transaction rate is high, third party costs are included in their transaction process	Bitcoin/E-cash transaction cost is less than other currency
3	Physical currency is not safe for users, it can harmful and sometimes the user can face the robbery problem. There is a lack of privacy and scalability	Digital currency is more secure and blockchain technology is used to increase the number of security and Special emphasis on privacy and scalability
4	Central banks are creating a vital role in monitoring the currency policy	Decentralized digital currency and generate peer-to-peer connection easy to use [8]

7 Advantages

i. Freedom of payment.
ii. Bitcoin wallet account is safe and there is no possibility to freeze the account by anyone.
iii. The transaction process is well organized, no one could rob the bitcoin/E-cash easily. Blockchain technology is used to protect the wallet for hackers or unauthorized access [23].
iv. The transaction cost is less than the other cost.
v. No third party is interrupted by their transaction process, so it is smooth [24].

8 Conclusion

From the above research work, it is concluded that the bitcoin is considered as a new platform to deploy digital currency. This research work explains the process, advantages, and disadvantages of bitcoin [1]. Bitcoin is not perfect but it remains useful for users. It is a safe transaction network [22]. Bitcoin transaction is more secured, and it does not flash users' personal information. The digital currency exchanges have now gained the access to convert bitcoins into fiat currency. Electronic cash provides strong security [2], where the bitcoin wallet cannot be hacked by the intruders. The transaction procedure is safe [25] and no third party can interrupt the transaction [24]. Nowadays, 200 million bitcoin wallets are available in the world and more than 400,000 people are using bitcoin every day [6]. Day-by-day the bitcoin wallet users are increasing with a continuous increase in its economic value, where it can help us to make our new digital lifestyle more secure and safe.

9 Future Work

Physical money and the physical transaction will be decreased while all the systems are digitalized, and people can use to do transactions digitally by using bitcoin or E-cash. In the future, the world transaction will be held by bitcoin [15]. In the future, interest in E-cash will be surveyed and this proposed approach will be modified to achieve a better user experience and satisfaction.

References

1. DeVries, P.D.: An analysis of cryptocurrency, bitcoin, and the future. Int. J. Bus. Manag. Comm. **1**(2), 1–9 (2016)
2. Jahan, I., Keam Psyche, K., Dutta, M.: An off-line E-Cash scheme based on group blind signature scheme. Int. J. Eng. Res. Dev. **12**(8), 46–53 (2016). www.ijerd.com. e-ISSN 2278-067X, p-ISSN 2278-800X
3. Raymaekers, W.: Cryptocurrency Bitcoin: distribution, challenges and opportunities. J. Payments Strat. Syst. **9**(1), 30–40 (2015)
4. Dutta, M., Psyche, K.K., Khatun, T., Islam, M.A., Islam, M.A.: ATM card security using bio-metric and message authentication technology. In: 2018 IEEE International Conference on Computer and Communication Engineering Technology (CCET), Beijing, pp. 280–285 (2018). https://doi.org/10.1109/ccet.2018.8542227
5. Eslami, Z., Talebi, M.: A new untraceable off-line electronic cash system. Electron. Comm. Res. Appl. 10(1), 59–66 (2011). https://doi.org/10.1016/j.elerap.2010.08.002, (http://www.sciencedirect.com/science/article/pii/S1567422310000657), ISSN 1567-4223
6. E-commerce (n.d.): Evolution or revolution E-commerce shifts into higher gear. E-commerce, 2 January 2014. http://ir.nielsen.com/files/doc_financials/Nielsen-Global-E-commerce-Report-August-2014.pdf. Accessed 2 Apr 2020
7. Crypto-Currency Market Capitalizations. http://coinmarketcap.com/. Accessed 20 Apr 2020
8. Twesige, R.L.: A simple explanation of Bitcoin and Block Chain technology. In: Conference Paper, January 2015. https://doi.org/10.13140/2.1.1385.2486
9. Luther, W.J., Olson, J.: Bitcoin is memory. J. Prices Mark. **3**(3), 22–33 (2013). https://ssrn.com/abstract=2275730, http://dx.doi.org/10.2139/ssrn.2275730
10. Harshita, S.T.: Implementation of fair-exchange and anonymous online E-cash protocol for E-commerce. Int. J. Inf. Technol. Comput. Sci. **8**, 66–74 (2016). https://doi.org/10.5815/ijitcs.2016.08.08
11. Karim, R., Rumi, L.S., Islam, M.A., Kobita, A.A., Tabassum, T., Hossen, M.S.: Digital signature authentication for a bank using asymmetric key cryptography algorithm and token based encryption. In: International Conference on Evolutionary Computing and Mobile Sustainable Networks (ICECMSN 2020) (2020)
12. Hossen, M.S., Tabassum, T., Islam, M.A., Karim, R., Kobita, A.A., Rumi, L.S.: Digital signature authentication using asymmetric key cryptography with different byte number. In: International Conference on Evolutionary Computing and Mobile Sustainable Networks (ICECMSN 2020) (2020)
13. Kazmirchuk, S., Anna, I. and Sergii, I.: Digital signature authentication scheme with message recovery based on the use of elliptic curves (2020). https://doi.org/10.1007/978-3-030-16621-2_26
14. Islam, M.A., Kobita, A.A., Hossen, M.S., Rumi, L.S., Karim, R., Tabassum, T.: Data security system for a bank based on two different asymmetric algorithm cryptography. In: International Conference on Evolutionary Computing and Mobile Sustainable Networks (ICECMSN 2020) (2020)

15. Team B: Understanding Bitcoin's Growth in 2015, 20 January 2016. https://blog.bitpay.com/understanding-bitcoins-growth-in-2015/. Accessed 5 July 2020
16. Tsang, P.P., Wei, V.K.: Short linkable ring signatures for e-voting, e-cash and attestation. In: Information security practice and experience, first international conference, ISPEC 2005, Singapore, April 11–14 (2005). https://doi.org/10.1007/978-3-540-31979-5_5
17. Sattar, A.: Analysis of offline E-cash schemes. Int. J. Adv. Res. **2**(8), 406–410 (2014)
18. Yulei, Y., Xiaolei, D., Zhenfu, C.: A trustee-based and efficient divisible e-cash scheme[J]. J. Comput. Res. Develop. **52**(10), 2304–2312 (2015)
19. Nakamoto, S.: Bitcoin: a peer-to-peer electronic cash system. https://bitcoin.org/bitcoin.pdf. Accessed 5 May 2020
20. Ren, L.: Proof of stake velocity: building the social currency of the digital age (2014)
21. King, S.N.S.: PPCoin: Peer-to-Peer Crypto-Currency with Proof-of-Stake, August 2012. http://www.peercoin.net/assets/paper/peercoin-paper.pdf. Accessed 8 May 2020
22. Dutta, M., Islam, M.A., Mamun, M.H., Psyche, K.K., Al Mamun, M.: Bank vault security system based on ınfrared radiation and GSM technology. In: Hemanth, D., Shakya, S., Baig, Z. (eds.) Intelligent Data Communication Technologies and Internet of Things, ICICI 2019. Lecture Notes on Data Engineering and Communications Technologies, vol. 38. Springer, Cham (2020)
23. Mohanty, S., Majhi, B., Das, S.: A secure electronic cash based on a certificateless group signcryption scheme. Math. Comput. Model. **58**(1–2), 186–195 (2013). https://doi.org/10.1016/j.mcm.2012.06.004, (http://www.sciencedirect.com/science/article/pii/S0895717712001276), ISSN 0895 7177
24. Chang, C.-C., Chen, W.-Y., Chang, S.-C.: A highly efficient and secure electronic cash system based on secure sharing in cloud environment. Sec. Commun. Netw. **9**(14), 2476–2483 (2016). https://doi.org/10.1002/sec.1517
25. Dutta, M., Psyche, K.K., Yasmin, S.: ATM transaction security using fingerprint recognition. Am. J. Eng. Res. (AJER) **6**(8) (2017). e-ISSN 2320-0847, p-ISSN 2320-0936

Smart Electrification of Rural Bangladesh Through Smart Grids

Dhrupad Debnath, Abdul Hasib Siddique, Mehedi Hasan, Fahad Faisal, Asif Karim, Sami Azam, and Friso De Boer

Abstract A smart grid is a new technology that integrates power systems with communication systems. It is an intelligent and efficient management system that has self-healing capabilities. The smart grid can be applied to manage networks that integrate different types of renewable resources for power generation. Bangladesh is currently experiencing severe power deficiency. Renewable energy sources such as solar power and biogas can play an important role in this scenario, especially in rural areas where electricity is even scarcer. By applying prototype concepts of smart grid, power generation from renewable resources and efficient load management can be achieved by a centralized control center. This will control the on-off sequence of the load and maintain the system stability. In this paper, different aspects of implementing a prototype of the smart grid in the rural areas of Bangladesh are discussed.

D. Debnath
Department of Electrical and Computer Engineering, University of Alberta, Alberta T6G 2G7, Canada
e-mail: d.debnath16@gmail.com

A. H. Siddique · M. Hasan
Department of Electrical and Electronic Engineering, University of Science and Technology Chittagong, Chattogram 4202, Bangladesh
e-mail: ahnion.ahs@gmail.com

M. Hasan
e-mail: mehedi.hasan01@northsouth.edu

F. Faisal
Department of Computer Science and Engineering, Daffodil International University, Dhaka 1207, Bangladesh
e-mail: fahad.cse@diu.edu.bd

A. Karim · S. Azam (✉) · F. De Boer
College of Engineering, IT and Environment, Charles Darwin University, Casuarina, NT 0810, Australia
e-mail: sami.azam@cdu.edu.au

A. Karim
e-mail: asif.karim@cdu.edu.au

F. De Boer
e-mail: friso.deboer@cdu.edu.au

P. Karuppusamy et al. (eds.), *Sustainable Communication Networks and Application*, Lecture Notes on Data Engineering and Communications Technologies 55,
https://doi.org/10.1007/978-981-15-8677-4_4

Keywords Smart grid · Rural area · Power demand · Renewable energy

1 Introduction

Capability in electrical power generation is a major driving force towards socio-economic development of a country. For a developing country like Bangladesh, the electricity capacity of power generation is far less than the demand [1, 2]. Therefore, to cope with the rapidly escalating demand for electricity, rigorous attention needs to be provided to generate electrical energy from renewable energy sources [3]. Rural areas of Bangladesh have limited access to the electrical power supply from the national grid compared to the urban areas [4]. People living in rural areas which do having electricity encounter frequent power shortages [4]. Solar photovoltaic (PV) systems are widely implemented in these areas [4]. To provide electricity in rural areas with renewable resources, a prototype of the 'smart grid' concept can be considered. This approach can lead to cost-effective deployment, efficient operation and good maintenance of the deployed system [3]. Some aspects of the smart grid systems can be implemented in this type of approach. This paper provides a discussion on how a basic framework of a smart grid system can be used to achieve control strategies and well-organized operation of a prototype system that incorporates renewable resources in rural areas of Bangladesh. The existing power system situation is discussed first as well as the current state of renewable resources of Bangladesh. A brief overview of the framework of a smart system is provided afterwards. Challenges of integrating renewable resources are highlighted. Finally, the different aspects of implementing the prototype using smart grids concept are discussed.

2 Existing Power Scenario of Bangladesh

Bangladesh, a low-income developing country [5], is highly vulnerable to setbacks arising from the ongoing electricity crisis. Natural gas, the main source of fuel for energy generation, is responsible for around 72% of the total commercial electricity consumption and around 81.72% of the total electricity generated [6, 7]. But, studies show indicate that the gas demand will increase up to 4,567 mmcfd by 2019–2020 [8] resulting in a shortfall of around 1,714 million cubic feet per day (mmcfd) [8]. Even if a slow growth rate of GDP is considered as 5.5% till 2025, Bangladesh will need about 19,000 MW of additional power each year [8].

Solving the ongoing electricity crisis was one of the major issues in the election manifesto of the current government [8]. The government focused on Quick Rental Power Plants which run on diesel fuel. Initially, it was able to reduce the gap of generation and demand for electricity, but the fuel cost made the cost of electricity considerably higher [9]. Due to the latest rise in oil prices in the international market, government subsidies for petroleum-based power plants went up. As a result, the price of electricity per unit was increased three times within just four months. Cost per KW-hr of electricity rose from TK 4.16 (USD 0.051) to TK 4.72 (USD 0.058) on December 1, 2011, then to TK 5.02 (USD 0.061) on February 1, 2012 and finally to TK 5.32 (USD 0.065) on March 1, 2012 [10, 11]. The average production cost of a unit of electricity is Tk 5.70 (USD 0.070), according to the Bangladesh Power Development Board (BPDB) [9].

As a developing country, it is not surprising that Bangladesh depends quite heavily on coal to produce electricity [1]. In the short term road map to meet energy demand by 2015, the Bangladesh government had contracted almost 2600 MW of coal fire power plants in Chittagong and Khulna [1]. A study of carbon emission shows that carbon dioxide emission rate (Metric tons per capita) increased very rapidly from 0.2 in 2005 to 0.3 in 2008 [3]. These developments current developments in power generating plants will make Bangladesh more vulnerable to high carbon emission.

System loss is another major problem for power systems in Bangladesh. This consists of the line loss, heat loss, unaccounted energy usage and electricity theft. System loss affecting public utilities is a persistent problem in Bangladesh's infrastructure. According to BPDB, the overall system loss calculated in 2008–2009 was 6.58% of net generation [1]. The monthly system loss for various months of the year of 2011 in Dhaka, based on information provided by Dhaka Electricity Supply Company Limited (DESCO) [2] is shown in Fig. 1.

The inefficiency of the overall transmission and distribution system, shortage of gas, unavailability of some power plants from time to time due to maintenance,

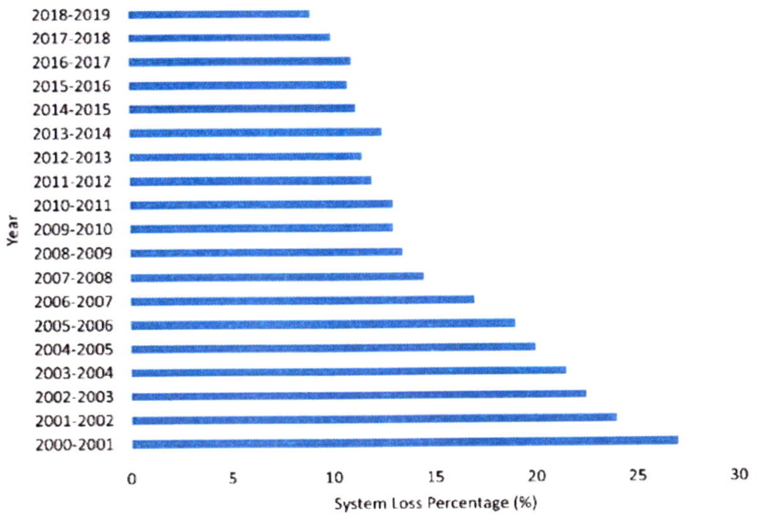

Fig. 1 System loss percentage in Bangladesh from 2000–2019 [2]

rehabilitation and overhauling, degenerated capacity due to aging of power plants are also some other problematic issues for the power system in Bangladesh [1].

The dependency on natural gas, oil and coal as well as abovementioned issues have changed the focus to renewable energy resources. Biomass, Biogas, Solar, Hydropower and Wind are the potential sustainable sources of energies. Energy security, as well as a cost-efficient and effective power supply to the off-grid rural areas of Bangladesh, can be provided if electricity can be harnessed from these renewable resources [3]. It is therefore imperative to study the renewable energy scenario of Bangladesh.

3 Renewable Energy Scenario in Bangladesh

To meet the existing deficit in power generation, renewable resources are becoming more attractive alternatives in Bangladesh. As a result, a policy for the effective utilization of renewable energy resources has been adopted by the Government of Bangladesh (GOB) [3]. The focus of this policy is to shift the considerable dependency on conventional fossil fuel-based thermal power plants. Global depletion of fossil fuels, increasing cost of purchasing and importing as well as a desire to move towards clean energy are the main driving forces behind it. So, the necessity of harnessing energy from renewable resources is essential. Among different types of available renewable resources in Bangladesh, the most potential ones which be explored rigorously by Bangladesh are discussed below.

3.1 Solar Energy

The abundance of solar radiation (daily 4.0–6.5 kWh/m^2) [12] has enabled the potential growth of solar-based energy resources in Bangladesh. The months of March to April and the months of December to January provide maximum and minimum solar radiation respectively [13]. Photovoltaic (PV) solar systems and Concentrating Solar Power (CSP) systems are the most common technologies. PV solar systems have been implemented extensively throughout the country, mostly in rural areas [3]. 801,358 Solar Home System (SHS), having a capacity of 36.5 MW, had been installed by January 2011 [3].

3.2 Biomass Energy

Bangladesh is an agricultural country. So, biomass is the most notable energy source in Bangladesh as biomass covers all types of organic matters which are available from a different type of crops. Biogas plants also use animal wastes from dairy and poultry farms. 70% of total final energy consumption is produced by biomass in Bangladesh [14].

3.3 Wind Energy

Wind energy generation has some prospects in the coastal areas of Bangladesh. The average wind speed available in coastal areas is 6.5 m/s and the density of wind power varies from 100 to 250 W/m^2 [15].

3.4 Hydro Energy

Bangladesh is a land of rivers. Approximately 1.4 trillion cubic meters of water per year flowing through different rivers. Even though the land is fairly plane, high current flows through major rivers for six months of the year which provides some locations with the prospect of 10KW to 5 MW power generation capacity [3].

Based on a study by Kaiser et al. [16], the relative contributions in terms of installed capacity in MWp for five renewable resources in Bangladesh are shown in Fig. 2. From the chart, it can easily be interpreted that solar and biogas are playing the most significant role. So it can be concluded that for rural areas, solar power can be the obvious choice for the main power generation source of a prototype grid where some technologies of smart grid concepts are implemented. For the successful implementation of this type of approach, the basic elements of the smart grid need to be reviewed.

4 Framework of Smart Grid

A smart grid is an intelligent network that uses digital and other modern technologies to observe and supervise the transportation of electricity from different generation resources to meet the dynamic electricity demands of the customers. Smart grid manages the requirements and capacities of all parts of the system including generators, grids, and customers as efficiently as possible. As a result, it minimizes costs and environmental impacts and maximizes system stability, reliability and resilience [17].

Smart grid helps to operate and proficiently manage the existing grid. The integration of a bi-directional communication system integrated with the power system is the most unique feature of the smart grid. Perhaps the most important feature of a smart grid is the ability to dynamically integrate the variable renewable resources which helps to reduce carbon emission and assists to meet the future power demand [18]. To maintain the balance of supply and demand, it also includes a storage system.

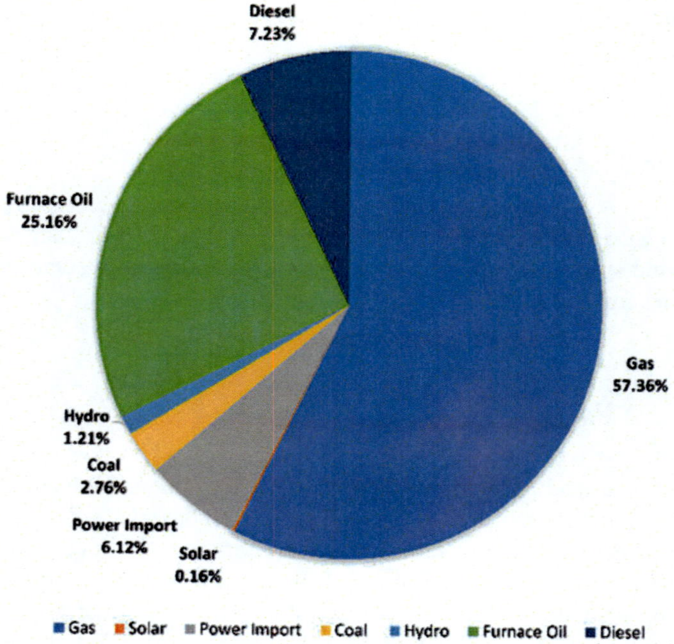

Fig. 2 Contribution of different implemented renewable energy technologies in Bangladesh in the year 2019 [2]

It is equipped with a real-time metering system that allows customer interaction facility, dynamic tariff system and demand-side management. Self-healing feature is an important characteristic of a smart grid. To maintain the stability of the system, the operators can manage the distributed resources to provide reactive power, voltage support and other ancillary services using the two-way communication infrastructure of the smart grid [19].

The technological improvements in the areas of communication, control and sensor technology are enabling the gradual implementation of a smart grid all over the world. The existing grids need to be updated with modern technological advancements for the transformation towards a smart grid. A standardized bi-directional communication system is the first area that needs to be updated. Integration of improved power electronics and measurement devices is also an integral part of a smart grid. An appropriate control method needs to be developed and implemented for the successful application of smart grids [20, 21]. The proposed prototype system borrows these features of smart grid technology along in addition to some other elements.

5 Smart Grid Prototype for Rural Areas of Bangladesh

Solar power is the most prominent source of renewable energy in Bangladesh [16]. For rural areas of Bangladesh, the Solar PV systems are most widely used. The goal of this smart grid prototype system is to provide electricity to a group of rural households of Bangladesh with the Solar PV system (Fig. 3).

5.1 Structure of Smart Grid Prototype

A stand-alone solar PV system consists of a solar panel, a charge controller circuit and a battery [22, 23]. As our existing system is an AC grid, the electricity obtained from the solar PV system is fed to the inverter before reaching to the load [23]. In case of a DC grid, the electricity can be directly provided to load.

Renewable source of energy, that is the solar PV system, will be connected to the grid through a charge controller circuit. Depending on demand and generation, the charge controller circuit will control the charge flow to the energy storage device and the load. During high demand, available electricity will be directed to the load. During excess generation, the controller circuit will charge the battery so that this energy can be used later. Also, the controller circuit will charge the battery if the

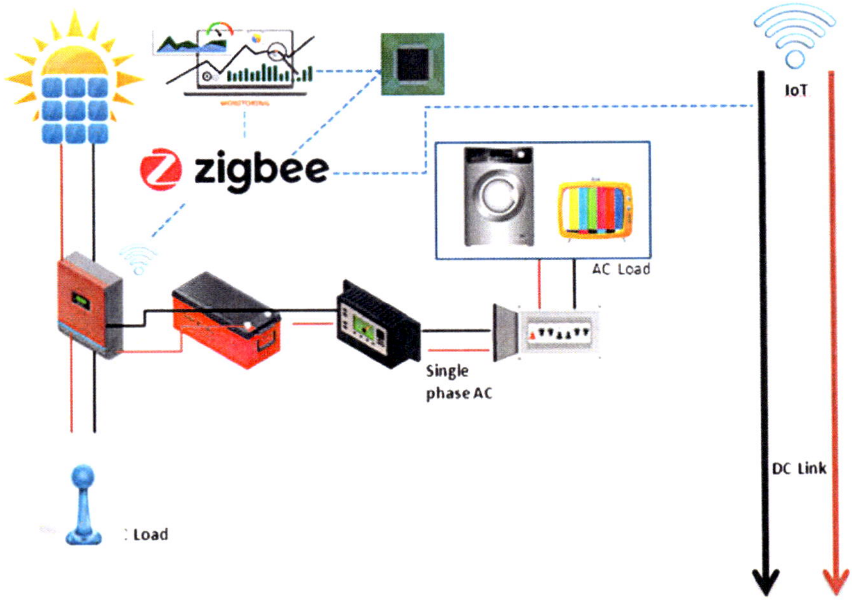

Fig. 3 Overview of a smart grid prototype for rural Bangladesh

charge level falls below the threshold value of the battery. When there is a generation greater than the rated value, the controller circuit can trip to protect the devices as well [24]. The electricity produced is connected to the distribution system. A Central Control Unit (CCU) controls the total system. It includes communication devices, monitoring devices and decision-taking system.

The communication with other devices can be maintained with a wired or wireless connection. Wireless communication systems can include Zigbee and IEEE 802 based standards for short-range and moderate range applications respectively [25]. Power Line Communication (PLC) system can be implemented for the wired system. These components will have a built-in modem. All the components will be communicating with the CCU using the modems. Communication is one of the most important parts of this system. So, it needs to be ensured that the communication links among the modems have minimal channel errors and thus high reliability.

The monitoring system will be composed of different measuring devices. These devices will measure different parameters such as voltage level, current, energy usage, generation, demand, etc. Since the loads are not sensitive, the measurements will be taken and transmitted to CCU periodically. This will result in a less complex system.

The decision-taking system of CCU will analyze these data. Based on it, it will take decisions that will be transmitted to respective components through modems. So, necessary actions will be taken corresponding to current parameters for efficient operation and management. For successful operation of the prototype, several other applications are also required.

5.2 Prototype of DC Grid

Since a solar PV system is applied, hence DC power will be generated. So, a prototype of the DC grid is required. The solar panel will be connected to a charge controller circuit which will be then connected to a DC grid through an energy storage device like a battery. The DC grid will be connected to loads through inverters which will convert the DC power to AC. All the inverters will be connected with CCU. Figure 4 shows the setup of the prototype DC grid. DC transmission system is preferred over AC transmission due to its higher efficiency. Only loss got is at the conversion of DC-AC. This is about 5–10% of the generated power. As micro inverters are used the loss will be towards the lower side of the band.

5.3 Synchronization of Voltage, Frequency and Phase

Synchronization of voltage, frequency and phase is an imperative aspect of this system. The inverters need to be synchronized before connecting them to the grid. A load and its associated inverter will be taken as reference node by the CCU. All other inverters will be synchronized corresponding to this inverter by CCU. The CCU

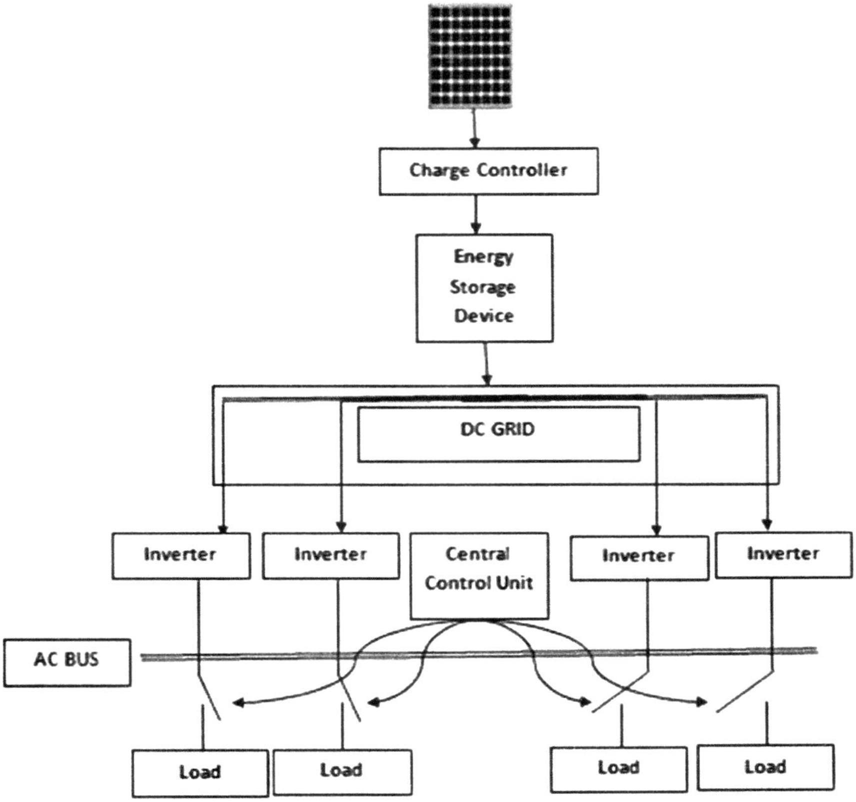

Fig. 4 Prototype of DC grid

will measure the voltage level. There are multiple inverters in this system each one connected to DC link. Each inverter is driving different load in the network. CCU controls the on-off sequence of these inverters according to the demand. Multiple inverter system has been proposed for this system to increase the reliability of the system. A central big inverter can go down even if one solar panel does not function properly whereas micro inverter diminishes this problem.

The control algorithm can be expressed as below.

- Initialize the reference inverter out of 'N' inverters and assign its index $i = 0$. Turn off all the switches at other loads.
- Measure the voltage (v), frequency (f) and phase ($\emptyset$) of each inverter and transmit to the CCU:
  $\{F\}_i = \{v, f, \emptyset\}_{\rm i}$ where $\rm i = 1, 2{,}3 \ldots \ldots \ldots (N - 1)$
- while $\{F\}_{\rm ref} \neq \{F\}_i$ do
  if $\{F\}_{\rm ref} < \{F\}_i$ then $\{F\}_i = \{F\}_i - \Delta\ \{\rm F\}_{\rm i}$
  where $\Delta\ \{F\}_i =$ step size

elseif $\{F\}_{\text{ref}} > \{F\}_i$ then $\{F\}_i = \{F\}_i + \Delta\,\{F\}_i$
else turn on the switch.
end while

- Transmit the decisions to the sensors at each load.

This control algorithm can also be applied if an inverter goes out of synchronization due to a major disturbance.

Voltage, frequency and phase of an incoming node and compare it with the reference node. Then CCU, using a control algorithm, will instruct the new incoming node to match its measured parameters till synchronization is achieved. This process is similar to the concept of micro-grids [26].

5.4 *Control of Harmonics*

The inverters can produce harmonics. To mitigate its effect, the CCU will periodically measure the voltages from the measuring devices connected at each inverter. Based on this, the CCU will perform a harmonic analysis and take appropriate measures to control the harmonics within an acceptable level [27].

5.5 *Managing the Prototype with Hybrid Energy Sources*

Diesel generators are the cheapest and the most easily available generators in Bangladesh [4]. It is widely used in rural areas for irrigation as well as supplying electricity during load shedding [4]. Since there is a lot of power shortage in rural areas, the people in these areas use diesel generators as back up source of electricity if they can afford it [4]. The prototype can have hybrid energy sources by including a diesel generator with renewable energy resources. This diesel generator can be used as a backup unit to provide electricity to the system when the load demand is higher than the capacity of solar PV sources. It can also be used to charge the battery when the battery voltage becomes low and there is no solar power available to charge the battery. Since a diesel generator is an AC source, it cannot be connected to the DC grid directly. So, a rectifier needs to be placed in between the DC grid and diesel generator. Traditionally, the diesel generator is turned on manually when it is required. But this can be done automatically as well by using a smart controller circuit. The smart controller can measure the voltage level of the solar PV system and load demand and based on that the smart controller can take decisions by itself whether the diesel generator should be turned on or not. Sensors placed on the battery will transmit the voltage level of the battery to the controller. The controller will compare this value with a threshold value. If the measured voltage of the battery is less than the threshold value, a smart controller will turn on the diesel generator. If the operating load goes beyond the rated value of the solar PV system, the measured value will be transmitted to the CCU which will switch off the loads. When there is low demand, like after midnight and when the charge level of the battery is also low, the diesel generator can be turned on to charge the battery. Since Biogas is also largely available as a renewable resource in rural areas [16], it can also be incorporated with the solar PV system in future.

5.6 Synchronization for AC Grid

The synchronization process for the prototype of the DC grid has been discussed previously. The similar approach can be followed to synchronize the inverters of multiple solar PV systems. These inverters need to be synchronized prior to their connection to the grid. Different operating parameters of these inverters will be transmitted to the CCU. The CCU will then compare these values with the reference node values and synchronize the inverters by instructing them to match the parameters.

5.7 Incorporating Irrigation Pumps

Bangladesh is an agricultural country. So, there is a lot of demand for irrigation pumps. This demand even gets even higher during the dry summer seasons. As a result, a lot of power outage takes place during the dry season when load demand in city areas, as well as rural areas, increase altogether. The irrigation pumps are mostly induction motor-based pumps [4]. As have known, the characteristic of the induction motor is to draw a large starting current. Solar PV systems are sometimes used to power these irrigation pumps [4]. So, the large starting current of the induction irrigation pumps can exceed the rated current values of solar PV systems [4]. The prototype system can accommodate this scenario. When the pumps are turned on, the sensors will transmit the signal to CCU. The CCU can monitor the current level and manage other operating parameters to accommodate the large staring current. It can also switch off other loads and turn on the diesel generator if required.

5.8 Protective Mechanism

Self-healing characteristic is an important aspect of smart grid systems. The prototype system can incorporate this idea. Since the prototype system has sensors all over which communicate with the CCU, the prototype system can sense the irregular operating conditions and take required remedial actions. The prototype system will be capable of detecting over-voltage or under-voltage conditions of the sources. When the demand gets too high, the CCU can manage this situation by adding more sources on the generation side. It can also manage the system by switching off some loads, which is also known as load side management. The prototype system will be equipped with the ability to provide reactive power support to the grid. To increase the protection of the total system, a coordinated protective relay system will be implemented as well. Frequency droop regulation technique should be implemented in the system to provide the real load (KW) sharing [28].

5.9 *Load-Energy Economy*

Demand-side management is an integral part of the prototype system. This helps to maintain the balance of generation and demand. Public awareness has to be created to ensure the efficient use of electricity. The people need to be educated and informed about when to use what type of devices. Users should be encouraged to change their habits to reduce the wastage of electricity. Sometimes, people keep electrical devices turned on even if they are not required. By creating public awareness, these types of behaviours can be changed. Also, users should be aware of peak demand periods. They should be encouraged to use less electronic devices during peak hours if possible. This will reduce and shift the peak demand and hence accommodate low generation during high demand [29].

5.10 *Local Energy Economy*

The local-energy economy will help the users under the prototype system to manage their load. When generation balances the demand, the price of electricity will be low [30]. But when demand is higher than a generation, power has to be bought from the grid or it has to be generated using the diesel generator. This is most likely to be at a higher price [30]. Also, excess generation from solar PV can be sold to the grid or used to charge the battery for later use. The difference in the price of generating power between the solar PV system and grid & diesel generator will encourage the more responsible application of demand-side management by the users [30].

6 Smart Grid and the Internet of Things (IoT)

The Internet of Things (IoT) is a breakthrough technology using which surrounding objects are connected through wired or wireless networks without any user intervention. The objects then become capable of communicating and exchanging information among themselves as well as to any servers that are available through local networks or public infrastructure such as the Internet. The information that

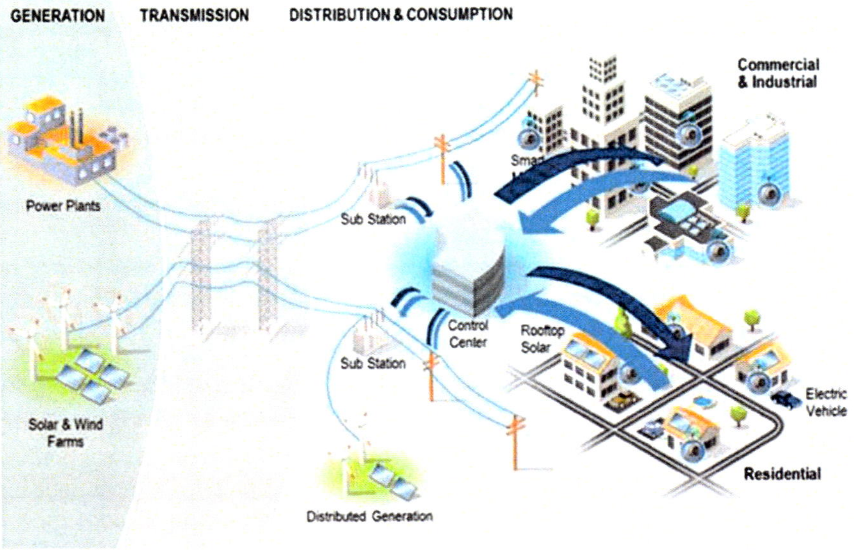

Fig. 5 IoT-based smart grid [32]

is collected and analyzed through multifarious IoT devices can offer advanced intelligent services to users [31].

In any Smart Grid setup, as shown in Fig. 5, application of IoT is vital, as it integrates almost in every sub-part of a Smart Grid, from power generation to distribution. Even at the users' end, various aspects of power consumption can be effectively quantified and enhanced, bringing a range of benefits to not only the power distributors but also to the user base.

Transformers are a key component in electricity distribution and transmission for recalibrating the source voltage for customer use. IoT enabled sensors can be installed internally to gauge operating parameters such as temperature, current, oil level fluctuation and even real-time environmental humidity and unit vibration [33]. All parameter values can be transmitted to monitoring stations through GPRS in real-time. This instantly alerts the monitoring stations in case of any emergency occurrences.

Further, sensor-enabled IoT based Power Meters, placed in any household that has basic home network-enabled, can collect many parameters from different devices in the household that uses electricity. This data can then be put to use for analyzing electricity usage, calculate billing information, make real-time decisions on load distribution, recognize detect unwanted malfunctions and even risks of probable outages. In addition, the power wastages can also be separately calculated, which can have a strong positive effect in reducing the overall monthly bill [34].

There were just a few of the examples of how IoT empowers a Smart Grid setup. IoT devices can further be placed into almost at any point of a Smart Grid and can amass a massive range of critical operating information. Such large datasets of information can also be processed using state of the art machine learning algorithms [35] to generate prediction patterns. Such advanced analytics can not only bring considerable improvement in service and cost savings but also can enable automatic fault-tolerant processes within various internal equipment [36].

7 Conclusion

The increasing price of electricity, rising power demand, and significant generation deficit is driving Bangladesh to move towards renewable resources. Different types of renewable resources are now being explored and implemented as an alternative source of energy, especially in rural areas. Instead of managing these systems centrally, it will be more effective to manage these resources on a small scale. A prototype of a smart grid as discussed in this paper can ensure the intelligent and autonomous management of supplying power from renewable resources to a few households in rural areas. It can also control the load side more efficiently and thus contribute to the solution of the energy crisis in Bangladesh.

References

1. Bangladesh Power Development Board: Annual Development Programme. http://www.bpdb.gov.bd/download/Annual%20Report-10.pdf
2. Bangladesh Power Development Board, Annual Report 2018–2019
3. Ministry of Power, energy and mineral resources, Government of the Peoples Republic of Bangladesh, Renewable energy policy of Bangladesh, Bangladesh. http://pv-expo.net/BD/Renewable_Energy_Policy.pdf
4. The Peoples Republic of Bangladesh: "Renewable energy policy of Bangladesh". Bangladesh, November 2008. http://pv-expo.net/BD/Renewable_Energy_Policy.pdf
5. Official Website of Rural Electrification Board (REB). http://www.reb.gov.bd/
6. The Database of World Bank. http://data.worldbank.org/country/bangladesh
7. Uddin, S.N., Taplin, R.: A sustainable energy future in Bangladesh: current situation and need for effective strategies. In: The 2nd Joint International Conference on Sustainable Energy and Environment (SEE 2006), Bangkok, Thailand (2006)
8. Islam, M.R., Islam, M.R., Beg, M.R.A.: Renewable energy resources and technologies practice in Bangladesh. Renew. Sustain. Energy Rev. **12**(2), 299–343 (2008)
9. Khan, S.: The search for alternatives. Star Weekend Mag. **8**(65) (2009). Accessed at: http://archive.thedailystar.net/magazine/2009/04/03/cover.htm
10. Staff Correspondent: Power tariff hiked again, The Daily Star. http://thedailystar.net/newDesign/news-details.php?nid=215311
11. Staff Correspondent: Fresh price hike for Power from March 1 in Bangladesh, BD News. http://www.bdnews.com/6617
12. Gofran, M.A.: Increase of electricity rate: impact on rural Bangladesh, The Daily Star. http://www.thedailystar.net/newDesign/news-details.php?nid=215655
13. Renewable Energy Research Center (RERC): University of Dhaka, Final report of Solar and Wind Energy Resource Assessment (SWERA) - Bangladesh project. http://ebookbrowse.com/swera-bangladesh-report-pdf-d199059998

14. REIN: Solar Energy. Renewable Energy Information Network, Bangladesh. http://www.lged-rein.org/knowhow.php?pageid=51
15. Azad, A.K.: A review on renewable power sources: prospects of Bangladesh and Scotland. EBook, St. Andrew's University, Scotland. http://pdfmio.com/download/renewablepower
16. Kaiser, M.A., Rahman, M.M., Sharna, S.A.: Wind energy assessment for the coastal part of Bangladesh. J. Eng. Appl. Sci. **1**(2), 87–92 (2006)
17. Baten, M.Z., Amin, E.Z., Sharin, A., Islam, R., Chowdhury, S.A.: Renewable energy scenario of Bangladesh: physical perspective. In: 2009 1st International Conference on the Developments in Renewable Energy Technology (ICDRET), pp. 1–5. IEEE, Dhaka (2009)
18. IEA: Technology Roadmap: Smart Grids. http://iea.org/papers/2011/smartgrids_roadmap.pdf
19. Strielkowski, W.: Renewable energy sources, power markets, and smart grids. In: Social Impacts of Smart Grids, pp. 97–151 (2020)
20. Strielkowski, W., Traditional power markets and an evolution to smart grids. In: Social Impacts of Smart Grids, pp. 9–54 (2020)
21. Smart Grid Concepts: Power System SCADA and Smart Grids, pp. 251–298 (2015)
22. Patel, M.R.: Wind and Solar Power Systems: Design, Analysis and Operation, 2nd edn. CRC Press/Taylor & Francis Group, London (2006)
23. Shafiullah, G.M., Oo, A.M.T., Jarvis, D., Ali, A.B.M.S., Wolfs, P.: Potential challenges: integrating renewable energy with the smart grid. In: 2010 20th Australasian Universities Power Engineering Conference, pp. 1–6. IEEE, Christchurch (2010)
24. Official Website of Rahimafrooz Limited. http://www.rahimafrooz.com
25. Halder, T.: Charge controller of solar photo-voltaic panel fed (SPV) battery. In: India International Conference on Power Electronics 2010 (IICPE 2010). IEEE, New Delhi, pp. 1–4 (2011)
26. Tse, N.C.F., Chan, J.Y.C., Lai, L.L.: Development of a smart metering scheme for building smart grid system. In: 8th International Conference on Advances in Power System Control, Operation and Management (APSCOM 2009), Hong Kong, China, pp. 1–5. IEEE (2009)
27. Fusheng, L., Ruisheng, L., Fengquan, Z.: Microgrid and distributed generation. Microgrid Technology and Engineering Application, pp. 29–46 (2016)
28. Hojo, M., Ohnishi, T., Adjustable harmonic mitigation for grid-connected photovoltaic system utilizing surplus capacity of utility interactive inverter. In: 2006 37th IEEE Power Electronics Specialists Conference, pp. 1–6, Jeju (2006)
29. Electric Power & Transmission & Distribution Forum, Droop setting of governors. http://www.eng-tips.com/viewthread.cfm?qid=11442
30. Energy Saving Trust, Power in Numbers: The Benefits and Potential of Distributed Energy Generation at the Small Community Scale. http://www.energysavingtrust.org.uk/…/Power%20in%20Numbers.pdf
31. Ahmed, A., et al.: An Intelligent and secured tracking system for monitoring school bus. In: 2019 International Conference on Computer Communication and Informatics (ICCCI), Coimbatore, Tamil Nadu, India, pp. 1–5. IEEE (2019)
32. IoT and Smart Grid. https://www.researchgate.net/profile/Mirza_Abdur_Razzaq
33. Pawar, R.R., Deosarkar, S.B.: Health condition monitoring system for distribution transformer using Internet of Things (IoT). In: 2017 International Conference on Computing Methodologies and Communication (ICCMC), Erode, pp. 117–122. IEEE (2017)
34. Santos, D., Ferreira, J.C.: IoT power monitoring system for smart environments. Sustainability **11**(19), 1–24 (2019)
35. Gunel, K., Ekti, A.R.: Exploiting machine learning applications for smart grids. In: 2019 16th International Multi-Conference on Systems, Signals & Devices (SSD) (2019)
36. Karim, M.A., Chenine, M., Zhu, K., Nordstrom, L. and Nordström, L.: Synchrophasor-based data mining for power system fault analysis. In: 2012 3rd IEEE PES Innovative Smart Grid Technologies Europe (ISGT Europe). IEEE, Berlin, pp. 1–8 (2012)

Dissimilar Disease Detection Using Machine Learning Techniques for Variety of Leaves

Varshini Kadoli, Karuna C. Gull, and Seema C. Gull

Abstract Agriculture is one of the main sectors in which about 70% of the population is dependent on, in India. The leaf disease detection would help the agriculturist in knowing whether the plants are affected by the disease before it could spoil the entire plantation. The method proposed helps in the early detection of leaf disease if the plant is affected. Advance machine learning has been used to determine the same by taking the diseased leaf image as an input. The system adopted with K-means clustering and the support vector machine classification technique after proper training is utilized for testing any plants' leaf diseases. The accuracy of the result obtained is in the range of 85–88%.

Keywords Image processing methods · Machine learning (ML) · Classification · Support vector machine (SVM) · Standard deviation (SD) · K-means clustering · Accuracy

1 Introduction

Developing world population has brought a ton of weight on rural assets. It is basic to get the greatest yield from harvest keeping in mind the end goal to support the populace and the economy. Plant diseases are the primary wellspring of plant harm which brings about financial and generation misfortunes in agrarian ranges. Inferable from troubled climatic and ecological conditions, an event of plant illnesses is on the ascent. There are different sorts of diseases in plants, the assortment of side effects,

V. Kadoli (✉)
K.L.E. Technological University, Hubli, India
e-mail: varshinikadoli35357@gmail.com

K. C. Gull
K.L.E Institute of Technology, Hubli, India
e-mail: karunagull74@gmail.com

S. C. Gull
Rani Channamma University, Belagavi, India
e-mail: seemacgull@gmail.com

P. Karuppusamy et al. (eds.), *Sustainable Communication Networks and Application*, Lecture Notes on Data Engineering and Communications Technologies 55,
https://doi.org/10.1007/978-981-15-8677-4_5

for example, spots or smirch emerging on the plant leaves, seeds and stanches of the plant. Keeping in mind the end goal to deal with these elements viably, there is a need to present a programmed strategy for plant observation that can examine plant conditions and applied information-based answer for recognition and groups them into different infections. Machine Learning is one of the distinguished methods for a proper structure to bolster this issue.

The proposed work includes the impression of image processing, where an input provided is an image, and the output might be an image or attributes of the image which helps to quantify input image. "An image is an array, or matrix, of square pixels arranged in rows and columns." As the image is a two-dimensional array, normally an advanced image is eluded as image components or pixels. Each pixel in the image is recognized by one or more numerical qualities that ultimately lead to being the characteristics of the image. For grayscale images, a solitary quality recognizes the force of the pixel. The intensity range can be of [0, 255] range. For color images, "image handling is a multidisciplinary field" [1]. It additionally concerns different regions, for example, machine learning, counterfeit consciousness, human vision research and so on.

1.1 Motivation

Due to the extreme climatic and ecological conditions, plants usually get affected by certain diseases. Leaf disease detection would help in identifying whether the plant is affected by the disease by its leaves. Advancement in technology, the way interacted with the environment, is also changing. This change can influence us to get better in each and every field including image processing technique. Exploring the connection between the physical and digital world may help the farmers. The method of detecting leaves diseases provides a helping hand to the agriculturists. It may give the farmers the best way of knowing about the health of the plant that they are yielding in the field. Machine learning is one such method that helps in detecting the disease in the leaves.

1.2 Problem Statement

To emerge automatic disease detection, a tool takes the leaf image as an input and enhances the image by applying different enhancement technique. Later, segmentation operation is applied to an enhanced image utilizing the *k*-means clustering algorithm. Features are extracted from the clustered images. Based on these features, the leaf is classified as either diseased or healthy using support vector machine. Thus, to find healthiness of a variety of leaves, a method is proposed as "dissimilar disease detection using machine learning techniques for a variety of plants' leaves."

2 Literature Survey

The proposed method has been obtained after going through the papers.

Apple leaf disease is one of the main problems that [2] have become the constraint in the growth of an apple. Apple leaf disease detection is done using image processing techniques and pattern recognition techniques. A color RGB input image of the apple plant is taken and then converted HIS, YUV and gray models. The background was then later removed using a threshold value and the spot segmentation was done using region growing algorithm (RGA). Around 38 classifying features were extracted based on color, texture and shape. The most valuable or the features considered important was extracted by combining genetic algorithm (GA) and correlation-based feature selection (CFS) to reduce the dimensionality of feature space and improve the accuracy of apple leaf disease detection. Finally, the SVM classifier is used to identify the disease that occurred in the apple leaf.

It explores the various imaging techniques. [3] These are developed mainly with a motto on deployment for monitoring methods for plants or crops. Classification of plants is done whether one is a healthy or diseased plant. It is expected to be accurate and the diseased plant should be detected as early as possible to help the agriculturists. Hyperspectral imaging has been used for the review. It is further used to provide additional information about the plant.

Machine learning is an advancement that has been in used in present days [4]. It is now developing and reduces the human effort and it comes under the field of computer vision. The diseases caused have a dangerous impact on agriculture. So a method is developed to detect the leaf disease. Normally, a plant is said to have a disease if it is affected by the pathogens such as bacteria, fungi and virus. Prediction for earlier detection has been used. This is done by using machine learning. Deep learning algorithms are also further used to detect the same. It acquires the data for the detection and the data is being stored in the cloud. It goes through the stages like image acquisition, image segmentation, feature extraction and classification. It also has to recognize the plant and hence the disease it has been affected. The output obtained is in the form of graphical visualization.

Spectral imaging has been in use within breeding because of the utility as a non-invasive diagnostic tool [5]. Canopy-scale measurements have resulted in low precision. It mainly tries for a prototype for the design that is a multi-spectral system for the study of plants. This analysis mainly takes machine learning into account for feature selection, disease detection to carry on with the classification. It further helps in the improvement of the system. The proposed model produced accurate results up to 92% when imaging was done for oilseed rape plants. False-color mapping of vegetation was used. The structure of the plant was further recorded using photometric stereo. The shape of the plants was also recorded. This allowed for the reconstruction of structure and leaf texture. The importance of this is to capture the structural information with the effect of reflectance and classification. This could be used in plant breeding with the quantization capability.

Evaluation in cluster analysis produces different clusters when used in diverse clustering techniques [6].

The distance between data vector x and centroid v is computed as using various measuring techniques.

The suggested depends on your dataset and output required that choose the proper distance measuring method with K-means to get the required output.

The feature set obtained is used for robust visual object recognition and identification [7]. The classifier used is SVM-based identification. The test case used here is human detection as a test case. The existing edge and gradient descent-based error detection are used for description. Histograms of oriented gradient (HOG) have been used for the test case as mentioned. This new approach has been an improvement in technology and hence gives a near-perfect classification of the pedestrian database including the pose variations of the human beings.

Euclidean distance:

$$D(x, v) = \sqrt{\sum_{i=1}^{n} (x_i - v_i)^2} \sqrt{\sum_{i=1}^{n} (x_i - v_i)^2 ss} \tag{1}$$

Manhattan distance:

$$D(x, v) = \sum_{i=1}^{n} (x_i - v_i) \left\| \sum_{i=1}^{n} |(x_i - v_i)| \right. \tag{2}$$

Chebyshev distance:

$$D(x, v) = \max|(x_i - v_i)| \tag{3}$$

Chi-square distance:

$$D(x, v) = \sqrt{\sum_{i=1}^{n} \frac{(x_i - v_i)^2}{(x_i + v_i)}} \sqrt{\sum_{i=1}^{n} \frac{(x_i - v_i)^2}{(x_i + v_i)}} \tag{4}$$

Raspberry Pi has been used as a controller to detect plant diseases and prevent them from spreading [8]. For image analysis, K-means clustering was done. According to the research, the leaf ailment was necessary to be recognized since it automatically detects symptoms of the plant diseases by K-means clustering in image processing. It has been useful for upgradation. It recognizes the diseases by using the particular picture and then providing the information by email, SMS. Automatic detection of diseases by its symptoms is done which would reduce the use of pesticides and hence increase the productivity in the farm.

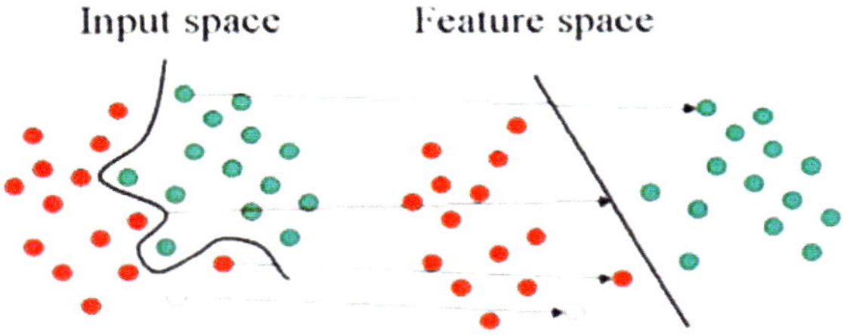

Fig. 1 SVM basic operation (Anon. 2011)

The authors explained the support vector machine [9] algorithm with the objective function.

$$C\sum_{i=1}^{m}\left(y^{(i)}\cos t_1\left(\Theta^T x^{(i)}\right)+\left(1-y^{(i)}\right)\cos t_0\left(\Theta^T x^{(i)}\right)\right)+\frac{1}{2}\sum_{j=1}^{n}\Theta_j^2 \quad (5)$$

where θ is the parameter vector, x is the feature vector, n is the number of features, m is the number of training sets, C is the regularization parameter, $\cos t_1\left(\Theta^T x^{(i)}\right)$ and $\cos t_0\left(\Theta^T x^{(i)}\right)$ are the costs when $y = 1$ and $y = 0$, respectively.

A basic representation of how it splits the data is shown in Fig. 1.

They said that for a given dataset, there may be multiple possibilities of hyperplanes but the SVM algorithm chooses the one that provides the greatest margin or the maximal margin between the classes. It is one of the sophisticated classification methods with high classification accuracy which made it so popular.

The process of classification of pomegranate leaf detection was done using image acquisition, image preprocessing and image segmentation, and thus, feature extraction was done [10]. The classification followed the method of support vector machine (SVM). Spatial filter, K-means clustering, GLCM were the additional techniques used to produce 90% result.

3 Proposed Methodology

The steps involved in the execution of the proposed work are image acquisition, preprocessing, feature extraction and classification which are explained in the following section.

3.1 *Image Acquisition*

"In image processing, the image acquisition can be defined as the process of retrieving the image from different types of sources." The hardware-based sources are used to retrieve the images. Performing image acquisition is always the first step in image

Fig. 2 Sample images of healthy and disease leaves

processing, because without an image, further, any processing is not possible. In the proposed method, the images are collected from farms, agricultural universities and books and data available on the website (which is publicly available) to prepare own database for healthy training of the system. The database contains the disease-affected leaves and healthy leaves.

The leaf has four different diseases. They are alternaria, anthracnose, bacterial blight, cercospora. The sample images of pomegranate plant leaves are shown in Fig. 2.

3.2 *Preprocessing*

The image set obtained from the dataset consists of noise. Noise may be dust, spores, water spots, etc. So images must be preprocessed to remove the noise. After preprocessing, the image should be resized to adjust the pixel values. Then it enhances the quality of the image with respect to contrast.

Try searching for the threshold that minimizes the intra-class variance [15], defined as a weighted sum of variances of the four classes:

$$\sigma_\omega^2(t) = \omega_0(t)\sigma_0^2(t) + \omega_1(t)\sigma_1^2(t) + \omega_2(t)\sigma_2^2(t) + \omega_3(t)\sigma_3^2(t) \tag{6}$$

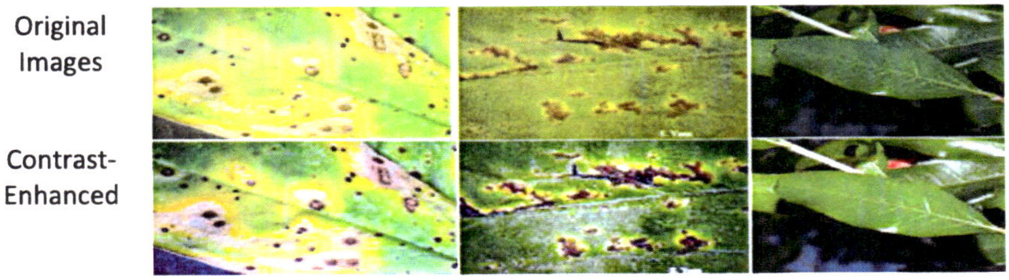

Fig. 3 Sample images after contract enhancement

Weights ω_0, ω_1, ω_2 and ω_3 are the probabilities of the four classes separated by a threshold t, and σ_0^2, σ_1^2, σ_2^2 and σ_3^2 are variances of these four classes. Then class probabilities and class means can be computed iteratively.

Apply morphological operations on the image (Fig. 3).

3.3 Feature Extraction

In the segmentation stage, the preprocessed image is given as an input. The segmentation means subdividing the whole image region into small regions. In the proposed method, k-means clustering algorithm is utilized for the segmentation process [11]. It is useful to extract the image structures.

In the proposed system, the k-means clustering intends to partition n perceptions into k-clusters in which every perception has a place with the cluster with the nearest mean, serving as a model of the cluster. Clustering is the process of portioning a group of data points into a small number of clusters. The following objective function (Equation 7) narrates the same.

$$\sum_{i=1}^{k}\sum_{x\in S_i}^{k}||x-\mu_i||^2 = \sum_{i=1}^{k}|S_i|\text{Var } S_i \tag{7}$$

where μ_i is the mean of points in S_i. This is equivalent to minimizing the pairwise squared deviations of points in the same cluster. For instance, the pixels in the image are clustered. Of course, this is a qualitative kind of portion. Figure 4 shows the example of the k-means clustering.

Thus, the steps involved in segmentation process using k-means clustering operation are

Input: Leaf image.
Output: Segmented clusters of leaf image.
Step 1: Find the centroid of the pixel.
Step 2: Divide the pixels into a cluster.
Step 3: Represent the clustered image.

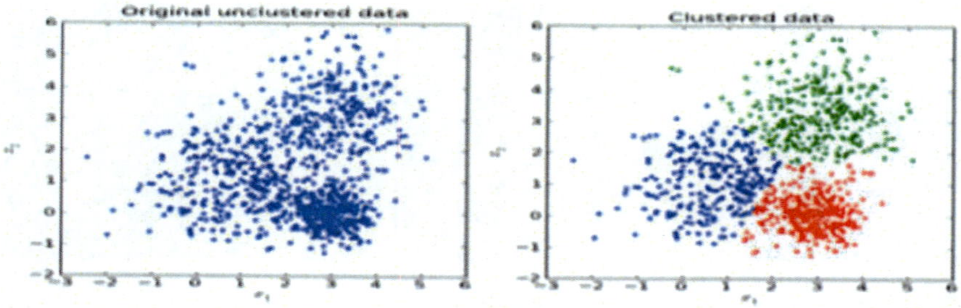

Fig. 4 *K*-means clustering. *Courtesy* https://qph.fs.quoracdn.net/main-qimg-770e75d8c69d0b77f49bb371cad50c3d

Step 4: Segmented output (Fig. 5).

In the proposed strategy, the segmented image is utilized to extract the characteristics of the input image [12]. To extract the region of interest, the estimation of these characteristics is essential [14].

The features extracted from the segmented images using GLCM are further utilized for the classification.

Thus, the process for statistical feature extraction is:

Input: Leaf image
Output: Statistical features
Step 1: Read the segmented leaf image
Step 2: Calculate the features by applying the GLCM technique
Step 3: Generate a feature vector.

In the proposed technique, statistical features like mean, standard deviation, entropy, skewness, kurtosis, correlation, etc. of segmented image ought to be superimposed to the sample image. All extracted features of the sample image are compared to the feature set of training images which were extricated and stored in characteristics or feature table (Fig. 6).

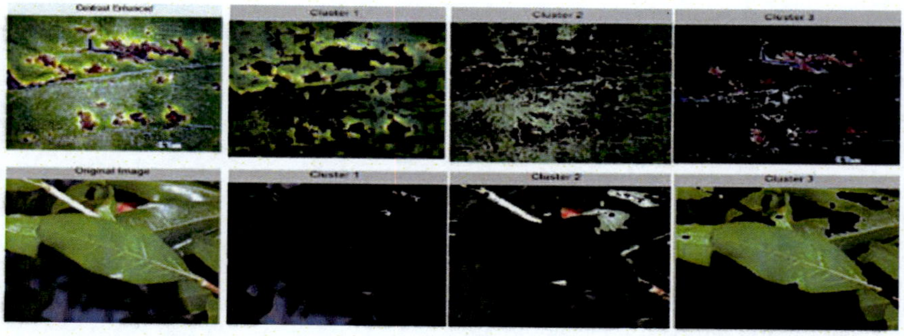

Fig. 5 Sample images after clustering

Mean	S.D	Entropy	RMS	Variance	Smoothness	Kurtosis	Skewness	IDM	Contrast	Correlation	Energy	Homogeneity
0.078876	0.978321	0.762589	0.974878	14.84385	47.81168099	1.709878	5.5747726	2150.696	1	15.597769	3.632011	255
0.466835	0.865708	0.796721	0.959196	14.15012	48.13958466	1.365848	4.3136177	1632.216	1	15.765401	3.674427	255
0.367586	0.910197	0.757318	0.962547	16.44411	51.41942893	1.667891	5.340374	2305.041	1	13.792639	3.402522	255
0.541238	0.751034	0.538239	0.922202	17.97166	37.66351987	2.582884	7.4036979	1306.813	1	10.495102	2.588339	255
0.512776	0.710321	0.894702	0.971681	17.1185	35.52045275	2.843172	10.450455	1162.225	1	27.603284	4.682011	255
0.697626	0.873892	0.487259	0.910412	31.56037	56.45960734	2.982981	8.114045	2844.325	1	4.4008361	1.612926	255
0.488618	0.958014	0.268706	0.940314	71.85277	83.0728783	5.120415	11.461609	5682.676	1	1.8269994	0.649677	255
0.430913	0.896565	0.765986	0.965598	17.43762	52.46393422	1.878881	5.7288992	2052.447	1	12.8361	3.273623	255
0.576072	0.909153	0.710405	0.958389	23.81361	60.20882544	1.673428	5.4362373	3230.313	1	6.958173	2.334859	255
0.746201	0.90982	0.52789	0.900746	40.04733	73.85749163	2.911928	7.5330257	4465.177	1	3.9931523	1.587334	255
0.88943	0.826304	0.818493	0.96507	16.41812	55.65340939	1.300243	4.3227964	2841.536	1	12.320378	3.301014	255
0.413971	0.970172	0.410585	0.972992	76.61838	97.98214573	3.843924	9.3641523	6332.289	1	1.5171447	0.599022	255
0.08629	0.94908	0.884848	0.993442	8.575495	35.45269477	0.717587	2.5151335	1110.396	0.999999	18.75243	4.105919	255
1.045052	0.816676	0.619172	0.920878	26.94916	60.87399316	2.481799	6.9559693	3471.252	1	6.6287909	2.20753	255
0.413097	0.845948	0.842361	0.976562	10.60321	41.47241151	1.194491	3.8288648	1594.445	1	22.346086	4.420793	255
1.006648	0.795249	0.796001	0.954265	16.59695	54.72505252	1.297603	4.6008753	2771.884	1	12.37107	3.282577	255
1.319761	0.864802	0.480227	0.919423	44.17804	76.8477228	3.321984	8.6894848	5494.231	1	3.411599	1.430894	255
0.27454	0.870993	0.859583	0.979117	9.741954	38.48165041	0.886391	3.5148447	1388.821	0.999999	19.345475	4.109354	175
0.565518	0.869229	0.673009	0.954929	21.60405	53.86948886	2.322427	6.2055053	2484.706	1	9.9313647	2.734996	255
0.262546	0.879222	0.798925	0.976501	13.27572	42.41367953	1.271065	4.3010873	1614.019	1	12.961897	3.281804	255
0.331847	0.866178	0.734424	0.957452	18.0845	41.43077451	3.817042	9.7439957	1158.825	1	22.607484	4.280236	255
0.841958	0.763881	0.704448	0.935609	18.29885	47.74902051	3.338588	7.1460656	2201.596	1	13.609027	3.268961	255
0.41152	0.981417	0.350451	0.94482	128.9061	118.6973325	3.6237	12.574377	10666.31	1	1.0988286	0.015921	255
0.667126	0.916899	0.480335	0.911658	41.22886	71.50431469	3.270195	8.1946567	4609.107	1	3.8336729	1.502789	255

Fig. 6 Characteristics or feature table

A mid-training phase the feature vector is stored as a text file and it is called as training data. The text file itself is utilized as a contribution for the classification. In the training stage, all the images are named utilizing the 0, 1, 2, 3 and 4 rotations. 0 represents the alternaria disease leaf. 1 represents the anthracnose disease leaf, 2 represents the bacterial disease leaf, 3 represents to cercospora disease leaf, 4 represents the healthy leaf. A mid the testing stage all the measurable features are extracted for the given image and stored as a text file and text data.

3.4 *Classification*

In the proposed strategy, the support vector machine (SVM) classifier is utilized. In a high or boundless dimensional space, a machine builds an arrangement of hyperplanes that are utilized for classification. Here, the margin is being looked to maximize between the data points and the hyperplane [13]. The loss function that will maximize the margin is given by

$$c(x, y, f(x)) = \{0, \text{ if} y * f(x) \\ \geq 11 - y * f(x), \text{ otherwise} \tag{8}$$

The hyperplane is accomplished the great partition if preparing information purposes of closest separation to any class having the biggest separation. If classifier having the larger margin, then generalization error will be low. In the support vector, the data points are near to the isolating hyperplane. By definition, "*x*" representing the data points of sort 1 and "–" representing the data points of 2. For a given training set, every value is named as having a place with one of class. It builds a model and new cases will be dispensed to one of the classifications. The illustrations are represented as points in space in Fig. 7.

Essentially, SVM can just resolve issues which are identified with double characterization. Presently, they have been extended to prepare multi-class issue. It utilizes the one after one strategy to fit all paired subclassifiers furthermore to locate the right class by choosing component to concede the multi-class order. The workflow of the proposed system is depicted in Fig. 8.

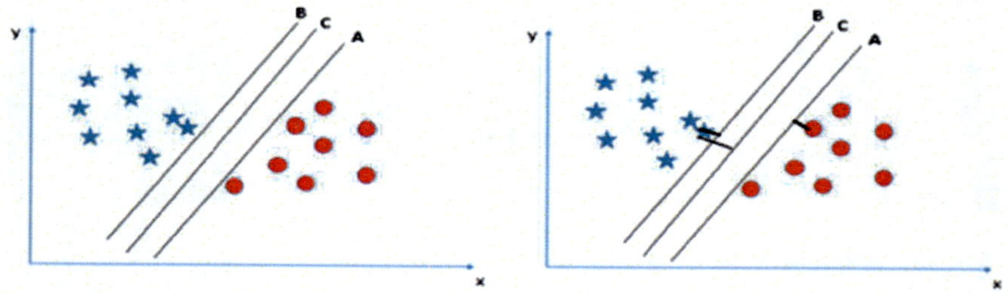

Fig. 7 Support vector machine

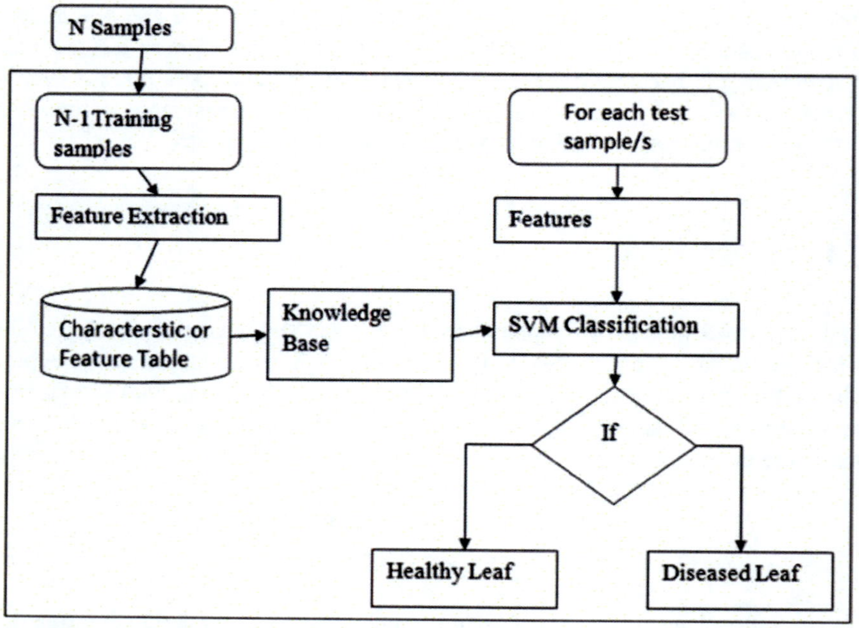

Fig. 8 Block diagram of classification

4 Experimental Results

The proposed method comprises of 125 pictures; the projected strategy utilizes the fundamental features like mean, standard deviation, skewness, kurtosis and many more. These features are put away in the feature or characteristic table as shown in Fig. 6. This table is acting as a reference of knowledge for the SVM classification.

The extracted characteristics or features of the segmented sample input image are given as an input for SVM classifier for the classification. In view of these qualities, it will classify the leaf as healthy and diseased. After classification, the results are directed with the assistance of performance parameters like specificity, sensitivity and accuracy. Taking into account, these values can without much strength legitimize the accuracy of the classification.

4.1 Performance Parameter

The dataset comprises of both diseased and healthy leaf images. The test results can be positive means; it predicts the image as a diseased leaf. On the off chance that the test result is negative means, it predicts the image as a healthy leaf. The different testing results are:

- True Positive—the segmented leaf is healthy or diseased and it is classified as healthy or diseased respectively.
- False Positive—the segmented leaf is healthy and is classified as diseased.
- True Negative—the segmented leaf is diseased and is classified as diseased.
- False Negative—the segmented leaf is diseased and is classified as healthy.
- *True Positive Rate*: It is additionally called as affectability. It decides the extent of real positive values that are accurately distinguished. It implies that the rate of the segmented leaf containing diseased is effectively classified as diseased.

$$\text{True Postive Rate} = \frac{\text{TP}}{\text{TP} + \text{FN}} \tag{9}$$

- *True Negative Rate*: True negative rate is likewise called as specificity. It decides the extent of negativity qualities which are accurately recognized. It implies that the rate of a healthy segmented leaf is effectively delegated healthy.

$$\text{True Negative Rate} = \frac{\text{TN}}{\text{TN} + \text{FN}} \tag{10}$$

4.2 Accuracy

It decides the factual measure of how well a classifier has effectively classified the images. The accuracy is the rate of genuine after-effects of both true positive and true negative qualities.

$$\text{Accuracy} = \frac{\text{TP} + \text{TN}}{\text{TN} + \text{TN} + \text{FP} + \text{FN}} \tag{11}$$

The classified results are shown in Fig. 9. The accuracy and error rates for SVM polynomial using a confusion matrix with a various ratio of training and testing dataset are depicted in Fig. 9. The accuracy of the stated SVM is ranging from 84 to 88%.

The scatter plots for the various training and testing dataset ratios with one feature (Mean vs. Correlation) to the confusion matrix are depicted in Fig. 10.

The affected areas of the various leaves with different types of diseases in percentage are depicted in Figs. 11, 12, 13 and 14.

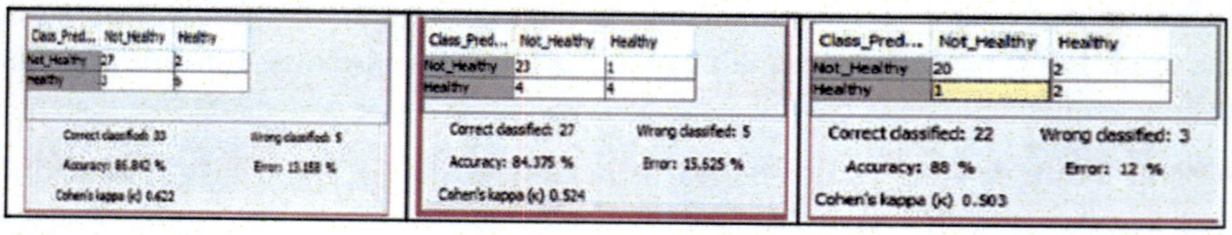

Fig. 9 Confusion matrices with accuracy and error rates for different training and testing ratios (70:30, 75:25 and 80:20, respectively)

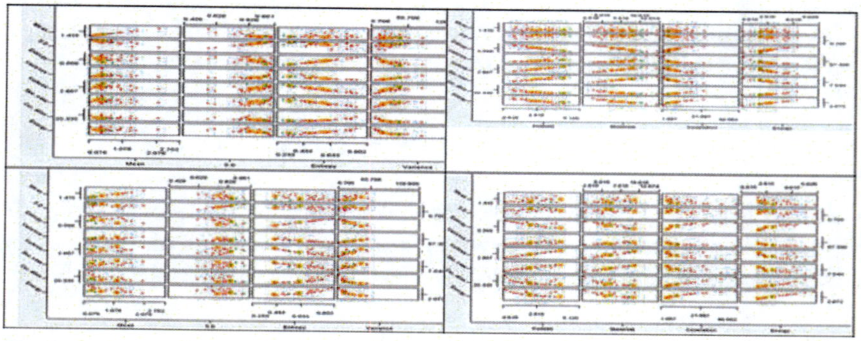

Fig. 10 Scatter plots with the misclassified highlighted for different training and testing ratios (75:25 and 80:20, respectively)

Fig. 11 Cercospora diseased brinjal leaf with 15.432% affected

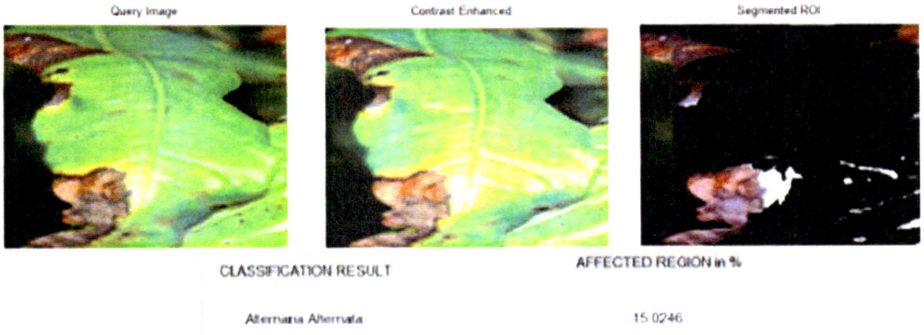

Fig. 12 Alternaria diseased mango leaf with 15.0245% affected

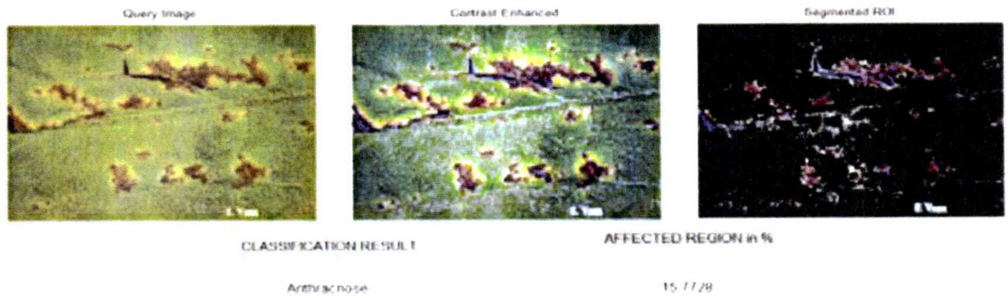

Fig. 13 Anthracnose diseased cotton leaf with 15.773% affected

5 Conclusion

This work proposes the development of diagnostic classifier for leaf images of the plant. Usually, plant images are having noise. Therefore, it is necessary to improve the contrast and suppress the noise present in the image, for identification of diseased leaf of a plant. The enhancement technique is used to improve the contrast of the images; it removes the noise present in the image. After the enhancement technique,

Fig. 14 Normal and healthy pomegranate leaf

the images are segmented using a *k*-means clustering algorithm. The segmented images are used for feature extraction and have utilized extracted relevant data about the segmented images for the classification. Support vector machines are used to analyze data and recognize patterns with the assistance of algorithms and accuracy of the result obtained is quantifiable up to 88%.

References

1. Twogood, R.E., Sommer, F.G.: Digital image processing. IEEE Trans. Nucl. Sci. **29**(3), 1075–1086 (1982)
2. Z. Chuanlei, Z. Shanwen, Y. Jucheng, S. Yancui, and C. Jia, "*Apple leaf disease identification using genetic algorithm and correlation based feature selection method*", 10:74–83, Jan 2017
3. Amy Lowe, Nicola Harrison, and Andrew P. French, "*Hyperspectral image analysis techniques for the detection and classification of the early onset of plant disease and stress Plant Methods*", 13(1):80, 2017
4. Malusi Sibiya and Mbuyu Sumbwanyambe, "*Deep Learning Models for the Detection of Plant Leaf Diseases: A Systematic Review*", Feb 2019
5. Charles Veys, Fokion Chatziavgerinos, Ali Alsuwaidi, James Hibbert, Mark Hansen, Gytis Bernotas, Melvyn Smith, Hujun Yin, Stephen Rolfe, and Bruce Grieve, "*Multispectral imaging for presymptomatic analysis of light leaf spot in oilseed rape Plant Methods*", 15, Dec 2019
6. Gull K.C., Angadi A.B. (2018) "A Methodical Study on Behavior of Different Seeds Using an Iterative Technique with Evaluation of Cluster Validity". In: Saini A., Nayak A., Vyas R. (eds) ICT Based Innovations. Advances in Intelligent Systems and Computing, vol 653. Springer, Singapore
7. N.Dalal and B. Triggs, "*Histograms of oriented gradients for human detection*", in 2005 IEEE Computer Society Conference on Computer Vision and Pattern Recognition (CVPR'05), vol 1, pages 886–893, 2005
8. M. Sankar, D. N. Mudgal, T. Varsharani Jagdish, N. W. Geetanjali Laxman, and M. Mahesh Jalinder, "*Green leaf disease detection using raspberry pi*", in 2019 1st International Conference on Innovations in Information and Communication Technology (ICIICT), pages 1–6, 2019
9. Karuna C. Gull, Sudip Padhye, Dr. Subodh Jain, "Automated Identification of Latent Attributes of Twitter users", International Journal of Emerging Trends & Technology in Computer Science (IJETTCS), Volume 6, Issue 3, May - June 2017, pp. 395-403, ISSN 2278-6856
10. Priti Badar and Suchitra.C," *Disease Detection of Pomegranate Plant Using Image Processing*", 5:7, May 2016

11. G. A. Wilkin and X. Huang, *"K-Means Clustering Algorithms: Implementation and Comparison",* in Second International Multi Symposiums on Computer and Computational Sciences (IMSCCS 2007), pages 133–136, 2007
12. M. C. Popescu and L. M. Sasu, *"Feature Extraction, Feature Selection and Machine Learning for Image Classification: A case study"*, in 2014 International Conference on Optimization of Electrical and Electronic Equipment (OPTIM), pages 968–973, 2014
13. Corinna Cortes and Vladimir Vapnik, *"Support-vector networks Machine Learning"*, 20(3):273–297, 1995
14. https://support.echoview.com/WebHelp/Windows_and_Dialog_Boxes/Dialog_Boxes/Variable_properties_dialog_box/Operator_pages/GLCM_Texture_Features.htm
15. https://en.wikipedia.org/wiki/Otsu%27s_method

Face Aging Through Uniqueness Conserving by cGAN with Separable Convolution

K. B. Sowmya, Mahadev Maitri, and K. V. Nagaraj

Abstract Inspired by the transition of style from Gatys and Goodfellow and the generative adversarial network (GAN), the use of conditional GAN (cGAN) is to achieve human age progression. cGAN is good at generating fake images, where labels are used as conditions which is why it is better than other GANs. For face aging, pre-processing of the original dataset is highly required. This makes it more computational. In terms of performance, "identity-preserving" technique is used to yield better results. Also, in this paper, a novel age order approach is presented by arranging reproduction misfortune and present the boundary λ which is even between huge age highlights and unobtrusive surface highlights. The trial results exhibited here, proposed cGAN, furnishes better maturing faces with personality protecting over other best in class age movement techniques.

Keywords Conditional generative adversarial network (cGAN) · Age estimation · Style transfer

1 Introduction

The human face is one of the features which best describes a person, and it is the only part and most complicated structure to define a person. The same structure of the human body is being used for many applications such as face recognition, used in public by government to find criminals or lost children and entertainment, smart unlock for devices, and many more. However, these applications will be not

K. B. Sowmya (✉) · M. Maitri · K. V. Nagaraj
Department of Electronics and Communication Engineering, RV College of Engineering, Bengaluru, India
e-mail: kb.sowmya@gmail.com

M. Maitri
e-mail: mahadevmm.ec16@rvce.edu.in

K. V. Nagaraj
e-mail: kvnagaraj.ec16@rvce.edu.in

P. Karuppusamy et al. (eds.), *Sustainable Communication Networks and Application*, Lecture Notes on Data Engineering and Communications Technologies 55,
https://doi.org/10.1007/978-981-15-8677-4_6

accurate as time passes since the face changes according to genes and the environment. Hence, it is most important to develop the cross-age verification [1] and recognition for a human face. In recent years, many age verification and estimation methods are proposed and witness various developments using neural network in the age estimation from a human face. The other method for age verification and recognition is to synthesis the human face for that age, i.e., face aging. Face aging is still a very difficult job for different reasons. First, the presented human faces that have different expressions and lighting conditions which pose major challenges in modeling aging patterns. From the survey, there are two methods to model the face aging: first, traditional face aging and second, deep learning method.

Current facial aging strategies can be divided into two: concept approaches and physical model approaches. Prototype methods are the one that decides the average face for a facial age group, and the variation between different average faces is the aging trend that would be useful for constructing the aging face. In contrast, the physical model describes the structure and shape of the face shifts in the form of muscle, hair color and crease, etc., with age. Such strategies, however, also include the same person's face aging sequences for a wide variety of ages. To overcome these problems, generative adversarial network (GAN) [2, 3] deals better with age progression. GAN was introduced to reconstruct the original (training) images. GAN hast two models joined adversarial: a generative model G and a discriminative model D. In recent years, the GAN approaches are helping the researcher to utilize this framework for many of their applications like image-to-image translation. Out of many, as a notable GANs, conditional generative adversarial network (cGAN) [4, 5] performs better than classic GAN since cGAN is trained with labels, i.e., condition. GANs have also been used to make improvements on human faces, such as changing color, inserting sunglasses or even quantitative aging. Nowadays, GAN gained more popularity because it can produce high-quality images from low quality. To produced images with different age, face features should remain intact and those features changed to get an image with the aged face. For that pre-trained classifier is required. More the accuracy of the discriminator better will be the image produced by the generator.

In this paper, mainly, face aging-based cGAN is proposed, which is focused on identity preserving. Because a common problem with previous GAN-based face modification methods is the fact that the identity of the original individual is always lost in a changed picture. In particular, contributions are as follows:

- Design an effective and efficient conditional GAN-based face aging system while preserving the identity of a person and cross-verifying with age estimation.
- Reduce the computational efficiency of the above cGAN while maintaining functionality and accuracy.

2 Related Work

2.1 Face Aging

The last couple of years have seen big attempts to tackle this issue [6], where aging accuracy and permanence of identity are widely seen as the two fundamental principles of its achievements. The conditional GAN-based approach for automatic face aging is suggested in this research. Unlike previous works that use GANs to change facial features for age progression, in this usage of identity preservation is done which is the high cost [6, 7].

Face aging, in real life, can be observed through a change in facial structures like skin's anatomy and facial muscles. These can be achieved by the prototype-based approaches. But those require an enormous amount of training data. But GAN does not require that much training, in this project training is only required for age prediction part. So much alteration in facial features will alter the image's credibility (the identity of the person at a particular age) [8]. So the only facial structure is changed to get face aging. Also, the resolution of the image plays an important role because facial structures may change in smaller increments which cannot be done in low-resolution image. Apart from biometric recognition [9], face aging is considered the most reliable personal identification. This is because of uniqueness in the facial features found in the face images.

2.2 Generative Adversarial Network

Generative adversarial network (GAN) framework was introduced by Ian J. Goodfellow in the year 2014 to generate similar data for the applications of artificial intelligence. In this framework, two models are trained, namely a generative model G and a discriminative model D. The functionality of these two models: where one measures the distribution of data and the other calculates the likelihood of how close the training data is to the output of the generator. When the model is training, G is attempting to increase the likelihood that D would make an error. When both models are assigned with arbitrary functions, there exists a unique approach with G reconstructing the data distribution and D attaining the probability equal to ½ everywhere. This is similar to a two-player game where one tries to win over the other [3].

2.3 Age Estimation

Existing age prediction systems attempt to calculate ages correctly or to classify age groups. The age is measured based on features derived that are length, width between eyes, nose and number of facial wrinkles and skin conditions. Conventional methods for extracting features include the local binary pattern operator, the biologically induced feature, and the Gabor filter. Recently, a deep-learning age prediction is suggested to learn filters and to extract features and learn ages. These are the active challenges in the field of deep learning technology. The techniques proposed are of two types, one which classifies the age into multiple age groups and accuracy is the measure of performance for classification method. On the other side, it predicts the age based on regression techniques, mean absolute error (MAE) and mean squared error (MSE). It is used to evaluate the method by means of performance measurement between the ground-truth age and predicted age [1].

For the face aging and age estimation [5, 10], there is numerous database available open-source such as MORPH, FG-NET, and IMDB-Wiki. The proposed work is evaluated on the IMDB-Wiki database which is discussed in Sect. 3.

3 Methodology

GAN has two main blocks—generator and discriminator. Generator G tries to generate image/data from a given input and noise. Discriminator D tries to estimate the results from the generator to predict it is real or fake. The training follows a two-player game with objective function $V(D, G)$ as shown in Eq. (1). The y_0 is conditional age, z is an input image to generator, and x is an image from the dataset.

$$\begin{aligned}\min_G \max_D V(D, G(z, y_0)) &= \mathbb{E}_{x\sim p_{\text{data}}(x)}\big[\log D(x, y_0)\big] \\ &+ \mathbb{E}_{z\sim p_z(z)}\big[\log(1 - D(G(z, y_0), y_0))\big]\end{aligned} \tag{1}$$

In the early stage of learning, the discriminator will reject the samples produced by the generator, as many iterations will increase the generator will produce better data which is not rejected by the discriminator. At this stage, the values of function $V(D, G)$ will be nearly half [3].

3.1 *Style Transfer*

Recently, a deep-learning age prediction is suggested to learn filters and to extract features and learn ages. Style transfer [11] goal is to take the input image's features and bind it with the set of styles defined. Latter image and the input image are combined to give an artistic style image. Optimization of content loss and style loss has to be done. These two losses depend on the pre-trained neural network's features. Using high-quality images will give better accuracy but the training of it will take more time [12] (Fig. 1).

In this, image along with expected age is given as input to GAN. It recognizes the facial features of the given input image and tries to convert into the expected aged

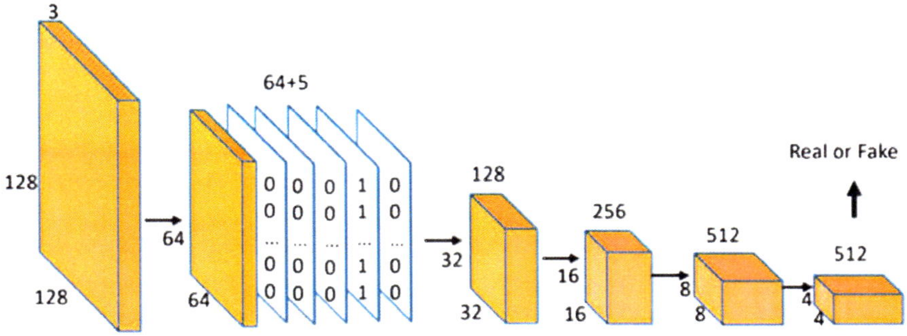

Fig. 1 Flow of data through stacked layers of a discriminator

image trying to keep as realistic as possible. The output of this image is given to discriminator, which again analyses the image and predicts age. If the predicted age and expected age matches, then the training will be completed. In our project, AlexNet is used inside discriminator which is already trained to predict the age. Generate has convolution layer, ReLu layer, convolution layer and ReLu layer, respectively. So overall, it is a system which accepts an image of a person that needs to be converted into an expected aged person [13]. With the multi-layer neural network, facial features for face aging are automatically identified using convolutional layers with kernels and structures the face for age progression. Some of the face aging features are wrinkles, tanning, etc. On top of these facial feature's transformations, the identity of the person is maintained throughout the network for face recognition.

3.2 Generator

The generator accepts image converts to a specific resolution image into give as input. There is 6 residual block in the generator. Before the first convolution, the input image and conditional feature are superimposed, and in a general term, age is considered as a condition and input image is of a person which is concatenated together. Invertible conditional GAN is adapted and used in this. High-quality input image will give better result compared to low-resolution images. First layer 4×4 convolution layer and LeakyReLu layer with conditional features, next three layers of 4×4 convolution layer, batch normalization layer and LeakyReLu layer with stride as two are used. For age classification, pre-trained AlexNet is used which shares a similar architecture compared to conv1 to pool5. In the end, three fully connected layers and one softmax layer. Overfitting can be prevented using dropout [4].

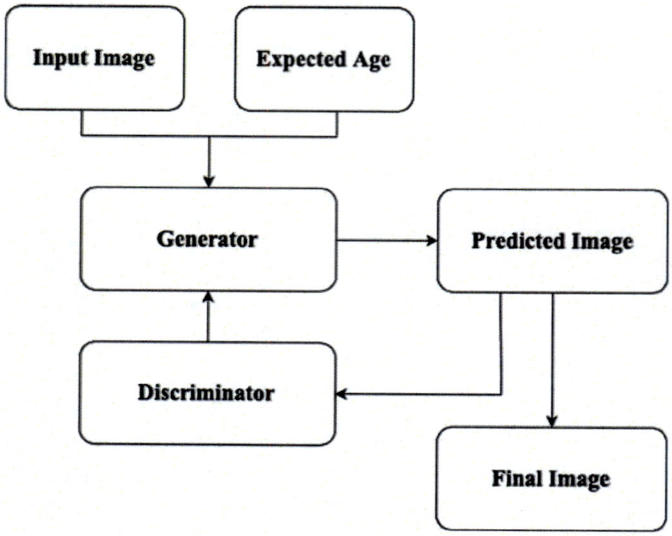

Fig. 2 Design of conditional GAN [14]

3.3 *Discriminator*

For age classification, pre-trained AlexNet is used which shares a similar architecture compared to conv1 to pool5. AlexNet uses eight layers. The first five layers are convolutional and followed by max-pooling layer, and the last layer is fully connected. It is a better version of convolutional neural network. After that two fully connected layers and one softmax layer is used. Overfitting can be prevented using dropout techniques, and it is the technique in which neurons are turned off with given probability. Each iteration uses different parameters from the sample. But it increases the training time. Figure 2 shows the architecture of the discriminator from the GAN [14].

3.4 *Age Classification*

To generate the image for a given range of age, pre-training of the classifier is needed. If an image generated from the generator does not fall in those ranges, penalties are given to the generator. The penalty will vary according to the age difference predicted by the discriminator. Similar to the softmax function used for a penalty, as the age difference between predicted age from discriminator and image from the generator increases the penalty value also increases [15]. Using a backpropagation algorithm generator changes the face of the image to correct the group.

4 Implementation

4.1 Dataset

The pictures are taken from the Cross-Age Celebrity Dataset (CACD2000) [16] including 163,446 photographs from 2,000 superstars assembled from the Internet. Pictures are drawn from mainstream web search tools utilizing the name and year of big names (2004–2013) as watchwords. Subsequently, computing the period of big names in the pictures by really taking away the time of birth from the year in which the image was taken, since there are for the most part white individuals in the CACD2000 database.

From the dataset, the range of age is from 16–62. So, for the categorization, the age range is divided into 5 which are 11–20, 21–30, 31–40, 41–50, 50+. In individual age categories, there are 8,656, 36,662, 38,736, 35,768, and 26,972 images, respectively.

4.2 Conditional GAN

Face aging is synthesizing the given image to a target age. For that discriminator is used. To fake, the aging generator tries but discriminator should not find that it is fake. So after a few iterations, generators will generate better images with facial structures. As discriminator can tell the images are real or fake with the given age conditional GAN [17] is used to get more realistic images. To optimize the network, matching aware discriminator is used as proposed in [18]. For identity, preservation mean square error (MSE) is used as a loss function.

5 Results

The dataset provided by CACD2000 is utilized in this work to face aging. The architecture for both generator and discriminator is designed efficiently such that without losing any features extracted from the existing method. The output from the generator, trained for all the images from CACD belonging to 5 age categories as described in Sect. 4.1, is 5 images belonging to each category created. From this work, the results are presented in Fig. 3 and the response time for the entire network are 576 ms. The outcome indicates that the proposed model is still competitive with existing ones, even when contrasting the aging of facial features.

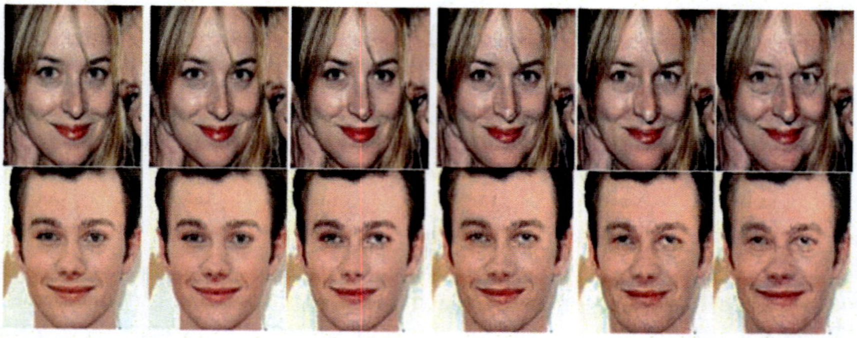

Fig. 3 Output of generator belonging to 5 categories of age

6 Conclusion

In this paper, a method for face aging using conditional GAN is proposed. Firstly, collecting the dataset and categorized into five as discussed in Sect. 4.1, second, designed the architecture of the cGAN efficiently and having same features, i.e., identity preservation of the person. Lastly, the cGAN is modified using counterparts of each layer designed earlier; for example, convolution is changed to separable convolution, making it much more efficient than existing methods, and mainly, focusing on the generator part which is key to GAN. Moreover, InfoGAN or dual conditional GAN can be used to produce better results by loading conditions as features, i.e., limiting the number of features and progression of age based on the categorized age group.

References

1. Nam, S.H., Kim, Y.H., Truong, N.Q., Choi, J., Park, K.R.: Age estimation by super-resolution reconstruction based on adversarial networks. IEEE Access **8**, 17103–17120 (2020)
2. Tang, X., Wang, Z., Luo, W., Gao, S.: Face aging with identity-preserved conditional generative adversarial networks. In: 2018 IEEE/CVF Conference on Computer Vision and Pattern Recognition, Salt Lake City, UT, pp. 7939–7947 (2018)
3. Goodfellow, I., et al.: Generative adversarial nets. In: Advances in Neural Information Processing Systems, pp. 2672–2680 (2014)
4. Antipov, G., Baccouche, M., Dugelay, J.-L.: Face aging with conditional generative adversarial networks. In: IEEE International Conference on Image Processing (2017)
5. Geng, X., Yin, C., Zhou, Z.-H.: Facial age estimation by learning from label distributions. IEEE Trans. Pattern Anal. Mach. Intell. **35**(10), 2401–2412 (2013)
6. Jia, L., Song, Y., Zhang, Y.: Face aging with improved invertible conditional GANs. In: 2018 24th International Conference on Pattern Recognition (ICPR), Beijing (2018)

7. Thengane, V.G., Gawande, M.B., Dudhane, A. A., Gonde, A.B.: Cycle face aging generative adversarial networks. In: 2018 IEEE 13th International Conference on Industrial and Information Systems (ICIIS), Rupnagar, India (2018)
8. Geng, X., Zhou, Z., Smith-Miles, K.: Automatic age estimation based on facial aging patterns. IEEE Trans. Pattern Anal. Mach. Intell. **29**(12), 2234–2240 (2007)
9. Jacob, Dr.: Capsule network based biometric recognition system. J. Artif. Intell. Capsule Netw. **2019**, 83–94. (2019). https://doi.org/10.36548/jaicn.2019.2.004
10. Fu, Y., Guo, G., Huang, T.S.: Age synthesis and estimation via faces: a survey. IEEE Trans. Pattern Anal. Mach. Intell. **32**(11) (2010)
11. Johnson, J., Alahi, A., Fei-Fei, L.: Perceptual losses for real-time style transfer and super-resolution. In: European Conference on Computer Vision, pp. 694–711. Springer (2016)
12. Gatys, L.A., Ecker, A.S., Bethge, M.: A neural algorithm of artistic style. arXiv preprint arXiv:1508.06576 (2015)
13. Antipov, G., Baccouche, M., Dugelay, J.: Face aging with conditional generative adversarial networks. In: 2017 IEEE International Conference on Image Processing (ICIP), Beijing (2017)
14. Chen, L., Hu, X., Zhang, Z.: Face aging with boundary equilibrium conditional autoencoder. IEEE Access **6**, 54834–54843 (2018)
15. Wang, S., Xia, X., Huang, Y., Le, J.: Biologically-inspired aging face recognition using C1 and shape features. In: 2013 5th International Conference on Intelligent Human-Machine Systems and Cybernetics, Hangzhou (2013)
16. Chen, B.-C., Chen, C.-S., Hsu, W. H.: Cross-age reference coding for age-invariant face recognition and retrieval. In: Proceedings of the European Conference on Computer Vision (ECCV) (2014)
17. Mirza, M., Osindero, S.: Conditional generative adversarial nets. In: Proceedings of Advances in Neural Information Processing Systems, Montreal, Canada (2014)
18. Reed, S., Akata, Z., Yan, X., Logeswaran, L., Schiele, B., Lee, H.: Generative adversarial text to image synthesis. In: Proceedings of The 33rd International Conference on Machine Learning, vol. 3 (2016)

A Wearable System Design for Epileptic Seizure Detection

V. Sangeetha, E. Shanthini, N. Sai Prasad, C. Keerthana, and L. Sowmiya

Abstract Epilepsy, a neurological disease, affects quite sixty-five million individuals all over the world. Epilepsy, a brain disorder, is characterized by intermittent seizures which are caused by disturbances in the electrical activity of the brain. These seizures will last from seconds to minutes and vary from an impaired consciousness up to severe convulsions of the whole human body. To reduce morbidity and mortality caused by epilepsy, real-time patient observation is essential to alert caretakers and to look after the emergency medications and give assistance. A wrist band monitors real-time biological parameters in cloud server whereas Myoware muscle sensor senses muscle contraction and SpO2 sensor measures pulse rate and oxygen level in the body. Databases are collected in the cloud server which helps doctor and caretakers to monitor patient's health regularly. An Android app is used to alert caretakers and alarm is used to alert the surroundings for emergency help and medication.

Keywords Epilepsy · Seizures · Wearable · Band

1 Introduction

Biomedical engineering or medical engineering is an area in which engineering-oriented concepts and technologies are applied in medicine and biological fields. Epilepsy is the neural disease which occurs due to infection in the brain. A person with epilepsy does not have a well-defined monitoring system to avoid a risky situation. Devices for epileptic patients are available but they are not affordable. Many types of epilepsy occur for a person according to their seizure effect in the brain. Hence, a beneficent monitoring system and a device are needed to take care of an epileptic patient. To develop a system for the detection of an epileptic seizure, the

V. Sangeetha (✉) · E. Shanthini · N. Sai Prasad · C. Keerthana · L. Sowmiya
Department of Electronics and Communication Engineering, Sri Ramakrishna Engineering College, Coimbatore, India
e-mail: sangeetha.v@srec.ac.in

N. Sai Prasad
e-mail: saiprasad2198@gmail.com

P. Karuppusamy et al. (eds.), *Sustainable Communication Networks and Application*, Lecture Notes on Data Engineering and Communications Technologies 55,
https://doi.org/10.1007/978-981-15-8677-4_7

biological parameters associated with the problem must be known. The basic parameters involved in detecting the epileptic seizure are muscle contraction, pulse rate of heart and oxygen level of the body. Muscle contraction means shrinkage or retrenchment of body muscle at the wrist or the bounces. The contraction occurs normally in our body, but sudden heavy muscle limpness leads to myoclonic seizures resulting in shock spasms and trouble in walking and speaking. The electrical activity in the brain during a seizure can also change our pulse and usually causes an increase in heart rate. It is the basic symptom of epileptic attack for epilepsy patients. Thus, measurement of pulse rate is very important for seizure detection. During most of the seizure attacks, the oxygen level in the blood was found to be reduced below 70%. Due to lack of oxygen level, the brain gets damaged and thus ending up in the loss of consciousness. By taking the biological parameters such as muscle contraction, pulse rate and oxygen in the body into account, prevention and detection of epileptic seizures are possible.

2 Literature Survey

In 2018, Dionisije Sopic proposed an e-glass, a wearable gadget [1] primarily based on four EEG electrodes for the detection and identification of epileptic seizures. Based on a warning from e-glass wearable system, it is feasible to inform caregivers to rescue the affected person and also to avoid loss of life due to the neurological disorders. Their wearable system has a sensitivity range of 93.8% and a specificity of 93.37%, which operates over 2.71 days on a single battery charge.

Jose Manuel Ramirez-Alaminos proposed a paper [2] that uses a low-cost epilepsy detection device to send an alert in case of an emergency. It may come across early signs of epilepsy seizures. The machine uses several parameters including the temperature, heart rate, and the extremities movement. The tool transmits the information about the occurrences of these seizures to a cell phone, where an application notifies the person or institution in charge of the patient. This application also keeps a record of these parameters and events which can be viewed later by the user.

In 2015, Imtiaz and team's paper [3] explained about the overall performance based totally on electricity intake. They exhibited a low strength equipment execution of epileptic seizure records choice algorithm with encryption and records transmission showing the alternate-offs across the exactness and the overall strength consumption of the machine. They showed that the total power saved by the machine through information selection may be carried out via transmitting less than 40% of data. They also manifested that a reduction of 29% of electricity is feasible while choosing and transmitting 94% of all seizure events and handiest 10% of background EEG.

A Pantelopoulous proposed a paper [4] on a preclinical demonstrator for the purpose of real-time seizure detection based on coronary heart rate. The gadget includes a body hub (a smart telephone having Bluetooth, low energy, and Android operating system) and an MIO Alpha Watch that is connected through Bluetooth. It can produce a nearby warning pursued with the aid of sending an SMS message while a seizure has been identified. The consumer has the chance to set the alert edge for easily the dimension of the pulse. It additionally detects when the device is disconnected or while the sensor is not in contact with the skin.

In this regard, Bunchfield et al. [5] explored a biomedical application based on a wireless sensor network (WSN). It investigates an application of biomedical sensor systems, which endeavors to screen patients for particular conditions in a totally non-invasive, non-intrusive way. This procedure utilizes an accelerometer to decide whether an individual's arm development is like that of an individual experiencing a seizure or intense cerebrum damage causing loss of cognizance or unconsciousness.

The outputs of the deployed algorithms have been tested on test subjects and confirmed few occurrences of false positives. In future tests, automated calibration primarily based upon fake positives will, in addition, reduce the occurrence of mistakes for a given test subject.

Lorincz et al. created a wearable wireless sensor-based platform [6] for understanding the patient movement analysis called mercury. This platform only employed motion sensors to detect if seizures were happening. Mercury also includes techniques for the consumption of energy and radio band to increase the lifetime of batteries.

Huige et al. developed the design [7] of sensor nodes in order to track human postures. Each of their nodes is equipped with an accelerometer and a magnetometer for motion detection. They also introduce energy-saving mechanism in order to get higher battery cycles.

In this paper [8] they examined the application of four accelerometers shortly called as ACM's which are connected to the limbs and surface electromyography electrodes connecting the upper palms for the detection of tonic-clonic seizures [9]. sEMG i.e surface electromyography electrodes can recognize the strain in the midst of the tonic period of tonic-clonic seizure, whereas ACM can recognize musical cases of the clonic period of tonic-clonic seizures [10–12]. AI strategies, which includes the highlight choice and least-squares bolster vector machine order, were utilized for recognizing tonic-clonic seizures from ACM and sEMG signals [13–15].

3 Proposed Work

The low-cost wearable system is used to monitor the epilepsy regularly and store the data in the cloud server. When a seizure occurs, it sends a message alert to the caretakers and it alerts the surrounding with the help of buzzer.

Figure 1 represents the block diagrammatic representation of our wearable system. The wrist band which the epileptic patient wears is embedded with the temperature sensor, SpO2 sensor, and muscle sensor. These sensors are controlled using the node MCU microcontroller.

The values of the sensors are transmitted using the Wi-Fi. At the receiving side, the Android app has been created using the MIT App Inventor 2 and the cloud has been created using ThingSpeak. The sensor values are stored in the cloud server.

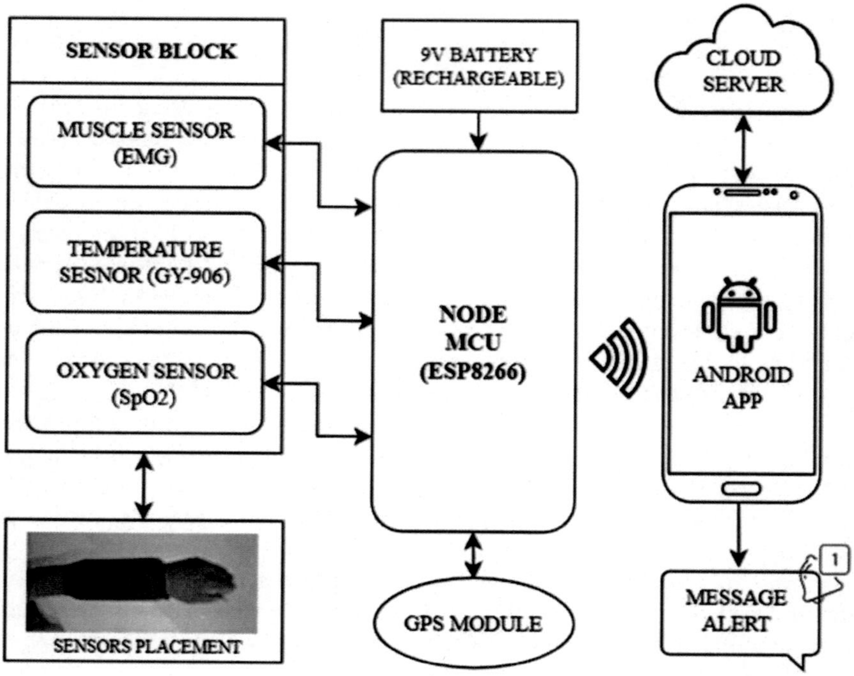

Fig. 1 Block diagram of the proposed system

These data can be viewed at any time with the help of Android app and ThingSpeak network by their caretakers. It is also helpful for doctors while giving treatment to patients. In case of any emergency or crisis, an alarm signal is sent to the caretakers with a location. If this condition prolongs for a long time, the alert with the buzzer is sent to nearby surroundings. Therefore, the patient is safe and secure at all time.

4 Technology Used

A wrist band monitors real-time biological parameters in cloud server whereas Myoware muscle sensor senses muscle contraction and SpO2 sensor measures pulse rate and oxygen level in the body. Databases are collected in the cloud server which helps doctor and caretakers to monitor patient's health regularly. During a seizure attack, the patient may become unconscious. An Android app is used to alert caretakers and alarm is used to alert the surroundings for emergency help and medication. The Android application collects the data from sensors and stores them in the cloud. At the receiver side, it acts as a medium to view the data and receive the alert message. The Android application also has access to maps which is used to find the location of the epileptic patient in case of emergency. The app acts as a path to store the data in

Fig. 2 EPISMART app using MIT App Inventor 2

the cloud so that it is accessible around the world. Using this app, the relative of the patient and family doctor can have regular monitoring of the epileptic patient. MIT App Inventer 2 platform is used to create the app which is termed as EPISMART app.

Figure 2 shows the login page of the EPISMART app created using MIT App Inventor 2. There is a login ID for each and every individual including the epileptic patient, caretakers, doctors, and common ID for a hospital.

5 Results and Discussions

The wrist band is used to monitor the health of the patient in normal conditions and if a seizure occurs it sends an alert message to the caretaker and alarms the buzzer for the immediate help from the surrounding.

Figure 3a, b shows the hardware components embedded within the wristband. The muscle sensor, temperature sensor, and pulse sensor are interfaced with the node MCU board. The node MCU board is dumped with the coding that gives the values of the sensors. Values obtained from these sensors are then sent to the EPISMART

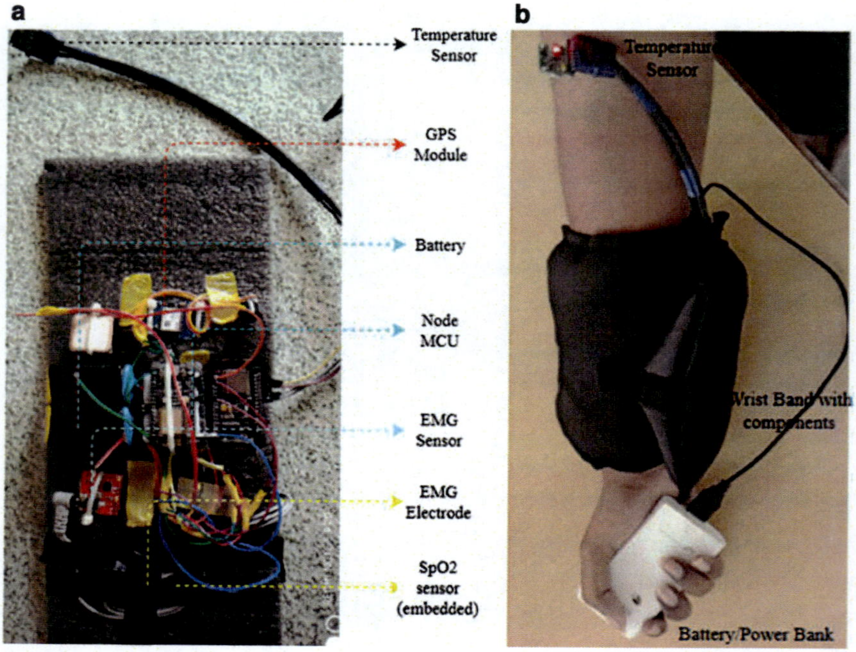

Fig. 3 **a** Product before stitching **b** Product after stitching

Android mobile application through Wi-Fi that is interfaced in node MCU board. The working of three sensors is controlled by node MCU.

Figure 3a shows the designed product before stitching as a band and Fig. 3b shows the product after stitching as a wrist band.

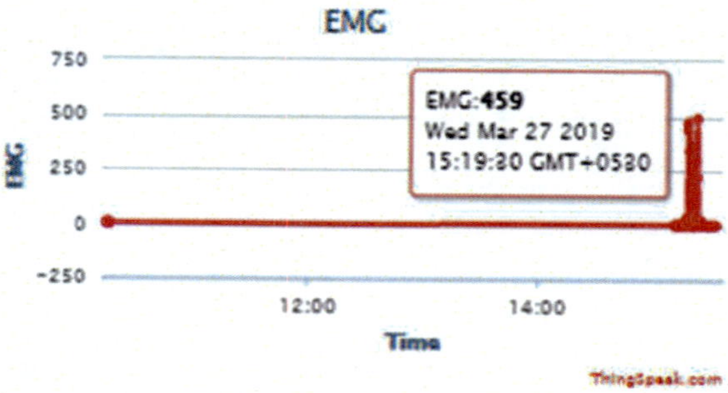

Fig. 4 Graphical representation of muscle contraction for epileptic patient

5.1 Graphical Representation of Obtained Muscle Contraction Value

Figure 4 shows the graphical representation of muscle contraction of the epileptic patient using EMG sensor. The threshold value for epileptic seizure is above 100. From the graph, it can be examined that the detected value is 459. Thus, the epileptic seizure is detected.

Figure 5 shows the muscle contraction of a person works out in a gym. Normally the muscle contraction value for a person working out in a gym is found to be below 100. From the graph obtained, the value was found to be 16, which indicates the person is normal.

Figure 6 shows the graphical representation of the EMG sensor for a normal adult. As stated earlier, the threshold value of epileptic seizure for a normal person is below 100. The graphical results state that the value obtained is found to be 4, which indicates the person is normal.

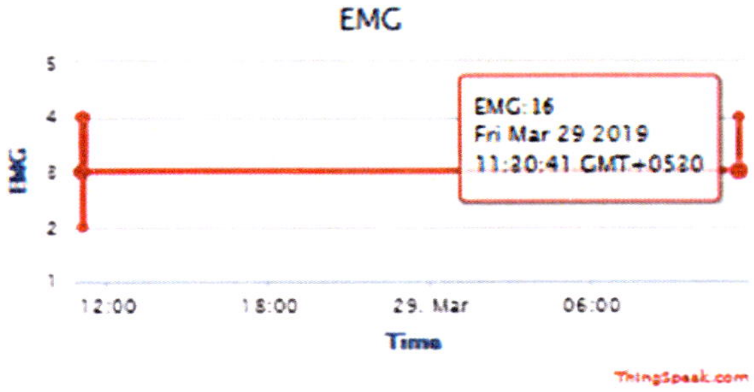

Fig. 5 Graphical representation of muscle contraction for person works out in a gym

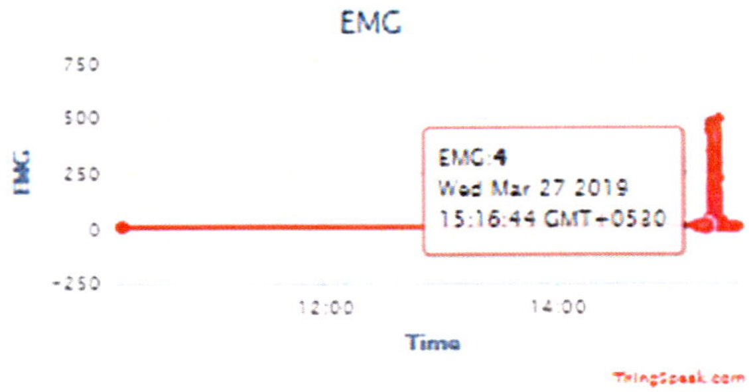

Fig. 6 Graphical representation of EMG sensor for normal adult

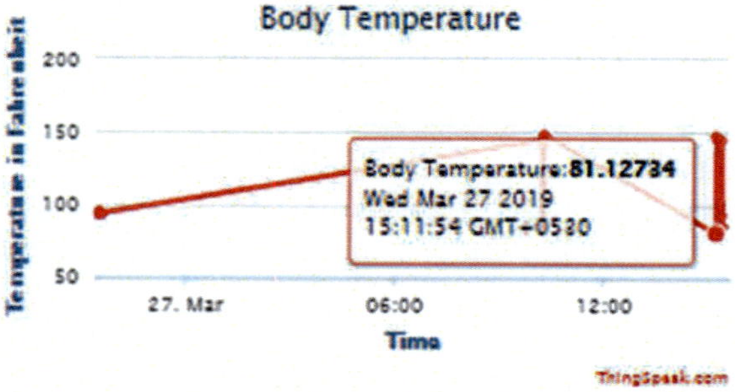

Fig. 7 Graphical representation of obtained values in temperature sensor

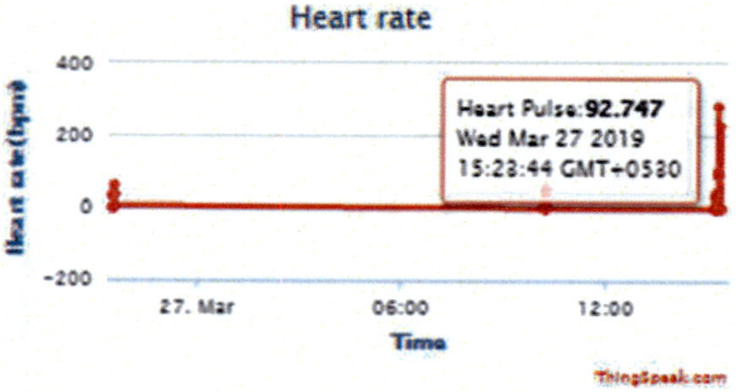

Fig. 8 Graphical representation of obtained values of heart rate

5.2 *Graphical Representation of Obtained Body Temperature Value*

Figure 7 shows the graphical representation of obtained values in the temperature sensor. Approximately, a normal person has a body temperature of value between 80 and 99 Fahrenheit. From the graphical representation, the detected value was found to be 81.12 F indicating the person is normal.

5.3 *Graphical Representation of Obtained Heart Rate and Pulse Oxidation Value*

Figure 8 shows the graphical representation of the obtained values of heart rate. A normal resting heart rate for adults is found to be 60–100 bpm. The detected value was found to be 92.47 bpm, which is indicating that the person is normal.

Figure 9 shows the graphical representation of obtained values in SpO2 sensor. The normal value of SpO2 sensor for a healthy person is found to be between 95%

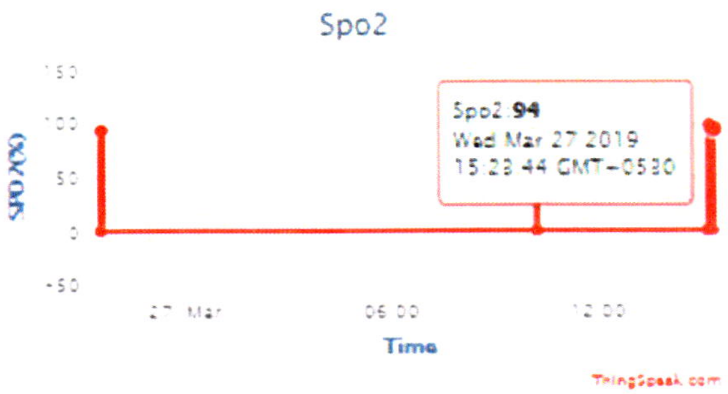

Fig. 9 Graphical representation of obtained values in SpO2 sensor

to 100% (oxygen level). From the graph, the detected value was found to be 94% stating that the person is normal.

6 Conclusion

The design of the low-cost wearable system will be very useful to epileptic patients. It is composed of a wrist band that contains muscle sensor, SpO2, and temperature sensor. In addition, it also contains node MCU that acts as both controller and Wi-Fi module which is used for wireless communication with the smartphone. The main feature is that Internet of things (IoT) technology has been used.

The patient history is stored in the cloud and the doctor can analyze it through the app at anytime and anywhere in the world. The information is delivered to the respective caretakers at an emergency condition (i.e.) they get an immediate alert message through the patient's wristband with his current location. The complete database is available at free of cost. The device helps to prevent the death of any patient who is unnoticed during a seizure attack. By this device, the patient can have continuous monitoring of his/her health. In case of sudden occurrence of epilepsy, a buzzer will sound as an alarm to get the immediate help and support from the nearby environment.

Many existing methods are bulky in nature and mostly kept in hospitals or homes to only detect the seizure attacks. But the proposed system is a handheld wearable device which is very helpful for patients anywhere they go. It alerts the caretakers at home as well to the registered hospital which is the research novelty of the proposed project.

References

1. e-Glass: a wearable system for real-time detection of epileptic seizures. In: 2018 IEEE 978-1-5386-4881-0/18/$31.00 @ International Symposium on Circuits And Systems (ISCAS)
2. Ramirez-Alaminos, J.M., Sendra, S., Lloret, J., Navarro-Ortiz, J.: Low-cost wearable bluetooth sensor for epileptic episodes detection. In: 2017 IEEE International Conference on Communications (ICC), Paris, pp. 1–6 (2017). https://doi.org/10.1109/icc.2017.7997413
3. Imtiaz, S.A., Logesparan, L., Rodriguez-Villegas, E.: Performance-power consumption tradeoff in wearable epilepsy monitoring systems. IEEE J. Biomed. Health Inform. **19**(3), 1019–1028 (2015). https://doi.org/10.1109/JBHI.2014.2342501
4. Pantelopoulos, A., Bourbakis, N.G.: A survey on wearable sensor-based systems for health monitoring and prognosis. IEEE Trans. Syst. Man Cybern. **40**(1), 1–12 (2010). https://doi.org/10.1109/TSMCC.2009.2032660
5. Bunchfield, T.R., et al.: Accelerometer-based human abnormal movement detection in wireless sensor networks. In: 1st ACM SIGMOBILE International Workshop on Systems and Networking Support for Healthcare and Assisted Living Environments, pp. 67–69, 11 June 2007
6. Lorincz, K.: Mercury: a wearable sensor network platform for high-fidelity motion analysis. In: 7th ACM Conference on Embedded Networked Sensor Systems, pp. 183–196, 4 November 2009
7. Huyghe, B., Vanfleteren, J., Doutreloigne, J.: Design of flexible, low-power and wireless sensor nodes for human posture tracking aiding epileptic seizure detection. In: 2009 IEEE SENSORS, Christchurch, pp. 1963–1966 (2009)
8. Milosevic, M., et al.: Automated detection of tonic-clonic seizures using 3D accelerometry and surface electromyography in pediatric patients. IEEE J. Biomed. Health Inform. **20**, 1333–1341 (2016)
9. Lin, S., Qomah, I., Lin, Y., Chiueh, H., Lin, C.: Design and implementation of a smart headband for epileptic seizure detection and its verification using clinical database. In: 2018 IEEE International Symposium on Circuits and Systems (ISCAS), Florence, pp. 1–5 (2018). https://doi.org/10.1109/ISCAS.2018.8351687
10. Padma, T., Kumari, C.U.: Sudden fall detection and protection for epileptic seizures. In: 2018 International Conference on Recent Innovations in Electrical, Electronics & Communication Engineering (ICRIEECE) (2018)
11. Moseley, B.D., Wirrell, E.C., Nickels, K., Johnson, J.N., Ackerman, M.J., Britton, J.: Electrocardiographic and oximetric changes during partial complex and generalized seizures. Epilepsy Res. **95**(3), 237–245 (2011)
12. MyoWare Muscle Sensor—SEN-13723—Spark Fun Electronics. Sparkfun.com. https://www.sparkfun.com/products/13723
13. Shafiul Alam, S.M., Bhuiyan, M.I.H.: Detection of seizure and epilepsy using higher order statistics in the EMD domain S. IEEE J. Biomed. Health Inform. **17**(2), 312–318 (2013). 2168-2194/$31.00©2013 IEEE
14. Tzallas, A.T.: Epileptic seizure detection in EEGs using time–frequency analysis. IEEE Trans. Inf. Technol. Biomed. **13**(5), 703–710 (2009). 1089-7771/$26.00©2009
15. Iasemidis, L.D., IEEE Member: Epileptic seizure prediction and control. IEEE Trans. Biomed. Eng. **50**(5), 549–558 (2003). 0018-9294/03$17.00©2003

Abatement of Traffic Noise Pollution on Educational Institute and Visualization by Noise Maps Using Computational Software: A Case Study

Satish K. Lokhande, Divya M. Motwani, Sanchi S. Dange, and Mohindra C. Jain

Abstract The aim is to study the impact of traffic noise pollution on an educational institution near Nagpur–Aurangabad Highway and propose a feasible noise assuaging measure. Optimizing the suitable barrier to attaining the desired attenuation, insertion loss (IL) for varying heights was computed using specialized software Predictor LimA. Further, software mapping of worked-out data was performed to generate noise contours for scenarios before and after the installation of the barrier enabling visualization of noise extent across the study area. Noise barrier design experiment concluded that a barrier of 4 m height along the college boundary could able to provide a maximum IL of 6 dB(A). Consequently, the subsequent noise level would be in the range of 57.9 to 62.9 dB(A). The developed noise maps displayed a considerable reduction in noise levels for the proposed 4 m height noise barrier.

Keywords Noise pollution · Noise barrier · Insertion loss (IL) · Transmission loss (TL) · Noise mapping · Cost effectiveness

---

S. K. Lokhande (✉) · M. C. Jain
Analytical Instrument Division (AID), CSIR-National Environmental Engineering Research Institute (NEERI), Nehru Marg, Nagpur 440020, Maharashtra, India
e-mail: s_lokhande@neeri.res.in

M. C. Jain
e-mail: mohindrajain@gmail.com

D. M. Motwani · S. S. Dange
AID, CSIR-NEERI, Nagpur, India
e-mail: m.divya.motwani@gmail.com

S. S. Dange
e-mail: sanchidange27@gmail.com

P. Karuppusamy et al. (eds.), *Sustainable Communication Networks and Application*, Lecture Notes on Data Engineering and Communications Technologies 55,
https://doi.org/10.1007/978-981-15-8677-4_8

1 Introduction

The emerging development of the cities has contributed to a much-unheeded form of environmental pollution, which is noise. The noise levels have increased considerably during the last hundred years due to human-made sources and are now doubling every ten years [1]. Increasing population, growing demand for vehicles, industrialization, rising construction works, lack of proper implementation, disobeying of noise rules, etc., are the major reasons for worsening of the problem. Noise pollution causes hearing loss [2], sleeplessness [3], irritability and stress, effects on work performance [4], reduced physical and mental abilities [5], increased blood pressure and uneven heart rhythms [1], etc. World Hearings Index (2017) ranked the noisiest cities of the world, in which Guangzhou, in China, topped the index followed by Delhi, the capital city of India, Cairo, and Mumbai [6]. Therefore, as per the ranking, undoubtedly, Indian people are at constant risk of being exposed to high noise levels, which may severely affect their health.

India needs a paradigm shift to deal with its noise management problems. With the rapid development of road network in India, vehicular traffic has risen to an average speed of 10.16% per annum over the last five years [7]. India is struggling with heterogeneous traffic conditions like congestion, mixed vehicle flow, changing weather conditions, road type and lack of traffic knowledge [8]. A study on Nagpur City concluded that residents are exposed to high noise levels of 70 dB(A) and above at almost all locations [8]. The underway construction of Metro in Nagpur has added to congestion of roads, thereby increasing the noise nuisance and discomfort [9]. However, due to growing awareness, Indian government institutions such as Maharashtra Metro Rail Corporation Limited (MMRCL) has come forward along with Central Road Research Institute (CRRI) to figure out places of Nagpur where noise is exceeding the permissible limits and reduction measures such as noise barriers and rubber pads are desired to be implemented [10].

Noise affects the students of educational institutions by disrupting a healthy educational environment and learning skills [11]. Ibrahim [12] studied the noise levels of an engineering college in Baghdad and mapped them with the help of SoundPlan software. The study indicated noise levels ranging from 55 to 80 dB(A) around the student gathering areas. Another study by D. Thattai [13] was carried out in a university campus of South India, comprising a medical facility. The obtained noise levels were violating the permissible limits which shows the need for the study to be carried out. Similarly, noise levels of three educational institutes of Chittagong city were studied by Mamun [14] besides these a questionnaire survey was also conducted. All the three studies revealed that the campuses are exposed to higher noise levels taking our attention towards the growing issue of noise levels, particularly in the educational institutions. As the educational institutions are usually located near the busy areas, it is necessary to provide appropriate measures to curb noise levels to provide a safe educational environment.

Noise barriers have proven to be very useful in mitigating noise pollution; however, due to lack of awareness and literature, they could not be appropriately put into practice in India. An effective noise barrier should be tall and long enough to block the overall noise levels through absorption, reflection, transmission and diffraction. Parameters, like effectiveness, maintenance, cost, acceptability, aesthetic, etc., must be given due consideration before designing a noise barrier. Besides this, weather conditions are also important aspects which can affect the barrier performance hence must be taken into account [15].

Noise mapping is an essential tool for getting a visual map of the sound contours of a particular area for management and control of environmental noise. In European countries, noise mapping is performed after every five years based on noise indicators (Lden) and (Ln) for industry, railway and road traffic sources [2]. However, the use of such maps and mapping practices is still limited in India. The current study aimed to access the noise levels of a college adjacent to a national highway and demonstrate it with the help of noise contours to get acoustic environment view of the area and proposes assuaging measures.

2 Study Area

An academic institution in Nagpur, India, was selected as the study area for the present work. The college is situated on Nagpur to Aurangabad Highway (NH-753A), just next to the traffic signal and before the petrol pump as shown in Fig. 1. Distance between college boundary and road is measured to be about 2.5 m, and the road is around 11.5 m from both sides. College boundary and road median are nearly 14 m apart. The college has two main buildings of height 15 m and 20 m, length 14 m and thickness of 77.0 m, respectively. Further, the college has a large campus along with

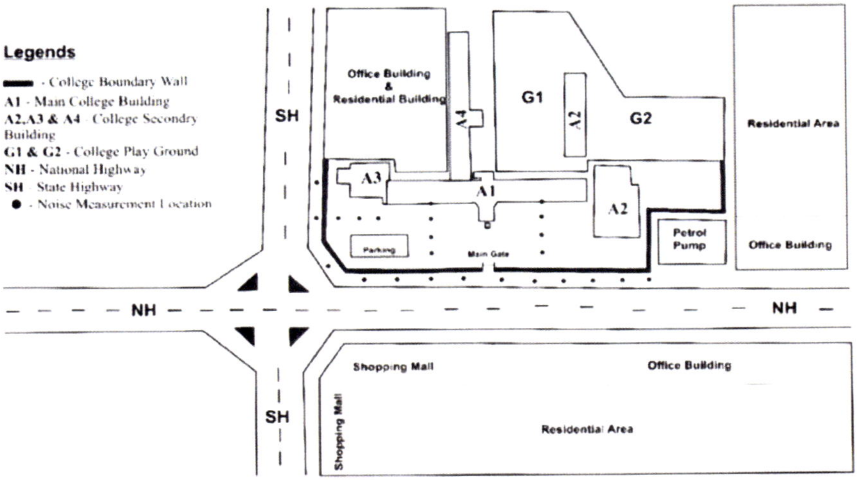

Fig. 1 College layout

parking area adjacent to the highway whereas the main building of the college is around 38 m from college boundary and about 40.5 m from the edge of the highway. The college is having an existing boundary of 2 m height that comprises of 1 m concrete construction and 1 m metal fencing.

3 Materials and Method

Noise measurements were performed using an integrated digital sound level meter (SLM) CK:172B Optimus Green (Cirrus, UK) having Fast, A-weighted response mode (LAF) with data logging of 125 ms time break. Noise measurements were carried out complying with ISO 1996-2:2007, which is an essential document of acoustics [16]. Before measurements, the sound level meter was appropriately calibrated using a standard calibrator and set fixed for the time duration of 20 min for monitoring purpose. The height of the source was 0.5 m, and the height of the receiver was 1 m. Average noise levels were recorded at every 1 m distance from the edge of the highway on 13 selected locations. The highway comprised of the vehicular flow of all types of vehicles, so the average speed on the road was calculated by classifying the vehicles into six categories namely bike, auto, car, truck, lorry and bus. The obtained average speed was 33–50 km/hr. Environmental noise influencing parameters such as temperature, humidity and wind speed were checked regularly before monitoring to ensure noise measurements carried out in suitable weather conditions.

Noise level measurements are performed at 24 different locations inside college premises at a distance of 10 m, 15 m, 30 m and 38 m from the college compound. Before the initiation of proper barrier design, optimization of the height of the barrier is crucial that can provide the desired attenuation relating to the geometric location of the barrier from the source and the receiver. For finalizing the adequate height of noise barrier for the present case study, assumptions of five height levels of 1 m, 2 m, 3 m, 4 m and 5 m were experimented on Predictor LimA software. Further, depending on the required insertion loss, a particular level of height has opted.

4 Noise Mapping

Noise mapping is a scientific method of representing noise levels of a particular area in the form of colour contours so that each colour stipulates the specific intensity of noise. Noise maps help in identifying and differentiating the high sound affected areas with other areas in the region.

The rapid development of GIS technology, fast computing capabilities, digital mapping, noise modelling tools, etc., in recent years and contribution of all has made it possible to develop 2D and 3D noise maps to understand noise problems precisely. Noise mapping is recommended to achieve the following objectives.

- Noise maps are easy to understand document for the general public compared with graphs and tabular formats
- Noise data can be mapped in terms of spatial–temporal variability
- Assist in planning for an upcoming development in the city
- Increase awareness amongst the common public as they get familiar to the extent of noise levels in there surrounding
- Noise developed for any city can become the benchmark for other cities for planning and mitigating noise pollution
- It helps the town planners better understand the urban noise problems and help them establish proper development plans for the city to set objectives for any noise issues.

Predictor LimA (LimA, 2016) (Type 7810) software is advanced software, which allows performing extensive calculations with precision. It uses ISO 9613 standard for calculations and subsequently predicting noise levels. Using the software noise, maps were generated to provide detail dissemination of highway road noise on the educational institute.

5 Results and Discussion

The results obtained based on the conducted noise assessment are engrossed in this section. Noise monitoring had performed in favourable weather conditions. Meteorological parameters such as temperature, humidity and the wind speed observed were 30 °C, 58.8% RH and 1.4 m/s, respectively. The investigation revealed that road traffic noise is the leading cause of high noise levels inside the college premises. The traffic signal and the petrol pump being closely located at two ends of the college front boundary make frequent halting and moving of vehicles that create engine noise and provoke regular honking of the horn. The noise generated by this contributes significantly and increases the overall noise level at the source.

Noise monitoring was performed at various locations along the highway and inside the college premises to realize the severity of the noise pollution. Noise monitoring conducted at 24 different locations inside college based on chosen distances from the college boundary specifies noise levels varies from 61 to 68 dB(A). Highway noise levels monitored at 13 positions starting from the petrol pump and ending on the traffic signal at 10 m distance from each other showed high variations. Amongst all the 13 measured noise levels, the highest noise level of 77.2 dB(A) found near a petrol pump and lowest of 73.4 dB(A), near to the signal. The overall average traffic noise level obtained during monitoring on the highway road was 75.4 dB(A). The impact of such high noise levels originating from the nearby highway on a college campus was 68 dB(A), which breached the permissible limits of Central Pollution Control Board (CPCB), New Delhi, for silence zones [17].

The college has an existing college boundary wall of 2 m height that comprises of 1 m concrete construction and 1 m metal fencing which is incapable of curbing the road traffic noise. Hence, the students and other staff of the college are at risk of exposure to the higher noise levels. In this case, a concrete barrier of greater height to shield the receiver from road traffic noise has been considered as the most suitable mitigation measure to curb the excess noise levels inside the college premises.

5.1 Assessment of Noise Barrier Height and Location

The conducted survey discovered that college is quite close to the highway, and the college boundary is along the paved road. Therefore, the barrier was proposed on the existing boundary wall. Consequently, the distance between the source and the receiver concerning the existing fence will remain unaltered. The validation of the monitored data and prediction of the noise levels for designing of a noise barrier was an essential aspect of the study. Many previous studies confirm that the acoustic performance of the barrier can be evaluated through the insertion loss of the proposed noise barrier. Consequently, with the help of specialized software, Predictor LimA, a barrier with varying heights of 1 m, 2 m, 3 m, 4 m and 5 m has been experimented and noise levels were predicted for each case. Insertion loss was derived by subtracting the measured noise levels inside the college premises at various locations with predicted noise levels computed for multiple heights of noise barrier [18]. Further, by comparing the noise reduction obtained for each case, the final barrier height is optimized as shown in Fig. 2.

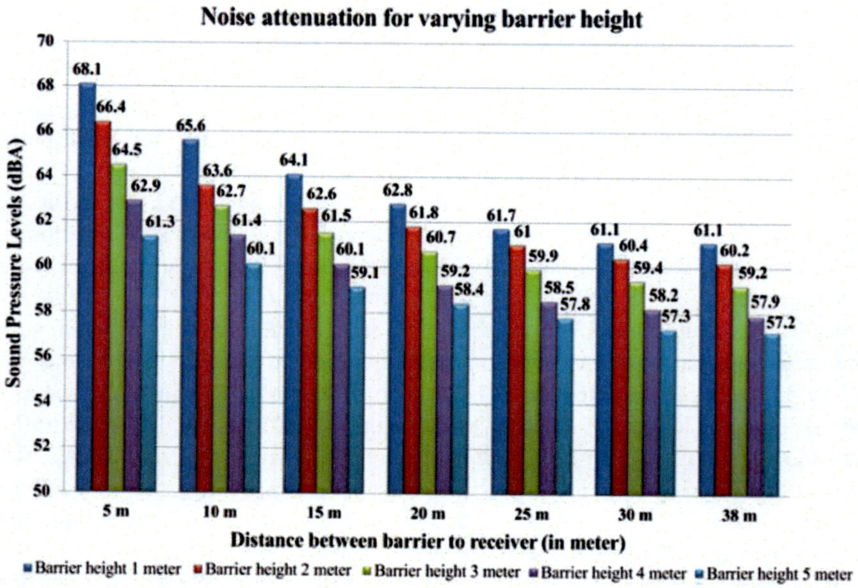

Fig. 2 Software predicted noise levels at different distances for varying barrier height

The predicted noise levels obtained with Predictor LimA software for 1 m height barrier signify the existing noise scenario of the college premises without barrier as it also has 1 m concrete boundary wall and predicted data confirms with the measured noise levels, thus validates the accuracy of the method. Besides, observed data reveals that noise reduction increases with an increased barrier height. Considering the above facts as well as the concerned cost of the barrier and aesthetic look, 4 m barrier height is suggested to mitigate the excessive noise levels.

Material choice is a crucial part of barrier construction and should be selected such that it provides required insertion loss plus an additional 10 dB(A) transmission loss (TL) [18]. The researchers have studied the use of waste materials such as palm tree pruning waste [19], demolished wastes [20], recycled ceramic wastes [21] as an alternative eco-friendly option for barrier construction. Also, vegetation could provide additional attenuation and also improve the aesthetic view of the area providing the feeling of the green area [15]. Different shapes of noise barriers such as wing, zig-zag, curved and flat type for their noise mitigation capacities have also been studied [22, 23]. Concrete is amongst the world's most widely used and adaptable materials for construction of noise barriers and hence recommended for the present case study [22, 24]. Proper designing of noise barrier requires consideration of both acoustical as well as non-acoustical parameters. They can be precast or cast in site; however, prefabricated barriers are relatively more expensive [24]. Implementing concrete barriers is advantageous due to their high life expectancy, which is generally 40–50 years and also their cost is 5% less than the wood barriers and 10% lower than aluminium barriers [4]. Hence, as per the performed experiment, a 4 m height of the barrier with concrete as a barrier material was able to provide noise reduction of 6 dB(A). The study also proposes plantation of sufficiently dense vegetation belt of 2 m width along the college boundary wall to attain additional attenuation.

Noise mapping is an effective scientific method and well-established practice that represents the noise levels of a particular area in the form of colour contour maps. The overall noise scenario before and after establishing noise barrier is visualized with the help of noise contour maps to indicate the most noise-prone areas were the extent of the college facing the national highway. Figure 3 shows noise dissemination of the study area for the existing 1 m boundary. It can be observed that the dominant noise levels appear near the main building of the college area ranges between 60 to 65 dB(A) as clearly delineated with brown colour contour, whereas noise levels along the boundary ranges between 65 to 70 dB(A) demarcated with violet colour contour.

Similarly, Fig. 4 shows the noise scenario of the study area after considering the proposed 4 m barrier as a noise abatement measure. The noise contour map depicted that there is a significant reduction in noise levels. Now, the dominant noise levels appear near the main building of the college area ranges between 55 to 60 dB(A) outlined with pink colour contour, whereas noise levels along the boundary ranges between 60 to 65 dB(A) demarcated with brown colour contour. Comparison analysis of both the maps concludes that establishing 4 m concrete barriers is an effective and reasonable mitigation strategy to reduce the noise levels in the educational institute to a significant extent. Although the suggested mitigation solution provides a significant

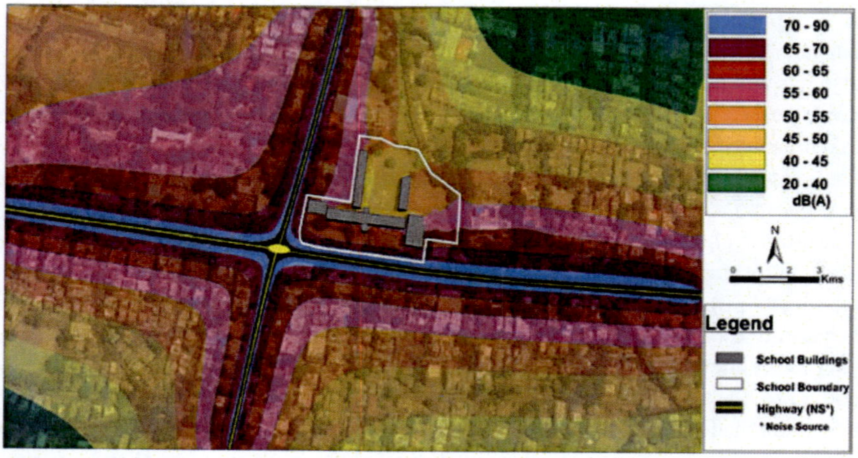

Fig. 3 Noise map of college having existing 1 m boundary (top view)

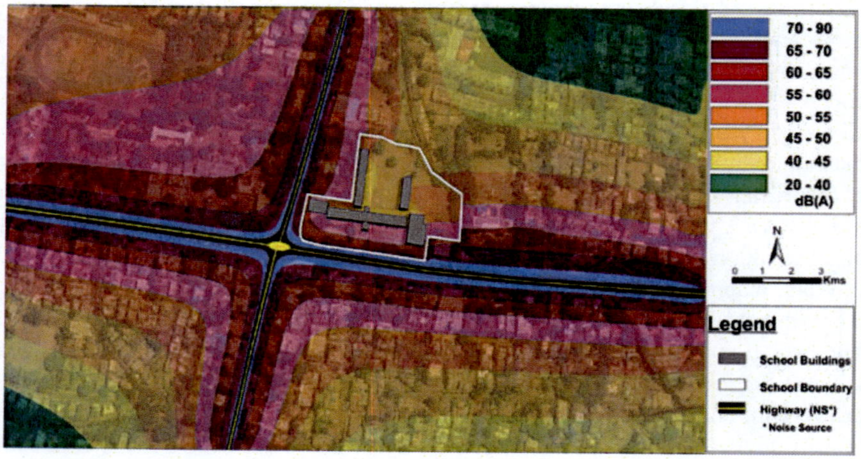

Fig. 4 Noise map of college with proposed 4 m barrier (top view)

noise reduction nevertheless, it could not able to attend the level below the permissible limits prescribed by CPCB owing to the highway, traffic signal and the petrol pump are at proximity to the institution. In this case, the implementation of 2 m vegetation belt can help achieve more noise reduction. Figure 5 represents the difference map between the noise levels before and after the implementation of the barrier of 4 m height and also defines the insertion loss achieved by the noise abatement strategy.

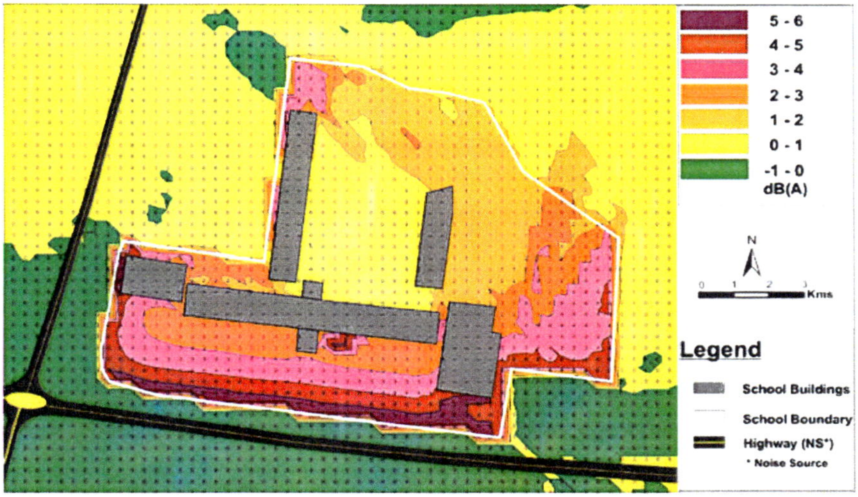

Fig. 5 Computing insertion loss of the noise barrier using the Predictor LimA Software

6 Conclusion

A constantly developing city like Nagpur with various constructions works and other development activities going on regularly is likely to face noise issues. The conducted study revealed that the teachers and students are at high risk as the college is exposed to noise levels, which exceed 50 dB(A) CPCB permissible limits. Unfortunately, the academic institution building faces the national highway creating its direct exposure to the noise sources. Obtained noise levels from the road were much higher; therefore, mitigation measures were of immediate concern for reducing these noise levels.

The average of noise levels obtained during monitoring on the highway was found to be 75.4 dB(A), whereas its impact on the college campus was about 60 to 65 dB(A). The experimental analysis concluded that a 4 m height barrier could provide noise reduction of 6 dB(A). Consequently, the predicted noise levels inside the college premises would be in the range of 55 to 65 dB(A). Besides this, the use of 2 m wide vegetation belt along the boundary wall inside the college premises can provide an extra reduction of noise levels and also help in enhancing the view of the barrier. Performed experimental analysis indicates that insertion loss increases with increased height of the barrier and decreases with increase in distance from the barrier. Investigation results validate the accuracy of the prediction data, and IL calculations as the measured noise levels for existing 1 m boundary wall and the computed noise levels for 1 m concrete barriers are similar.

Obtained data were mapped to view the noise dissemination of the study area, and it indicates that the college parking area and the college buildings are amongst the most noise prone regions. The observations from the noise map concluded that a proposed 4 m barrier was able to provide a substantial reduction in noise as compared to existing 1 m concrete wall thereby proving to be a suitable measure to combat the noise levels. Use of concrete as a material for noise barrier construction is recommended to combat the excessive noise level. The present work promotes the use of contour noise maps to get a clear idea of the present and futuristic noise levels of the study area. Though noise barriers are useful, they are moderately expensive and are not entirely efficient are amongst its limitations. New concepts and innovative research in the field of acoustics are much needed, which will prove helpful in dealing with the increasing nuisance of noise pollution. The present study is one of its kind as it has studied, analysed and mapped the noise levels and suggestive measures have also well provided. Also, this study can be a valuable reference for future projects dealing with areas coming under silence zones and prove helpful.

Acknowledgements The authors wish to thank CSIR—National Environmental Engineering Research Institute, Nagpur, India, for giving this opportunity and facilitating us with all the necessary requirements for this paper. This article has been checked for similarity index through iThenticate software with KRC No: CSIR-NEERI/ KRC/2019/SEP/AID/1.

Conflict of Interest The authors declare that they have no conflict of interest.

References

1. Hunashal, R., Patil, Y.: Assessment of noise pollution indices in the city of Kolhapur India. In: International Conference on Emerging Economies—Prospects and Challenges India (ICEE 2012), vol. 37, pp. 448 – 457 (2012). Procedia—Soc. Behav. Sci.
2. Manwar, V., Mandal, B., Pal, A.: Application of noise mapping in an Indian opencast mine for effective noise management. In: 12th ICBEN Congress on Noise as a Public Health Problem (2017)
3. Bluhm, G., Nordling, E., Berglind, N.: Road traffic noise and annoyance-an increasing environmental health problem. Noise Health **6**(24), 43–49 (2004)
4. Ahac, H., Lakušic, D.V.: Croatian experience in road traffic noise management-concrete noise barriers. Rom. J. Transp. İnfrastruct. **3**(1) (2015) https://doi.org/10.1515/rjti-2015-0019
5. Petrovic, R., Petrovic, Z.N.: Design of noise protection in urban areas- case study of an elementary school. In: Twenty Third National and Fourth İnternational Conference Noise and Vibration (2012). https://www.researchgate.net/publication/268270921
6. Gray, A.: These are the cities with the worst noise pollution. World Economic Forum 2017. https://www.weforum.org/agenda/2017/03/these-are-the-cities-with-the-worst-noise-pollution/. Accessed 9 Dec 2018
7. Vij, G., Agarawal, M., Gupta, M.: Assessment of traffic noise and development of noise models for national highway-06, passing through Chhattisgarh, India. Ind. J. Sci. Technol. **9**(47) (2016). https://doi.org/10.17485/ijst/2016/v9i47/103966
8. Lokhande, S.K., Pathak, S.S., Kokate, P.A., Dhawale, S.A., Bodhe, G.L.: Assessment of heterogeneous road traffic noise in Nagpur. Arch. Acoust. **43**(1), 113–121 (2018). https://doi.org/10.24425/118086
9. The Times of India. Nagpur Metro. The Times of India Newspaper (2018). https://timesofindia.indiatimes.com/topic/nagpur-metro. Accessed 31 May 2018
10. Roy. A.: Metro to conduct noise pollution study (2018). https://timesofindia.indiatimes.com/city/nagpur/metro-to-conduct-noise-pollutionstudy/articleshow/62963089.cms. Accessed 15 June 2018

11. Bhosale, B., Late, A., Nalawade, P., Chavan, S., Mule, M.: Studies on assessment of traffic noise level in Aurangabad city, India. Noise Health **12**(48), 195–198 (2010). https://doi.org/10.4103/1463-1741.64971
12. Ibrahim, S.: Noise mapping of the campus of the college of engineering/the university of Al-Mustansiriyah. J. Environ. Earth Sci. **5**(4), 108 (2015)
13. Thattai, D., Sudarsan, J.S., Sathyanathan, R., Ramasamy, V.: Analysis of noise pollution level in a University campus in South India. In: IOP Conference Series: Earth and Environmental Science (2017). https://doi.org/10.1088/1755-1315/80/1/012053
14. Mamun M., Shil A., Paul A., Paul P.: Analysis of Noise Pollution Impacting Educational Institutes near Busy Traffic Nodes in Chittagong City. International Conference on Green Architecture (ICGrA). (2017)
15. Pigasse, G., Kragh, J.: Optimised Noise Barriers; A state of the Art Report; Report 194, December 2011
16. ISO 1996-2:2007. Acoustics – Description, measurement and assessment of environmental noise – Part 2: Determination of environmental noise levels
17. Central Pollution Control Board: The noise pollution (Regulation and control) rules (2000). http://cpcb.nic.in/displaypdf.php?id=Tm9pc2UtU3RhbmRhcmRzL25vaXNlX3J1bGVzXzIwMDAucGRm
18. Klingner, R.E., McNerney, M.T., Busch-Vishniac, I.J.: Design guide for highway noise barriers. Conducted for the Texas Department of Transportation in cooperation with the U. S. Department of Transportation Federal Highway Administration, by the Centre for Transportation Research Bureau of Engineering Research the University of Texas at Austin (2003)
19. Gil-Lopez, M.-M., Verdu -Vazquez, M.-R.B.: Acoustic and economic analysis of the use of palm tree pruning waste in noise barriers to mitigate the environmental impact of motorways. Sci. Total Environ. (2017). http://dx.doi.org/10.1016/j.scitotenv.2017.01.162
20. Jeffrey: Construction and Demolition Waste Recycling a Literature Review. Dalhousie University's Office of Sustainability (2011)
21. Arenas, C., Vilches, L.F., Leiva, C., Alonso-Fariñas, B., Rodríguez-Galán, M.: Recycling ceramic industry wastes in sound absorbing materials; Materiales de Construcción 66(324) (2016). https://doi.org/10.3989/mc.2016.10615
22. Fleming, G.G., Knauer, H.S., Lee, C.S., Pedersen, S.: FHWA highway noise barrier design (2000)
23. Lee, K., Park, C.S., Kim.: Reduction effects of shaped noise barrier for reflected sound. J. Civil Environ. Eng. **5**, 2 (2015). http://doi.org/10.4172/2165-784X.1000171
24. Bjelic, M., Vukicevic M., Petrovic A.: Analysis of materials used for production of noise protection barriers. In: Noise and vibration 23rd National and 4th International Conference, pp. 17–19 (2012)

RainRoof: Automated Shared Rainwater Harvesting Prediction

Vipul Gaurav, Vishal Vinod, Sanyam Kumar Singh, Tushar Sharma, K. R. Pradyumna, and Savita Choudhary

Abstract A large proportion of rainwater harvesting systems is not optimized for maximum water conservation. Along with requiring tedious timelines to gain approval, the rising urbanization necessitates the adoption of shared rainwater harvesting, a system that uses a shared rainwater store for the surrounding community. The process of estimating the feasibility of such a system is complex and requires a lot of manual inspection. This paper presents a machine learning and computer vision-based solution that automates the entire process by implementing rooftop image segmentation, depth estimation, rainfall prediction, and optimal tank placement. Rainfall prediction is done using a seasonal autoregressive moving average (SARIMA)-based rolling forecasting model with a mean absolute error of 85.69. The Mask R-CNN segmentation model is enhanced by the Canny edge algorithm and contour mapping to calculate the catchment volume of the rooftop. The system can thus provide the feasibility of an installation and estimate the break-even period based on the computed metrics.

Keywords Shared rainwater harvesting · Rooftop segmentation · Depth estimation · Optimal tank placement · Rainfall prediction · Computer vision · Deep learning · Machine learning

1 Introduction

Water scarcity has emerged as an immense problem around the world and a tropical country like India, which is blessed with sufficient rainfall scattered across various states, is still facing water shortages. In India, rainwater harvesting is not feasible for every housing settlement typically due to the large upfront costs and the convoluted government procedures in place, involving tedious paperwork and inspections carried

V. Gaurav (✉) · V. Vinod · S. K. Singh · T. Sharma · K. R. Pradyumna · S. Choudhary
SVAG Research Lab, Sir M. Visvesvaraya Institute of Technology, Bangalore, India
e-mail: vgaurav3011@gmail.com

S. Choudhary
e-mail: savitha_cs@sirmvit.edu

P. Karuppusamy et al. (eds.), *Sustainable Communication Networks and Application*, Lecture Notes on Data Engineering and Communications Technologies 55,
https://doi.org/10.1007/978-981-15-8677-4_9

out by only qualified engineers, thus rendering it inefficient [1]. It requires the feasibility study for the given region which involves various factors ranging from rainfall received by that region, daily water consumption rate, potable water availability, and catchment area estimation [2]. Our proposed system is motivated to solve the problem of feasibility estimation and the associated deployment procedures by automating the process requiring expensive and time-intensive skilled inspection by providing a machine learning-based end-to-end tool. The proposed solution only requires a scaled satellite image of the user's rooftop and the tool performs enhanced image segmentation to obtain the segmented roof dimensions. With the computed area, the depth estimation module combined with the region-specific rainfall prediction system calculates the total rooftop catchment volume. This value is then extrapolated to compute the mean rainfall volume based on the surface runoff coefficient of the rooftop material to obtain the total predicted water accumulated in the system. Based on a detailed statistical model herewith explained in Sect. 3, the non-drinking water usage estimation is used as the verification metric to validate the feasibility of the shared rainwater harvesting system. Our image segmentation pipeline consists of a Mask R-CNN model whose accuracy is enhanced by the Canny edge detection algorithm thus rendering a much more efficient automated tool. Using these state-of-the-art machine learning models, our system reduces the arduous manual procedure taking several months to a simple and intuitive online system that computes the entire data in a few minutes on-demand.

Our tank placement prediction algorithm is a unique set of methods used to predict the perfect location to create the tank. This is also based on scaled satellite imagery and the depth elevation map available from the open geocoding directory. The satellite image is first subjected to a Gaussian blur. Next, the volume of the tank is used to obtain tank dimensions for the required capacity. This computed value along with the image is subject to the lowest elevation mapping where the tank is predicted to be placed at the geographical minima for that region. With this freely computed data, the community can raise a quote or a tender directly with any company without having to go through the hassles of recruiting agents to perform manual inspection and offer their estimates over a long time period.

2 Related Work

Redmon et al. [3] studied that object detection is carried out by using classifiers that evaluate its various locations and scales in the image to constructing bounding boxes that can be processed by the classifier. This can be treated as a regression problem instead of localization where class probabilities can be assigned to each region in the proposed box and further use them for rooftop detection. Singhal et al. [4] studied the process of automating building detection in aerial images using Canny edge detection which emphasizes on manipulating the straight-line borders of building as edges and color invariances present in the image that can be used for segmenting out the building. Chen et al. [5] experimented with convolutional neural networks for rooftop segmentation and found them better than the other approaches because of its capacity to extract high-dimensional feature maps from the manual design. Yu Yi studied the possibility of measuring the area of irregular objects based on pixel ratio and perspective projection [6] which can be extended for use in measuring the area of rooftops from aerial satellite images. Ashutosh Saxena et al. experimented with various conceptual approaches to determining depth from single monocular images [7]. Parmar et al. [8] studied the nonlinear relationships present in the rainfall data and experimented with various models to find out that artificial neural networks are the most optimal approach to predict rainfall. Eni and Adeyye [9] studied the seasonal variations of rainfall, in general, using the autoregressive integrated moving average (ARIMA) model which was found to be suitable for the seasonal variations of rainfall data. Hajani et al. [10] analyzed the cost benefits of the rainwater harvesting system and factors affecting them in the urban regions and also studied various methods of regression to predict the cost with a greater accuracy and confidence score. Choudhary et al. [11] studied the rainfall patterns in India with various experimentations of models to determine that artificial neural networks and random forest regression are most suitable for predicting rainfall for India. Umapathi et al. [12] studied the life cycle cost (LCC) for a complete rainwater harvesting system in India which included the profit feasibility study of any rainwater system investment. Balasubramani et al. [13] designed a geospatial system to construct underground maps that can be used for urban infrastructure planning and utilized for various applications. Kim et al. [14] proposed a deep learning-based approach for underground object detection which can be used when buried obstacles are encountered. Umapathi et al. [15] proposed a monitoring-based investigation of rainwater collection systems in groups with non-conventional end users for rainwater. Martinez [16] reviewed the methodologies related to rainwater harvesting and studied the characteristics related to each hydraulic structure which guarantees the runoff collection in the basic area. Rahimi [17] proposed a cost-efficient system with long service life and optimization of storage tanks based on the balance among the rainwater volume and non-potable demands. Norman [18] reviewed the research gap between the geospatial technology in estimating the rainwater harvesting systems quality and study the implication of roofing material and roofing surface conditions.

3 Methodology

The approach involves six major phases which include detection and segmentation of rooftops, depth estimation, and catchment area volume estimation, predicting non-drinking water consumption and rainfall propensity for that region. In addition, the optimal tank placement for shared buildings is attempted to determine. The feasibility is determined using rainfall prediction and cost calculation based on the water consumption rate. The aim was to develop a complete product that can study rainwater harvesting (Fig. 1).

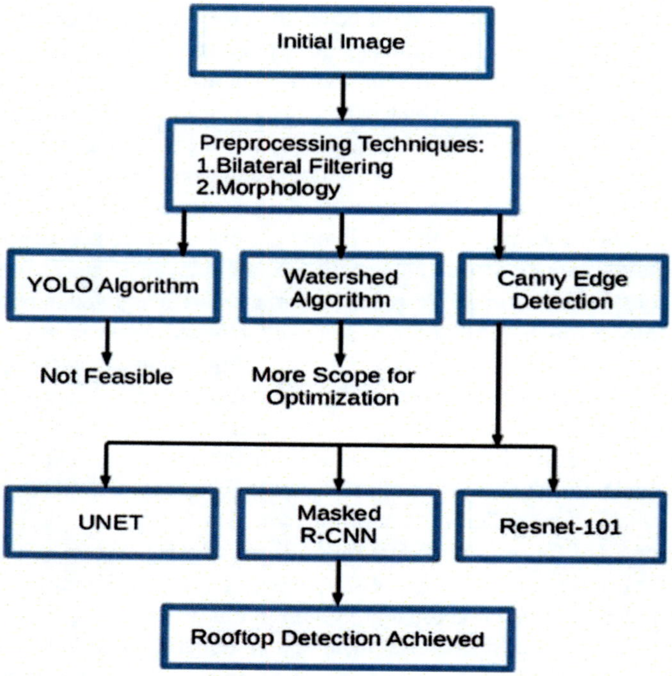

Fig. 1 Rooftop detection procedure

3.1 *Rooftop Segmentation*

The satellite images are preprocessed with bilateral filtering and morphological transformation to obtain sharper and better boundaries in the images. The YOLO algorithm performs well with few clear resolution satellite images but suffers from the misclassification of rooftops which is detrimental to the utility of the approach. Thus, the image is preprocessed further with watershed transformation and Canny edge detection to obtain segmented results. The former suffers from over-segmented results compared to the latter and thus, the images are preprocessed using Canny edge detection and then annotated to create a custom dataset.

3.2 *Instance Segmentation*

The state-of-the-art models are tried to obtain segmentation masks of rooftops experimenting with U-Net [19], one of the finest networks in segmenting biomedical images, ResNet [20] used for segmenting and image recognition, and Mask R-CNN [20] specialized in instance segmentation. Out of these, Mask-RCNN produced the best results. The Mask-RCNN is modified to obtain a custom network with a few

Fig. 2 Input image for rooftop detection

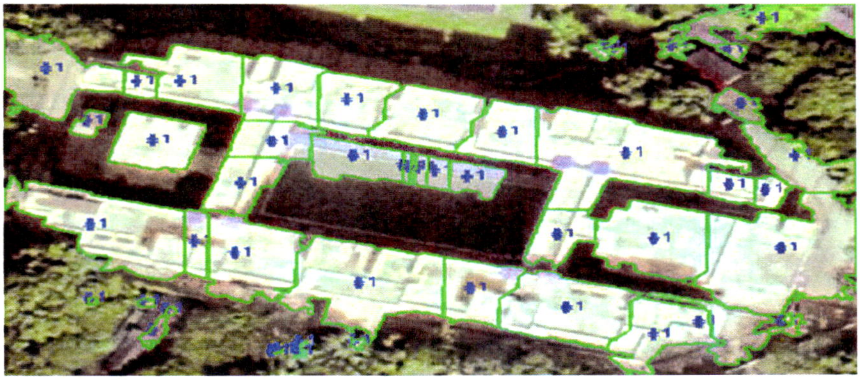

Fig. 3 Watershed segmentation result

changes to its architecture to tune the receptive field required by the satellite images (Figs. 2, 3 and 4).

3.3 Roof Catchment Area Estimation

The concept of pixel per metric ratio and contour lines is used. Contours are plotted in the segmented result and sorted from left to the right. The larger contours are accounted for the estimation of area and smaller ones are discarded as noise to obtain the specific region of interest. The midpoint of each region is computed, and the Euclidean distance is calculated for each set of midpoints and further divided it by pixel per metric to obtain the final measurement of the rooftop catchment area (Fig. 5).

Fig. 4 Canny edge detection result

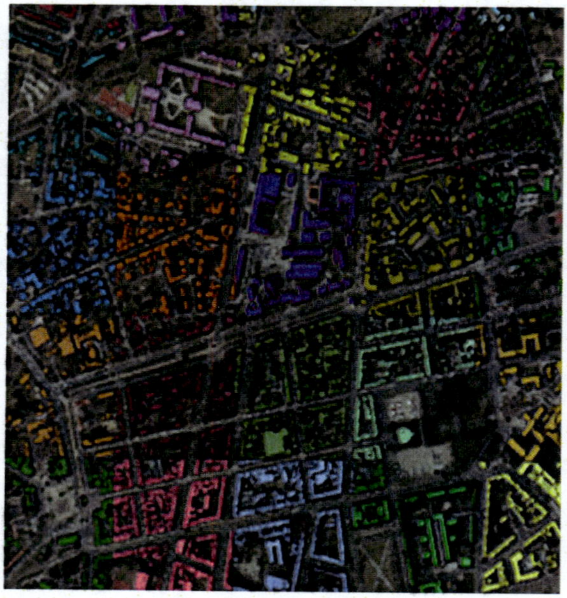

Fig. 5 Mask R-CNN instance segmentation of rooftops

3.4 Lowest Elevation Point Estimation

A contour map is plotted to identify the areas with the least elevation around the terrain of the buildings. An array which consists of evenly spaced elevation values referred to as digital elevation modeling is used for approximate depth estimation. It computes the least elevation point by segregating the image into boxes and then calculates the local minima for every box. The global minima obtained through the

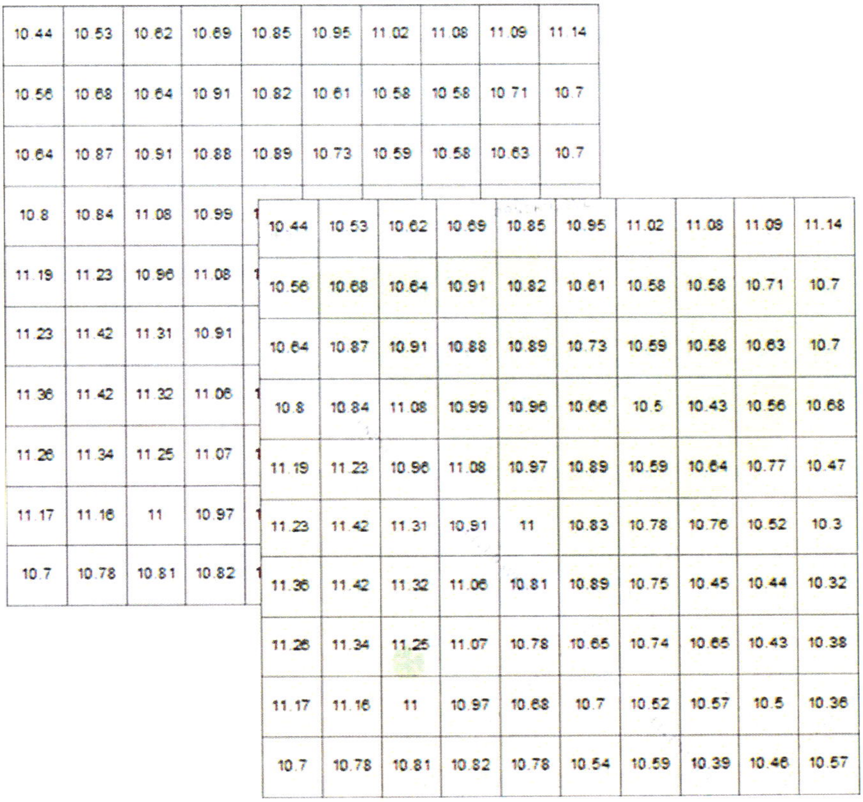

10.44	10.53	10.62	10.69	10.85	10.95	11.02	11.08	11.09	11.14
10.56	10.68	10.64	10.91	10.82	10.61	10.58	10.58	10.71	10.7
10.64	10.87	10.91	10.88	10.89	10.73	10.59	10.58	10.63	10.7
10.8	10.84	11.08	10.99	10.96	10.66	10.5	10.43	10.56	10.68
11.19	11.23	10.96	11.08	10.97	10.89	10.59	10.64	10.77	10.47
11.23	11.42	11.31	10.91	11	10.83	10.78	10.76	10.52	10.3
11.36	11.42	11.32	11.06	10.81	10.89	10.75	10.45	10.44	10.32
11.28	11.34	11.25	11.07	10.78	10.65	10.74	10.65	10.43	10.38
11.17	11.16	11	10.97	10.68	10.7	10.52	10.57	10.5	10.36
10.7	10.78	10.81	10.82	10.78	10.54	10.59	10.39	10.46	10.57

Fig. 6 Least elevation estimation using digital elevation model

above procedure is taken as the lowest elevation point for the given area matrix. The multiple least elevation points are taken and they select the one which has no obstacle at that point and use this depth to estimate the position of the tank. This approach provides us with the optimal placement of the tank (Figs. 6 and 7).

3.5 *Break-Even Analysis and Underground Mapping*

Our approach to estimate depth is based on unsupervised monocular depth estimation [21] which can perform depth estimation on a single 2D image without any true labels. Monocular depth estimation uses a left–right disparity consistency and a reconstruction loss to determine the depth. Solving for correspondence between left–right images gives the disparity. Depth is inversely proportional to the disparity. The model receives the stereo pairs of images and outputs one of them after reconstruction. The perfect stereo images from satellite data were unable to obtain and hence, resorted

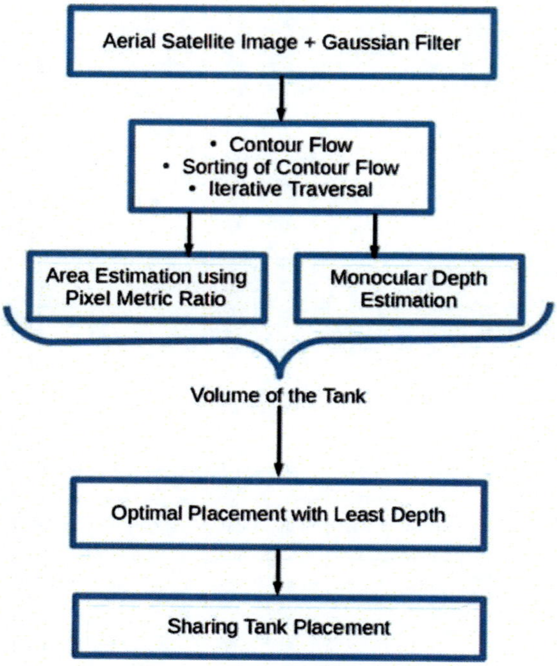

Fig. 7 Optimal tank placement algorithm

to using real-time predicted rainfall depth as the depth metric as in Sect. 3.6. The semi-supervised approach to monocular depth estimation was implemented with reference to [22]. Upon computing, the volume of water collected on the roof and correcting for the satellite scale factor obtained the net water collected. With the computed volume, the predicted non-water drinking usage is used to get the consumption value. Note that this metric is influenced by a multitude of factors and hence, approximate values from waterportal.org have been used. The construction of a rainwater harvesting tank is not only dependent on the catchment area and depth but also the external factors such as terrain, availability of space, and the obstacles underground. For this, we need to take care of the piping infrastructure and gas pipelining underground. This entails the use of ground penetration radar (GPR) to detect underground buried obstacles [14]. For GPR-based object detection, the strength of the signals becomes essential to determine the depth up to which needed to place the underground storage tank. There are three main phases in ground penetration radar-based underground object detection: Phase 1 is the data collection phase using the GPR system, Phase 2 is the data processing, and Phase 3 is the classification phase using CNNs and their feature maps. Different architectures and library independent models can be used for transfer learning rather than pretrained weights. Phase 1 entails the use of the GPR system with the required signal strength to obtain scale-invariant image scans of the land features. Phase 2 is the statistical analysis of the scanned data from Phase

Table 1 Average non-drinking water consumption per person

Category	Liters/person
Drinking and cooking	7
Bathing	20
Sanitation	40
Washing clothes	25
Washing utensils	20
Miscellaneous	23
Total	135

1 and includes data preprocessing and data reconstruction. The data is prepared to be used by the classification phase of the pipeline. Phase 3 uses a convolutional neural network (CNN) architecture with residual connections (ResNet) to classify the scanned images for the presence or absence of obstacles that would impede the storage tank placement. Based on this, a calculator tool has been implemented that estimates the break-even period based on the cost of investment and the consumption details given for a shared roof rainwater harvesting system (Table 1).

The feasibility of a rainwater harvesting model is determined by the per capita drinking water and non-drinking water consumption in that region and the predicted rainfall to be received by that region. The catchment area estimation is a multiple regression problem. The total estimated cost depends on the mean material pricing for laying the piping structures and the estimated cost of the tank. The required information is obtained from the rooftop catchment area prediction and the depth is estimated. The surface runoff coefficient determines the amount of water that would drain away from the catchment surface. This amounts for a portion of water that cannot be captured in the tank and thus must be proportionally excluded from our estimation. Figure 8 tabulates the runoff coefficients for various types of catchment surfaces.

Type of Catchment	**Coefficients**
Roof Catchments	
- Tiles	0.8- 0.9
- Corrugated metal sheets	0.7- 0.9
Ground surface coverings	
- Concrete	0.6- 0.8
- Brick pavement	0.5- 0.6
Untreated ground catchments	
- Soil on slopes less than 10 per cent	0.0 - 0.3
- Rocky natural catchments	0.2 - 0.5
Untreated ground catchments	
- Soil on slopes less than 10 per cent	1.0 - 0.3
- Rocky natural catchments	0.2 - 0.5

Fig. 8 Runoff coefficients for various catchment surfaces

Table 2 Dataset for climatic changes and rainfall patterns

Column	Description	Observation
1	City	Statewise name of cities
2	Precipitation amount	Rainfall in the area (in mm)
3	Air temperature	Kelvin
4	Average temperature	Kelvin
5	Dew point temperature	Kelvin
6	Maximum air temperature	Kelvin
7	Minimum air temperature	Kelvin
8	Relative humidity	Percentage
9	Specific humidity	Gram of water vapor per kilogram of air
10	Normalized difference vegetation index	Number

3.6 Rainfall Estimation

The amount of rainfall is used as a depth metric to calculate the break-even analysis of a region. The data obtained from the Government of India's official rainfall data consists of the aforementioned parameters (Table 2).

When negative binomial regression is applied to the set of input features, we found that the humidity (both relative and specific) and air temperature were the most significant features in predicting the precipitation amount in all the states across India. Random forest regression handles the correlations between the temperatures and humidity parameters particularly well and thus can achieve acceptable accuracy. However, we noted that there are many seasonal patterns in the dataset. These seasonal patterns were handled in a much better fashion when the data is passed as input to a simple three stacked layered LSTM [23] along with recurrent neural networks that handle the seasonality in the data. The recurrent neural network calculated the mean absolute loss for the training data to be 0.033 and test mean absolute loss to be 0.029 [11]. This makes it possible to predict the amount of rainfall that would occur in the region for studying the overall feasibility of the system design.

The GOI provides time series monthly rainfall data as well which had the following variations from 1910 to 2000. It had a highly seasonal nature with peak rainfall observed only during the months from June to September. To cater to these seasonal changes, the data has experimented with seasonal autoregressive integrated moving average (SARIMA) which correctly predicts the time period of rainfall ranging from June to September, which is the valid onset of monsoon in India. This method proves to be better than autoregressive integrated moving average (ARIMA) and much more adaptable to the data specifically for Indian rainfall patterns with a mean absolute error of 85.69 mm compared to 94.56 mm in case of ARIMA (Figs. 9 and 10).

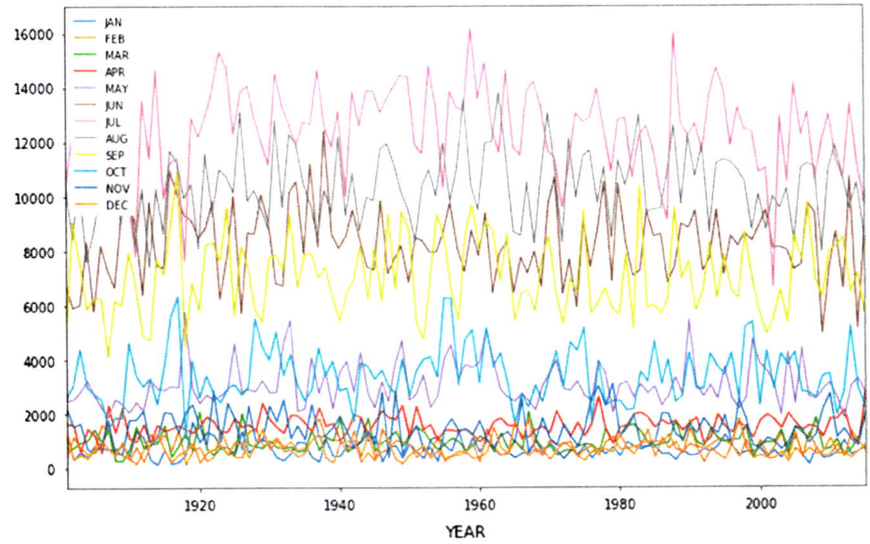

Fig. 9 Time series analysis of annual rainfall across India

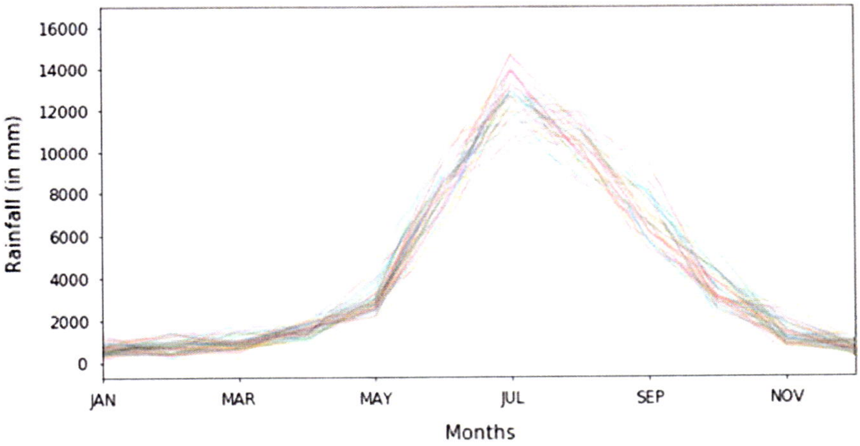

Fig. 10 Seasonal patterns of annual rainfall in India

4 Formulas

The Canny edge detection mentioned in the paper is formulated as follows [24]:

The operator is used to define the gradient amplitude and direction of a smooth data array $I(x, y)$. $P_x[i, j]$ and $P_y[i, j]$ are direction partial derivatives in the respective directions.

$$P_x[i, j] = (I[i+1, j] - I[i, j] + I[i+1, j+1] - I[i, j+1])/2 \quad (1)$$

$$P_y[i, j] = (I[i, j+1] - I[i, j] + I[i+1, j+1] - I[i+1, j])/2 \quad (2)$$

The gradient amplitude is calculated as:

$$M(i, j) = \sqrt{f_x(i, j)^2 + f_y(i, j)^2} \quad (3)$$

and gradient direction depends on:

$$\theta(i, j) = \arctan\big(f_y(i, j)/f_x(i, j)\big) \quad (4)$$

The pixel per metric ratio is used for the calculation of the size of objects in a two-dimensional image, and the formula as follows:

$$\text{pixel per metric} = \text{object width}/\text{scale factor} \quad (5)$$

The capacity estimation can be done by performing a modified version of the famous Shannon channel capacity equation [25]:

$$C = W \log\left(1 + \frac{P_S}{P_N}\right) \quad (6)$$

Seasonal autoregressive integrated moving average is SARIMA (p, d, q, P, D, Q) as follows:

$$\phi_p(L)\tilde{\phi}_P(L^S)\Delta^d\Delta_S^D y_t = A(t) + \theta_q(L)\tilde{\theta}_Q(L^S)\epsilon_t \quad (7)$$

$\phi_p(L)$ is autoregressive order
$\tilde{\phi}_P(L^S)$ is seasonal autoregressive order
Δ^d is differencing
Δ_S^D is seasonal differencing
y_t is time series
$A(t)$ is constant
$\theta_q(L)$ is moving average order
$\tilde{\theta}_Q(L^S)$ is the seasonal moving average order
ϵ_t is noise.

Multiple linear regression is formulated as follows:

$$y_i = \beta_0 + \beta_1 x_1 + \beta_2 x_2 + \cdots + \beta_n x_n + \epsilon \quad (8)$$

y_i dependent variable.
x_i independent variable.

β_i parameter.
ϵ error.

5 Results and Discussion

The product successfully automated the procedure of rainwater harvesting inspection which was earlier highly time-consuming and expensive as well. The water consumption rate is studied with both drinking and non-drinking estimations taken into account. Based on these parameters, a profit calculator was successfully implemented and the optimal placement of the tank for a shared housing settlement was achieved using contour lines and monocular depth estimation. Rainfall estimation was carried out by the recurrent neural network whose mean absolute loss for the training data is 0.033 and the test mean absolute loss is obtained as 0.029 which proved to be better than other regression techniques. The rainfall probability was determined by the time series analysis experimenting with ARIMA and SARIMA which gave 94.56 and 85.69 as mean absolute error making SARIMA more suitable for deployment. The comparison of actual and forecasted results by the SARIMA model is displayed in Fig. 14. The end product involves an end-to-end progressive Web application based on the Flask Framework in which the user uploads the satellite image of the region as shown in Fig. 11. The deep learning network is run on the image from backend and the segmented result is produced as in Fig. 12. Based on the rainfall data predicted, the water usage, and number of clusters, a profit calculator using multiple linear regression runs as the backend of the profit calculator to predict the time period needed for the return of investment. The current model was developed for the shared rainwater harvesting problem. The solution enumerates several phases to bring about the most optimal solution. The first of which is the estimation for the volume of the tank from the segmentation of rooftop images and estimating its area from satellite images. The next phase includes the digital elevation model for depth estimation used along with the ground penetration radar (GPR) for optimal placement of the tank as shown in Fig. 16. Apart from estimating the location of the tank, computing the cost and break-even analysis is essential. Thus, rainfall prediction is necessary and is used to calculate the volume of water filling the tank. The prediction is done using SARIMA models with rolling forecasting. The feasibility of the system is determined by the break-even analysis computed as a result of all the above-mentioned phases (Figs. 13, 14 and 15).

6 Conclusion

The system currently uses rainfall estimation as a depth metric and the future scope of the product will be to design a suitable algorithm for depth estimation from satellite images to calculate the actual capacity of each rooftop as a potential rainwater

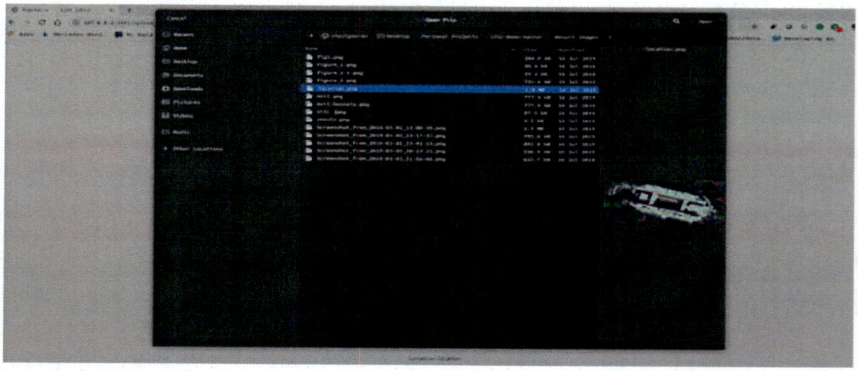

Fig. 11 Scaled satellite image input to the image processing pipeline

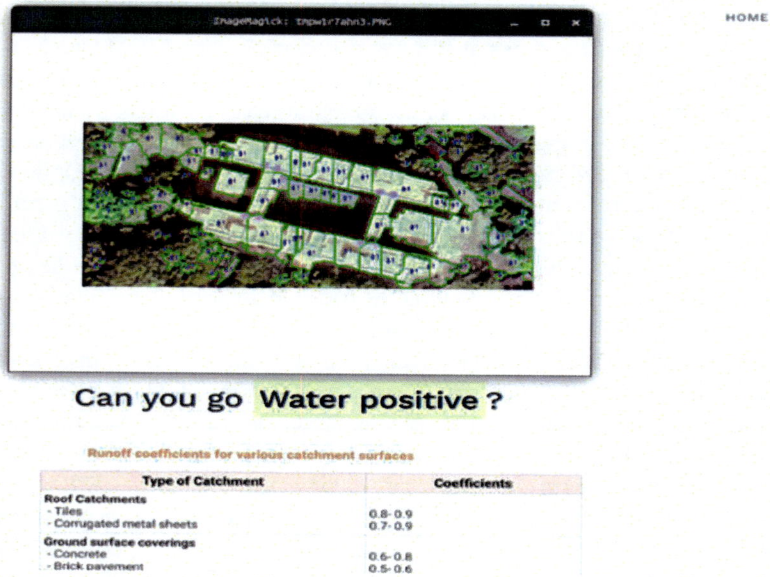

Fig. 12 Image segmentation and area computation output

harvesting storage. Currently, the system works on the concept of pixel per metric ratio and more efficient implementation to calculate the catchment area from the satellite images is also needed. These works can efficiently increase the feasibility of the product and aid in automating shared rainwater harvesting universally.

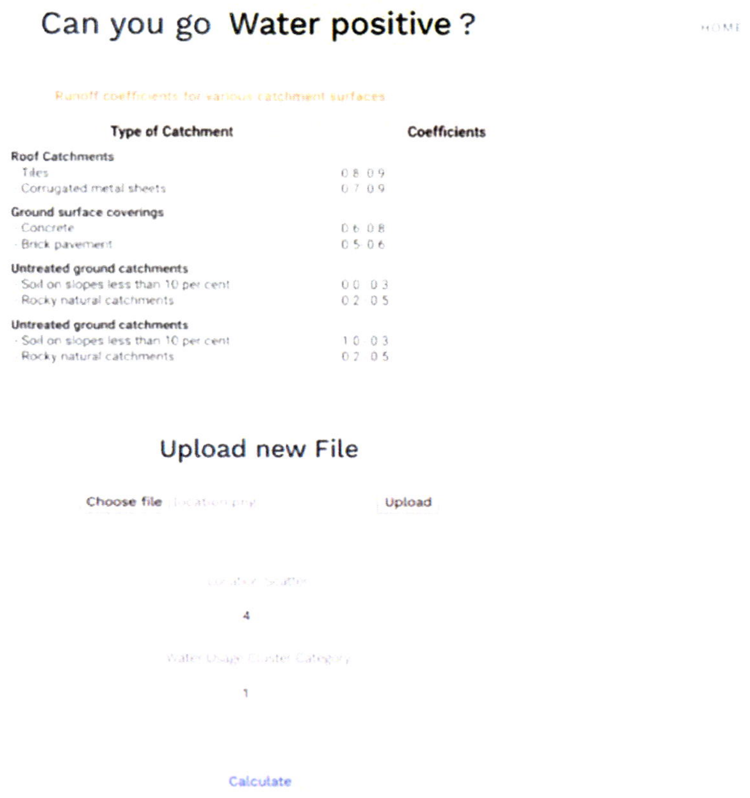

Fig. 13 Break-even analysis of the shared rainwater harvesting system

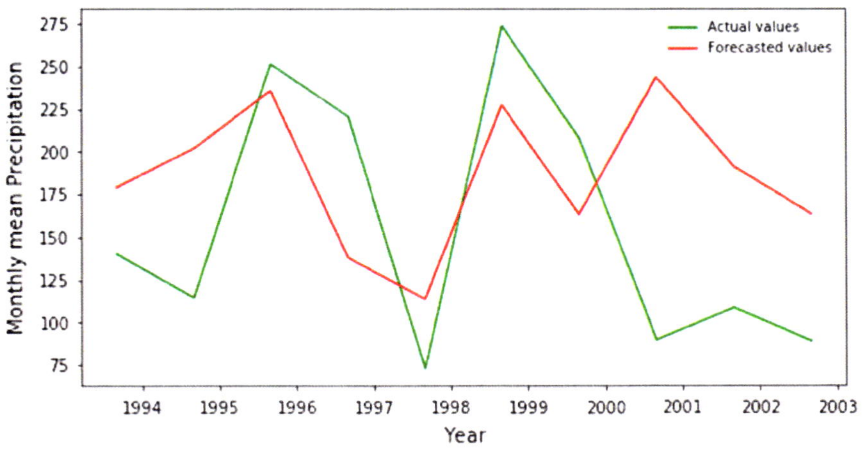

Fig. 14 SARIMA rainfall prediction results

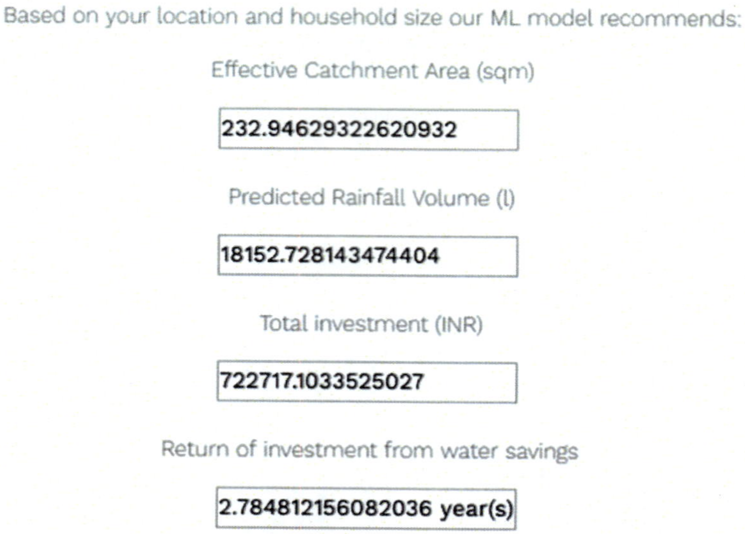

Fig. 15 Final prediction results and break-even analysis

Fig. 16 Final tank placement prediction

References

1. Campisano, A., Butler, D., Ward, S., Burns, M.J., Friedler, E., DeBusk, K., Fisher-Jeffes, L.N., Ghisi, E., Rahman, A., Furumai, H., Han, M.: Urban rainwater harvesting systems: research, implementation and future perspectives. Water Res. **121**, 386 (2017)
2. Berwanger, H., Ghisi, E.: Investment feasibility analysis of rainwater harvesting in city of Itapiranga. Int. J. Sustain. Human Dev. 114 (2014)

3. Redmon, J., Divvala, S., Girshick, R., Farhadi, A.: You only look once: unified, real-time object detection. In: IEEE Conference on Computer Vision and Pattern Recognition (2016)
4. Singhal, S., Radhika, S.: Automatic detection of buildings from aerial images using color invariant features and canny edge detection. Int. J. Eng. Trends Technol. (IJETT) **1**(8) (2014)
5. Chen, Q., Wang, L., Wu, Y., Wu, G., Guo, Z., Waslander, S.L.: Aerial imagery for roof segmentation: a large scale dataset towards automatic mapping of buildings. ISPRS J. Photogrammetry Remote Sens. **147** (2019)
6. Yi, Y.: Research on superficial area measurement method of irregular object-based on image processing. In: International Conference on Electronics, Network and Computer Engineering (ICENCE) (2016)
7. Saxena, A., Chung, S.H., Ng, A.Y.: Learning depth from single monocular images. In: Neural Information Processing System (2018)
8. Parmar, A., Mistree, K., Sompura, M.: Machine learning techniques for rainfall prediction: a review. In: International Conference on Innovations in Information Embedded and Communication Systems (2017)
9. Eni, D., Adeyye, F.J.: Seasonal ARIMA modelling and forecasting of rainfall in Warri Town, Nigeria. J. Geosci. Environ. Prot. **3**, 91–98(2015)
10. Hajani, E., Rahman, A.: Reliability and cost analysis of a rainwater harvesting system in peri-urban regions of greater Sydney. Water 6, 945–960 (2014). https://doi.org/10.3390/w6040945
11. Choudhary, S., Gaurav, V., Sharma, T., Pradyumna K.R., Vishal V.: Forecasting dengue and studying its plausible pandemy using machine learning. Int. J. Res. Advent Technol. **7** (2019)
12. Umapathi, S., Pezzanati, D., Beecham, S., Whaley, D., Sharma, A.: Sizing of domestic rainwater harvesting systems using economic performance indicators to support water supply systems. Water **11**(4), 783 (2019). https://doi.org/10.3390/w11040783
13. Balasubramani, B.S., Belingheri, O., Boria, E.S., Cruz, I.F., Derrible, S., Siciliano, M.D.: GUIDES-geospatial urban infrastructure data engineering solutions. In: ACM SIGSPATIAL International Conference on Advances in Geographic Information Systems (2017)
14. Kim, N., Kim, K., An, Y.-K., Lee, H.J., Lee, J.J.: Deep learning-based underground object detection for urban road pavement. Int. J. Pavement Eng. (20 Dec, 2018)
15. Umapathi, S., Pezzanti, D., Beechamz, S.: Sizing of domestic rainwater harvesting systems using economic performance indicators to support water supply systems. Water (2019). ISSN 2073-4441; CODEN: WATEGH
16. Martinez, L.: Design criteria for planning the agricultural rainwater harvesting systems: a review. Appl. Sci. (2019). ISSN 2076-3417; CODEN: ASPCC7
17. Rahimi, O.: Analysis of rooftop rainwater harvesting in Kabul's new city: a case study for family houses and educational facilities. Open J. Civ. Eng. (2018). https://doi.org/10.4236/ojce.2018.82013
18. Norman, M.: Review of remote sensing and geospatial technologies in estimating rooftop rainwater harvesting (RRWH) quality. Int. Soil Water Conserv. Res. **7**(3), 266–274 (Sept 2019)
19. Ronneberger, O., Fischer, P., Brox, T.: U-Net: convolutional networks for biomedical image segmentation. arXiv:1505.04597v1 [cs.CV] (2015)
20. He, K., Zhang, X., Ren, S., Sun, J.: Deep residual learning for image recognition. In: IEEE Conference on Computer Vision and Pattern Recognition (2016)
21. Godard, C., Aodha, O.M., Brostow, G.J.: Unsupervised monocular depth estimation with left-right consistency. In: Computer Vision and Pattern Recognition (CVPR) (2017)
22. Godard, C., Aodha, O.M., Firman, M., Brostow, G.: Digging into self-supervised monocular depth estimation. In: International Conference on Computer Vision (ICCV) (2019)
23. Sherstinsky, A.: Fundamentals of recurrent neural networks (RNN) and long short term memory (LSTM). arXiv: 1808.03314v4 (2018)

24. Li, J., Ding, S.: A research on improved canny edge detection algorithm. In: Applied Informatics and Communication. ICAIC 2011. Communications in Computer and Information Science, vol. 228. Springer, Berlin (2011)
25. Yaghmaee, F., Jamzad, M.: Estimating watermarking capacity in gray scale images based on image complexity. EURASIP J. Adv. Sig. Process. 851920 (2010)

A Smart Biometric-Based Public Distribution System with Chatbot and Cloud Platform Support

Shashank Shetty and Sanket Salvi

Abstract Public distribution system (PDS) is a hierarchy of a government system in India consisting of government, machinery of government (MoG), broker, and the beneficiary as a consumer with below poverty line (BPL), who is provided with the food and non-food items at a subsidized rate. Radio-frequency identification (RFID) and fingerprint reader-based automatic ration shop also called a fair price shop is a practical approach in the public distribution system for efficient and accurate distribution of ration across the needy without any manual intervention. Automatic rationing for public distribution using the fingerprint sensor can be used to prevent irregularities as well as fraudulence. It involves less manual work, and faster results are seen. The entire mechanism is Arduino-based, thus making it an easy approach. Automatic rationing prevents non-transparent procedure followed by the shopkeeper, who may sell the ration for a higher price to someone and forge a false entry in the records. RFID tag used and the fingerprint details carry the user's information and the microcontroller connected to the reader will authenticate the users. After successful two-layered verification, the quantity of the ration allotted will be automatically distributed among the consumers. The automated question–answer system with Chatbot assistance is provided for improved customer service and engagement. The smart ration distribution system (SRDS) manager is supported by a cloud platform for visualization and monitoring of the system. The novelty of the work is to low-cost design and cloud platform integration with Chatbot assistance.

Keywords Public distribution system · RFID · Biometry · Chatbot · Internet of things (IoT)

S. Shetty (✉)
Department of Computer Science and Engineering, NMAM Institute of Technology, Karkala, Karnataka, India
e-mail: shashankshetty06@gmail.com

S. Salvi
Department of Information Science Engineering, Nitte Meenakshi Institute of Technology, Bangalore, Karnataka, India
e-mail: sanketsalvi.salvi@gmail.com

P. Karuppusamy et al. (eds.), *Sustainable Communication Networks and Application*, Lecture Notes on Data Engineering and Communications Technologies 55,
https://doi.org/10.1007/978-981-15-8677-4_10

1 Introduction

From ATMs to smartwatches, technology has made life so much simpler than ever before. However, the advantages of technological advancements are best-made use by people with higher living standards and one who can afford it. For a developing country like India, where the majority of the population is below poverty line, it is more crucial than technological advances reach to these people in some or other form. There are several schemes made available by Govt. of India for the section of people who are below the poverty line. However, its reachability and transparency remain an issue of concern. Thus, technology can be used to address such issues and increase outreach by providing ease of access and simple to use applications, and also providing a system to help govt. schemes.

Traditionally, vending machines are used to provide goods or items on demand. Delivering fixed ration to an individual based on biometric and RFID, one can save upon several resources and maintain transparency throughout the process of ration distribution. Such a system can work with minimal human intervention and allow round the clock working capability. It also paves the way for a new type of smart ration distribution system, where ration can be provided like current ATM Systems. On the other hand, with increasing penetration of Chatbots in various fields, integration of Chatbot [1] with the SRDS would offer a simpler and more human way of interaction to manage SRDS resources.

The rest of the paper is organized as follows: Sect. 2 highlights the previously carried out work in the domain of distribution systems; Sections 3 and 4 focus on our proposed system design and implementation, respectively, followed by a discussion on results and conclusion.

2 Related Work

The planning commission performed a comprehensive survey on targeted public distribution system (TPDS) intending to validate the efficacy of the system implemented and found that the 58% of subsidized ration distributed by the government do not reach the families with the below poverty line [2]. It was due to the non-transparent procedures of the ration shop vendor, error in identification, or unethical process followed in the system. Basanta et al. [3] in their study point out the inability of the PDS due to the corruption in the government created system and proposed measures to overcome these issues at the microlevel by revamping the system. It is revealed that there are considerable complications and gaps in the implementation of PDS [4]. Several efforts are made to solve the above-discussed issues.

Advantages of various technologies like IoT [5–8] and cloud computing [9] can be employed to build more secure and easy-to-use public distribution systems. Valarmathy et al. [10] proposed an automated ration distribution machine replacing the conventional system and ration card with the GSM and RFID support to avoid any

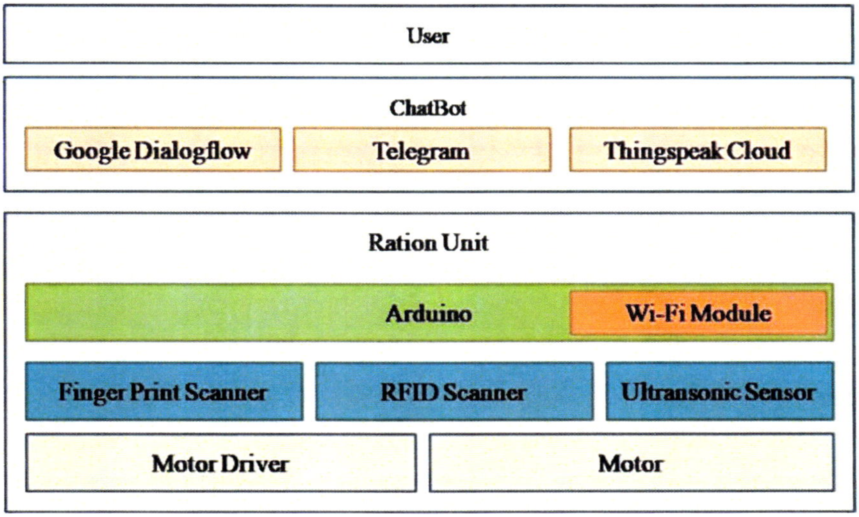

Fig. 1 Block diagram of proposed smart ration distribution system

wastage of food grains or misconduct in the system. Kowshick [11] introduces the public dispensing machine with the RFID reader as an alternative for ration cards and utilizes the cloud for storing customer information. Madur [12] proposes RFID as a smart card and ARM7-based ration distribution system. Priya et al. [13] devised an E-PDS to prevent any irregularities using a smart card and RFID. Vaisakh et al. [14] designed and developed a ration dispensing system using the Internet of things with GSM and fingerprint sensor to validate the customer identification.

These works provided significant insight into various design principles and conceptual understanding required for building our proposed system.

3 Proposed System

The proposed system is divided into three layers, namely ration unit, Chatbot, and user as shown in Fig. 1. Under the ration unit, the components are subdivided into sensors, actuators, communication module, and controller. Components such as motor driver and motor come under actuator while fingerprint scanner, RFID reader, weight sensor, and ultrasonic sensor come under sensors. Actuators and sensors are connected to a controller which has a Wi-Fi communication module. By using this Wi-Fi module, the proposed monitoring and control operations can be performed remotely. The Chatbot layer has three subcomponents, namely Google Dialogflow,[1] Telegram[2] and

[1] https://cloud.google.com/dialoglow.

[2] https://telegram.rg/.

Thingspeak Cloud.[3] Google Dialogflow provides an easy to make Chatbot service, which uses Google Cloud Platform for training the query response system of the designed Chatbot. Telegram provides messaging services and Chabot APIs in the form of webhooks. ThingSpeak cloud delivers data storage and visualization services for IoT devices. And finally the topmost layer, the user which implies an individual with managerial access to the SRDS.

To enhance the existing ration distribution system, a system is proposed with simplicity and security of transactions which are of prime importance. A two-layered security check is proposed, i.e., first, the fingerprint is checked to see if the user has been enrolled and then the RFID card number is checked against the fingerprint provided by the user, and only if the match is found, the ration will be provided to the user. Keeping into consideration the increasing size of the database, the customer authentication details are saved over the cloud. If one RFID card is given to a family, the ration for a stipulated time can be got only once in the proposed machine. If all the fingerprints of the family members are registered under one RFID card, then any member of the family can collect their family's ration. This eliminates the disadvantage of restricting a single person of the family for collecting their family's ration.

The ultrasonic sensor and weight sensor continuously monitor the available ration in the containers and keep updating the same over ThingSpeak cloud platform from where the SRDS manager is notified. The manager also gets notified about the family or person who has received or not received the monthly ration.

4 Implementation

The implementation of the system is divided into hardware and software. Hardware implementation discusses the design and detailed workflow of different data throughout different hardware components, whereas software implementation discusses design, integration, and data flow in the Chatbot.

4.1 Hardware Implementation

Figure 2 shows a detailed circuit diagram of the proposed system. The sequence diagram of the overall system is shown in Fig. 3. The proposed system uses an Arduino Uno with external Wi-Fi ESP8266 module to leverage the features such as low-cost availability, community support, and a huge range of libraries available for integration. The implementation of each module is discussed as follows:

[3] https://thingspeak.com/.

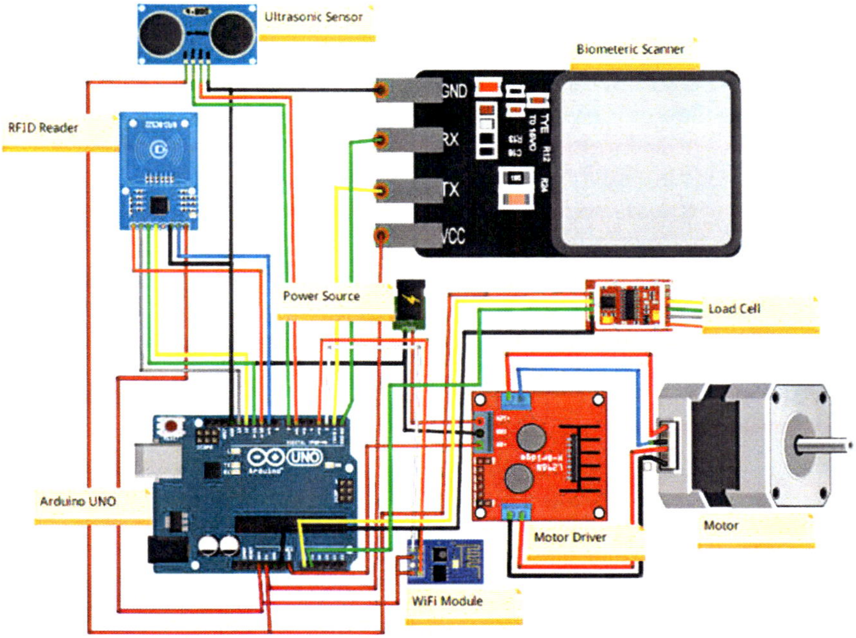

Fig. 2 Detailed circuit diagram of proposed system

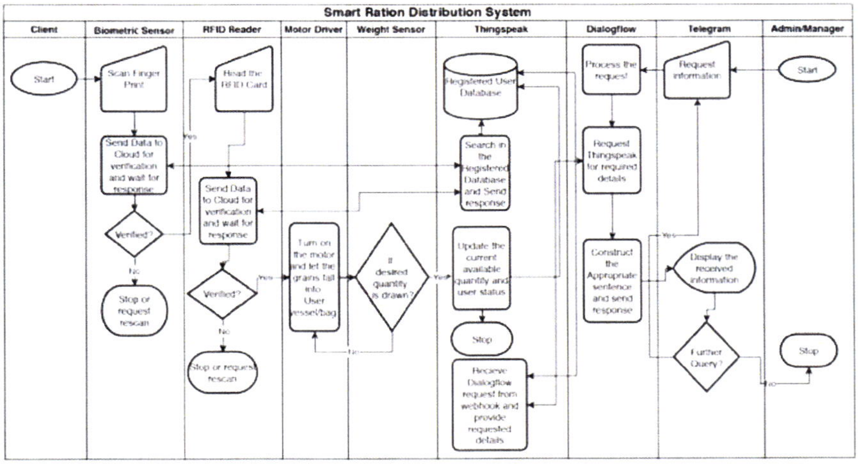

Fig. 3 Sequence diagram of the overall system

- **Module 1—Fingerprint scanner:** Fingerprint scanner GT-511C3 is deployed for scanning the fingerprint of the user. The consumer needs to register the fingerprint to obtain a ration. When the new consumer tries to enroll for the system, the message will be prompted to place their finger on the scanner for recording their fingerprint. After a successful scan, the message will be displayed in the result box, and the fingerprint scanner will be ready for the next scan by incrementing the ID number; also, the entry would be updated over the cloud. To verify, whether a user or user's fingerprint has been registered and to check the ID generated for a particular user's fingerprint, place a finger on the scanner and select the "Identify (1: N)" button. The scan results are displayed in the result box. Timeout message will be prompted if the finger is not positioned on the scanner for a significant period and a cancel button can be employed to stop the identify check.
- **Module 2—RFID authentication:** The information of the ration for a particular consumer is incorporated in the RFID tag. Every consumer is provided with the authorized RFID tag. After the successful fingerprint verification, the RFID tag is scanned against the RFID reader unit. RFID tag comprises of unique information (i.e., a combination of characters and digits), which is collected when the RFID tag is scanned. The unique code is matched against the already recorded data, and if it matches, then the procedure continues or else prompts for the authorized RFID tag to be scanned again.
- **Module 3—Working of valve using motor:** After the successful two-way authentication (i.e., the fingerprint and the RFID scan), the valve is opened through the motor and the process of ration distribution begins. As per the information stored against the consumer, the prescribed quantity of food is provided.
- **Module 4—Level indicator:** The container used has radius 6 cm and height 100 cm. This stores around 10 Kgs of ration. The levels marked at ThingSpeak are at the scale of 10. The formula to calculate the remaining quantity of ration is shown below:

$$R_{\text{Ration}} = 3.14 \times r^2 \times (10 - h) \times d_i \tag{1}$$

where R_{Ration} is the total quantity of ration remaining inside the container in Kgs, r is the radius of the container used, h is the distance between lid and the stored item in decimeters, and d_i is the density of the item.

This is a fail-safe for weight sensor. Thus, weight sensor values and ultrasonic sensor values are updated over the cloud for monitoring the ration quantity.

4.2 Software Implementation

The Chatbot is created in three different stages, which include the usage of three different platforms. Its implementation is discussed accordingly as follows:

- **Automated Question–Answer System (Google Dialogflow)** Google Dialogflow is a cloud-based service provided by Google for designing custom question–answering systems. It includes terms such as agent, intent, entity, and actions. An agent is a full-fledged Chatbot which contains all relevant questions and answers, for example, movie ticket booking agent, pizza ordering agent, etc. An intent is named according to the context of the question and answers. For instance, if the intent name is "Ask User Name", it would include questions such as "What is the name of the User?" and "Who is the user?". It also includes a set of expected replies. At the back-end, Google trains the system to understand and predict other different ways of asking the same question. For questions or answers which include multiple names for the same object an entity is associated, which is a group of words implying the same object. Actions are used to classify a set of intents. A webhook is also provided such that the system can be directly integrated with any of the popular existing platforms. Based on all these concepts, a set of predefined questions and its expected answers is defined for the proposed SRDS.
- **Chatbot interface (Telegram)** The webhook provided by the Google Dialogflow requires a dedicated connection with Telegram. To achieve this, a default Chatbot called "Bot Father" from Telegram is used. Bot Father provides a template for Chatbot and its authentication token. This authentication token is then shared with Google Dialogflow to provide Dialogflow's agent over Telegram.
- **IoT cloud storage platform (ThingSpeak)** ThingSpeak is an IoT cloud storage, visualization, and data analytics platform by Mathworks. ThingSpeak provides its APIs for communication between IoT device and cloud. A dedicated channel is created at ThingSpeak, and its read–write APIs are used in the IoT device, i.e., the Arduino Uno of SRDS. Using this API, SRDS sends data as well as receives data from ThingSpeak. On the other hand, the Google Dialogflow webhook takes input from the ThingSpeak channel as and when requested by Telegram Client. For instance, if the query is "How much ration is available in SRDS?", then the Dialogflow will get the latest updated value of Weight Sensor from ThingSpeak, add additional text response around it and reply "The currently available ration is X Kgs" where X is the value taken from ThingSpeak weight sensor.

5 Experimental Setup and Results

The implemented system is tested against two registered users with a preset quantity of 1 kg ration for each. Figure 4 shows various stages of user validations and ration distribution. Figure 4a, indicates the user getting authenticated based on the registered fingerprint. Figure 4b shows the second level of authentication where the user is authenticated based on the RFID card. Figure 4c shows the distribution of the ration post verification of provided details.

At SRDS manager, the manager can view the ration level on ThingSpeak cloud as well as receive messages from the Chatbot. Figure 5 shows the responses that a manager can view. Figure 5a shows ThingSpeak channel feeds for two fields, one

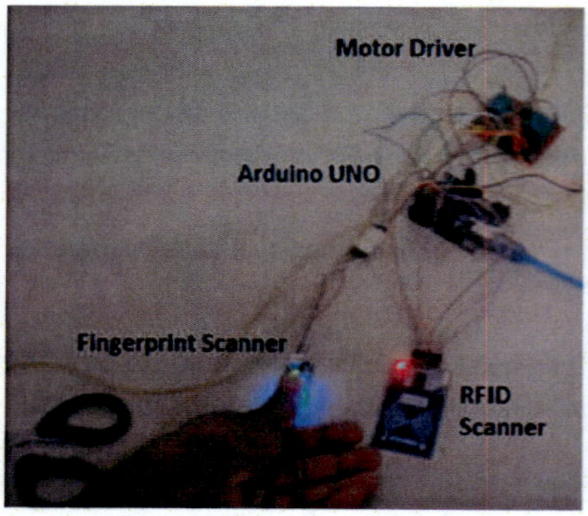

(a) Finger Print Scanner

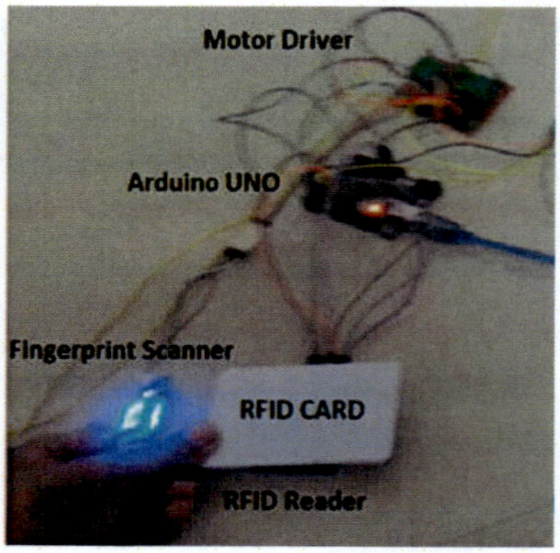

(b) RFID Tag Reader

(c) Ration Collection Unit

Fig. 4 Various stages of user validations and ration distributions

for rice and other for wheat. Ten levels are indicating 10 decimeters of the container height and using Eq. 1, the remaining ration is calculated, and the same is conveyed to the manager. Figure 5b shows the Chatbot–bot interface and interactions with the SRDS system Chatbot.

6 Conclusion

Smart ration distribution system (SRDS) is a transparent ration vending machine that enables consumers to obtain ration without any manual interventions or any involvement of intermediaries and keeps track of accountability of the food grains distributed, thus preventing the fraudulence and corruption involved in the public distribution system and providing the food and non-food items to the needy in the subsidized price. Two-way authentication is incorporated that includes fingerprint and RFID scanner to eliminate unauthorized users from accessing the rations

(a) Thingspeak Channel Feed (b) Ration Vending Chatbot

Fig. 5 Smart ration distribution system manager view

allotted to the consumers with below poverty line. After successful authentications, the allotted ration can be collected by the consumer's time-to-time easily and securely. The simple user interface with Chatbot support helps consumers to interact with the system more systematically. The regular monitoring, storage, and visualization facility are provided to the SRDS manager through ThingSpeak cloud platform. The current prototype developed operates on the solid food items and in future, our work would be focused on designing a touchless public distribution system that works on liquids as well.

Our proposed system uses registration of the entire family with one RFID card as well as many fingerprints of members from the same family. This allows the system to work in the situation where due to certain circumstances if one member cannot receive the ration other members from the family can get it. However, it is important to maintain the RFID card and report in case it is lost or damaged. The system can further be improved by adding facial recognition systems to avoid usage of any additional component.

References

1. Salvi, S., Shetty, S.: AI based solar powered railway track crack detection and notification system with chatbot support. In: 2019 Third International conference on I-SMAC (IoT in Social, Mobile, Analytics and Cloud) (I-SMAC), pp. 565–571 (2019)
2. Public Commission: Performance Evaluation of Targeted Public Distribution System (TPDS). Working Papers id: 6773, eSocialSciences (2015). https://ideas.repec.org/p/ess/wpaper/id6773.html

3. Kumar, B., Mohanty, B.: Public distribution system in rural India: implications for food safety and consumer protection. Procedia—Soc. Behav. Sci. **65**, 232–238 (2012). https://doi.org/10.1016/j.sbspro.2012.11.116
4. Balani, S.: Functioning of the public distribution system: an analytical report. Working Papers id:5628, eSocialSciences (2014). https://ideas.repec.org/p/ess/wpaper/id5628.html
5. Salvi, S., et al.: Cloud based data analysis and monitoring of smart multilevel irrigation system using IoT. In: 2017 International Conference on I-SMAC (IoT in Social, Mobile, Analytics and Cloud) (I-SMAC), pp. 752–757 (2017). https://doi.org/10.1109/I-SMAC.2017.8058279
6. Shetty, S., Salvi, S.: SAF-Sutra: a prototype of remote smart waste segregation and garbage level monitoring system. In: 2020 International Conference on Communication and Signal Processing (ICCSP), Chennai, India, pp. 0363–0367 (2020). https://doi.org/10.1109/ICCSP48568.2020.9182408
7. Raj, J.S., Ananthi, J.V.: Automation using IoT in greenhouse environment. J. Inf. Technol. **1**(01), 38–47 (2019)
8. Pasumpon, A.P.: Artificial intelligence application in smart warehousing environment for automated logistics. J. Artif. Intell. **1**(02), 63–72 (2019)
9. Shetty, S., Shalini, P.R., Sinha, A.K.: Article: a novel web service composition and web service discovery based on map reduce algorithm. In: IJCA Proceedings on International Conference on Information and Communication Technologies ICICT, no. 4, pp. 41–45, October 2014
10. Valarmathy, S., Ramani, R., Akhtar, F., Selvaraju, S., Ramachandran, G.: Automatic ration material distributions based on GSM and RFID technology. Int. J. Intell. Syst. Appl. **5** (2013). https://doi.org/10.5815/ijisa.2013.11.05
11. Karthik, K.: Cloud-based ration card system using RFID and GSM technology. Int. J. Eng. Res. Technol. **2** (2013), 328–334
12. Madur, A.N., Nayse, S.: Automation in rationing system using arm 7. Int. J. Innov. Res. Electr. Electron. Instrum. control Eng. **1**(4) (2013)
13. Priya, B.G., Nikumbh, S.: E—public distribution system using smart card and GSM technology. In: 2017 International Conference on Intelligent Sustainable Systems (ICISS), pp. 244–249 (2017)
14. Vaisakh, A.K., et al.: IoT based intelligent public ration distribution. In: 2019 International Conference on Communication and Electronics Systems (ICCES), pp. 1894–1897 (2019)

Performance Evaluation of Clustering Techniques for Financial Crisis Prediction

S. Anand Christy, R. Arunkumar, and R. Madhanmohan

Abstract In the present days, financial crisis prediction (FCP) is winding up progressively in the business advertises. As organizations gather an ever-increasing number of data from day-by-day activities, they hope to draw valuable decisions from the gathered data to aid on practical assessments for new client demands, e.g., client credit classification, certainty of return that was expected, and so forth. Banks as well as institutes of finance have connected diverse mining methods on data to upgrade their business execution. With all these strategies, clustering has been measured as a major strategy to catch the usual organization of data. Be that as it may, there are very few examinations on clustering methodologies for FCP. In this work, we assess two clustering algorithms, namely *k*-means and farthest first clustering algorithms for parsing distinctive financial datasets shifted out off time periods to trades. The evaluation process was conducted for datasets Weislaw, Polish, and German. The simulation results reported that the *k*-means clustering algorithm outperforms well than farthest first algorithm on all the applied dataset.

Keywords Partitioning clustering · Clustering · Financial dataset · Density-based clustering

S. A. Christy (✉)
Department of Computer Applications, St. Joseph's College of Arts and Science(Autonomous), Cuddalore, India
e-mail: scholarchristy83@gmail.com

R. Arunkumar · R. Madhanmohan
Department of CSE, Annamalai University, Chidambaram, India
e-mail: arunkumar_an@yahoo.com

R. Madhanmohan
e-mail: madhanmohan.mithu@yahoo.com

P. Karuppusamy et al. (eds.), *Sustainable Communication Networks and Application*, Lecture Notes on Data Engineering and Communications Technologies 55,
https://doi.org/10.1007/978-981-15-8677-4_11

1 Introduction

Despite the availability of numerous financial datasets, they do not afford to contribute to a valuable decision making. Worldwide rivalries, dynamic markets, as well as fast advancement in the data and correspondence innovations are a portion of the significant difficulties within the present financial industry. For example, financial organizations are in consistent requirements for more data analysis, that is winding up additional extensive as well as complex. The measure of data accessible is always expanding; our capacity to process it turns out to be increasingly troublesome. Productive revelation of helpful knowledge from these datasets is in this manner turning into a dispute and a huge monetary need.

Then again, data mining is the way toward removing helpful, frequently beforehand obscure data, supposed knowledge, out of very big dataset and utilized in the various application such as the fraud detection, client maintenance, market analysis, and so forth. As of late, financial crisis prediction (FCP) has turned out to be exceptionally powerful and gainful in breaking down financial datasets [1]. Be that as it may, mining financial data present unique difficulties [2]: heterogeneity, complexity confidentiality, external factors as well as dimension. The FCP dispute is to discover the patterns rapidly when they are substantial and also to perceive the time when the patterns are not any more powerful. Additionally, planning a fitting procedure for finding significant knowledge in financial data is an extremely complex assignment.

In this work, we study two clustering algorithms for examining distinctive financial datasets for an assortment of utilizations: credit cards extortion detection, investment exchanges, stock market, and so on. We talk about the favorable circumstances and inconveniences of every strategy in connection to better comprehension of internal organizations of financial datasets in addition the capacity of every clustering technique in this unique situation. At the end, the motivation behind this examination is to give an outline of how fundamental clustering techniques were connected on FCP.

2 Related Works

In cluster analysis, the dataset with the closer similarity is clustered together [3]. In any case, the groups are not characterized ahead of time. Regularly, clustering analysis is a valuable beginning stage for different purposes, for example, summarization of data. A Cluster of Data Articles can be measured as a type of Data Compression. Diverse spaces may be able to imply grouping strategies to analysis data, for example, science, information retrieval, medicine, and so forth. In the business and fund, clustering may be able to be utilized, for example, to fragment clients into various gatherings for extra analysis and showcasing exercises. Clustering is a usual process in data compression; here are very few applications of the finance that utilizes this system contrasted with classification as well as affiliation analysis. Another clustering strategy is thickness depended [3] that avoids segmenting example space using the mean-center point; however, rather thickness-based information is utilized, by that it is tangled, unpredictable shaped, however, very much disseminated dataset be able to be grouped effectively.

OPTICS. OPTICS gives a point of view to investigate the size of density flanked groups. OPTICS makes an increased ordering of cases in light of the density distribution. This cluster ordering might be able to be utilized by a wide scope of density flanked grouping, for example, DBSCAN.

In any case, OPTICS requires few priori, for example, neighborhood radius (ε) as well as a least amount of items [4].

DBSCAN [5, 6] depends on density connected rate from subjective center objects that comprises the objects with the MinPts it records the participation of the cluster initially further points can be referred in [4].

"Since reachability plot is unfeeling to the information parameters," [7] recommend which the qualities ought to be sufficiently "large" to give up a decent outcome by means of no unclear illustrations as well as reachability plot look not barbed. Investigations demonstrate which MinPts utilizes ranges somewhere in the range of 10 and 20 dependably get great outcomes with sufficiently vast ε. Quickly, reachability plot is an exceptionally natural intends to derive the indulgent of the financial data of thickness flanked organization and its usual dimensions are utilized.

[8] employed an online developing methodology for identifying of financial articulations' peculiarities. The online advancing strategy [9] is a vibrant method for grouping data stream. This technique progressively builds the quantity of clusters by figuring the separation among illustrations and existing cluster focuses. In the event that this separation is higher than edge esteem, another cluster is made and instated by the case [10, 11] used hierarchical agglomerative clustering [3] way to deal with dissect securities exchange data. The authors proposed a productive metric for estimating the similitude between clusters, an ideal issue for various leveled agglomerative grouping techniques. The creators additionally specified that some pre-preparing systems, for example, mapping, dimensionality decrease, and standardization ought to likewise be connected to enhance the execution. Besides, they utilized precision–recall technique [10] to expand the cluster quality.

3 Clustering Algorithms

3.1 K-means Clustering

K-means [12, 13] be a familiar clustering algorithm that performs grouping of n data points to *K* clusters by minimizing the data points from *K* cluster heads in an iterative manner. *K*-means is a sort of unsupervised learning technique regularly utilized for taking care of the notable clustering issue. This procedure method tracks a less complex and less demanding path for arranging a known dataset utilizing a specific count of clusters (assuming *k* clusters) fixed apriori. The essential rule is to characterize *k* clusters, one for each group. The arrangement of clusters ought to be completed shrewdly because of the way that shifting area prompts diverse outcomes. Subsequently, the best technique for setting them is to keep them quite far from each other. The succeeding advance takes each point has a place with a given dataset and partners it to the nearest cluster. At the point when none of the fact of the matter is inadequate, the underlying advance is done and an early gathering age is likewise done.

At that occasion, there is a prerequisite to perform recalculation of *k* new centroid as barycenter of the clusters got from the preceding step. Subsequent to *k* new centroid is resolved, another coupling procedure will happen between the comparable dataset clusters and the nearest new cluster. A circle will be made. Thus, it is seen that the k focuses adjust their area well-ordered till none of the progressions exist or it can named as the clusters do not move any more. Toward the end, this strategy means to minimize an aim function recognized as squared error function and is represented as:

$$J(V) = \sum_{i=1}^{c} \sum_{j=1}^{c_i} \left(\|x_i - v_j\|\right)^2 \tag{1}$$

where $\|x_i - v_j\|^2$ represents the Euclidean distance of x_i and v_j, count of data points (c_i) cluster (ith) and *count given by* "*c*".

3.2 Farthest First Clustering

This clustering approach is introduced by Hochbaum and Shmoys in 1985 [8, 14]. It is similar to *k*-means in which it opts the centroid and assigns the entities in the cluster; however, with maximum distance, initial seeds are value that is at more distance to the mean values. In this situation, assigning clusters is dissimilar, and at initial cluster, a link with high session count will be received, for instance, at cluster-0 more than in cluster-1, etc. This algorithm requires fewer adjustments [8]. Firstly, it takes point *Pi* selects the preceding point *Pi* maximum distance. *Pi* is centroid, and the points are *p*.1, *p*.2, ..., *p.n* otherwise entities member of dataset belong to cluster out off Eq. (2).

$$\text{Minimum}\{\text{maximum dist}(p.\text{i}, p.1), \text{maximum dist}(p.1, p.2)\ldots\} \tag{2}$$

4 Results and Discussion

The method has used three datasets, and they are the "German, Weislaw, and the Polish " with the instance with the maximum variation from 240.01 to 100.01 containing 20 to 64 characteristics and were two types non-bankrupt and bankrupt used (Table 1).

Table 2 and Figs. 1, 2, and 3 show the cluster formation of *K*-means as well as farthest first algorithms. The cluster 0 in addition to cluster 1 in table represents the number of instances grouped in cluster 0 and cluster 1 as bankrupt and non-bankrupt. For better clustering performance, the number of instances in the cluster 0 and cluster 1 should be closer to the number of instances under bankrupt and non-bankrupt. From the table, it is shown that among the 700 bankrupt instances in German dataset, the *K*-means algorithm clusters 713 instances where the farthest first algorithm clusters 869 instances. These values show that the *K*-means clustering performs well and attain

Table 1 Description of dataset used

Dataset	Source	Instances	Characteristics	# of class	Bankrupt/non-bankrupt
German	UCI	1000.01	22	2.0	701/301
Weislaw	UCI	240.02	34	2.0	120/120
Polish	UCI	7028	65	2.0	114/130

Table 2 Cluster assignments of various dataset using *K*-means and farthest first algorithms

Dataset	*K*-means		Farthest first	
	Cluster-0	Cluster-1	Cluster-0	Cluster-1
German	713	287	869	131
Weislaw	128	112	196	44
Analcat data	25	25	27	23

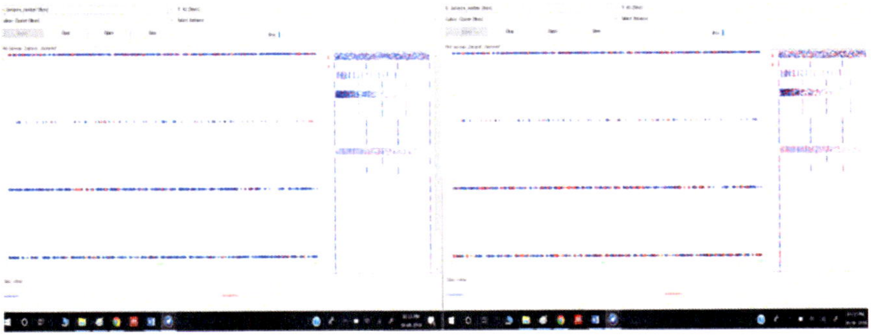

Fig. 1 Visualization of clustering on German dataset

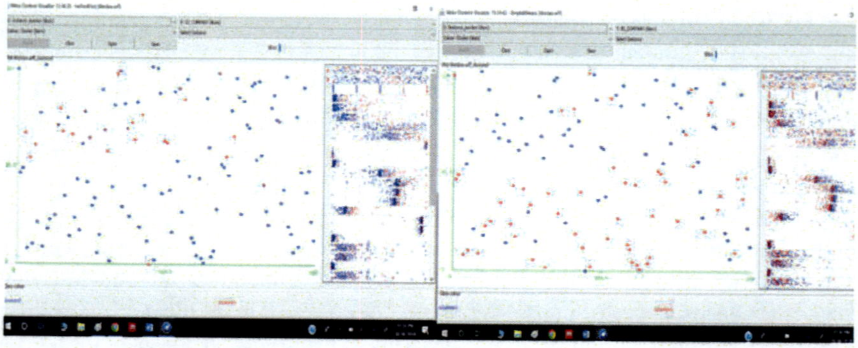

Fig. 2 Visualization of clustering on Weislaw dataset

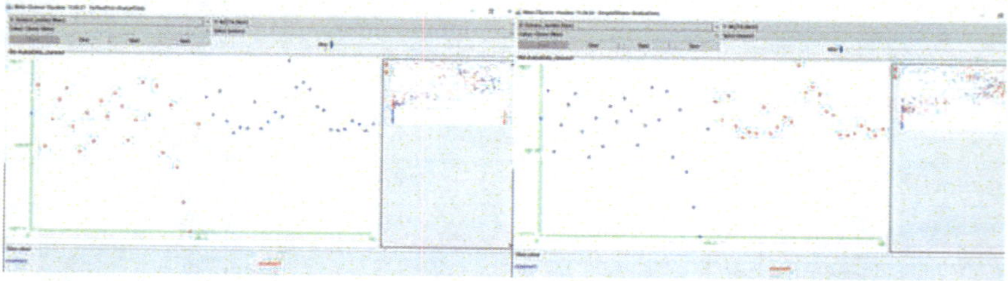

Fig. 3 Visualization of clustering on Analcat dataset

superior clustering than compared one. In the same time, for Weislaw dataset, the *K*-means clustering algorithm clusters 128 instances as bankrupt and 112 instances as non-bankrupt. These values imply that *K*-means algorithm clusters are almost closer to the original ones. Similarly, for Analcat dataset, *K*-means algorithm is superior to farthest first algorithm in several dimensions.

5 Conclusion

FCP has turned out to be exceptionally powerful and gainful in breaking down financial dataset. Similar to classification, cluster analysis groups comparative data objects into groups, in any case, the classes or clusters were not characterized ahead of time. Regularly, clustering analysis is a valuable beginning stage for different purposes, for example, data summarization. In this work, we assess two clustering algorithms, namely *k*-means as well as farthest first clustering algorithms, for parsing distinctive financial datasets shifted out off time periods to trades. The validation process reported that the *k*-means clustering algorithm outperforms well than farthest first algorithm on all the applied dataset.

References

1. Weigend, Data mining in finance: report from the Post-NNCM-96 workshop on teaching computer intensive methods for financial modeling and data analysis. In: Fourth International Conference on Neural Networks in the Capital Markets NNCM-96, 1997, pp. 399–411.
2. Kasabov, N.: Evolving Connectionist Systems, pp. 40–42. Springer, London Berlin Heidelberg (2003).
3. Tan, P-N., Steinbach, M., Kumar, V.: Introduction to Data Mining, pp. 150–172. Addison Wesley (2006)
4. Ankerst, M., Breunig, M.M., Kriegel, H-P., Sander, J.: OPTICS: ordering points to identify the clustering structure. In: ACM SIGMOD International Conference on Management of Data, 1999. pp. 49–60.
5. Ester, M., Kriegel, H-P., Sander, J., Xu, X.: A density-based algorithm for discovering clusters in large spatial databases with noise. In: 2nd International Conference on Knowledge Discovery and Data Mining (KDD-96). 1996. pp. 226–231
6. Ramesh, S., Smys, S.: A software-based heuristic clustered (SBHC) architecture for the performance improvement in MANET. Wirel. Pers. Commun. **97**(4), 6343–6355 (2017)
7. Quinlan, J.R.: Learning first-order definitions of functions. J. Artif. Intell. Res.. **5**, 139–161 (1996)
8. Sharma, N., Bajpai, A., Litoria, R.: Comparison the various clustering algorithms of weka tools. Int. J. Emerg. Technol. Adv. Eng. **2**(5) (2012). ISSN: 2250-2459.
9. Han J., Kamber, M.: Data Mining: Concept and Techniques. Morgan Kaufmann Publishers, 2nd Edn (2005).
10. Manoharan, S.: Performance analysis of clustering based image segmentation techniques. J. Innov. Image Process (JIIP) **2**(01), 14–24 (2020)
11. Raj, J.S.: QoS optimization of energy efficient routing in IoT wireless sensor networks. J. ISMAC **1**(01), 12–23 (2019)
12. Richa Loohach and Dr. Kanwal Garg Effect of distance functions on simple K-means clustering algorithm. Int. J. Comput. Appl. **49**, (2012).
13. Wittman, T.: Time-Series Clustering and Association Analysis of Financial Data (2002). Available: https://www.math.ucla.edu/~wittman/thesis/project.pdf.
14. Bensmail, H., DeGennaro, R.P.: Analyzing imputed financial data: a new approach to cluster analysis (2004). Available: https://www.frbatlanta.org/filelegacydocs/wp0420.pdf.
15. Omanovic, S., Avdagic, Z., Konjicija, S.: On-line evolving clustering for financial statements' anomalies detection. In: International Symposium on Information, Communication and Automation Technologies, ICAT 2009. XXII, 2009, pp. 1–4

Optimization of Job Scheduling with Dynamic Bees Approach

Harsimrat Singh and Chetan Marwaha

Abstract Cloud-based job scheduling is used to achieve high throughput. The problem of load balancing hampers the performance of the cloud. To end this, the bees life algorithm is used to achieve optimization in job scheduling. This paper proposes unique dynamic scheduling using the bees algorithm with the shortest job first approach to achieve a high degree of load balancing. The effectiveness of cloud computing depends greatly upon the performance of the cloud scheduler. The scheduling algorithm used achieves high performance, since waiting time for jobs is reduced. The overall algorithm is divided into phases. The first phase is the initialized section in which fog and execution nodes are defined. In the second phase, job fetching and uploading are accomplished. In the third phase, jobs are sorted according to the shortest burst time. In the last phase, bees are deployed to locate the best possible processor for job execution. The result of the dynamic bees algorithm shows effective load balancing as compared to the static bees algorithm. Result in terms of execution time, waiting time, execution speed, and memory allocation shows improvement by 2–3%.

Keywords Cloud computing · Job scheduling · Optimization algorithms · Dynamic bees life algorithm

1 Introduction

Cloud computing platforms attract mass community due to the affordable and vast availability of resources [1]. Heavy traffic toward cloud resources could leads to deadlock, and waiting time for other users increases significantly. This problem causes loss of confidence and hence degrades the quality of service [2]. In broker-driven

H. Singh (✉) · C. Marwaha
Department of Computer Engineering and Technology, Guru Nanak Dev University, Amritsar, India
e-mail: harsimratsingh1996@gmail.com

C. Marwaha
e-mail: marwaha.chetan@gmail.com

P. Karuppusamy et al. (eds.), *Sustainable Communication Networks and Application*, Lecture Notes on Data Engineering and Communications Technologies 55,
https://doi.org/10.1007/978-981-15-8677-4_12

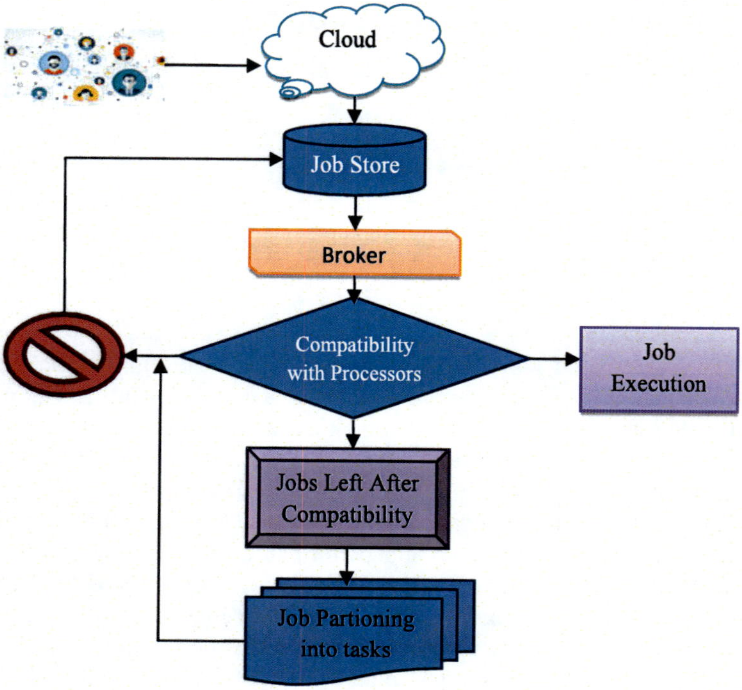

Fig. 1 Dynamic bees approach for job execution

VM selection policy, the fittest machine is selected and the job is allotted no criteria is specified regarding the release of resource causing a high probability of deadlock [3]. Resource provisioning thus requires scheduling based on multiple parameters. An algorithm based on multiple criteria is optimization algorithm and satisfies the need for avoiding a deadlock situation. The structure of the cloud satisfying requirements of different clients is given in Fig. 1. The job is submitted to the cloud by the registered user. Jobs form batches and picked by the broker. The broker is responsible for checking the compatibility of a job against the processor resources. Job sorting is also performed by the broker. Once job sorting is performed by the broker and compatibility check is performed, job allocation is performed [4]. After the execution of the job, the result in the form of execution time is returned. The job compatibility, if not satisfied, then jobs are portioned into the task and stored within the job buffer. The process is repeated until all the jobs are executed successfully. The objective of the proposed work is dynamic job scheduling and is performed. To compute various parameters like execution time, waiting time, execution speed, memory allocation, compare the proposed work with static bees approach [5]. To achieve job execution, the best possible resources are searched for. This is accomplished using job scheduling algorithms [6]. Job scheduling algorithms based on single metrics are not effective in today's era. Multicriteria algorithms such as genetic algorithm, PSO, and bees algorithms are capable of handling the problem of multiobjective jobs. All of these algorithms are discussed in Sect. 2. The problems within these algorithms are discussed in Sect. 3. Section 4 gives the mathematical foundation and proposed work. Section 5 gives the performance analysis and result, and Sect. 6 gives the conclusion and future scope. Section 7 gives references.

2 Multiheuristic Approach to Job Execution

The multiheuristic approach is based on multiple conditions. This means, multiple conditions if satisfied, then the job is executed on the current processor. The job execution based on multiple multiheuristic approaches is given in the tabular form.

2.1 Genetic Algorithm-Based Job Execution

The genetic approach is one of the oldest mechanisms to execute a job based on multiple objectives. The genetic approach is divided into phases. The first phase corresponds to population selection, the second phase corresponds to mutation, in the next phase objective function is evaluated and at last, the crossover is performed. The population selection is based on the satisfaction of objective function. Figure 2 shows the sequence of phase execution in a genetic approach.

The genetic approach is used by many researchers to execute jobs with multiple objectives. This is given in Table 1.

2.2 Particle Swarm Optimization (PSO)

Particle swarm optimization is a multiheuristic approach in which swarm intelligence is used. Allocation of resources in the best possible way is accomplished using this mechanism. Figure 3 demonstrates the steps of this approach.

The techniques demonstrating the use of PSO in job scheduling is given in Table 2.

2.3 Bees Algorithm for Job Scheduling

Bee's life algorithm is revolutionary in updating the mechanism by which a genetic-based approach executes the job. The phases associated with bee's life algorithm are given in Fig. 4.

The research utilizing the bees algorithm is given in Table 3.

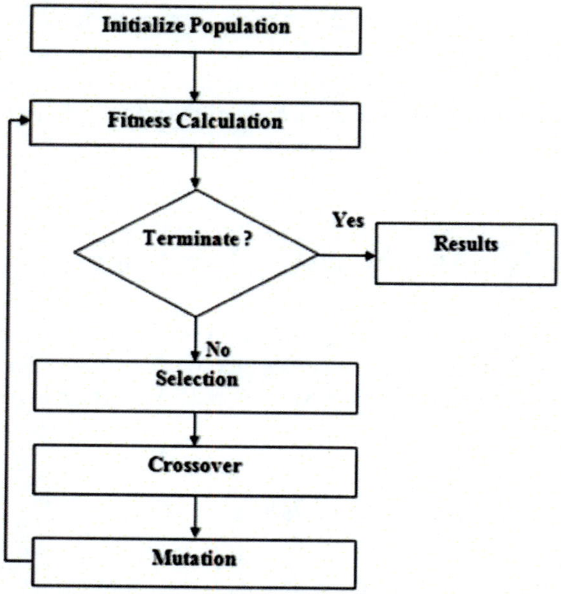

Fig. 2 Genetic algorithm phase execution

Table 1 A genetic algorithm for the execution of job

Authors	Year	Multiobjectives	Merits	Demerits
Ling and Leung [7]	2007	• Load balancing • Cost-effective scheduling	All the jobs are executed then the algorithm is allowed to finish.	The convergence rate is poor
Pavez-Lazo and Soto-Cartes [8]	2011	• Load balancing • Cost-specific allocation	Load balancing is achieved effectively	The convergence rate is high
Garg and Singh [9]	2015	• Grid allocation • Reducing execution time	Allocating grid to the cloudlet effectively and workflow scheduling is achieved with high performance	Load balancing is an issue
Ferrarotti et al. [10]	2016	• Job execution time • Load balancing	The load is balanced with high performance	The rate at which execution takes place is slow

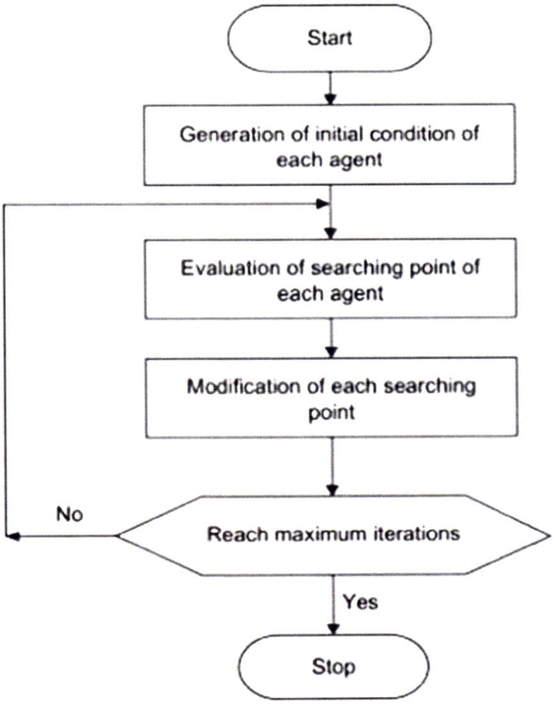

Fig. 3 PSO steps for job scheduling

Table 2 PSO utilized in job scheduling

Authors	Year	Multiobjectives	Merits	Demerits
Pacini et al. [11]	2015	• Load balancing • Cost-effective scheduling	Three-level approach for securely executing the job	The convergence rate is poor
Smanchat and Viriyapant [12]	2015	• Workflow job scheduling • Cost-specific allocation	High traffic of cloudlets can be executed easily	The convergence rate is high
Mohtasham et al. [13]	2016	• Execution time • Waiting time	Allocating resource to the cloudlet effectively and cloudlet scheduling is achieved with high performance	Load balancing is an issue
Devi and Poovammal [14]	2016	• Overlap community detection • Execution time	Load balancing is effective	Execution time and waiting time can be reduced effectively

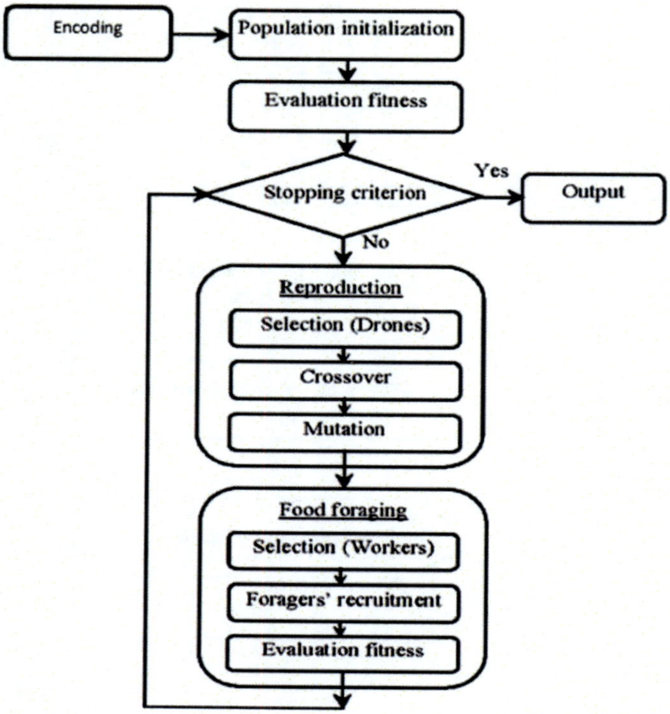

Fig. 4 Bees algorithm for job scheduling

Table 3 Bees life algorithm for job execution

Author	Year	Multiobjective	Merits	Demerits
Lee [15]	2015	• Makespan • Flowtime	Makespan and flowtime are reduced	Convergence and load balancing is a problem
Kimpan and Kruekaew [16]	2016	• Makespan • Flowtime	Execution time is reduced with a multiheuristic approach	Rate of convergence and load balancing is an issue
Bitam et al. [17]	2018	• Minimizing execution time • Job execution	Makespan and flowtime are reduced	Low convergence rate due to unfinished processes and waiting time is high
Vishwakarma and Jain [18]	2019	• Cloud security • Job execution	Minimizing execution time is achieved using honey bot approach	Load balancing is an issue not achieved through this approach

3 Problem Definition

The problem from the bees life algorithm indicates that load balancing is achieved but waiting time is a problem. The execution speed is also high in the existing bees approach. The job is matched against the compatible processor and if the match does not occur, the cloudlet is added within the waiting queue. The process will not come out of the queue until the processor with the compatibility is located. The process will be in an indefinite loop and may not execute at all. This causes a low convergence rate associated with the bee's life algorithm. The discovered problems from the bees life are listed as under (Fig. 5)

- The static approach causes unfinished or daemon processes.
- Low convergence rate due to unfinished processes.
- Job allocation based on first come first serve approach.
- High execution time if compatibility is violated.

3.1 Difference in Static and Dynamic Bees Approach

In this section we briefly differentiate between the state and dynamic bees approach is given in Table 4.

Table 4 Difference between the static and dynamic bees approach

Parameters	Static approach	Dynamic approach
Number of tasks	Fixed per virtual machine	Variable depending upon the load
Convergence	Takes a large amount of time	Takes less time
Load balancing	It is achieved but to a lesser extent	It is achieved to a better extent
Cost	More virtual machines used to lead to higher cost	Less virtual machines usage caused the least cost
Waiting time	It is high as task partitioning is fixed	It is less as task partitioning is variable

4 Mathematical Foundation and Proposed Work

Some metrics are evaluated in the dynamic bees algorithm to prove worthy of study. These metrics should be effective enough to state the validity of the proposed approach. The mathematical foundation for the proposed approach is given as under

4.1 Mathematical Foundation

The approach uses the objective function. The objective function is based on cost, load balancing, and compatibility check. The objective function is given as under Eq. 1.

$$y = ax^2 + bx + c \tag{1}$$

Here '*y*' is the objective function value which is to be maximized. '*x*' indicates the confidence level or degree of importance associated with each factor. '*a*' is the obtained value of load balancing, '*b*' is the cost factor, and '*c*' is the compatibility factor.

The waiting time indicates the amount of time the cloudlet needs to wait after arrival before the resource is allocated. The waiting time is given as under Eq. 2.

$$W_t = T_t - B_t \tag{2}$$

'*T*' indicates the turnaround time, '*B*' indicates the burst time, and '*W*' indicates the waiting time. The execution speed is the parameter indicating the fast convergence. The execution speed is measured with the response time. This is given as under Eq. 3.

$$\text{Speed} = P_t - A_t \tag{3}$$

'*P*' is the time during which process first get the CPU and '*A*' is the arrival time. The turnaround time is another metric that is estimated with the formula. The turnaround time is given as under Eq. 4.

$$\text{TAT} = C_t - A_t \tag{4}$$

'*C*' gives the completion time of the job, and '*A*' indicates the arrival time.

4.2 Proposed Work

The proposed work is focused upon the job partitioning and execution approach using the bees life algorithm. Scheduling is based on a time scale depending upon the process burst time. Deadline-oriented approach is used for determining the best possible job for execution over the processor. The phases of this approach are given as under.

– **Job Sorting**

This phase receives the job and then performs the sorting operation based upon the burst time associated with the job. The job burst time if lowered is placed in the first place within the batch. The job sorting operation is based on the shortest job first approach. This sorting operation is performed before assigning a broker to the job. The algorithm for this phase is given as under

- Initialize the process and add it into the ready queue.
- Assign CPU to process having the lowest burst time and execute it.
- If two processes have similar Burst time then SJF executes the process which is arrived first is executed.
- Repeat the steps above until the queue become empty.
- Calculation of waiting time and turnaround time of the individual process is done.
- Calculate the average waiting time and average turnaround time.

– **Resource Sorting**

Bubble sort has been applied to receive the resources which are highest in number. After that, the highest resources will be allotted to the jobs. The resource sorting operation is based on a bubble sort approach. The algorithm for resource sorting is given as under.

Algorithm- Resource Sorting
Input: Resource List, N(number of resources)
Output- Sorted List

- Set index i = 0
- Set index j = 0
- For i = 0 to N
  For j = 0 to N-i-1
  If ($Resource_i$ > $Resource_{i+1}$)
  Temp = $Resource_i$
  $Resource_i$ = $Resource_{i+1}$
  $Resource_{i+1}$ = Temp
  End of if
  End of For
  End of For
- Output Job list

– Job partitioning

The job partitioning is performed by receiving the jobs and then checking against the pool of resources. The job requirement if matches with the processor resources then the job is entered into the active queue. The jobs that are present within the queue are evaluated again after all the jobs present within the active queue are finished. The job is partitioned into tasks to make the compatibility of jobs with the processor. These portioned jobs are submitted again to the queue for resource allocation. The algorithm for the job partitioning is given as under

Algorithm-Job Partitioning
Input: Job List, i(index), N (Number of Jobs), CE(Computing elements or processors), CE_R(CE resources), Top=0
Output: Task List

- Input Job list from first phase (Job_i)
- Set Top=0
- For i = 0 to N
  If Job_i. Requirments > CE_i
  Waiting_list [Top] = Job_i
  Top = Top+1
  Else
  CE_i = Job_i
  End of if
  End of for
- Top = i
- Check Waiting Queue
  If (is empty (Waiting_list))
  Call Bees Algorithm
  Else
  L = waiting_list [Top]
  Top = Top-1
  End of if

$Jobs_i$ = L/CE_R
Goto Job Sorting algorithm

The model for the first two phases is given as under

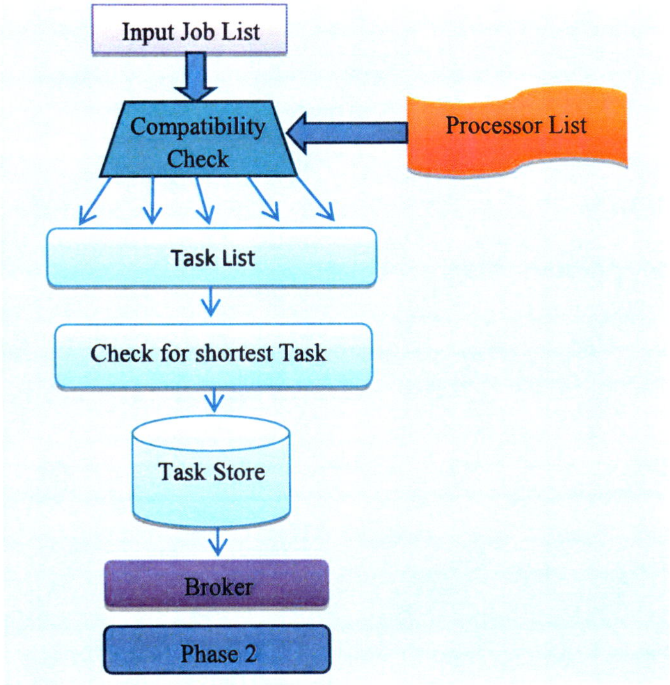

Fig. 5 Proposed model for the first two phases

The compatibility check mechanism corresponds to the number of resources possessed by a virtual machine and requirements of the process. The process requirements must be matched with the resources possessed by VM otherwise compatibility check fails. Thus, the number of resources is a parameter used to check the compatibility.

– **Dynamic bees approach**

This is the final phase of the proposed approach. The bees are deployed to execute the jobs. This approach receives jobs after sorting and task partitioning. The bees are deployed to locate the nearest food source. The bees hunt down the food source and the job is executed. The fittest resource is searched, and if resource from the processor is left over, the core is partitioned to utilize the resource further. Since the fittest processor is selected hence load balancing is effectively achieved. The algorithm for this approach is given below;

Algorithm bees
Input: Job list, resource list
Output: Waiting time, speed, memory allocation

- Deploy bees corresponding to the number of resources

- If (is found ($Resource_i$))

  Execute job
  End of if

- If (Resource_Finished)

  Deploy Scout bees for resource searching
  End of if

- If (Finished (Job_i))

  Return waiting time, memory allocation, speed
  End of if
  The model for the dynamic phase is given in Fig. 6.

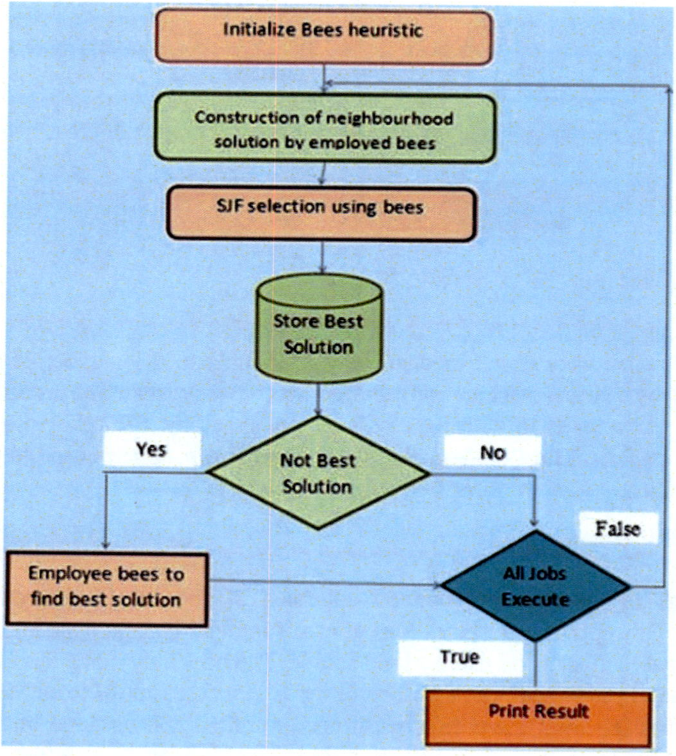

Fig. 6 Bees algorithm for job scheduling

First, all the jobs are uploaded and stored within the text file. Fog nodes and execution nodes are also created. The number of jobs should be directly proportional to jobs uploaded. In case these jobs are not equivalent to execution nodes, then process wait for execution and waiting time increases. Inputted jobs are checked against the processor resources. This will build a fittest processor sort list. After this, depending upon the burst time of the job and processor resources, the task list is prepared. This task list is sorted according to the shortest job first approach. These tasks are again stored within the task list. This task list is submitted to the broker for execution. This bee corresponding to the task list is formed. These bees look for the best possible processor for the evaluation of the task. After this initial solution is prepared and building a complete solution, bees are randomly scattered. The result is again obtained. This result is compared to the previous solution. The best possible solution is retained.

5 Performance Analysis and Result

The performance of the dynamic bees approach is much better as compared to the static bees approach. The result is presented in the form of waiting time, execution time, memory allocation, and speed. The number of fog nodes and execution nodes is varied to determine the variation in the result. The result is given in Table 5. The number of fog nodes and execution nodes is kept at a constant 5 value.

The plot for Table 5 is given in Fig. 7. The execution speed shows improvement by a margin of 10%.

Table 5 Speed comparison

Number of jobs	Speed (static bees) ms	Speed (dynamic bees) ms
10	282	293
20	512	532
30	712	722
40	900	912
50	950	970
60	1023	1072
70	1200	1278
80	1300	1356
90	1396	1450
100	1520	1560

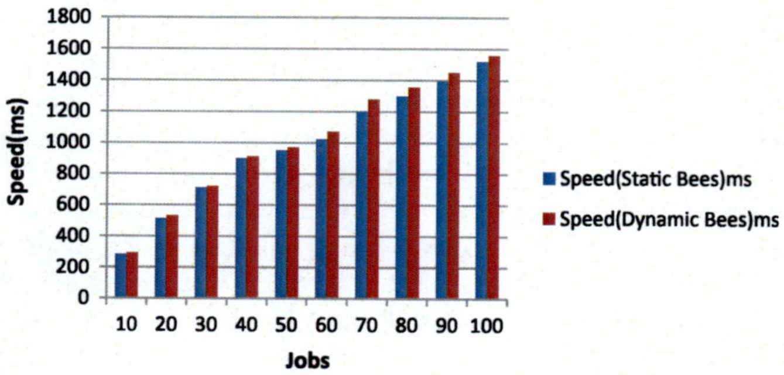

Fig. 7 Speed plot for dynamic and static bees approach

Task completion speed is given in Table 5. This speed measures the efficiency of the proposed system. The allocation could be early but due to interruption, the processor may be prompted hence task completion is considered in this case. The waiting time also shows significant improvement. The waiting time is shown in Table 6. The improvement by a margin of 7% is achieved.

The plot showing the betterment of the proposed approach in terms of load balancing and latency is given in Fig. 8.

Table 6 Waiting time or latency comparison

Number of jobs	Waiting time (static bees) ms	Waiting time (dynamic bees) ms
10	2490	1293
20	5120	2532
30	7112	3722
40	9020	4912
50	9530	5970
60	10,223	6172
70	12,020	7178
80	13,030	8156
90	13,496	9150
100	15,520	10,560

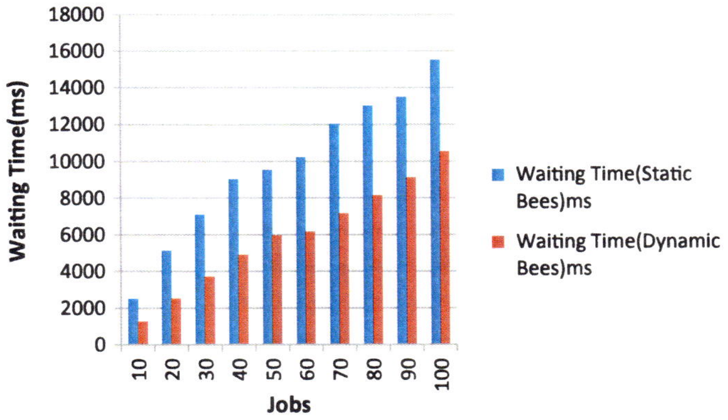

Fig. 8 Waiting time for dynamic and static bees approach

The memory allocation increases with the number of processes. The process burst time directly impacts the memory consumption. The performance of the proposed mechanism is certainly better as the process does not last within memory for a longer time and hence less memory is consumed. Table 7 shows the performance of the proposed mechanism in terms of memory consumption.

The plot for Table 7 is given in Fig. 9.

Table 7 Memory consumption comparison

Number of jobs	Memory (static bees) KB	Memory (dynamic bees) KB
10	9.56	5.33
20	12.433	6.87
30	14.675	8.656
40	17.786	10.34
50	20.434	12.343
60	23.2234	13.454
70	30.434	15.4343
80	32.345	17.343
90	33.344	20.343
100	35.3434	21.2323

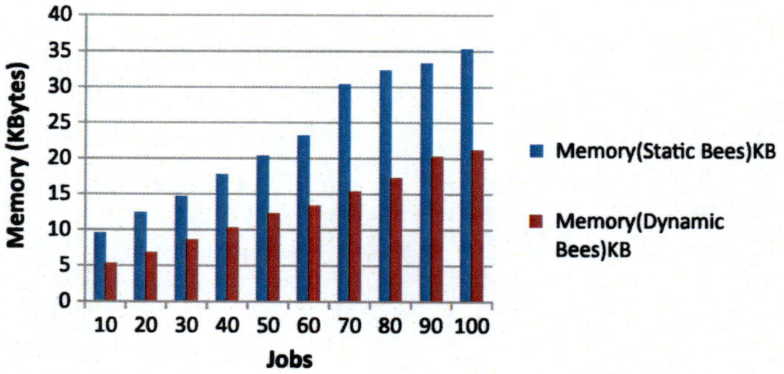

Fig. 9 Memory consumption by dynamic and static bees approach

The execution time indicates the time of completion associated with the process or job. The arrival time must be subtracted from job completion time to obtain the actual execution time of a job. Thus when all the job waiting time is added together, then the execution time of a batch is obtained.

The plot for the execution time is given in Fig. 10 (Table 8).

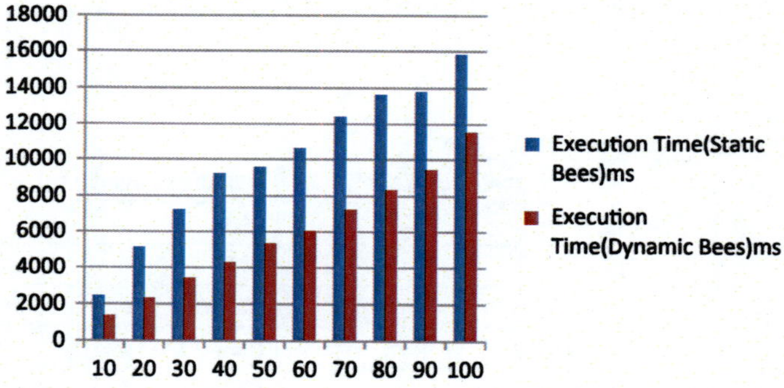

Fig. 10 Execution time comparison

Table 8 Execution time for dynamic and static bees algorithm

Number of jobs	Execution time (static bees) ms	Execution time (dynamic bees) ms
10	2440	1393
20	5140	2332
30	7232	3422
40	9230	4312
50	9580	5370
60	10,623	6072
70	12,420	7278
80	13,630	8356
90	13,796	9450
100	15,820	11,560

6 Conclusion and Future Scope

Static bees scheduling algorithm is used to achieve optimization in terms of waiting time. The jobs are executed on machines that are not sorted. The machine executes the job on the first come first serve basis. This mechanism although executes all the jobs but execution time increases. To tackle the issue, a dynamic mechanism is needed. The proposed mechanism not only sorts the jobs but also sorts virtual machines. The execution of a job is much faster as compared to static bees algorithm. The waiting time, turnaround time, and execution time are reduced greatly using the proposed system. The order of scheduling is broker-driven. The broker is a software agent that is used to select the optimal VM on receiving the job. The job is presented to the system in batches. The execution of a job if not possible on the current machine then the machine is divided into cores. The proposed mechanism is termed dynamic since the processor is selected based on job requirements.

The dynamic bees approach with job sorting and partitioning mechanism achieves better load balancing and hence the result is improved. In the first phase, job sorting is achieved. To accomplish this, a bubble sort mechanism is applied. The primary parameter used for this purpose is the job burst time. The next phase is used to check the compatibility of the job with available resources. The complexity violation leads to task partitioning. The task partitioning approach allows all the jobs to be executed within the prescribed deadline. Bees approach allows the execution of the job list and results in terms of execution time, memory, waiting time, and speed is presented. The resulting improvement by the margin of 2–4% is achieved.

In the future, similar job identification with a benchmark dataset can be used. The additional parameters can also be accommodated within the dynamic bees approach.

References

1. Yao, G., Ding, Y., Member, S., Hao, K.: Using imbalance characteristic for fault—tolerant workflow scheduling in Cloud systems. IEEE Access, vol. 9219, no. c (2017). https://doi.org/10.1109/tpds.2017.2687923
2. K.V.: A stochastic development of cloud computing based task scheduling algorithm. J. Soft Comput. Paradig. (1), 41–48 (2019). https://doi.org/10.36548/jscp.2019.1.005
3. Han, R., et al.: PCS: predictive component-level scheduling for reducing tail latency in cloud online services. Proc. Int. Conf. Parallel Process., vol. 2015-Decem, pp. 490–499 (2015). https://doi.org/10.1109/icpp.2015.58
4. Switalski, P., Seredynski, F.: Scheduling parallel batch jobs in grids with evolutionary metaheuristics. J. Sched. **18**(4), 345–357 (2014). https://doi.org/10.1007/s10951-014-0382-0
5. Mangla, N., Singh, M.: Effect of scheduling policies on resource allocation in market oriented grid. In: 2012 International Conference on Computing Sciences, pp. 212–216, Sept 2012. https://doi.org/10.1109/iccs.2012.30
6. Singh, H., Marwaha, C.: Fog computing: terminologies, architecture and challenges. Int. J. Innov. Sci. Mod. Eng. **6**(3), 7–13 (2019). https://doi.org/10.35940/ijitee.c1162.116319
7. Ling, S.H., Leung, F.H.F.: An improved genetic algorithm with average-bound crossover and wavelet mutation operations. Soft. Comput. **11**(1), 7–31 (2007). https://doi.org/10.1007/s00500-006-0049-7
8. Pavez-Lazo, B., Soto-Cartes, J.: A deterministic annular crossover genetic algorithm optimisation for the unit commitment problem. Expert Syst. Appl. **38**(6), 6523–6529 (2011). https://doi.org/10.1016/j.eswa.2010.11.089
9. Garg, R., Singh, A.K.: Adaptive workflow scheduling in grid computing based on dynamic resource availability. Eng. Sci. Technol. Int. J. **18**(2), 256–269 (2015). https://doi.org/10.1016/j.jestch.2015.01.001
10. Ferrarotti, F., Schewe, K.D., Tec, L., Wang, Q.: A new thesis concerning synchronised parallel computing—simplified parallel ASM thesis. Theor. Comput. Sci. **649**, 25–53 (2016). https://doi.org/10.1016/j.tcs.2016.08.013
11. Pacini, E., Mateos, C., Garino, C.G.: A three-level scheduler to execute scientific experiments on federated clouds. IEEE **13**(10), 3359–3369 (2015)
12. Smanchat, S., Viriyapant, K.: Taxonomies of workflow scheduling problem and techniques in the cloud. Futur. Gener. Comput. Syst. **52**, 1–12 (2015). https://doi.org/10.1016/j.future.2015.04.019
13. Mohtasham, A., Filipe, R., Barreto, J.: FRAME: fair resource allocation in multi-process environments. In: Proc. Int. Conf. Parallel Distrib. Syst. - ICPADS, vol. 2016-Janua, pp. 601–608 (2016). https://doi.org/10.1109/icpads.2015.81
14. Devi, J.C., Poovammal, E.: An analysis of overlapping community detection algorithms in social networks. Proc. Comput. Sci. **89**, 349–358 (2016). https://doi.org/10.1016/j.procs.2016.06.082
15. Lee, X.: Heuristic task scheduling with artificial Bee colony algorithm for virtual machines. In: IEEE Conference Publication (2015)
16. Kimpan, W., Kruekaew, B.: Heuristic task scheduling with artificial Bee colony algorithm for virtual machines. In: Proceedings—2016 Joint 8th International Conference on Soft Computing and Intelligent Systems and 2016 17th International Symposium on Advanced Intelligent Systems, SCIS-ISIS 2016, pp. 281–286, Dec. 2016. https://doi.org/10.1109/scis-isis.2016.0067
17. Bitam, S., Zeadally, S., Mellouk, A.: Fog computing job scheduling optimization based on bees swarm. Enterp. Inf. Syst. **12**(4), 373–397 (2018). https://doi.org/10.1080/17517575.2017.1304579
18. Vishwakarma, R., Jain, A.K.: A honeypot with machine learning based detection framework for defending IoT based botnet DDoS attacks. In: Proceedings of the International Conference on Trends in Electronics and Informatics, ICOEI 2019, pp. 1019–1024, Apr. 2019. https://doi.org/10.1109/icoei.2019.8862720

Enhancing Cloud Security Using Secured Binary-DNA Approach with Impingement Resolution and Complex Key Generation

Jasmine Attri and Prabhpreet Kaur

Abstract Cloud computing is used by the mass community. Categories of users using cloud resources are indifferent. Cloud security is thus the area of concern. This paper focuses on the security aspect by modifying the BDNA procedure. Although DNA-based encryption is considered one of the safest mechanisms for managing data within the cloud, it has a flaw of a Key Clash that is rectified using a random number generator and hashing mechanism within BDNA. The impingement occurs when the key generated with DNA has the same location as the earlier key location. The employed mechanism is termed as chain-based BDNA to improve security further also with impingement handling. It considered chaining-based BDNA and BDNA approach to tackling the problem of impingement with keys. The parameters considered are execution time in key formation, reliability, number of impingements. BDNA is based on binary encryption and to enhance the security further, excess three codes are merged within the proposed mechanism. The proposed system is implemented using Netbeans8.1 java-based platform along with cloudsim4.0. The overall result of encryption is least as time consumption is reduced and highest in terms of reliability as compared to the BDNA approach.

Keywords Cloud security · BDNA · Chaining · Impingement

1 Introduction

Sander and Tschudin [1] Cloud security is an issue surrounding the mass community and required to be tackled to increase descent traffic towards cloud resources [2]. To this end, cloud security procedure is employed both at client and service provider end. Cloud security consists of a set of policies that yield optimize performance by

J. Attri (✉) · P. Kaur
Department of Computer Engineering and Technology, Guru Nanak Dev University, Amritsar 143005, India
e-mail: attrijasmine@gmail.com

P. Kaur
e-mail: prabhpreet.cst@gndu.ac.in

P. Karuppusamy et al. (eds.), *Sustainable Communication Networks and Application*, Lecture Notes on Data Engineering and Communications Technologies 55,
https://doi.org/10.1007/978-981-15-8677-4_14

rejecting unauthorized access to resources. Although cloud services and apps attract people, however, security issues could put cloud resources, system information, and data management at risk. To tackle the issue cloud security mechanisms must be enforced both at client and service provider end.

NIST [3] Cloud security mechanisms required for handling mass storage both at the client at the server end. There are three service models requires handling distinct requirement from the user. Primary work associated with the proposed work deals with software as a service [4]. The encryption mechanism in the general situation employed in cloud computing is transposition cipher. In this mechanism, if data "ABC" is to be transmitted than according to the transposition mechanism "CAB" can be transmitted [5]. This encryption mechanism is weak. To tackle the issue random interchange mechanism is employed to make the key complex. Both public and private key mechanisms are used for encoding and decoding of keys. During the advancement in the encryption process, DNA encryption is developed for making information more and is secure. In the cloud, this technique is used to enhance user trust in the service provider.

DNA Encryption is the process of perplexing genetic information. The computational method that is used is genetic sequencing. This mechanism plays a critical role in communication security. There are several reasons for using DNA encryption.

- This technology is one of the rapidly growing mechanisms that found its application in the cloud environment also.
- It is quite possible with DNA encryption to create complex crypto algorithms.
- It becomes possible to design algorithms for unbreakable problems.
- Less storage is required to handle DNA-based encryption as compared to DES and RSA algorithms.
- The power source is not required during the computation of DNA.
- Millions of bits can be stored in a few milligrams of data in the real world.

Kchim [6] Several modifications to DNA encryption are suggested to improve the efficiency of the encryption process. DNA encryption process is elaborated in Fig. 1

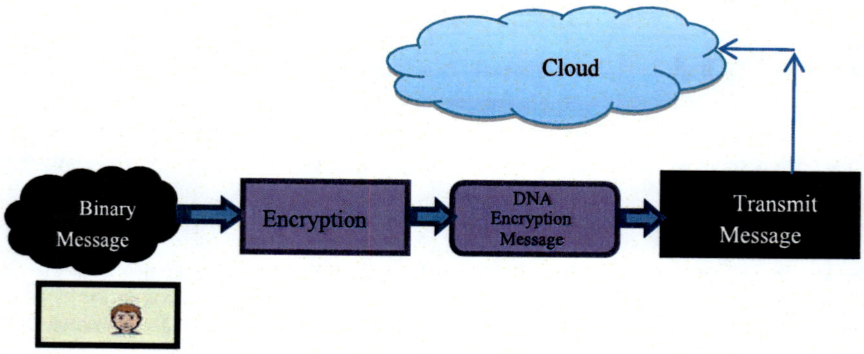

Fig. 1 Binary Encryption process under DNA encryption

A comparison of literature is presented in the literature survey along with merits and demerits.

The objective of the proposed methodology is as given below

- To handle the large file with least execution time
- To avoid impingement problem to handle data loss
- To achieve a better and secure key for encryption
- To compare the result of BDNA approach with the proposed approach.

The general structure of this paper is considered as under. Section 2 gives detail of studied literature of the existing mechanism associated with encryption for secured data transmission. Section 3 gives the problems and corresponding solutions to the problems. Sections 4 gives the methodology of proposed work along with a description of each phase, Sect. 5 gives performance analysis and result, Sect. 6 gives the conclusion and future scope and last section gives the references.

2 Literature Survey

This section presents the encryption mechanism that is used to secure the cloud. The security of cloud computing is considered at distinct levels. At the first level, the least secure mechanisms will be considered and continue towards more secure mechanisms of security toward cloud computing

2.1 Level 1 Security Mechanisms

Level 1 security mechanisms include security systems that can be further improved or in other-words are least secure. Comparison Table 1 provides insight into the level1 security.

Random number generator is missing in almost all the studied literature given in Table 1. Also, the security of data is at stake with the use of public key mechanisms.

2.2 Level 2 Security Mechanisms

Level 2 security is advanced as compared to level 1 security levels. The random number generator is accommodated within most of the mechanisms to provide security among encryption mechanisms.

These mechanisms are more secured as compared to strategies without a random number generator. The mechanism employed could be further enhanced by employing strategies to tackle the issue regarding key generation.

Table 1 Level 1 security standards

Technique	Security metrics	Merits	Demerits	Considered in
RSA	Key Size Execution time	The key that is formed depends upon random prime numbers	Public Key formed can be hacked easily that is transmitted over the public medium	Mahalle and Shahade [8], Seo et al. [9]
DES	Key Size	Key size is small and is a building block for most secure encryption standards in the modern era	Encryption standards are poor using a 56-bit key size	Security [10], Gupta et al. [11]
Homomorphic Encryption	Key Size Execution time	Key size is reduced but complex by the use of random number generator	Encryption standards use both public as well as private key mechanisms that make the entire mechanism complex	Wang et al. [12], Alabdulatif et al. [13], Miguel [14]

2.3 *Level 3 Security Mechanism*

Level 3 security mechanisms more secure as compared other two levels. To manage the abnormalities within the dataset, pre-processing mechanisms are employed. Also, to manage the storage, deduplication is utilized. The mechanisms ate discussed in Table 2.

There is a problem of impingement within the discussed literature. Impingement resolution can be resolved by the use of chaining in the proposed system. The problem discovered in BDNA approach that is discovered to be most secured than the rest of the literature is given in Table 3.

The proposed system accommodates BDNA with excess 3 code rather than binary codes. The phases of BDNA is described as under

2.4 *Phases of BDNA Approach*

BDNA encryption algorithm uses the symmetric key generation algorithm that is DES [7]. A Block cipher of 64 bits is used in the case of the DES algorithm. 56 bits form 64 bits are used for encryption and the rest of the bits are used for padding and checking parity. 16 rounds lead to the generation of the final key. The key is maintained using a security framework. The local database is used to maintain and

Table 2 Level 2 Encryption mechanisms

Considered in	Technique	Description	Advantage	Disadvantage
Zhang et al. [15]	Dynamic DNA coding	It first generates the Chaotic sequence of the image pixels. Then, it uses DNA encoding to encrypt the image pixels	It gives better security to image and can effectively resist statistical, shear, and another attack	Reliability is low
Gautam et al. [16]	BCD-based coding	In this, the BCD code is used to encode the data and the key is used once. It also used 1's complement and 2's complement for encoding	It gives efficient and performs well for text crypto. Throughput of the system is high	CPU time is more
Abdalla et al. [17]	Playfair algorithm using XOR	It converts image pixels into the matrix and then calculated key value using XOR	It makes it hard for the attacker to perform a frequency analysis based on the used pixel digraphs	Complexity is high

Table 3 Level 3 Security mechanisms for the security of cloud

Considered in	Technique	Description	Merit	Demerit
Chandel et al. [18]	Homomorphic encryption	This encryption determines the complexity of key and space conservation is applied	Less space but the more complex key	Space conservation may cause additional overhead
Sohal and Sharma [2]	Security of cloud storage using BDNA approach	This approach is of encryption based on DNA with high security	Complex key generation is included	Impingement problem is not tackled

store secret keys. The framework manages the keys using access rights of the users. The steps involved in the existing BDNA mechanism is given as under

- A transposition cipher is obtained from 64-bit plain text. This transposition is obtained by changing the position of the bits within the plain text.
- 16 festal rounds are applied on 64-bit keys with distinct functions.
- 32 bits generated from 16 rounds form the first half of the key.
- Substitution and permutations are performed to generate the next half of the key.

- The finalized key is generated by applying the XOR operation.

The second phase consists of AES encryption. This key generation is applied to the output generated from the first phase of the key generation. The steps involved in the encryption is given as under

- Substitute bytes: in this phase, the bits from the transmitted text or plain text is replaced with the key codes.
- Shifting Rows: the rows from the generated ciphertext are shifted randomly to generate a complex matrix.
- Mix Columns: the same shift operation is also applied to columns of the generated matrix.
- Add round keys: the generated matrix is applied with XOR operation to generate finalize key.

The final phase consists of applying the first and second phases except the mixing of columns phase. The generated key is then stored at the server end. The phases are applied to the uploaded file. The problem of impingement is not handled even at the file uploading phase. This problem is tackled within the proposed model. The result obtained from AES, DES, DNA, and BDNA is given in this section.

It is observed that BDNA is best in terms of execution time. The file size is directly proportional to the execution time (Fig. 2).

The key size must be complex and using the BDNA key is more secure as compared to AES, DES, and DNA encryption mechanism. The primary reason for security enhancement is the assembling of multiple encryption mechanisms.

The key size indicates complexity and security. With the BDNA approach key size is significantly enhanced. The plot is given in Fig. 3.

The complexity of the key in the BDNA approach is more as compared to other approaches. The problem of impingement is handled by the use of a hash-based mechanism described in the next section.

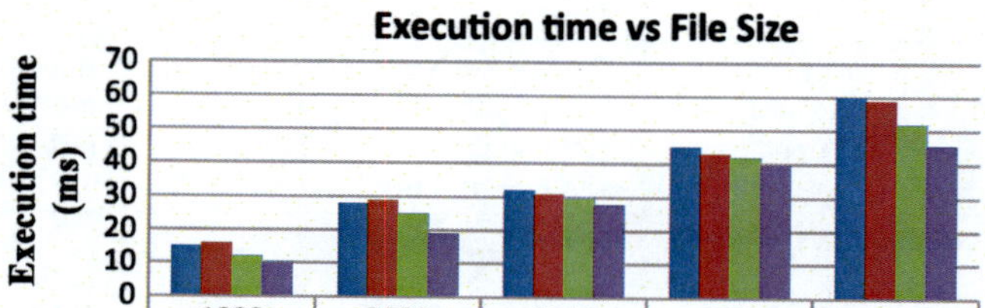

	1000	2000	3000	4000	5000
AES(ms)	15	28	32	45	60
DES(ms)	16	29	31	43	59
DNA(ms)	12	25	30	42	52
BDNA(ms)	10	19	28	39.98	45.63

Fig. 2 Execution time comparison with the file size

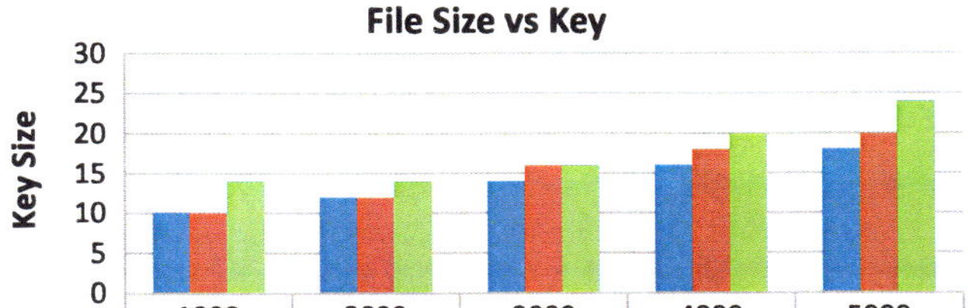

	1000	2000	3000	4000	5000
AES(bits)	10	12	14	16	18
DES(bits)	10	12	16	18	20
DNA(bits)	14	14	16	20	24

Fig. 3 Key Size comparison with the file size

3 Proposed System

The problem of impingement is significant in BDNA encryption along with binary encoding operation. This problem can be rectified by the use of chaining and excess 3 encryption. The overall procedure of the proposed system starts with the initialization of the queue that will store the generated key.

3.1 Handling Impingements

The impingement is generated as the same key is generated for multiple data values. In general, the division method is followed within the BDNA approach for generating a key. The procedure of key generation is described through example 1.

Example 1: Consider a data 2309 to be transmitted towards the destination then the highest prime number of two digits is assumed, i.e., 97 and 2309 is divided by 97. The obtained remainder is 80. But it's possible that remainder generated from other data can also be 80. This led to an impingement.

To tackle the issue, the queue is maintained for storing the key values. Generated keys can be multiple for given data. These keys can be stored at the same location using slot policy. Slot policy means the same location is divided into parts. Each part is capable of storing one key value. The process is elaborated in Fig. 4.

Fig. 4 Representation of Keys using key

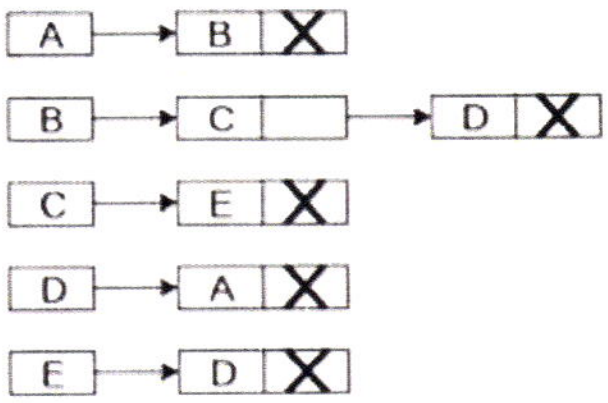

The keys in Fig. 4 are represented with "A", "B", etc. The keys in the same row are similar to a distinct index attached to them. This process resolves the impingement to a great extent.

3.2 Excess 3 Encoding

Binary encoding scheme employed in the BDNA approach is less secure since plane encoding can simply be hacked and security is at stake. To resolve the issue binary encoding scheme is further made secured using excess 3 encodings. The length of the key is increased and hence security is also enhanced. The mechanism used for enhancing security is described using example 2.

Example 2: The example considering the content "A" to be transmitted is, first of all, converted into ASCII code values. "A" is converted to 96. The binary equivalent for the same are 101000 but in excess 3 the same data is represented as 11001001.

This encryption is relatively difficult to hack and hence security is enhanced. The key generation is followed by this step to make it more complex in the BDNA approach.

3.3 Proposed Model

The proposed model converts the generated key with the BDNA approach into a more secured form by the application of an excess 3 encoding scheme. The model also incorporated pre-processing by handling impingement at an early stage and hence reliable key generated in the least execution time. Figure 5 gives an in-depth mechanism of a considered approach.

The proposed model works in phases. The pre-processing mechanism is considered that convert the data file into normalized form by eliminating impingement from generated keys. The procedure that is used for impingement resolution is known as chaining. After the impingement is resolved, the excess three mechanisms are applied to the generated key through the DNA approach. Also, the Binary code mechanism is replaced with excess 3 code in the proposed approach to enhance the security further. The worth of the proposed system is proved in the next section considering different parameters.

Difference between the proposed algorithm and HASH and SHA algorithms

Parameter	HASH	SHA	Proposed Algorithm
Key generation	Folding method Division Mid square	Public key Private Key	Excess 3 Folding
Complexity	High	Comparatively low	Extremely High

(continued)

(continued)

Parameter	HASH	SHA	Proposed Algorithm
Security	Less Secure with collision problem	Security is low due to public key encryption	Highly secured with excess 3 BDNA encryption
Key length	Depends upon data used	128 bits	255 bits

4 Performance Analysis and Results

The proposed system makes the changes in the BDNA approach at two distinct levels. First of all, the key generation is modified by accommodating chaining with the queue. Also, binary codes at the encryption phase are replaced with the excess three mechanisms to ensure more secure encryption as compared to a binary encoding scheme. The result is given in terms of execution time and key length. The size of the file being uploaded is varied and the result is obtained. The result is given in Tables 4 and 5.

The execution time with different file sizes with the proposed approach is minimized. The resulting improvement of nearly 10% is observed that Is significant and plots for the same is given in Fig. 6.

The complexity of the key is also enhanced by the use of excess 3 code and Channing mechanism. Key complexity is increased significantly which is demonstrated in Table 6.

The key length in the case of BDNA is a maximum of 32 bits. The proposed model is based upon a larger key size since multiple data entries are contained within the same index values. This is also demonstrated in Fig. 7.

The results indicate that by accommodating chaining and excess three mechanism security is enhanced and result in terms of execution time is improved.

5 Conclusion and Future Scope

DNA-based encryptions are considered to be most secured that is based upon human DNA structure. The DNA-based encryption although is secured but prone to impingement problems. The problem exists as a division method is employed in key generation. This means distinct data items can yield the same keys that result in impingement. To tackle the issue, the queue is placed at the encryption end. In case the same key is generated that has to be stored at the same index value is placed within the queue. Also, to improve the security, the excess 3 code is accommodated in BDNA rather thana binary coding scheme. This scheme ensures that a more secured key

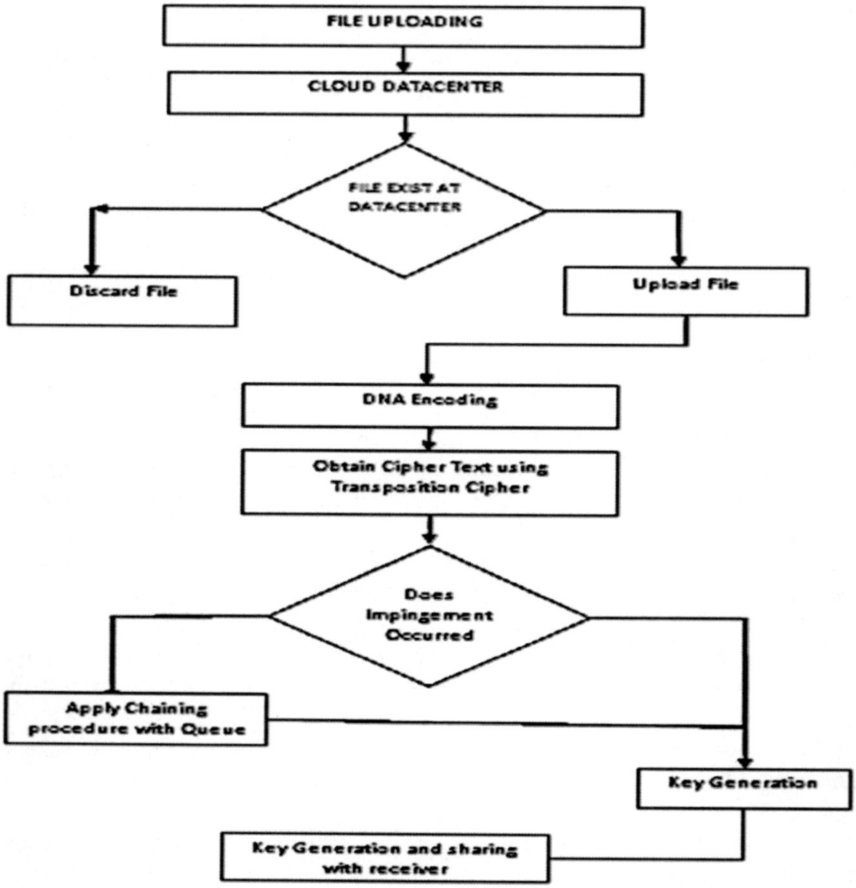

Fig. 5 A proposed model with impingement resolution and excess 3 encoding

is generated that has increased length as compared to a binary encoding scheme. The result due to least impingement is improved in terms of execution time and complexity. The key becomes more complex and execution time is improved by the margin of 10%.

In the future, the security mechanism of BDNA with excess 3 code can be demonstrated within ECG signals. The application is most suitable within the Internet of Things with medical provisioning applications.

Table 4 BDNA problems

Properties	Magnitude and description	Discovered Problem/s
Used Phases	Five • Uploading • File Checking • Encryption • Downloading • Decryption	Large file uploading causes increased propagation delay
Security mechanism used	Binary-based DNA encryption mechanism	Division method in the studied literature is exposed to impingement
Length of formed key	Length of a key has a maximum size of 32 bits	Size or length of the key is small
Consumed time	Generating key consume the least amount of time	The propagation time is high in case of redundancy exist within the file
What's in future?	Phases with key and uploading downloading standards along with impingement detection	–

Table 5 Execution time comparison of BDNA and Proposed approach

File length (Bytes)	BDNA (ms)	Proposed BDNA (ms)
1000	10	9
2000	19	11
3000	28	15.5
4000	39.98	18.98
5000	45.63	21.45

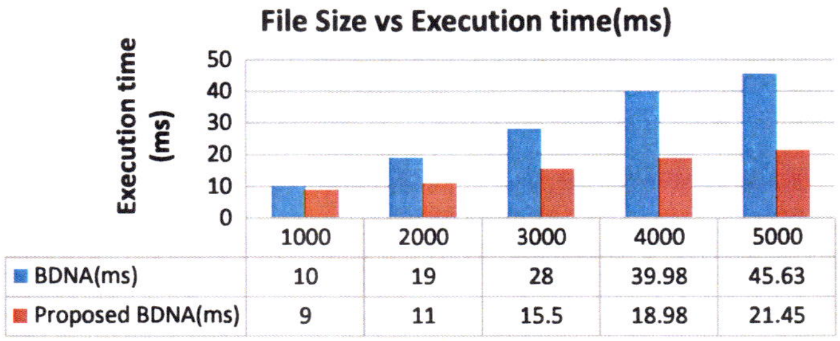

Fig. 6 Execution time comparison with BDNA and the proposed approach

Table 6 Key size with the different file size

File length (Bytes)	BDNA (bits)	Proposed BDNA (bits)
1000	14	64
2000	14	64
3000	32	128
4000	32	128
5000	32	256

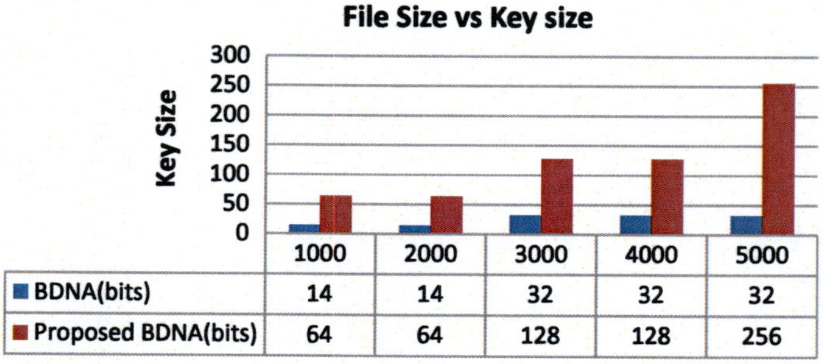

Fig. 7 The key size for BDNA and the proposed approach

References

1. Sander, T., Tschudin, C.F.: Protecting Mobile Agents Against Malicious Hosts, pp. 44–60 (1998). https://doi.org/10.1007/3-540-68671-1_4
2. Sohal, M., Sharma, S.: BDNA-A DNA inspired symmetric key cryptographic technique to secure cloud computing. J. King Saud Univ. Comput. Information Sci. (2018). https://doi.org/10.1016/j.jksuci.2018.09.024
3. NIST: The NIST definition of cloud computing. Observatorio Económico EEUU, no. BBVA Research, pp. 1–8 (2016). https://doi.org/10.6028/NIST.SP.800-145
4. Sridhar, S., Smys, S.: Intelligent security framework for IoT devices. In: International Conference on Inventive Systems and Control (ICISC-2017) Intelligent, 1–5 (2017)
5. El-Yahyaoui, A., Ech-Chrif El Kettani, M.D.: Data privacy in cloud computing. In: 2018 4th International Conference on Computer and Technology Applications, ICCTA 2018, 25–28. Institute of Electrical and Electronics Engineers Inc (2018). https://doi.org/10.1109/CATA.2018.8398650
6. Kchim, S.K.: Data Privacy in Cloud Computing. IEEE Conference Publication (2019)
7. Singh, N.N.: Encryption Schemes of Cloud Computing: A Review. IEEE Conference Publication (2018)
8. Mahalle, V.S., Shahade, A.K.: Enhancing the data security in cloud by implementing hybrid (Rsa & Aes) encryption algorithm. In: 2014 International Conference on Power, Automation and Communication, INPAC 2014, no. I, pp. 146–149 (2014). https://doi.org/10.1109/INPAC.2014.6981152
9. Seo, H., Liu, Z., Großschädl, J., Kim, H.: Efficient arithmetic on ARM-NEON and its application for high-speed RSA implementation. Security Commun. Netw. **9**(18), 5401–11 (2016). https://doi.org/10.1002/sec.1706
10. Security, Computing: Addressing Data Security Challenges in the Cloud The Need for Cloud, No. July (2010)

11. Gupta, S., Jain, S., Agarwal, M.: Ensuring data security in databases using format preserving encryption. In: Proceedings of the 8th International Conference Confluence 2018 on Cloud Computing, Data Science and Engineering, Confluence 2018, pp. 214–218 (2018). https://doi.org/10.1109/CONFLUENCE.2018.8442626
12. Wang, C., Wang, Q., Ren, K., Lou, W.J.: Ensuring data storage security in cloud computing. In: Iwqos: 2009 IEEE 17th International Workshop on Quality of Service (2009). 37-45\n302. https://doi.org/10.1109/IWQoS.2009.5201385
13. Alabdulatif, A., Kumarage, H., Khalil, I., Atiquzzaman, M., Yi, X.: Privacy-preserving cloud-based billing with lightweight homomorphic encryption for sensor-enabled smart grid infrastructure. IEEE ACCESS (2017). https://doi.org/10.1049/iet-wss.2017.0061
14. Miguel, R.: HEDup: secure deduplication with homomorphic encryption. In: 2015 IEEE International Conference on Networking, Architecture and Storage (NAS), pp. 215–223, IEEE (2015). https://doi.org/10.1109/NAS.2015.7255226
15. Zhang, J., Hou, D., Ren, H.: Image encryption algorithm based on dynamic DNA coding and Chen's hyperchaotic system. Math. Problems Eng. (2016). https://doi.org/10.1155/2016/6408741
16. Gautam, S., Mishra, S., Shrivastava, M.: An enhanced encryption technique using BCD and bit complementation. Int. Res. J. Eng. Technol. (IRJET) **4**(1), 896–900 (2017). https://irjet.net/archives/V4/i1/IRJET-V4I1158.pdf
17. Abdalla, F.M., Mohammed, K., Babiker, A.: Modified Playfair Cipher for Encrypting Images **3**(3), 1063–1068 (2018)
18. Chandel, S., Ni, T.Y., Yang, G.: Enterprise Cloud: its growth & security challenges in China. In: Proceedings—5th IEEE International Conference on Cyber Security and Cloud Computing and 4th IEEE International Conference on Edge Computing and Scalable Cloud, CSCloud/EdgeCom 2018, pp. 144–52 (2018). https://doi.org/10.1109/CSCloud/EdgeCom.2018.00034

A Survey of Blockchain Technology Applications and Consensus Algorithm

E. Indhuja and M. Venkatesulu

Abstract Blockchain technology applications are growing fast in today's digital world. Blockchain technology is a decentralized and immutable data format. Blockchain consists of a distributed ledger of information called blocks. However, to use blockchain effectively, it is necessary to consider solving the security and privacy problems. This technology is applied in various fields, for example, finance, healthcare, cybersecurity to name a few. The consensus layer is the most important for the blockchain architecture, in which the consensus protocol is confirmed to how the new block is added to the blockchain network. A consensus algorithm is used to solve mathematical puzzles and increases the overall security for each transaction. Here, a survey is presented on the state-of-the-art of blockchain technology, its applications to different fields and classification of the consensus algorithms.

Keywords Blockchain technology · Healthcare · Consensus algorithm · Security · Privacy

1 Introduction

Blockchain technology is a decentralized and immutable data store in the network. This network creates a block to store the information. It can be created to store the patient's health-related information. Once the block is created the information can never be altered. These applications used in healthcare is mainly patient-centric [1], attribute-based security algorithms to avoid external attackers [2]. Each block contains related information. What is Blockchain? Logically, it is a chain of blocks which contain specific information (database), but securely and genuinely grouped

---

E. Indhuja (✉) · M. Venkatesulu
Department of Information Technology, Kalasalingam Academy of Research and Education, Anandnagar, Krishnankoil, Tamilnadu 626126, India
e-mail: indhjait16@gmail.com

M. Venkatesulu
e-mail: m.venkatesulu@klu.ac.in

P. Karuppusamy et al. (eds.), *Sustainable Communication Networks and Application*, Lecture Notes on Data Engineering and Communications Technologies 55,
https://doi.org/10.1007/978-981-15-8677-4_15

in a network (peer-to-peer). It is a collection of nodes chained with each other and that the whole network operates in a decentralized manner.

The medical data sharing using blockchain provides secure transfer by employing encryption and decryption algorithms. Different types of models are used in health records to improve privacy and access control [3]. In traditional methods, hospitals maintain the medical records manually and collect the patient health report [4]. This type of method can only store a limited amount of information and cannot store a large amount of medical data [5]. Verifying the integrity of healthcare records while storing and transferring, use permissioned blockchain model. The security problems are solved using cryptographic algorithms and the EHR transactions are verified. Each and every block in the network contains a hash value. Previous block hash value is taken in the current block so the transactions are more secure [6].

The decentralized approach support managing a very large amount of data and also provide the audit, interoperability and accessibility abilities. The miners need to unravel the protection puzzles for extracting the data stored within the electronic health care record [EHR]. It is also a decentralized append-only ledger that's managed by a peer-to-peer network of nodes. Instead of being managed by one trusted entity, the nodes during a blockchain network use consensus algorithms to independently validate transactions and add them to the ledger. Once the information stored, it is not easy to change the content and hash value. These mechanisms are very useful for healthcare data systems to automate data-access policies and log instances to exchange within the blockchain for participants for later view and verification.

2 Overview of Blockchain Technology

A blockchain is a decentralized network, in the network everyone can easily access the information. This network contains a sequence of blocks, each block store the transaction information for every transaction process. A blockchain, in a simple term, a time-stamped series of immutable records of information and each block maintains particular records and hash value. Each transaction on a block is secured with a digital signature that proved its authenticity. This technology allows all the network participants to succeed in an agreement, commonly referred to as consensus. All the information stored on a blockchain is recorded digitally and features a common history that is out there for all the network participants. This way, the probabilities of any fraudulent activity or duplication of transactions are eliminated without the necessity of a third-party.

Each blockchain contains n number of blocks to be added to the network. A block contains its own distributed ledger. The ledger maintains its copies of the record for every transaction. Ledger stores the data in a block and the nonce value is created by each block for every transaction. The nonce value randomly generates the block and the hash value helps encrypt the data and ensures security during the transaction [7]. Figure 1 shows that the blockchain Architecture. A block of data is linked to the previous block hash value and thus provides security to the information processed.

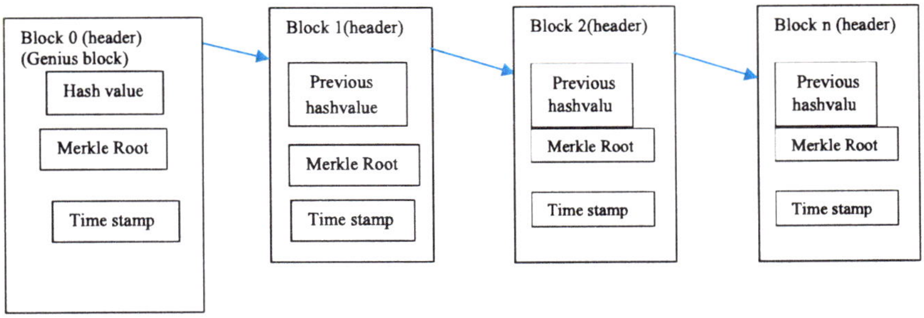

Fig. 1 Blockchain architecture

The main advantage of using the hash value is that it not easy to understand by humans and input is through functions containing numbers, letters, and outputs a fixed-length file.

$$\text{Hash (block)} \leftarrow \text{Hash (H1} + \text{H2} + \text{H3} + \ldots + \text{Hn)} \quad (1)$$

$$y = h(x) = [10x \bmod 1] \quad (2)$$

Where H1, H2, H3, Hn are the hash values of the previous blocks in the blockchain network, and also h(x) is the overall hash value of the block. The hash value is of specific length and uses different functions and calculate the values using hashing algorithms. The Hash value is used to maintain the integrity of the data and each hash value is unique. That is, two data sets do not have the same hash value (Table 1).

Blockchain technology uses a hashing algorithm in each block and takes into the input set, the previous block hash value. The cryptographic hash is a digital signature and digest of some amount of data in the block. SHA (secure hashing algorithm), SHA 2, SHA 256, MD 5.SHA 256 algorithms are used in the blockchain (Fig. 2). It generates hash values of 256 bits length, every time the algorithm is used and blockchain also uses the Merkle root tree [8]. A secure hash function is one of the

Table 1 Types of blockchain

Public blockchain	Private blockchain	Federated blockchain
Distributed network to access the blocks	Centralized network to access the blocks	Decentralized network to access the blocks
All participants to read and write and audit the blocks	Selected participants to read and audit the blocks	Selected participants to read and audit the blocks
Lower throughput	Higher throughput	Higher throughput
Low Transparency	High Transparency	High Transparency
Example: Bitcoin, Factom, Etherum, Blockstream	Example: Multichain, Blockstack,	Example: R3, Hyperledger

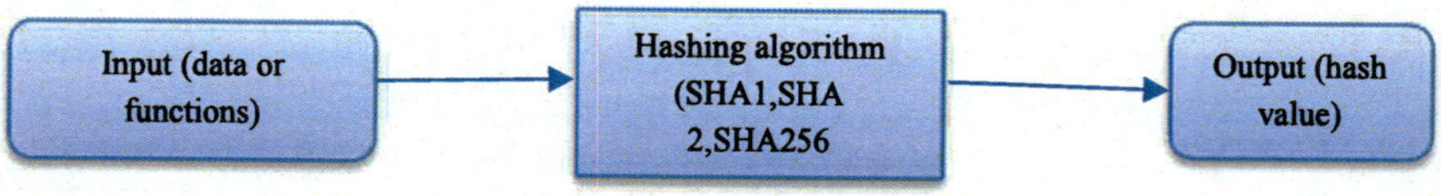

Fig. 2 Hash value generation

types of cryptographic functions issued by the national institute of standards and technology. The SHA 256 algorithm is based on the SHA 2 algorithm, this algorithm created a 256-bit message digest. The algorithm includes the two processes first one is message preprocessing and another process is the main loop.

$$\text{Merkle root value (MRV)} = \text{hash A} + \text{hash B} \tag{3}$$

$$\text{Hash A} = \text{Hash (HA} + \text{HB)} \tag{4}$$

$$\text{Hash B} = \text{Hash (HC} + \text{HD)} \tag{5}$$

Merkle root tree is a data structure that allows a large amount of data to be securely shared. Merkle tree is used to determine all transaction process and the user can easily identify the process. Figure 3 shows that the Merkle tree structure and its working process. Hashing a pair of nodes is to create the tree structure, for example, HA, HB nodes create a tree and also find the Merkle root value (MRV). Each node has its hash value for the transaction process. The Merkle tree groups all of the data inputs into pairs. Each leaf node has a hash of transactional data and the non-leaf node has a previous hash value. Merkle trees only require an even number of leaf nodes. If the transaction is odd, the last hash value will be duplicated once again to create an even number of leaf nodes in the tree structure. Merkle tree reduces the data size and

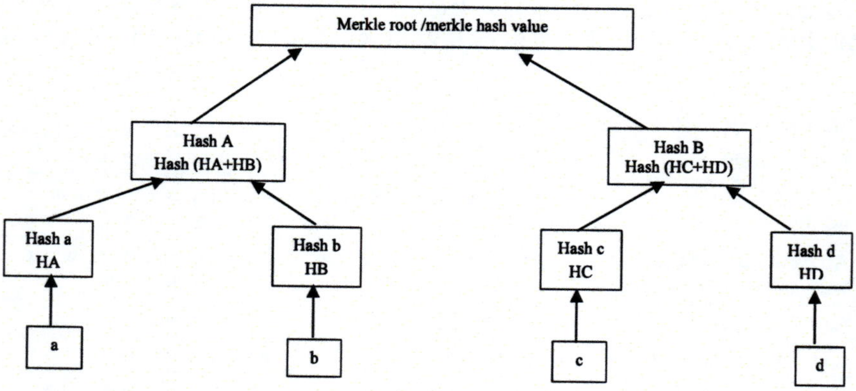

Fig. 3 Merkle tree structure

verifies the transactional validity to ensure the integrity of the data and also reduce the memory requirement. Merkle tree summarizes all the data transactions and stores in the block header. The computations should be fast enough for the transactional process.

3 Types of Blockchain

The blockchain contains different types of blocks in the network. The transaction data is shared between individuals, organizations, and industries using these types of blocks. The blocks store and manage the data. The user performs the different activities depending upon the following types of blocks

A. Public blockchain (permissionless blockchain)

B. Private blockchain (permissioned blockchain)

C. Federated blockchain (public permissioned)

3.1 Permissionless Blockchain

The public blockchain network is easily accessible to everyone, anyone can read and write the information and it is a decentralized network type. The user can publically access the block and easily join the network without any authentication, a single person cannot control the overall network and everyone acts as the authority of the blockchain, equally, divide the roles and responsibilities in the public chain network. Permissionless chain is used in the bitcoin and etherum.

3.2 Permission Blockchain

A private blockchain allows access for only authorized persons in the centralized network. An organization or industry uses this type of network. It is a centrally controlled 1 network and provides authentication, identification, and verification process for the users. The data transaction is very confidential and securely transfer the data from one organization to another organization. The blockchain environment uses the hyper ledger fabric tool (Table 2).

Table 2 Comparison of consensus algorithms

Consensus algorithm	Types of blockchain	Introduced the year	Power of tolerated	Scalability	Double spending	Attacks	Examples
POW	Public	2009	<25% computing the power	Strong	Yes	Yes	Bitcoin, etherum, smart contracts
POS	Public	2013	<51%	Strong	No	No	Next coin
DPOS	Public	2016	<51%	Strong	No	No	Bit shares, cryptocurrency
LPOS	Public	2016	<51%	Strong	No	No	Bitcoin
POET	Private	2018	<51%		No	No	Hyper ledger sawtooth
PBFT	Private	2015	<33%	Weak	No	No	Smart contract
SBFT	Private	2014	<33%	Weak	No	No	Smart contract, cryptocurrency
DBFT	Private	2016	<33%	Weak	No	No	Smart contract, cryptocurrency
POA	Public	2016	<50%	Strong	No	No	Bitcoin, cryptocurrency
POI	Consortium	2015	<50%	Strong	No	No	Internet of things application
POC	Public	2014	<30%	Strong	No	No	Bitcoin, Etherum

3.3 Federated Blockchain

A public permissioned chain is a combination of a public and private chain. This type of blockchain is formed by a group of individuals sharing the information. In the Consortium blocks, some nodes act as the private blocks allow only authorized users to access the data and some nodes act as the public blocks allow to all the users without any restrictions, finally form the federated blockchain network [9]. Federated blockchain is a decentralized network. The Federated blockchain is used in the R3, EWF.

4 Types of Consensus Algorithm

A consensus algorithm is a decision-making process where a group or individual can construct the support for the decision, the maximum number of nodes construct the same decision to approve the actions. The common agreement of the participants constitutes the current state block of the ledger. This algorithm is more reliable and ensures trust for every participating node and unknown nodes [10]. The consensus algorithm gives equal rights to all the participants.

Concepts used in Blockchain Technology:

Proof of Work (POW)
Proof of stake (POS)
Delegated Proof of stake (DPOS)
Leased proof of stake (LPOS)
Proof of elapsed time (POET)
Practical Byzantine Fault Tolerance (PBFT)
Simplified Byzantine Fault Tolerance (SBFT)
Delegated Byzantine Fault Tolerance (DBFT)
Proof of activity (POA)
Proof of importance (POI)
Proof of capacity (POC).

4.1 Proof of Work (POW)

The proof of work is the first consensus algorithm used in the blockchain network. The algorithm is used to select the miner to generate the next sequence of blocks in the network. In the transaction between two blocks, the algorithm selects the miner block which solves the complex mathematical puzzle first and acts as the miner block and gets the reward for the solution [11]. The node which solved the mathematical problem in the chain network is allowed to add the new block within the 10 min for each transaction. It eliminates attacks and ensures security for the transactions. The miner needs to communicate with the other nodes and this process is done in a more centralized manner and time-consuming [10].

4.2 Proof of Stake (POS)

The proof of stake is the alternative process of the POW. In the POW consensus algorithm, every node solves the complex puzzles for the mining process whichever node first finds the solution acts as the miner, and every node gets validate by adding the new block in the chain network. In the POS algorithm, the miners are allowed to solve the problems based on the coin of stake. The node mining process is entirely random. Nodes having more number of coins in their account in the network are qualified for being a miner. After the process is completed, all nodes verify the voting system for choosing the validators. The POS algorithm saves energy consumption. The community bond of stakeholders is not strongly connected and it is a decentralized process The POS algorithm reduces up to 51% of attacks and is energy efficient.

4.3 Delegated Proof of Stake (DPOS)

The delegated proof of stake is a variation of proof of stake. The blocks of witness or validators depend upon the vote of stakeholders. The transaction process is faster and efficient, decentralized using the DPOS algorithm. The miner/validators of the nodes are called the delegated nodes. It is the traditional contract between the witness and the validators. The witnesses generate the block of information for every transaction. Sometimes the witness does not create a new block within a particular time, and the stakeholder can vote for generating the next block. The witness can access the overall block, and qualify for transaction fee. The DPOS algorithm permits transparency in the database.

4.4 Leased Proof of Stake (LPOS)

The leased proof of stake is a type of proof of stake. LPOS algorithm introduces the wave's platform for the blockchain network. The wave platform ensures lower power consumption compared to the others. This process allows the blocks to create a centralized community in a decentralized manner [12]. The smallholders also involve the blocks and change the staking for the transaction. The user creates the token and uses it for better security.

4.5 Proof of Elapsed Time (POET)

The POET algorithm is one of the best consensus algorithms. This algorithm uses the permissioned blockchain network. The users can access any block to get permission and after that, the transaction process can be continued. The permission network decides on the mining rights and voting system. This algorithm consumes more power. The participant who completes all the mining process creates a new block in the network. POET algorithm mainly focuses on power consumption and data transparency.

4.6 Practical Byzantine Fault Tolerance (PBFT)

The practical Byzantine Fault Tolerance (PBFT) algorithm is designed to work efficiently to use the asynchronous process. This algorithm is used in the distributed network to reach the consensus where some of the nodes in the network fail to respond to the incorrect information. The main purpose of this algorithm is to ensure the safety of the failure nodes and reduce the faulty nodes [13]. All the nodes in the network are arranged in some specific order. One node is selected as the primary node and other nodes work in the backup plan inside the network to communicate with others. This algorithm does not provide any reward for the miners.

4.7 Simplified Byzantine Fault Tolerance (SBFT)

In this algorithm, the first block generally collects all the transaction details and validating. The batching process is completed for the transaction after the new block is created in the network. Here, a single validator handles the bundle of transactions and follow certain rules for validating the transaction [14]. After validating and adds their digital signature for security purposes. The algorithm identifies the faulty block keys and is immediately rejected.

4.8 Delegated Byzantine Fault Tolerance (DBFT)

The DBFT (Delegated Byzantine Fault Tolerant) algorithm is predicated on the PBFT (Practical Byzantine Fault Tolerance) algorithm. PBFT algorithm can solve distributed network consensus effectively, but the more nodes join consensus, quickly the performance drops, because of the time complexity. For this reason, NEO (New Economy Movement) proposed an algorithm named dBFT, which mixes the characteristics of dPoS. By voting on the blockchain, it decides the name list of consensus nodes for the next round, namely authorizing a couple of nodes to succeed in consensus and make a new block, the opposite nodes will act as ordinary nodes to receive and verify blocks. DBFT algorithm is used to achieve faster transactions compared to other algorithms. Prevents malicious attacks during transactions. This algorithm first confirms 100% of all transactions and allows creating a new block within the 15–20 s. The network supports large-scale commercial applications and the throughput is higher compared to others. It does not require high energy.

4.9 Proof of Activity (POA)

The proof of activity consensus algorithm is a combination of POW and POS algorithms. The transaction between blocks is allowed in the network for mining to solve complex mathematical problems. The miners have two things in the blocks; the first one is the stored header information and the other one is a reward for the process [15]. It is more secure to transfer the data and requires less storage space. The attacker cannot easily attack the blocks because a large amount of transactions is required to add the block in-network and cannot change any information without authorization.

4.10 Proof of Importance (POI)

Proof of importance consensus algorithm was first introduced by NEM (New Economy Movement).POI algorithm determines which network nodes are eligible to add the blocks in the network. The transaction process is known as the harvesting process. In the transaction exchange for harvesting, a block and nodes can collect the transaction fees within that block. Accounts with a better importance score will have a better probability of being chosen to reap a block. The NEM protocol requires that an account holds a minimum amount to be eligible for harvesting [16]. This algorithm uses the transaction graph and the priority is assigned to every node for the transaction.

4.11 Proof of Capacity (POC)

The proof of capacity consensus algorithm is based on the proof of work. The transaction between the nodes to create a new block uses two steps, namely mining and plotting the blocks in the network. The plotting process happens on the hard drive disk and blocks create the nonce values. Mostly the nonce value uses the proof of work and it is difficult to generate the hash values, miners first solve the nonce value for mathematical puzzles after creating the new blocks for the transaction process. The process needs to be completed by the miners to get rewarded.

5 Applications of Blockchain Technology

Blockchain technology is used in different domains mostly in the financial-related areas. Mainly this type of network used a distributed ledger and securely maintain all the information used to hash and cryptography algorithms. Some of the applications of the blockchain technology are listed out below. Figure 4 represents the various application domains in blockchain technology.

5.1 Healthcare

Blockchain technology is used in the healthcare industry. Personal health records are encrypted and stored in this network. The authorized user can access the medical record using the private key. The healthcare chain network maintains the patient's medical records and clinical test results, lab reports stored in the network. The decentralized technology is used to create a network of patient's data and doctors can access the patient's health data faster. This technology is used in supply chain management,

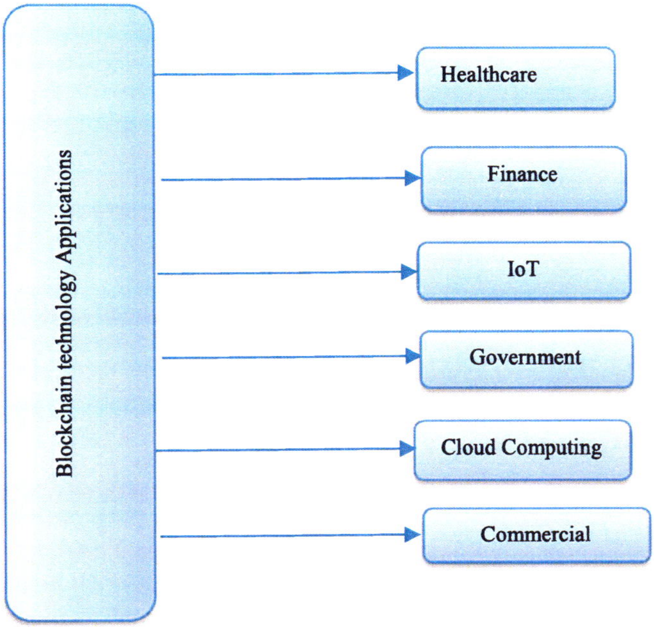

Fig. 4 Blockchain applications

especially in the paramedical industries. It is guaranteed full transparency in the process. The ledger maintains the overall drug supply process from the start to the end until the drug reaches the consumer. This technology is used to store securely the medical data information in the database in an encrypted format.

5.2 *Finance*

Financial sectors used blockchain technology for safe and secure transactions. The digital transaction used the cryptocurrency Bitcoin and Etherum. The distributed ledger stored the own copy of the records all transaction details [17]. This network is immutable and highly secure as the data is encrypted using encryption algorithms so the hackers cannot easily change any records. The only authorized user can access the data using the digital id. Many banks use this network and provide a digital identity for the user and can easily identify the fraud, secure transactions. This technology usage reduced transaction costs and ensure security during data transfer. The transaction process is fast, most accurate and inexpensive. This technology increased the trust in the public [18].

5.3 Internet of Things (IoT)

The Internet of things is the connection of smart devices. Most of the smart applications are based on centralized system connections. The centralized system maintains all the devices and some of the drawbacks are a single point of failure, trust management, security issues, and so on [19]. Blockchain technology is a decentralized network and trusts evolved though consensus protocol used ledger in the network. This technology enables smart devices in IoT to reduce trust problems and improved efficiency [20]. Use of digital ledger for the process of peer to peer transactions used the IoT can easily find the attacks and encrypted the data for secure transaction. This technology provides scalability and security for smart devices [21].

5.4 Supply Chain Management

Supply chain management is the management of the following of the product and all services process to reach the final product of the consumer. It is growing rapidly, involving all the stakeholders, and tracks the products in the chain. This technology was emerging as a strong, detangling all information and transactions, communication, exchange of the product in the supply chain system. This technology is used in the supply chain for procurement, provenance and traceability, digital payments, contract logistics, manufacturing. Blockchain provides several facilities for companies to transport goods. It can be used to queue up the events in the supply chain, allocating the correct place for different shipping containers.

5.5 Government

The government sector uses the blockchain technology to securely manage all confidential data and transactions. This technology is used at the national, state, and local government levels for maintaining citizen's information with improved security and privacy [22]. In the current voting system for the government, each vote is attributed to one id and also has the possibility of creating s fake id. Blockchain usage ensures more security in the voting system so that the attackers do not change any information in the voting systems and does not create the fake id [23]. Record management, identity management, voting, taxes, are some of the used in blockchain technology.

5.6 *Cloud Computing*

Cloud computing uses the traditional database to store information. A combination of blockchain technology and cloud computing technology creates many opportunities for efficient data storage and sharing. The decentralized ledger maintains the transactions and makes its copies of the data stored in the cloud server [24]. Point to point encryption is done to securely transfer the data. A Blockchain network enables a secure way of transmitting the data from the cloud server. This technology adds the Proof of Past Function (POP) and works with a delay function. POP provides the cloud computing network with a timestamp and ensures the safety of the data [25].

5.7 *Commercial*

Blockchain Technology is also used in the media and entertainment fields. The commercial applications of the blockchain network create specific policies and smart contracts. The media digital rights management can be built on the blockchain technology. The distributed ledger provides traceability, visibility, transparency to the Media and Entertainment fields. The media and entertainment industry is mostly a contract-based business and this technology provides the smart contract. Smart contracts provide accurate, faster, and cost-effective transactions for contract management in the media industry. The blockchain technology used in the media industry eliminates fraud, reduce cost, and time.

6 Conclusion

Blockchain Technology is a decentralized and distributed ledger, maintains all transactions, securely transfer the information. It is used in various applications to execute transactions in a trustful environment and eliminates the single point of failure. In this survey, the blockchain architecture and working principles of the network are explained. The classification of consensus algorithms, their use, and applicability are also presented. Finally, the paper explained the major applications of blockchain technology. In future work, the performance of the consensus algorithms is intended to verify and also the security aspects.

Acknowledgements The first author thanks the management of Kalasalingam Academy of Research and Education for providing scholarship for carrying out this research work.

References

1. Nakamoto, S.: Bitcoin: A Peer-to-Peer Electronic Cash System (Online) (2008). Available: https://bitcoin.org/bitcoin.pdf
2. Haber, S., Stornetta, W.S.: How to time-stamp a digital document. J. Cryptol. **3**(2), 99–111 (1991)
3. Hardin, T., Kotz, D.: Blockchain in health data systems: a survey. In: 2019 Sixth International Conference on Internet of Things: Systems, Management, and Security (IOTSMS)
4. H¨olbl, M., Kompara, M., Kamiˇsali´c, A., Zlatolas, L.N.: A systematic review of the use of blockchain in Healthcare. Symmetry **10**(10) (2018). https://doi.org/10.3390/sym10100470
5. Moser, M., Bohme, R., Breuker, D.: An inquiry into money laundering tools in the Bitcoin ecosystem. In: Proceedings of the eCrime Researchers Summit (eCRS), IEEE, pp. 1–14 (2013)
6. Dev, J.A.: Bitcoin mining acceleration and performance quantification. In: Proceedings of the 2014 IEEE 27th Canadian Conference on Electrical and Computer Engineering (CCECE), IEEE, pp. 1–6 (2014)
7. Conti, M., Kumar, E., Kumar, S., Lal, C., Ruj, S.: A survey on security and privacy issues of bitcoin. IEEE Commun. Surv. Tutorials **20**, 3416–3452 (2018)
8. Dorri, A., Steger, M., Kanhere, S.S., Jurdak, R.: BlockChain: a distributed solution to automotive security and privacy. IEEE Commun. Mag. **55**, 119–125 (2017)
9. Li, Z., Kang, J., Ye, D., Deng, Q., Zhang, Y.: Consortium blockchain for secure energy trading in industrial Internet of Things. IEEE Trans. Industrial Informatics **14**, 3690–3700 (2018)
10. Ongaro, D., Ousterhout, J.K.: In search of an understandable consensus algorithm. In: Proceedings of 2014 USENIX Annual Technical Conference, Philadelphia, PA, pp. 305–319 (2014)
11. Bentov, Lee, C., Mizrahi, A., Rosenfeld, M.: Proof of activity: extending bitcoin's proof of work via proof of stake. In: ACM SIGMETRICS Perform. Eval. Rev., vol. 42, no. 3, pp. 34–37 (2014)
12. Yli-Huumo, J., Ko, D., Choi, S., Park, S., Smolander, K.: Where is current research on blockchain technology?—A systematic review. PLoS One **11**(10) (2016)
13. Castro, M., Liskov, B.: Practical byzantine fault tolerance. In: Proceedings of the Third Symposium on Operating Systems Design and Implementation, no. February, USENIX Association, pp. 173–186 (1999)
14. Nguyen, G.-T., Kim, K.: A survey about consensus algorithms used in blockchain. J. Information Process. Signal 7, February 2018
15. Bentov, Lee, C., Mizrahi, A., Rosenfeld, M.: Proof of activity: extending bitcoin's proof of work via proof of stake. ACM SIGMETRICS Perform. Eval. Rev., vol. 42, no. 3, pp. 34–37 (2014)
16. Nguyen, G.T., Kim, K.: A survey about consensus algorithms used in blockchain. J. Inf. Process. Syst. **14**(1), 101–128 (2018)
17. Mukhopadhyay, U., Skjellum, A., Hambolu, O., Oakley, J., Yu, L., Brooks, R.: A brief survey of cryptocurrency systems. In: Proceedings of the 2016 14th Annual Conference on Privacy, Security and Trust (PST), IEEE, pp. 745–752 (2016)
18. Laskowski, M., Kim, H.M.: Rapid prototyping of a text mining application for cryptocurrency market intelligence. In: Proceedings of the 2016 IEEE 17th International Conference on Information Reuse and Integration (IRI), IEEE, pp. 448–453 (2016)
19. Christidis, K., Devetsikiotis, M.: Blockchains and smart contracts for the internet of things. IEEE Access **4**, 2292–2303 (2016)
20. Zeadally, S., Das, A.K., Sklavos, N.: Cryptographic technologies and protocol standards for Internet Things. Internet Things (2019). https://doi.org/10.1016/j.iot.2019.100075
21. Devibala, A.: A survey on security issues in IoT for blockchain Healthcare. In: 2019 IEEE International Conference on Electrical, Computer, and Communication Technologies (ICECCT)

22. Hou, H.: The application of blockchain technology in e-government in china. In: Proceedings of the 2017 26th International Conference on Computer Communication and Networks (ICCCN), IEEE, pp. 1–4 (2017)
23. Sullivan, C., Burger, E.: E-residency and blockchain, computer law & security review. Int. J. Technol. Law Pract. (2017). https://doi.org/10.1016/j.clsr.2017.03.016
24. Esposito, C., De Santis, A., Tortora, G., Chang, H., Choo, K.-K.R.: Blockchain: a panacea for healthcare cloud-based data security and privacy. IEEE Cloud Comput. **5**, 31–37 (2018):
25. Sharma, P.K., Chen, M., Park, J.H.: A software-defined fog node based distributed blockchain cloud architecture for IoT. In: IEEE Access, vol. 6, pp. 115–124 (2018)

FPGA Implementation of Turbo Product Codes for Error Correction

M. G. Greeshma and Senthil Murugan

Abstract This paper aims to design and implement the very efficient turbo product codes decoding technique for errorless signal reception. The turbo product codes (TPC) are finding a wide range of applications in the communication field where ever highest data rate is required is selected for the design. The message is encoded to form the TPC such that a $k \times k$ message will be converted into an $n \times n$ TPC. The signal is transmitted through the AWGN channel. Iterative Chase-Pyndiah type 2 decoding algorithm is followed in the decoding section. This decoding technique can decode the transmitted codeword with very much less probability of errors as well as have a very low decoding complexity and is also efficient in achieving the highest data rate. The (n, k, d) hamming codes are selected for the design, because of its properties like the highest code rate, ability to correct the single-bit error and identify multiple bit errors. The hardware is implemented in the Xilinx FPGA platform.

Keywords Turbo product codes · Chase decoding · Error correction · Bit error rate · İterative decoding · SISIO decoding

1 Introduction

Error detection and error correction play an important role in communication, especially where ever high code rate data transmission is required. Whenever a signal is passed through a channel, the noise will be added to the transmitted signal and the resulting signal at the receiver side will be noisy. Any change in the received signal from that of the transmitted one can be noted as a noise. This includes changes in the position of the bits, missing data bits, or even partial or complete loss of data. To overcome this, usually, encoding will be done at the transmitting end and similar

M. G. Greeshma (✉) · S. Murugan
Department of Electronics and Communication Engineering, Amrita Vishwa Vidyapeetham, Amritapuri, India
e-mail: greeshmamg03@gmail.com

S. Murugan
e-mail: senthilmurugan@am.amrita.edu

P. Karuppusamy et al. (eds.), *Sustainable Communication Networks and Application*, Lecture Notes on Data Engineering and Communications Technologies 55,
https://doi.org/10.1007/978-981-15-8677-4_16

decoding will be performed at the receiver end. Encoding means the addition of some parity bits to the data bits and the decoding of the same should be performed. It's possible that the bits may or may not be decoded correctly. But for an efficient data transmission scheme, the correct decoding of data is very necessary. There are many kinds of codes namely the linear codes, block codes, convolutional codes, cyclic codes, etc. The common types of channel codes are block codes and convolution codes. These two codes have some variations. Block codes accept *k* number of data bits and thus render n number of coded bits. This is done by adding an *n-k* number of redundant bits or simply the parity bits to the k number of data bits. Generally, they can be represented as *(n, k, d)* block codes, where n is the codeword length, k is the number of data bits, and d is the minimum hamming distance. Turbo product codes (TPCs) are formed by strong linear block codes which consist of many simple linear block codes. Like Turbo Codes they do not depend on an interleaver, however, they classify under serially concatenated codes and therefore have no issue with the error floor problem of flattening the curve. It is possible to use a SISO decoder to produce the outputs of the soft decision decoder. The maximum a posteriori probability (MAP) algorithm, Chase-Pyndiah iterative decoder, the MAP algorithm based on symbols and the Soft Output Viterbi algorithm are some of the soft decoding algorithms. The modulation technique used here is BPSK modulation and the modulated signal is transmitted over the AWGN channel. For soft decision decoding, the iterative Chase-Pyndiah decoding algorithm has been simulated to lower the bit error rate.

2 Literature Survey

In this section, the various research articles and papers that formed the foundation of references are briefly discussed. Turbo Product Codes and the various specialties associated with it while using it for the error detection and correction purpose is very important in the field of communication.

Turbo product codes (TPC) finds its wide range of applications in the communication field wherever there is required highest code rate data transmission, code words of larger length, and good performance for error-free data transmission [1]. The chase decoding algorithm of the Turbo Product codes presented, in this paper, which selects some less reliable positions and these positions create the corresponding test sequences. Mainly two different methods for reducing the Chase-Pyndiah decoding algorithm are proposed in this paper. The first scheme aims at reducing the number of the least reliable selected positions by eliminating those having comparatively negligible probabilities of error. On the other hand, the second method aims at reducing the number of least reliable positions by eliminating unnecessary positions algebraically. The analysis of the bit error rate (BER) output of different 2-D Turbo Product Codes is carried out in this study [2]. The TPC decoder is designed for the hard input hard output data which is impaired by the AWGN and multipath fading effects. The BER performance of the TPC is investigated for both the sequential and non-sequential decoding techniques. The calculation indicates that the decoding of the non-sequential code in the Raleigh fading stream can offer almost double coding benefits compared to the AWGN streams, and the gain can also be interpreted as being inversely proportional to the code rate and word length [3]. This research paper focuses on the encoding and decoding of the Turbo Product Codes based on the soft input and soft output scheme. For a BPSK modulated signal when passed over an AWGN channel the TPC will perform faster calculations, simple decoding, and attains a faster convergence. The observation is that on an increasing number of iterations the performance of the Turbo Product Codes is increasing significantly. The same operations when repeated with gradient decoding algorithm show a greater compromise between the complexity of chase decoding and performance. The main focus of this research is to perform the iterative turbo decoding of the Turbo Product Code based on its single parity check component (SPC) [4]. For the Turbo Product Code formed with hamming code, the SPC decoder performs almost similar to the duel code MAP algorithm and performs better than the typical block Turbo decoder n the Raleigh fading channel [5]. This paper aims to introduce a decoder that has zero degradation of performance and can reduce the complexity. The reduced implementation complexity of the BCH codes has led to the selection of the same. The weightage and reliability factors are prefixed without any normalization. The decoder proposed in this research can scale with test pattern parameter say P which is a function of the iteration number. There is a wide range of standards are available for TPC shortening [6]. The research aims to study the various aspects available for the purpose. For this purpose, the component codes chosen are shortened hamming codes. The selection is based on the IEEE 802.16 standard to obtain distinct block sizes for the encoded codes. To model the shortened TPC, the undetermined error probability of component codes error is estimated. The notable advantages of this work are the reliability factors are not at all needed as in Pyndiah's paper, so the decoding complexity is surprisingly reduced. This is possible by avoiding the normalization operations of the complete codes for each of the iterations.

The work is carried out to study the reliability and ambiguity of the LDPC codes with the TPC [7]. TPC is a short length code of length within 2048 bits whereas LDPC is semi-random type. Study proves that the LDPC has a slightly better performance compared to TPC but the disadvantage is that number of iterations required for observing such a difference. 2D TPC formed from hamming codes performance are evaluated in this study [8]. The study concludes that at higher SNR regions TPC shows outstanding performance than any other codes under the same conditions of parameters, test patterns, and decoding algorithms. The idea is to reduce the number of test patterns considered for the Chase decoding algorithms [9]. Taking into account the relationship between syndromes and number of errors, TPCs are classified into distinct conditions such that only the TPC with a lower number of errors is decoded. The simulations indicate that around 50% of the TPC is not necessarily decoded without affecting the performance. To design a hybrid decoder, adopting SISO and HIHO decoding techniques to decode the row and column vectors in each iteration of a TPC [10]. This will be done based on the syndromes for rows and columns, respectively. Simulations indicate that decoding complexity is surprisingly reduced and the BER performance is also improved slightly. The main idea of this research is to study the reason for selecting the Turbo Codes especially their limits on the channel capacity [11]. Concatenated coding and iterative decoding are two fundamental ideas focused on this research. This can be mainly interpreted as the turbo principle. The prime focus of encoding is to the addition of some parity bits to the message bits and then transmits them through the channel for efficient data transmission and reception [12]. The same kinds of operations are followed while recovering or decoding the signal at the receiver. Communication channels have different variety of errors that will get added with the information passed through the channel [13]. This may create some variations in the information bits received or may even result in the complete loss of data. For efficient and reliable data transmission the information transmitted through the channel should be always less than or equal to the channel capacity, thus satisfying Shannon's channel capacity theorem [14]. In the script, Hamming codes can detect two-bit errors in the codeword and can correct a one-bit error without detecting the uncorrected error bit [15]. The requirement of optimized performance and energy-efficient error correction in wireless communication systems leads to an iterative error correction turbo code [16]. The iterative decoding scheme results in computational complexity, delay in decoding and high power consumption also according to the channel conditions the number of iterations required for decoding also varies. Therefore an early termination of the iteration at specific times is very necessary to limit the computational complexity without any degradation in performance. To achieve the near-optimum performance for the block turbo codes on high noise levels, two novel performance-enhancing decoders can be used [17]. One of the decoders which are based on PSO (particle swan optimization) can provide very much less decoding latency for block turbo codes having higher block lengths. On the other hand, the decoder based on SVM (support vector machine) can adapt to the varying channel properties. This specialty makes the decoder efficient for designing application-specific decoders.

3 Turbo Product Codes Encoding

Claude Berrou first developed the Turbo Coding in 1993. But the method has a disadvantage of error floor problem. Scientists have then introduced the Turbo Product Codes to overcome the limitation. Serial/parallel concatenation of two or more constituent codes will form the TPC. The constituent codes may be either Hamming codes or BCH codes etc. The basis of forming the TPC is,

(a) Using Block Codes instead of systematic/non-systematic codes.
(b) SDD (Soft decision decoding) is used rather than HDD (hard decision decoding).
(c) Combing shorter block codes develop long codes of low decoding complexity.
(d) Iterative decoding technique.

Here TPC is created by serially concatenating two hamming codes (n, k, d). Figure 1 shows the block diagram of TPC encoding.

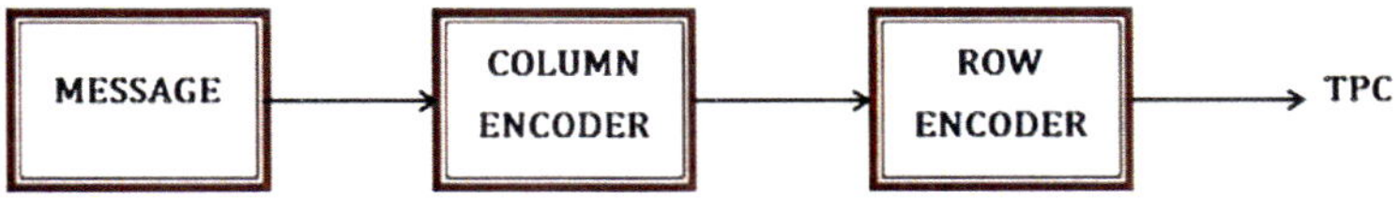

Fig. 1 Turbo product code encoder

3.1 How to Form TPC?

Consider two (15,11) hamming codes and let us name these as C_1 and C_2 having the parameters codeword length n_1 and n_2, data length $k_1$and k_2 and minimum Hamming distance d_1 and d_2, respectively. The steps are as follows;

(a) Form the 11 × 11 messages
(b) Perform column encoding of the message by adding parity bits to the 11 × 11 message to get the 15 × 11matrix
(c) Perform row encoding by adding parity bits to the 15 × 11matrix to get the 15 × 15 matrix

Arrange the column and row parity bits on positions to obtain the TPC as shown in Fig. 2.

Fig. 2 Structure of turbo product code

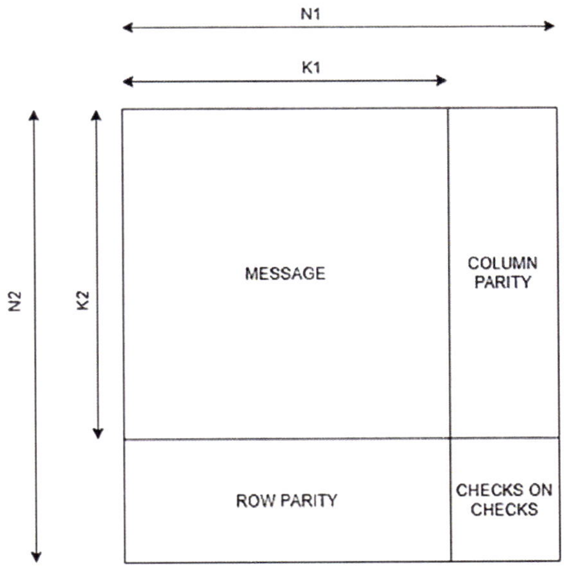

3.2 *Parity Calculation*

d1, d2, d3 … d11 represents the bit positions of the message. P1, P2, P3, P4 are the parity bits, respectively.

$$P1 = d1 \oplus d2 \oplus d3 \oplus d4 \oplus d6 \oplus d8 \oplus d9 \quad (1)$$

$$P2 = d2 \oplus d3 \oplus d4 \oplus d5 \oplus d7 \oplus d9 \oplus d10 \quad (2)$$

$$P3 = d3 \oplus d4 \oplus d5 \oplus d6 \oplus d8 \oplus d10 \oplus d11 \quad (3)$$

$$P4 = d1 \oplus d2 \oplus d3 \oplus d5 \oplus d7 \oplus d8 \oplus d9 \quad (4)$$

3.3 *The Decoder*

The block diagram for the proposed decoding technique is shown in Fig. 3. It includes a row decoder and a column decoder alongside a memory block. The decoder is designed using the MATLAB as well as FPGA platforms and is implemented using Xilinx FPGA "xc7z020clg484-1". The AWGN channel is modelled in the MATLAB platform and the output from the channel is obtained as the RX inputs to the decoder section. α is the weighting factor which is used to eliminate the unreliability which

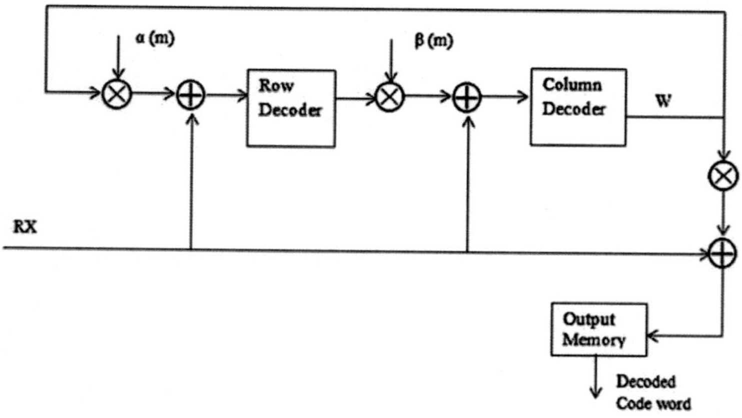

Fig. 3 Proposed model of decoder

might be associated with the extrinsic information '*W*' especially during initial iterations. β is called the reliability factor is used to transform the hard decision output to soft decision output. Both the α and β values range from 0 to 1.

3.4 Step-by-Step Decoding Procedure

One complete iteration is divided into 2 half iterations in order to reduce the decoding complexity. Step 2 to Step 14 details the 1st half iteration and the same will be repeated in the next half-cycle with the updated RX.

Step 1: Get the Received signal matrix **RX** (15×15) from the AWGN channel. For each of the rows, vectors perform the following operations.

Step 2: Find the absolute value of RX.

Step 3: Form ***p*** least reliable positions.

Step 4: Form error patterns **E** with respect to ***p***.

Step 5: For each error patterns thus formed in step 4 perform **RX** $\oplus$ **E** and form the candidate codeword ***CCW***.

Step 6: Perform matrix operation between each row of the **RX** and columns of the transpose of parity check matrix **PT**. Obtain the 3-bit syndrome as the result of this step.

Step 7: Locate the syndrome position in the **PT** and form the error pattern E' by placing 1 at the located position and $0'$s at the other positions.

Step 8: Perform **RX** $\oplus$ **E**$'$ and obtain **TX**.

Step 9: Check for equality between ***CCW*** obtained in step 5 and **TX** in step 8.

Step 10: If ***CCW*** is not the same as that of **TX** repeat step 5.

Step 11: If ***CCW*** is same as that of **TX** decode it as the corrected codeword ***Ccorrect***.

Step 12: Convert the ***Ccorrect*** to BPSK ***C***.

Step 13: Calculate extrinsic information.
$W = \beta\ (Iteration) * D - RX$ Where $D = |R - C|^2$.

Step 14: Update RX by $RX = RX + \alpha(\text{iteration}) * W$.

3.5 Flow Chart of the Proposed Method

In this section, the detailed flowchart representing one complete iteration which is subdivided into two half iterations is shown. The number of iterations can be specified. Whatever be the number of iterations, each will have a subdivision of two half iterations. This reduces the decoding complexity by performing the decoding operation one by one just like slicing a big piece into smaller pieces. This not only reduces the decoding complexity but also the hardware utilization improved since it requires only half the components because of the half iteration technique and of course the speed is also improved because of low complexity. Figure 4 shows the flow chart of the proposed decoding technique.

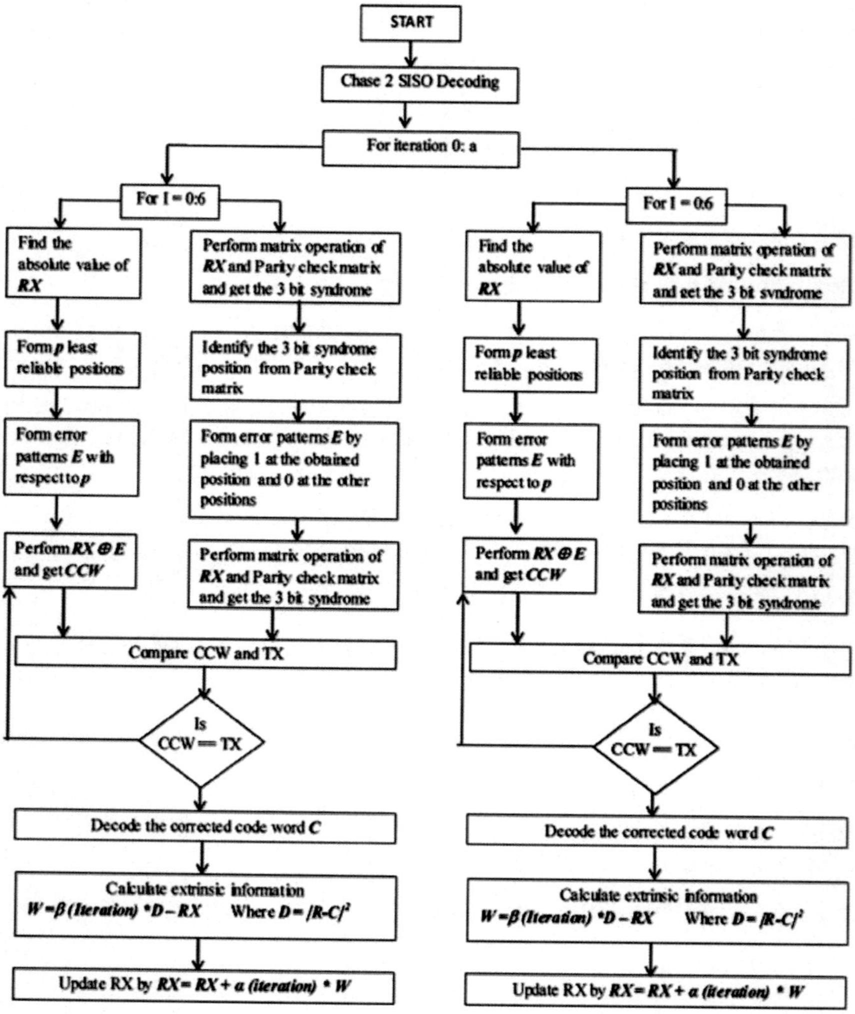

Fig. 4 Proposed decoding flowchart

4 Simulations and Results

This part details about the simulations and results. The proposed method of decoding can decode the received codeword correctly in an efficient and fast manner. Whatever be the bit position which got wrongly decoded the proposed decoder can decode the code words correctly. Figures 5 and Fig. 6 shows simulations done for the (15, 11) hamming code for different data streams with 8 and 10 iterations, respectively. Figure 7 shows the Chase decoding of (15, 11) code. The (15, 11) code having a code rate of 0.733(code rate $r = k/n$) When simulated for 8 iterations show an

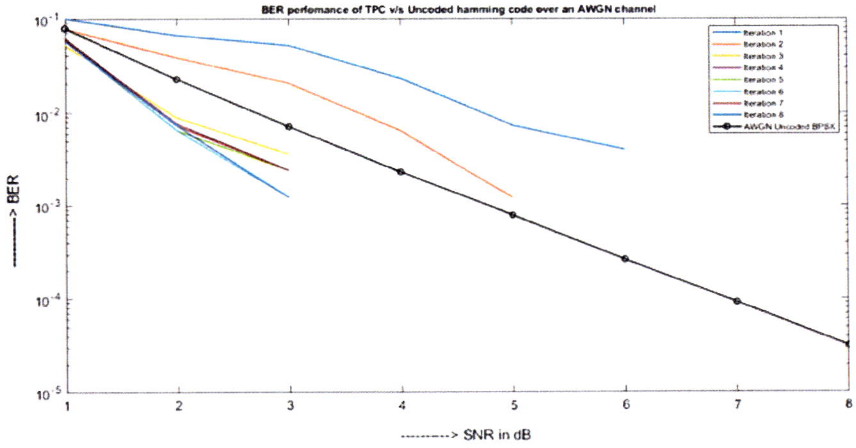

Fig. 5 (15,11) code with 8 iterations

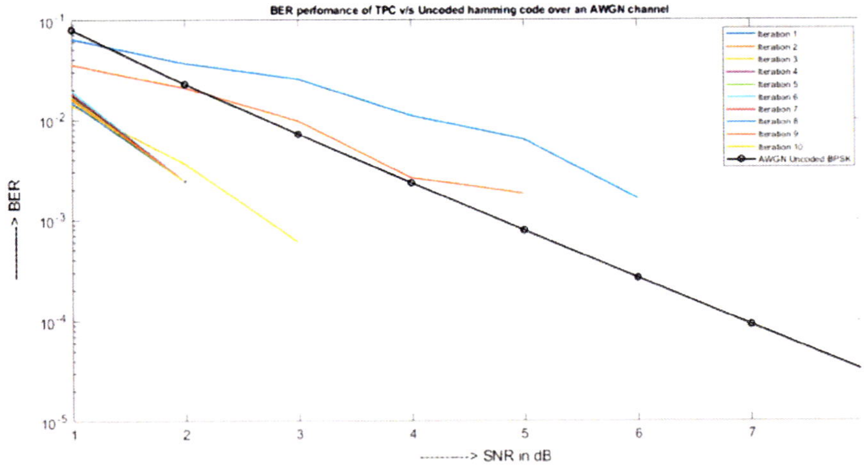

Fig. 6 (15,11) code with 10 iterations

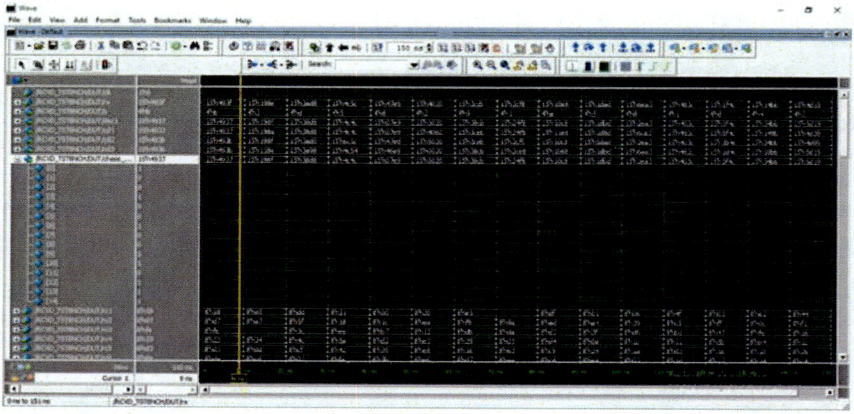

Fig. 7 Chase decoding of (15, 11) code

improvement of 1.7 dB for BER of 10^{-3} over the uncoded bpsk and for 10 iterations shows an improvement of 2 dB for the same BER over the uncoded signal. From this, it's clear that as the number of iterations increases the decoding efficiency also improves. The decoding technique is implemented on "xc7z020clg484-1" for a clock of 10 MHz and a time period of 10 ns. Figure 8 shows the utilization report and Fig. 9 shows the timing report whereas Fig. 10 shows the power summary. Table 1 details the implemented design characteristics.

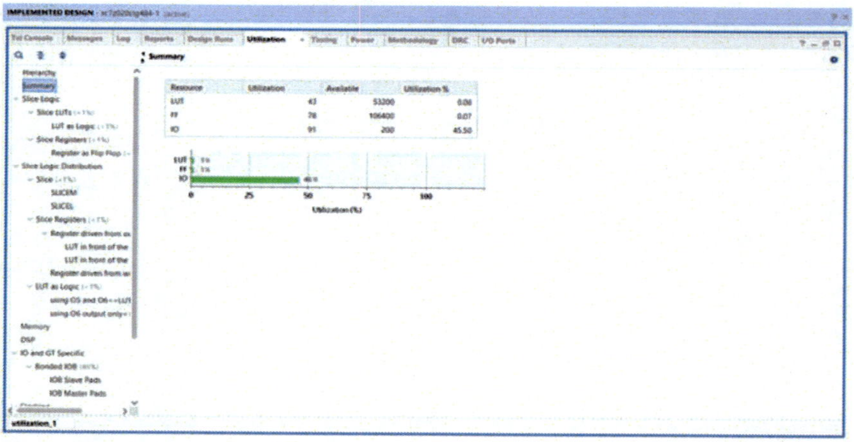

Fig. 8 Utilization Report

Fig. 9 Timing Report

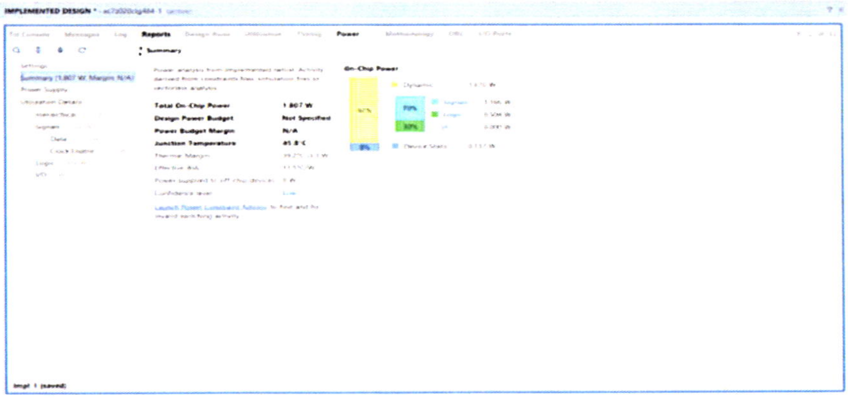

Fig. 10 Power Summary

Table 1 Implemented design characteristics

Parameter	Features	Observations
Area utilization	LUT	0.08%
	FF	0.07%
	IO	45.50%
Timing requirements	Setup time	96.49 ns
	Hold time	0.205 ns
	Pulse width	49.5 ns
Power summary	Dynamic power	1.6 W
	Static power	0.13 W

5 Conclusion

Turbo Product Codes prove its efficiency in error detection and error correction and find its wide range of applications in the field of communication. Turbo Product Codes based on hamming codes are very suitable for those areas wherever the highest code rate signal transmission is required. TPC based on Hamming codes is well enough to be used for error-free signal transmission in a large variety of communication applications. The proposed method proves to be efficient in errorless decoding by a factor of around 1.7 and 2 dB improvement in SNR for the same BER of 10^{-3} in comparison to the uncoded bpsk signal.

References

1. Cho, W.S.J.: Reduced complexity Chase-Pyndiah decoding algorithm for turbo product codes. In: IEEE Workshop on Signal Processing Systems, Beirut, Lebanon, pp. 210–215 (2011)
2. A.-D. A. J. A.-Q. M. A. Muaini, S. A. A.: Ber performance of non-sequential turbo product codes over wireless channels. In: IEEE GCC Conference and Exhibition (GCC), pp. 93–96 (2011)
3. Wang, Y.F.F.-G., Tang, Y.: The iterative decoding algorithm research of turbo product codes. In: Proceedings of ICACIA, pp. 97–100 (2010)
4. Y. L. G. . O. S. . M. S. . H. C. Kee: Turbo decoding of product code based on SPC component decoding. In: International Conference on Communications, Circuits and Systems (2004)
5. Argon, S.W.M.C.: An efficient chase decoder for turbo product codes. IEEE Trans. Commun. **52**, 896–898 (2004)
6. Xu, C.: Shortened turbo product codes: encoding design and decoding algorithm. IEEE Trans. Vehicular Technol. **56**, 495–509 (2007)
7. Landolsi, M.A.: A comparative performance and complexity study of short-length LDPC and turbo product codes. In: 2nd International Conference on Information Communication Technologies, vol. 2 (2006)
8. Hel, Y., Ching, P.: Performance evaluation of adaptive two-dimensional turbo product codes composed of hamming codes. In:Proceedings of the 2007 IEEE International Conference on Integration Technology (2007)
9. Guo Tai Chen, L.C.: Test-pattern-reduced decoding for turbo product codes with multi-error-correcting eBCH codes. IEEE Trans. Commun. **57**, 307–310 (2009)
10. Erl-Huei Lu, P.-Y.L.: A syndrome-based hybrid decoder for turbo product codes. In: International Symposium on Computer, Communication, Control and Automation, vol. 2, pp. 208–210 (2010)
11. Burr, A.: Turbo Codes: the ultimate error control codes. Electronics Commun. Eng. J. **2**, 155–165 (2011)
12. Prasad, N.-R.V.: OFDM for Wireless Communications Systems (2004)
13. Rappaport, S.: 2002 Wireless Communications Principles and Practice (2011)
14. Shannon, C.E.: A mathematical theory of communication. Bell Syst. Tech. J. **379–423**, 623–656 (2004)
15. Wicker, S.B.: Error Control Systems for Digital Communications and Storage
16. Salija, P., Yamuna, B.: Implementation of turbo code with early iteration termination in GNU radio. J. Telecommun. Electronic Comput. Eng. **9**, 53–59 (2017)
17. Sudharsan A., Vijay Karthik, V.B., Vaishnavi, V.C., Abirami J.S.D.S., Yamuna, E.B.: Perfomance enhanced ıterative soft ınput soft output decoding algorithms for block turbo codes. J. Telecommun. Electronic Comput. Eng. **8**

FetchZo: Real-Time Mobile Application for Shopping in COVID-19 Pandemic Situation

Sudhish Subramaniam and Subha Subramaniam

Abstract In this paper, FetchZo real-time application is developed for the shopping purpose in COVID-19 pandemic situation. Coronavirus is a serious virus in India, which spreads via close contacts. Maintaining social distancing becomes more important to avoid coronavirus infection. Shopping for daily needs is a great issue with social distancing and have come up with FetchZo application, which will locate your nearest shop and update you with many people currently present in the shop. This sustainable application provides security with login credentials for the shopkeeper as well as the customer. The shopping list also can be updated by the shopkeeper thus providing a real-time application for this pandemic COVID 19 situation.

Keywords COVID 19 · Coronavirus · Shopping application · Social distancing

1 Introduction

In the critical situation of COVID-19, the only way to prevent coronavirus is by social distancing. Keeping this in mind an application has been developed, which would ease shopping of essential items during a lockdown situation. The lockdown has forced everyone to stay at home. This brings up the major problem of availability of essential items such as rice, wheat, medicines, and other groceries. To help people staying at home with the essential items this model has been proposed. At the initial stage, the model takes input from the shopkeepers about their stock availability. The user is entitled to use the application to get the nearest shop, having the required

S. Subramaniam (✉)
VIT, Vellore, India
e-mail: sudhish.subramaniam2018@vitstudent.ac.in

S. Subramaniam
SAKEC, Mumbai, India
e-mail: subha.subramaniam@sakec.ac.in

P. Karuppusamy et al. (eds.), *Sustainable Communication Networks and Application*, Lecture Notes on Data Engineering and Communications Technologies 55,
https://doi.org/10.1007/978-981-15-8677-4_17

item. The model also calculates the number of people in each shop and hence gives the optimum shop which has the least number of people and is the closest among all.

2 Literature Review

Aarogya Setu (The scaffold for freedom from infection) is an Indian open-source COVID-19 "Contact following, Syndromic application, and Self-appraisal" computerized administration, fundamentally a portable application, created by the National Informatics Center under the Ministry of Electronics and Information Technology [1]. The application arrived at in excess of 100 million introduces in 40 days. On 26 May, in the midst of developing protection and security concerns, the source code of the application was made open. On second April 2020, India propelled the Aarogya Setu versatile application.

Application for aiding and expand the endeavors of constraining the spread of COVID19, to empower Bluetooth based contact following, use of likely hotspots and dispersal of pertinent data about COVID19. The application has more than 114 million clients as of 26th May, which is more than some other Contact Tracing Application on the planet. The Application is accessible in 12 dialects and on Android, iOS, and KaiOS stages. Residents the nation over are utilizing Aarogya Setu to secure themselves, their friends and family, and the country. Numerous adolescents likewise call Setu as their Bodyguard.

The key mainstays of Aarogya Setu have been straightforwardness, protection, and security and in accordance with India's arrangement on Open Source Software, the source code of Aarogya Setu has now been made open-source [2, 3].

3 Overview of Fetchzo Application

The application comprises two parts, namely the shopkeeper side and the customer side. At the shopkeeper's side, he/she can update his stock details and the model stores in the database. FetchZo algorithm runs on the database to find the exact location and of the shop [4].

At the customer's end, the customer's location is accessed and he/she is asked to select the required items with the specified quantity. This is again uploaded to the database, on the customer table [5].

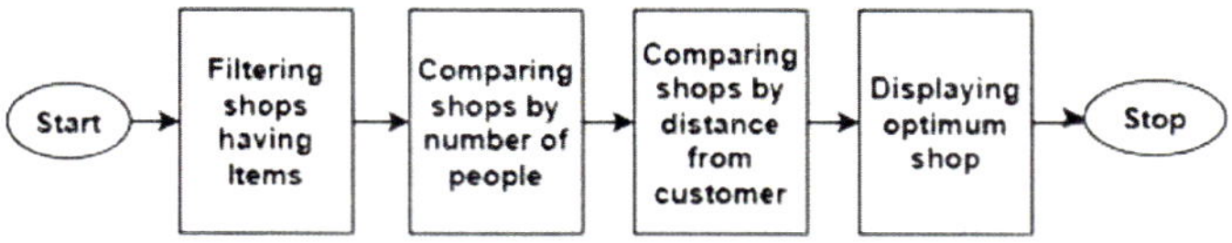

Fig. 1 Data Flow diagram of FetchZo application

FetchZo application fetches the location of the customer and reaches the nearest shops in map. The distance between the customer and the shops are calculated and displayed on the map [6]. The FetchZo application turns on Bluetooth of customer device during installation of the application. The Bluetooth algorithm calculates the number of people in the shop displays it on the map [7, 8].

Using the three parameters availability of the item, distance of the shop from the customer and the number of people in the shop respectively, FetchZo application displays the optimum shop on the screen [9, 10].

In a pandemic situation, people still have to stand in long queues for the items. To overcome this problem the model has a token button. Once the optimum shop is being displayed a counter token button is displayed which will be updated with runtime. It sends a message to the shopkeeper about the customer and his details and his time of the visit to the shop. The shopkeeper can check the list of required items and can be ready with the items as per the token time. This reduces manual token systems which are existing in all our Indian cities. It provides social distancing which can prevent shopkeepers and customers from infections and direct contact.

Only dmart, Star and Supermarkets are surviving which attracts large queues and crowds. By installing this FetchZo application, small shop vendors can also survive without fear. This will help in maintaining social distancing in the shops and avoiding the crowd. Surely the application will be very helpful to both the customer and the shopkeeper keeping the COVID-19 pandemic in mind.

Figure 1 depicts the data flow diagram of FetchZo application. It initially filters nearby shops having a stock from the customer side. It counts the number of people present in the shop and compares it with other shops. It selects the shop with a minimum number of people. It takes distance as a parameter and compares it with the existing shops. This FetchZo application gives social distancing as the highest priority in this COVID-19 situation. It takes shop distance from the customer side as the second parameter and displays the optimum shop.

4 Implementations of Fetchzo Application

As shown in Figs. 1 and 2 the application flows in two parts, customer side and shopkeeper side.

At the shopkeeper's side (Fig. 2), the application starts with the Bluetooth scan. It detects the devices around it and redirects the page to the shopkeeper login. Here it checks if the account is pre-existed, if yes then login else it creates an account. Once the shopkeeper has logged in, he/she has to enter the stock details which is updated to the database. Now the location of the shopkeeper's device is automatically tracked and updated to the database.

At the customer's side (Fig. 3), the application starts the Bluetooth scan and stores the nearby devices. After the login procedure, it displays the essential items required during the lockdown. It asks for the quantity of the item selected. Once an item is selected the items with quantity gets added to the cart. The items are uploaded to the database, the algorithm checks for the items in the nearby shops of the customer's location. The shops with the essential items needed are filtered. Now the algorithm

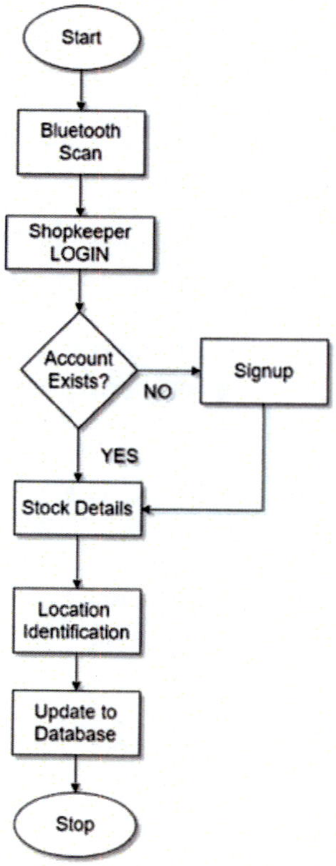

Fig. 2 Shopkeeper side flowchart

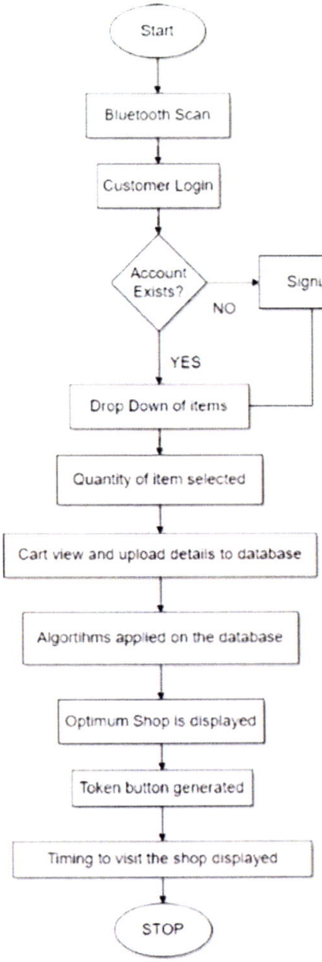

Fig. 3 Customer side flowchart

checks for the number of people present in each shop. It also checks the distance of the shop from the customer's location. With the number of people in the shop having priority, the optimum shop is displayed in the application. The token number is generated and the optimum time to visit the shop is also assigned [11, 12].

4.1 FetchZo Algorithm

- Detecting the nearest shop with the given location
- Calculating the least number of people in the shop
- Providing a token for avoiding the queues

4.2 Detecting the Nearest Shop with the Given Location

Shops register in the application with their respective locations and the details. The shop owners will have to upload their stock details. At the customer side, the user will register his location and his details, this data will be recorded in the database. In the database with the entries of both the tables. Those shops are filtered out which have the required items. The location of the customer is compared with the location of the filtered-out shops in the shop owner's table. By this, the nearest shop can be found.

4.3 Calculating the Least Number of People in the Shop

This is done with the help of Bluetooth Low Energy (BLE). The shop owner's phone becomes the host and it detects the number of people inside the shop using Bluetooth. This information gets uploaded in the database and hence it helps in evaluating the number of people in his shop.

4.4 Attaching a Token

To avoid long queues in the shops a counter button is made which is connected to the database. With the increase in the counter buttons, the model allows a time starting from morning 8 o clock. The application can have its online payment system (example UPI) which will help the customer to do online payments, keeping digital India and social distancing in mind. Once the online payment is done after verification of available items, it becomes convenient for the customer to reach the shop and collect shopping items. This ensures little contact and maintains social distancing.

5 Results and Snapshots

5.1 At Customer Side

Snapshots of the FetchZo application in real-time are shown below. Figure 4 depicts application initial page. This page gives the option to login as a customer or the shopkeeper. Snapshots of the application as shown below Fig. 5 depicts the cart view of the customer.

Snapshot of the application in Fig. 6 shows the application, detecting the location of the customer. Figure 7 shows the application, displaying the number of shops around the customer.

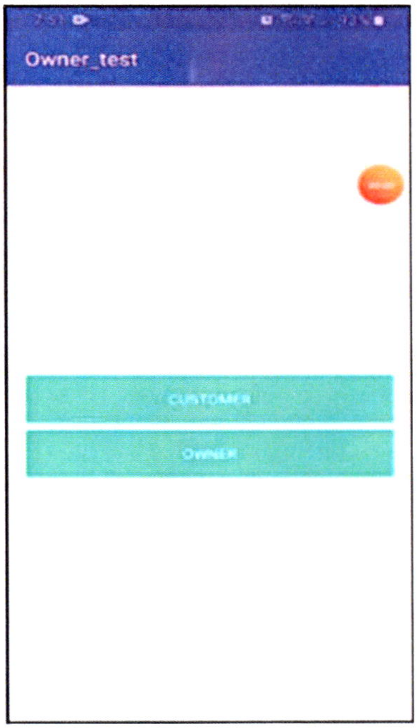

Fig. 4 Login page

Fig. 5 Cart-customer side

5.2 *Shop Keeper Side*

The stock details of the owner are being updated in the database.

Figure 8 shows the login page at the owner's side. As shown in Fig. 9, application asks to add the items to the owner's side.

After displaying the image, it gives the option to enter the item and the quantity of the item to be added in the shopkeeper database. Figure 10 depicts the image of this. Figure 11 shows the items added and shown on the application page. As shown in Fig. 12a, all the items in the cart are uploaded to the database. These are the images from the application developed using Android Studio. The software is user friendly and allows the user to develop android applications.

As depicted in Fig. 12a, b the login information and authentication of both the customer and owner are saved in the database.

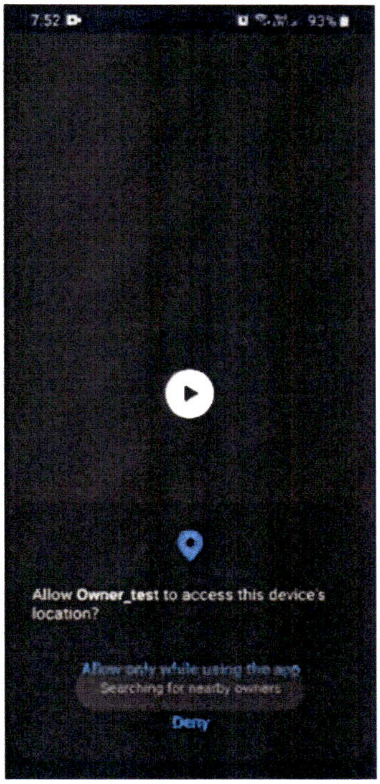

Fig. 6 Location of the customer

Figure 13 shows the saved page of the location. The location of the owner is being analyzed and prepared for comparing with the location of the customer. Figure 14 depicts the calculation of the distance between the shops and the customers.

6 Conclusion

As the COVID 19 situation has become very fierce and the situation is getting complex day by day, would like to present this FetchZo application model so that it will ease the situation and help people in shopping. In future, the model will be surely useful for the day-to-day essential items. This will not only increase the economy but also would prevent the spreading of coronavirus and will help to avoid long queues in the shops. Getting the required items, finding the nearest shop, avoiding contact with people hence following social distancing, avoiding cash payment, these are some of the major advantages of the application. It can be concluded that FetchZo model(application) will surely help people in the lockdown situation of COVID-19.

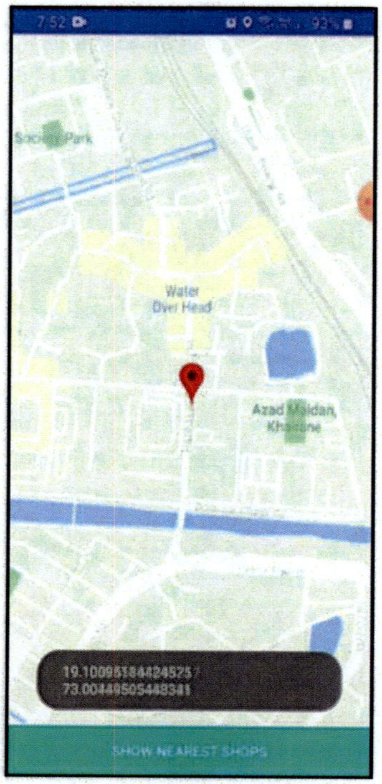

Fig. 7 Displaying the nearby shops with distance

This application will make shopping easier and also will help India fight against COVID-19 by social distancing and avoiding contact.

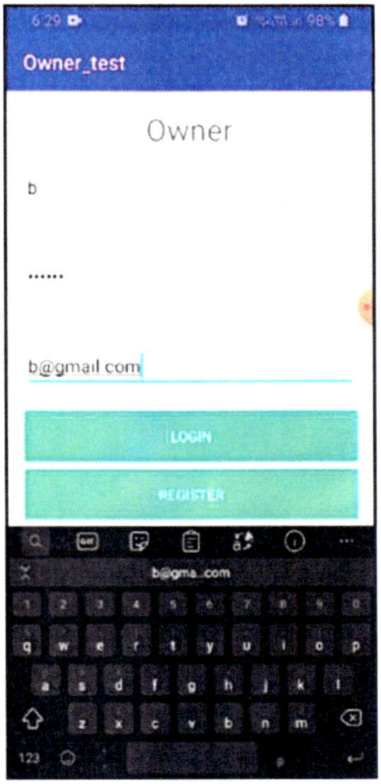

Fig. 8 Login page of Owner

Fig. 9 Add Items

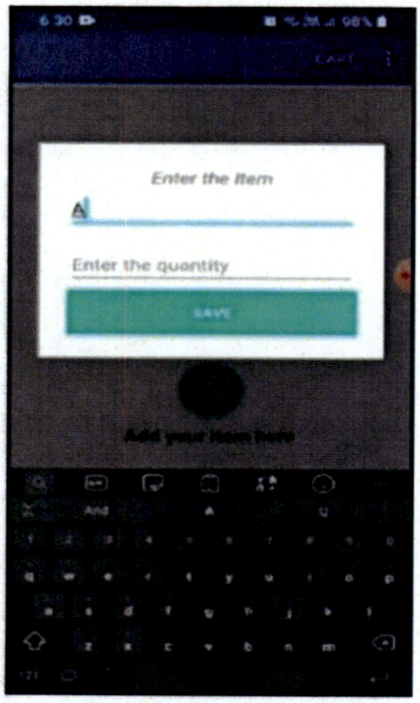

Fig. 10 Adding Items-Owner Side

Fig. 11 Items-Owner Cart

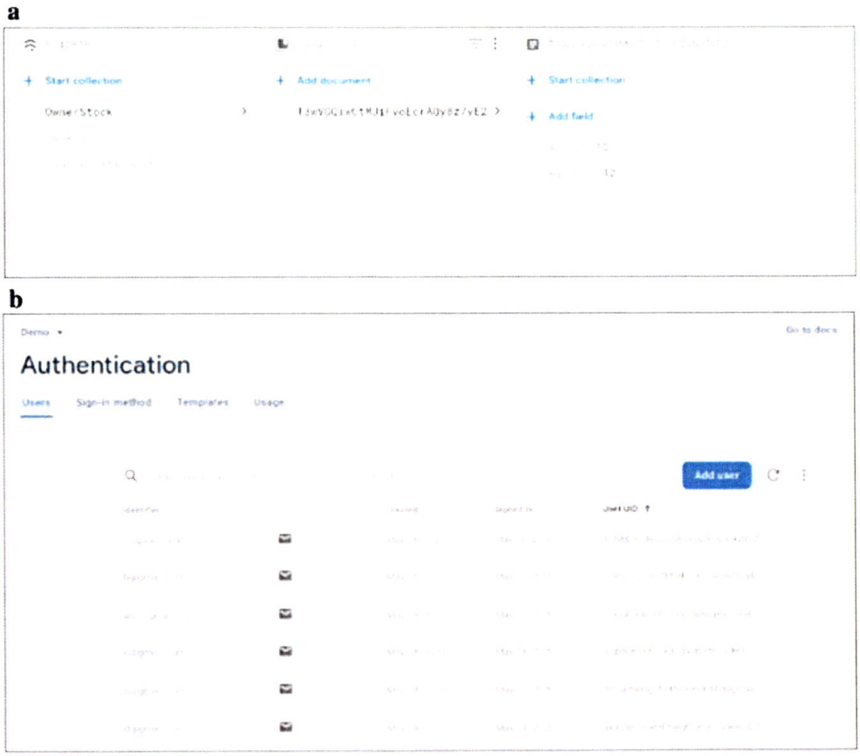

Fig. 12 **a** The Items of the cart are uploaded to the Database, **b** Customer and Owner Database

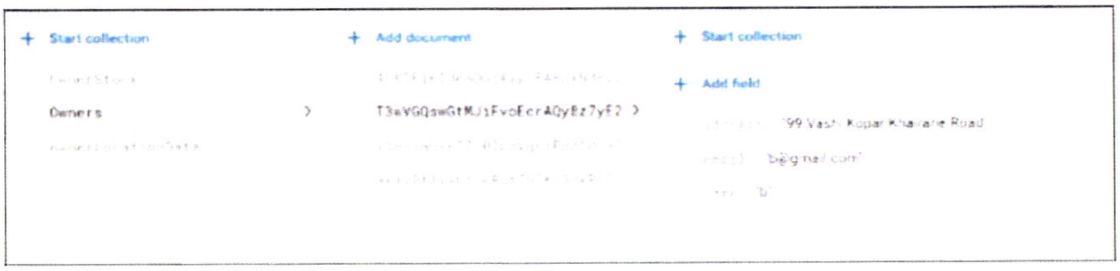

Fig. 13 Saved Location Details of FetchZo Application

Fig. 14 Calculation of distance between the shops and the customers

References

1. nic-delhi/AarogyaSetu_Android, nic-delhi, 27 May 2020, Retrieved 27 May 2020
2. Jump up to:a b "Govt launches 'Aarogya Setu', a coronavirus tracker application: All you need to know". Livemint, 2 April 2020. Retrieved 5 April 2020
3. Aarogya Setu Mobile Application. MyGov.in. Retrieved 5 April 2020
4. Sridhar, S., Smys, S.: A hybrid multilevel authentication scheme for private cloud environment. In: 2016 10th International Conference on Intelligent Systems and Control (ISCO), pp. 1–5, IEEE (2016)
5. Palanisamy, R.: The impact of privacy concerns on users' trust, attitude and intention of using a search engine: a conceptual model and research issues. Int. Acad. Res. J. Business Manage. **2**(3), 31–43 (2013)
6. MohdAnis, T.B., Subramaniam, S.: Operating system: the power of android. Int. J. Sci. Res. (IJSR) (Online) **03**(11), 1914–1919 (2014)
7. Marcario, G., Torchiano, M., Violante, M.: An invehicle infotainment software architecture based on google android. In: SIES, pp. 257–260. IEEE, Lausanne, Switzerland July 2009
8. Nogueira, L., Pinho, L.M.: Time-bounded distributed qos-aware service configuration in heterogeneous cooperative en-vironments. J. Parallel Distrib. Comput. **69**(6), 491–507 (2009)
9. Higuera–Toledano, M.T., Issarny, V.: Java embedded real time systems: an overview of existing solutions. In: Proceedings of the Third International Symposium on Object-Oriented Real-Time Distributed Computing (ISORC 2000), pp. 392–399. IEEE Computer Society, Washington, D.C, USA (2000)
10. Guerra, R., Schorr, S., Fohler, G.: Adaptive resource management for mobile terminals—the actors application. In: Proceedings of 1st Workshop on Adaptive Resource Management (WARM10), Stockholm, Sweden, April 2010
11. Maia, C., Nogueira, L., Pinho, L.M.: Experiences on the implementation of a cooperative embedded system framework. CISTER Research Centre, Porto, Portugal, Tech. Rep., Jun '10
12. Ahmad, M.S., Musa, N.E., Nadarajah, R., Hassan, R., Othman, N.E.: Comparison Between Android and iOS Operating System in terms of Security. School of Computer Science, Faculty of Information Science & Technology, Universiti Kebangsaan Malaysia, 43600 Bangi, Selangor, Malaysia 2013 8th International Conference on Information Technology in Asia (CITA)

An Emerging Trust-Based Security on Wireless Body Area Network

R. Sudha

Abstract Recent approach in wireless technology leads to WBAN which promises ordinary ambulatory health monitoring for an prolonged period of time and afford real-time advise of the patient's position to the doctor. The drop of privacy is one of the main problem in WBAN. Authentication is the main step against security. Enhanced verification scheme prohibits the networks from pretenders and disturbing users meritoriously. Here proposes a trust-based authentication protocol and its simulation result establishes that they outperform the actual systems in terms of enhanced trade-off among anticipated security holds and computational difficulty.

Keywords Authentication · Security · Onion routing · Digital signature · WBAN

1 Introduction

Security in any wireless technology especially in wireless body area networks is highly needed. Authentication process is solitary of the preliminary steps for security employed to put off from the unconstitutional users and pretenders. Authentication schemes vary as per the nature of wireless body area network. For such networks, there is a need of specific lightweight authentication schemes. Security begins with a transaction of wanted security suite between the two conveying the gatherings, hub, and center point. The security choice sets off a security relationship between the two gatherings. To actuate a pre-shared or creating another mutual ace key. Security affiliation conventions are done in view of the key trade arrangements.

The process region of a WSN is extremely huge and can be used in ecological observing, manage temperature and moisture, vehicle transfer control, checking of human body organs, and among others. Figure 1 exemplifies a situation of WBANs in the medical area where patients that are being observed can be in a hospital, at

R. Sudha (✉)
Associate Professor & Head, Department of Computer Science, PSG College of Arts & Science, Coimbatore, India
e-mail: sudha279@yahoo.com

P. Karuppusamy et al. (eds.), *Sustainable Communication Networks and Application*, Lecture Notes on Data Engineering and Communications Technologies 55,
https://doi.org/10.1007/978-981-15-8677-4_18

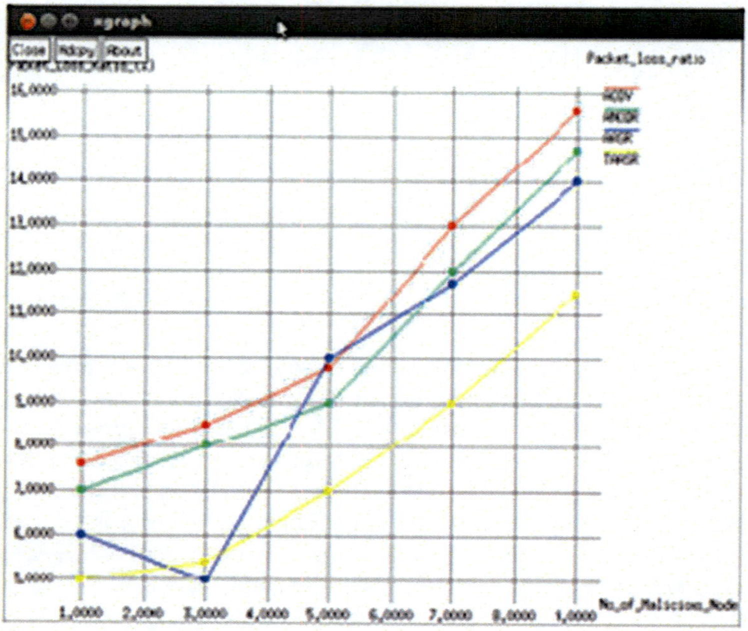

Fig. 1 Comparison based on packet loss ratio

home, or anyplace besides performing an goings-on routine. Sensing data are send to health experts in the course of the Internet.

The fundamental security requirements in WBAN are described below [1].

1. Data Confidentiality. To protect the data from a revelation, the system necessitates data discretion
2. Data Authentication. Applications together with mutually medical and non-medical relevance demand data authentication. Symmetric technique can be used in a WBAN to attain data authentication. This method shares the secret key to work out Message Authentication Code for all data.
3. Data Integrity. This is compulsory as an rival can modify the data that is broadcasted over an anxious channel. Deficiency of data integrity system paves a way to the opponent to amend the information before it reaches the destiny.

2 Related Work

In order to obtain the issues in existing wireless body area networks, a vast survey has been made based on wearable gadgets and wireless body area networks. Sensors play a vital role in body area networks. In few research models mentioned that sensors must be constructed based on cryptographic parameters since it is major challenge in WBAN to meet necessary security measures [2, 3]. In few research models, biometrics is considered as an important factor since it describes the relation of the individual in an unique manner and helps to secure the system communication between wireless body area networks [4]. Various security mechanisms are evolved in wireless body area networks based on symmetric cryptosystem. The system restricts the access to the assets if it faces any issues. However, it lags to provide security against physical arrangements in some cases [5].

Other than the intricacy of detecting component, hub's key administrations in WBAN offer each part abundance. On the different, some exploration misuses the awry cryptosystem in portable and specially appointed systems even have been expected, and attempted to appear at the peculiar uniqueness of WBAN [6, 7]. One misery about the topsy-turvy cryptosystem is a source imperative inconvenience; however, current work has demonstrated that performing ECC devours a considerable measure a lesser measure of memory and processing power [7]. These investigates tended to a span of controlled body area networks and avoid the discussion about sensor systems implantation. Wireless body area networks work based on the arranges of connected elements in human body and the communication is possible from any place. Recent framework discussed in [8, 9] provides communication process in inter- and intra-wireless body area networks in telemedicine application. Using wireless module like Bluetooth for short-range and ZigBee for long-range communication, the proposed model dispatches the data from sensors to the sink. Sathesh [1, 10] discussed the importance of rack remote sensors and developed the wireless body area network model similar to Tmote sky design.

An end-to-end mhealth application for patient monitoring is discussed [11] as European MobiHealth venture which progress based on GPRS and UMTS. Using assorted sensors, the vital signs such as electrocardiogram, circulatory strain, pulse are continuously monitored in mobihealth setup. Communication through the sensors is performed thorough exceptional gadget which works based on single-hop ZigBee or Bluetooth module. The fundamental issues in health applications are considered in terms of insurance, dependability of correspondence assets, and quality of service ensures. The French undertaking BANET [12] means to develop a structure, copies, and instruments to mode improved remote correspondence frameworks proposing the greatest differentiation of WBAN-based applications, contained by the customer regular reasoning, restorative, and game fields. The investigation reports the issues in expansion channels in WBAN and its MAC conventions along with substitute remote systems. Falck et al. [13] discussed about the German BASUMA venture and its wireless body area networks based on the framework. The design has UWB in its front end along with IEEE 802.15.3 as MAC decorum. This system also utilizes time allocations based on opening schedules by getting the period disputes through CSMA/CA communication process.

In literature [14], an economic bendable wireless body area network for urbanized utilization. The objective of the research present in terms of developing location-based application which helps to obtain results in the on field testing. Sensors such as WiMoCA-hubs are used in the experimental process and the sensors could be placed anywhere since it is characterized by MEMS accelerometers, and the performance will not be deviated. The exploration of wearable sensors is focused in IBBT IM3

[15] which helps to analyze the wearables of the patients. In this process, a heat beat is used to detect and send to medical specialists and provide solution to the patient. Based on cryptosystems, the security of the wireless body is network and is analyzed in literature [16]. Cryptosystems are used to validate the risks in WBAN and provide strong security to the system. From the survey, it is observed that almost all the models have limitations in its security factors. Very few models are analyzed the issues related to security in mobihealth applications. Based on the observed limitations, the proposed models are formulated in the following section.

2.1 Trust and Security Issues

Data security is for the most part in view of two elements: trust and security. In WBAN, every hub is engaged with the bundle exchange ought to guarantee that their neighboring hubs are trustful and secure. The component which verify that the data about the source is really who it professes to be. The signatures and encryption instruments should require an arrangement to check by any hubs the wellsprings of that data. Security and trust are firmly commonly subordinate element that cannot be distanced.

2.2 Proposed Methodology

We indicate a WBAN by B and make the accompanying suspicions.

2.2.1 Group Signature

Gathering signature is a strategy for enabling individuals from a gathering to sign namelessly in a WBAN steering convention. Gathering signatures can be seen as customary open key marks with extra protection highlights. This approach is to run a gathering key assention convention toward the start of each schedule vacancy and utilize the subsequent gathering key as the normal parameter and versatile. The more productive approach is to utilize a gathering key understanding convention with a specific end goal to concede to the normal parameter and gathering chief to create and convey this beginning worth. Gathering signature conspire has bunch administrator, who is reaction for including new individuals and denying mark of individual hubs in obscurity are given to a gathering supervisor.

Open Key GB+: key this is normal to every one of the hubs of a gathering B.

Private Key GN1−: key which gives security for the information of individual hub N1 in a gathering B where N1 ∈ B.

Hub N1 uses GN1 private key to sign a message, and this signed message can be unscrambled by means of people in general key GB+ by alternate hubs in B, which keeps the obscurity of N1 [17].

2.2.2 Onion Routing

The fundamental work of onion steering convention is foundation of association and taking into consideration unidentified correspondence. Amid Route ask for the messages are dully encoded the data while sent source to goal hubs of onion switches [18]. While in Route-Request has each transitional hubs known as onion switches seizes a layer of encryption and uncover directing data when pushes the message from goal to the source hub. This system safeguards these go-between hubs about knowing the cause, goal, and substance of the message. To pass an instant message, the directing onion is an information structure which frames concealed layer by encryption for sending an instant message with back-to-back layers of encryption. In the meantime while back warding an instant message, it unscrambles their relating layer and the first plaintext message visible just to sender and beneficiary. It is end-to-end encryption and decoding process between the source and the goal in ill-disposed condition.

3 Proposed Work

Many trust administration plans have been proposed to assess trust esteems and the vast majority of the trust-based conventions for secure directing computed trust esteems in light of the qualities of hubs carrying on appropriately at the system layer. Trust estimation can be application subordinate and will be diverse in light of the outline objectives of proposed plans. The trust administration measurements incorporate overhead (e.g., control parcel overheads), throughput, bundle conveyance proportion, bundle dropping rate, and postponement.

3.1 Public Key Cryptography

For onion directing, the messages are scrambled and unscrambled utilizing PKC. In this paper, we have utilized elliptic curve Diffie Hellman key trade calculation.

3.2 Public Key Encryption

The secrecy of correspondence is achieved by public key encryption (PKE) amid transmission.

Here, the sender utilizes the general population key of the beneficiary to scramble the substance of the message. The enciphered message is then transmitted to the recipient and the collector would then be able to utilize their own coordinating private key to decode the message.

3.3 Elliptic Curve Diffie Hellman Key Exchange Algorithm

ECDH is utilized for the motivations behind key understanding. Assume two hubs, n1 and n2, wish to trade a mystery key with each other. N1 will create a private key dA and an open key QA = dAG (where G is the generator for the bend). Essentially, n2 has his private key dB and an open key QB = dBG. In the event that n2 sends its open key to n1 then ready to figure dAQB = dAdBG. Also if n1 sends its open key to n2, at that point he can compute dbQA = dAdBG. The mutual key is the x co-ordinate of the computed point dAdBG. Any busybody would just know QA and QB, and cannot ready to appraise the common mystery key.

3.4 Digital Signature

The objective of an advanced mark conspire is to guarantee the sender of the message cannot disavow a message that they sent. Accordingly, the motivation behind computerized marks is to guarantee the non-denial of the message being sent. This is helpful in a down to earth setting where a sender wishes to make an electronic buy of offers and the recipient needs to have the capacity to demonstrate who asked for the buy.

3.5 Protocol Design

The personality data is appointed to every hub in instatement stage or when the hub will be designed.

3.6 Trusted Anonymous Route-Request

In the proposed conspire, initially, every hub will be arranged with the steady trust esteem 50 utilizing hub trust work. The proposed convention can choose the better way (trusted and briefest) utilizing trust esteem and the quantity of bounces. At the point when the RREQ and RREP message are created in the system, every hub add its own trust an incentive to the trust aggregator on these course revelation stage. Every hub likewise refreshes its own particular steering table. The accompanying recipe can be utilized to assess the trusted and briefest way.

Entirety of trust esteems Where, Sum of trust esteem $= \sum$ trustvalue (I).

3.7 Trusted Routing Procedure

The directing calculation can be actualized in view of the current on-request impromptu steering convention like AODV. The primary directing methods can be condensed as takes after:

1. During course disclosure, two stages are performed
   (a) First before sending the RREQ, the trust estimation of each neighboring hub is introduced to 50
   (b) Second source hub communicates a RREQ parcel in the arrangement to find confided in neighbors.
2. If a transitional hub gets the RREQ bundle, it checks the RREQ by utilizing its gathering open key, and includes one layer best of the key-scrambled onion. It checks the trust estimation of the neighbor hub and in light of the trust factor, the hub chooses its next neighboring hub for RREQ bundle exchange. This procedure is rehashed until the point when the RREQ parcel achieves the goal or terminated.
3. Once the RREQ is gotten and checked by the goal hub, the goal hub collects a RREP bundle in the arrangement and initiates communication with source hub.
4. On the invert way back to the source, each moderate hub approves the RREP bundle and updates its directing and sending tables.
5. Then it expels one layer on the highest point of the key-encoded onion, and keeps broadcasting the refreshed RREP in the arrangement.
6. When the source hub gets the RREP parcel, it verifies the bundle, and updates its steering and forwarding tables. The course revelation stage is finished.
7. The source hub begins information transmissions in the set up course in the arrangement. Each intermediate node advances the information bundles by utilizing the route pseudonym.

4 Experimental Result

The Random Way Point Mobility Model portrays the development of hubs. The respite time is set to 10 s. what's more, most extreme speed set to 5 m/s. The recreation time is set to 100 s. furthermore, nodes are similarly dispersed in 800 × 800 m zone. Subsequent to getting the estimations of every execution metric as indicated by every convention, the diagram has been plotted to demonstrate the examination between AODV, ANDOR, AASR and TAASR.

As appeared in Fig. 1, the proposed TAASR has most astounding capacity to distinguish packet dropping ratio with the assistance of its trust administration approach and it beats the current strategies AODV, ANDOR and AASR. AASR accomplishes 5% more prominent misfortune proportion than TAASR in normal.

As appeared in Fig. 2, while there is increment in number of malevolent hubs, the normal throughput of four protocols diminishes clearly. Throughput of the proposed TAASR is higher than the staying existing protocols.

Figure 3 depicts the analysis of end-to-end delays. In this TAASR invest energy in the security handling in their course revelation, its postponement is higher than AODV. On the off chance that ANODR is under a substantial assault, it will dispatch new course disclosures for the broken courses, which present more deferrals in normal. Contrasted with the assaulted ANODR, AASR, and AODV, the proposed TAASR decreases the need of re-steering because of its trust-based validation and onion directing which results in 20 ms less of deferral in normal.

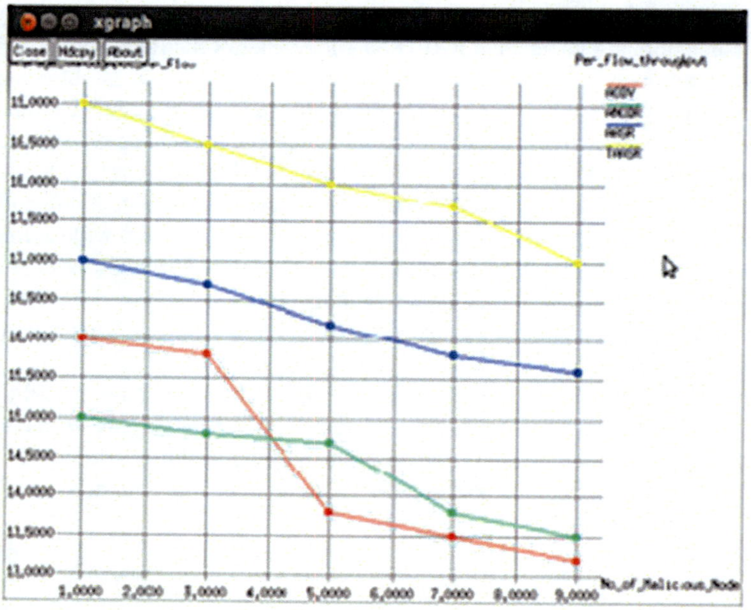

Fig. 2 Comparison in light of throughput

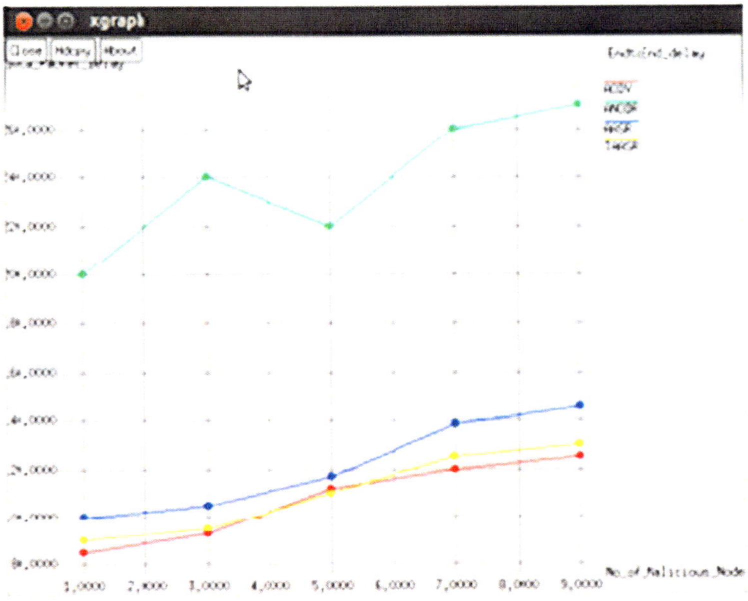

Fig. 3 Comparison in light of end-to-end delay

Performance Evaluation of Mobile Scenario Under Adversarial Environment:

To mimic the adversarial conditions, the aggregate nodes are set into 20% of its actual value, i.e., 9 nodes, as malignant nodes. The system portability is changed from 1 to 5 ms and record the execution consequences of the four protocols (Fig. 4).

In spite of the execution variety, TAASR dependably accomplishes the most elevated throughput because of its trust-based taking care of nature. This can be clarified by its capacity in protecting the bundle dropping assault.

The bends of the end to-end delay are appeared in Fig. 5. Because of the extra security preparing time in RREQ flooding, AODV, ANODR, and AASR have longer postponements than TAASR, while AASR has 20 ms less of deferral than TAASR in normal (Fig. 6).

5 Conclusion

The trust and trust relationship among nodes can be spoken to, figured and consolidated utilizing a thing feeling. In our TAASR proposed protocol, nodes can participate together to get a target sentiment about another node's reliability. They can likewise perform confided in directing practices based on the trust relationship. With a conclusion limit, nodes can adaptably select the method to execute cryptographic process. Consequently, the computational overheads are lessened without the need of asking

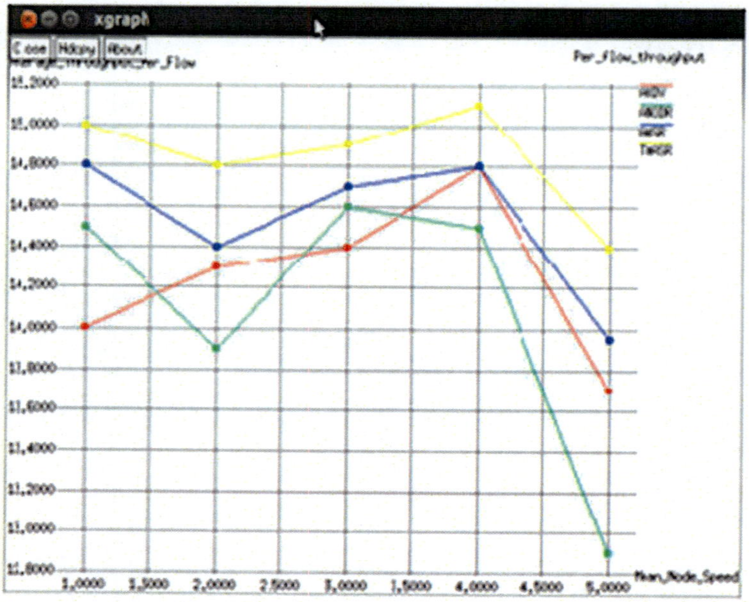

Fig. 4 Performance correlation in light of throughput

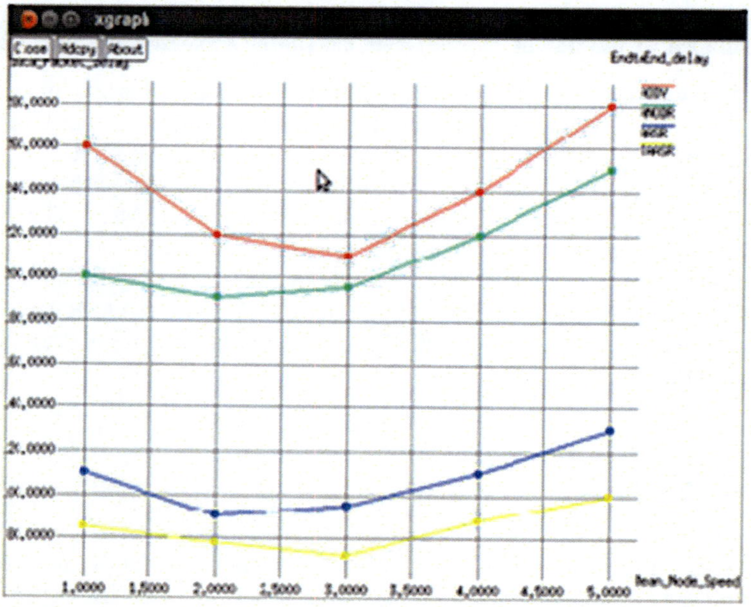

Fig. 5 Performance correlation in light of end-to-end delay

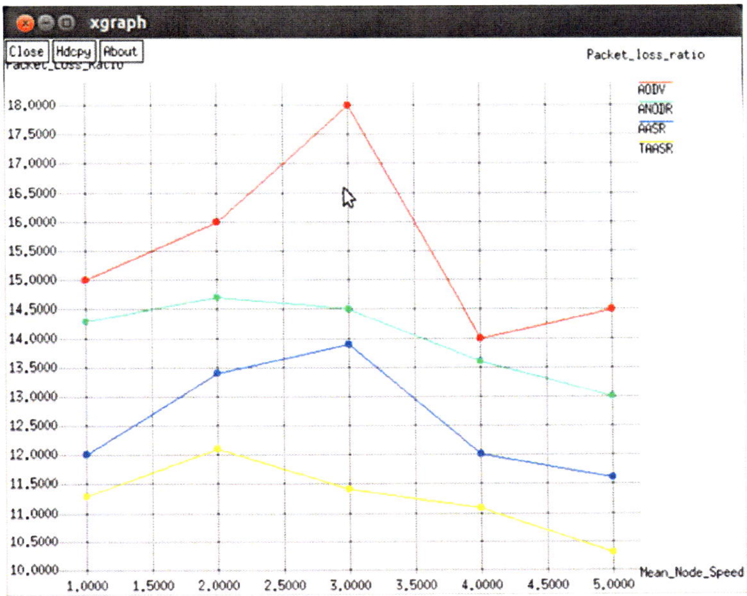

Fig. 6 Performance correlation in light of packet loss ratio

for and checking declarations at each directing activity. Our TAASR steering convention is an all the more lightweighted; however, more adaptable security arrangement than other cryptography and confirmation plan. It utilizes trust esteems to support parcel sending by keeping up a trust counter for every node. On the off chance that the trust counter esteem falls underneath a limit, the comparing halfway hub is noxious node. In this proposed conspire, approved node has high throughput and parcel conveyance proportion can be enhanced altogether with diminishing normal end-to-end delay by expanding trust esteem.

References

1. Sathesh, A.: Optimized multi-objective routing for wireless communication with load balancing. J. Trends Comput. Sci. Smart Technol. (TCSST) **1**(02), 106–120 (2019)
2. Poon, C.C.Y., Zhang, Y.T., Bao, S.-D.: A novel biometrics method to secure wireless body area sensor networks for telemedicine and m-health. Commun. Mag. IEEE **44**, 73–81 (2006). https://doi.org/10.1109/MCOM.2006.1632652
3. Sudha, R., Devapriya, M.: A survey on wireless body sensor networks for health care monitoring. In: Int. J. Sci. Res. (IJSR) **3**(9), 1574–1578 (2014)
4. Sudha, R., Devapriya, M.: Enhanced bio-trusted anonymous authentication routing technique of wireless body area network. In: Biomedical research 2016 in Special Issue: S276–S282, September 2016
5. William, C., Tan, C.C., Wang, H.: Body sensor network security: an identity-based cryptography approach. In: Proceedings of the ACM conference on wireless network security (WiSec '08),

pp. 148–153. ACM Press (2008). https://doi.org/10.1145/1352533.1352557
6. Lim, S., Oh, T.H., Choi, Y.B., Lakshman, T.: Security issues on wireless body area network for remote healthcare monitoring. IEEE Int. Conf. Sens. Netw. Ubiquitous Trustworthy Comput. **2010**, 327–332 (2010). https://doi.org/10.1109/STUC.2010.61
7. Sharmilee, K.M., Mukesh, R., Damodaram, A., Subbiah Bharathi, V.: Secure WBAN using rule-based IDS with biometrics and MAC authentication. In: 2008 10th IEEE international conference on ehealth networking applications and services, pp. 102–107. IEEE (2008). https://doi.org/10.1109/HEALTH.2008.4600119
8. Otto, C., Milenkovic, A., Sanders, C., Jovanov, E.: System architecture of a wireless body area sensor network for ubiquitous health monitoring. J. Mobile Multimedia **1**(4), 307–326 (2006)
9. Jovanov, E., Milenkovic, A., Otto, C., de Groen, P.C.: A wireless body area network of intelligent motion sensors for computer assisted physical rehabilitation. J. NeuroEng. Rehab. **2**(1), 16–23 (2005)
10. Moteiv [online] https://www.moteiv.com
11. van Halteren, A.T., Bults, R.G.A., Wac, K.E., Konstantas, D., Widya, I.A., Dokovski, N.T., Koprinkov, G.T., Jones, V.M., Herzog, R.: Mobile patient monitoring: the mobihealth system. J. Inf. Technol. Healthcare **2**(5), 365–373 (2004)
12. Bhalaji, N.: QOS and defense enhancement using block chain for fly wireless networks. J. Trends Comput. Sci. Smart Technol. (TCSST) **1**(01), 1–13 (2019)
13. Falck, T., Espina, J., Ebert, J.P., Dietterle, D.: BASUMA—the sixth sense for chronically ill patients. In: International workshop on wearable and implantable body sensor networks (BSN), pp. 57–60, Cambridge, MA, USA, 3–5 April 2006
14. Farella, E., Pieracci, A., Benini, L., Rocchi, L., Acquaviva, A.: Interfacing human and computer with wireless body area sensor networks: the Wimoca solution. Multimedia Tools Appl. **38**(3), 337–363 (2008)
15. Pandian, M.D. :Enhanced network selection and handover schema for heterogeneous wireless networks. J. ISMAC **1**(01), 160–171 (2019)
16. Jang, C.S., Lee, D.G., Han, J.W., Park, J.H.: Hybrid security protocol for wireless body area networks. Wirel. Commun. Mobile Comput. **11**, 277–288 (2011). https://doi.org/10.1002/wcm.884
17. Boneh, D., Boyen, X., Shacham, H.: Short group signatures. In: Proceedings of CRYPTO, pp. 41–55 (2004)
18. Zhang, Y., Liu, W., Lou, W., Fang, Y.: MASK: anonymous on-demand routing in mobile ad hoc networks. IEEE J. Wirel. Commun. **5**(9), 2376–2385 (2006)

Preventing Fake Accounts on Social Media Using Face Recognition Based on Convolutional Neural Network

Vernika Singh, Raju Shanmugam, and Saatvik Awasthi

Abstract In today's world, most people are intensely dependent on online social networks (OSN). People use social sites to find and make friends, to associate with people who share comparable intrigue, trade news, organize the event, exploring passion. According to a Facebook review, 5% of monthly active users had fake accounts, and in the last six months, Facebook has deleted 3 billion accounts. According to the Washington Post, Twitter has suspended over 1 billion suspect accounts over a day in recent months. Detection of a fake profile is one of the critical issues these days as people hold fake accounts to slander image, spread fake news, promote sarcasm that has attracted cybercriminals in. There are numerous machine learning methodologies, such as supervised learning and SVM-NN, and they are produced for the effective detection of a fake profile. In this paper, convolutional neural networks are proposed with many artificial neural network algorithms like face recognition, prediction, classification, and clustering for the efficient identification of account being real or fake and elimination of fake profile account. Furthermore, the study is grounded on the fact of the face recognizing of the user and performing feature detection and time series prediction. If the user account detected fake, it would not be created.

Keywords Social media analytics · Online social media · Face recognition · Convolutional neural networks · Age prediction

V. Singh (✉) · S. Awasthi
Galgotias University, Greater Noida, India
e-mail: vernikasingh8@gmail.com

S. Awasthi
e-mail: saatvikawasthi1998@gmail.com

R. Shanmugam
School of Computing Science & Engineering, Galgotias University, Greater Noida, India
e-mail: dr.sraju@yahoo.com

P. Karuppusamy et al. (eds.), *Sustainable Communication Networks and Application*,
Lecture Notes on Data Engineering and Communications Technologies 55,
https://doi.org/10.1007/978-981-15-8677-4_19

1 Introduction

There are increasing numbers of people in the current generation with social online networks like Facebook, Twitter, Instagram, LinkedIn, and Google + [1]. Social networks allow people with common interests or collaborative reasons. It provides them with access to numerous services for example messaging, posting comments on their cyberwalls which are public, commenting on other users' profiles post, and exchanging the masking of identity for malicious purposes has become progressively prevalent over the last few years. People rely heavily on OSNs to remain in contact, to organize trade news, activities, and perhaps even e-business [2]. People rely heavily on online social networks (OSNs) to create and share individual personal profiles, email, images, recordings, audios, and videos, and to find and make friends that have attracted cybercriminals' interest in carrying out a variety of malignant activities. Government associations are used (OSNs) as a forum for effectively providing government-driven services to people and educating and informing them about different situations. This heavy utilization of social networks results in immense measures of data being disseminated and organizations use social networks to promote, advertise, and support their business online. Fake profile accounts show that people do not represent them as a real person. Such an account is manually opened by a person after that actions are automated by bot. Fake profile account is categorized into a Sybil account and duplicate account. A duplicate account applies to a user's account and maintained by the user other than their main account. Fake accounts are classed into user classified (reported) or unauthorized (unwanted) groups of accounts. User malicious account records show individual profiles made by a client from a company or non-human element. Alternatively, undesirable accounts are however user identities that are configured to be used for violation of security and privacy.

The social networking site database for Facebook records a statistic of 4.8% for duplicate accounts, the number of user-misclassified accounts is 2.4%, and the number of unauthorized accounts is 1.5% [3]. In 2019, Facebook announced the deletion of 2.3 billion fake profile accounts. This is almost twice as many as 1.2 billion accounts withdrawn in the first quarter of 2018. The Facebook Compliance Report shows that as much as 5 million of its monthly active users are fake and that there is a growing number of attacks. Facebook is estimated to have more than two billion monthly active users and one billion active users each day in the major online social network reports. Accordingly, only 5% of its active monthly users are false in Facebook reports [4]. It is very convenient today to make false accounts. Nowadays, fake profile accounts can be purchased on the Web at an extremely cheaper cost; furthermore, it can be delivered to the client using publicly supporting. Now it is easier to purchase followers online from Twitter and Instagram. The goal behind Sybil account formation is to defame someone else's image, digital terrorism, terrorist propaganda, fear-based oppressor publicity, campaigns for radicalization, distribution of pornography, fraud and misinformation, popularity shaping, division of opinions, identity insecurity. In this paper, the promptly accessible and designed methodologies are evaluated that are utilized for the fruitful detection of identifying fake accounts on Facebook utilizing AI models, of human-created detection that is accomplished by observing the attitude and the needs of people by examining their experiences. Detection is accomplished by observing the attitude and the needs of people by examining their communication and interaction with each other. Then, there is a need to give a fake account to our machine learning model to allow the algorithm to comprehend what a fake account is. Convolutional neural network (CNN) and feature classification algorithms are being used. Our methodology is to differentiate on the reality between true user accounts and fake accounts by checking them at the time of creation and not allowing the fake profiles to be created.

2 Problem Identification

2.1 Online Social Network (OSN)

Social media play a vital role in our life as 45% of the population around the world spend at least one hour daily on social networks to share news, post pictures, tweeting, commenting, liking, etc. [1]. Companies use the online social network to advertise and promote their business online. Government organization uses social media as a platform to deliver government services to citizens in an efficient manner and to educate and inform them in various ways. The highly dependent nature of billions of people over social networks has attracted the interest of cybercriminals to carry out malicious activities.

Computer-mediated communication (CMC) is a basic portion of our every day lives as, a result it has ended up essential for the populace, it is never going stalwart, as the society is never returning to a more physical environment, to considering characters on CMC stages like social media and gatherings. The idea of identity is expressly distinguished in this consider because it has to be overhauled logically as a critical theme of the social media wrangle about. As expressed prior, a few analysts have examined this issue, but determined three-discusses are required since it is connected to "self" and "self" as demonstrated overnight. Inquiry about concerned with the identity, picture, and online gatherings has numerous possibilities and is especially meriting of thorough consideration owing to social organizing designs predominant over thousands of a long time. Amid the final era, numerous more youthful people effectively lead complicated genuine organizing, which may not reflect their current or offline presence. It has ended up a matter of concern. As of late specified viewpoint is a fair one in setting up untrue characters in social media. In, however, another neighborhood, faux accounts are made in advanced organizing and reasonable consideration was done that has moreover procured.

2.2 Fake Profile

Fake profiles are created to defame someone's image by using an individual name and pictures, and terrorist propaganda, sharing fake news, and distribution of pornography. It has a high impact on younger age who are highly dependent on social networks. Most people try to portray an image of themselves who they are not and try to communicate with their dear ones and hate ones to harm them mentally which lead to serious problems such as depression or suicide. Fake accounts can be categorized into duplicate account which is maintained by a user in addition to their principal account whereas false account further breaks down into user-misclassified accounts which represent the user personal account created as a business account and undesirable accounts which are created to carry out malicious activities.

Social networking site, Facebook, has estimated that 4.8% profiles are double accounts, 2.3% profiles are unsolicited, and 1.6% are undesirable accounts [3]. The Facebook company has started removing 2.2 billion accounts that were detected fake in the Q1of 2019. That is double the number detected in Q4 of 2018 where 1.2 billion fake profiles were deleted. According to Facebook's Enforcement Report, it is 6% monthly active users are fake and the increase is due to the rise in automated attacks.[5] The creation of fake accounts and buying Twitter and Instagram followers along with putting likes online on posts are very easy nowadays. It can be bought online at a very less cost and can be given to the customer via crowdsourcing services.

The reason behind the creation of fake accounts is mostly among these, i.e., either to defame another person, terrorist propaganda, spreading rumors and false news, influencing popularity, polarizing opinions, identity theft, radicalization campaigns, cyberbully, dissemination of pornography and fraud, and online extremism.

2.3 Approach to Solve

The paper finds out the methods and the measures currently available for the detection of counterfeit accounts using machine learning technologies. Detection is carried out by learning the behaviors of individuals and requires a detailed analysis of the activities of social media interactions with other accounts. To identify other fake accounts and to predict fake accounts for potential use, this machine learning algorithm is training with the latest collection of identified fake profiles. Neural networks and algorithms are used for classification. Our solution uses distinct characteristics of the platform and restricts a platform to use an identity over the so-called Internet.

3 Literature Survey

This section summarizes some of the related work done in the field of detecting fake profile accounts using machine learning models.

3.1 Sarah Khaled, Neamat El-Tazi and Hoda M. O. Mokhtar

In 2018, a new algorithm is proposed to detect fake accounts and bots on Twitter. Classification algorithms of machine learning have been utilized, and they focus on technologies adopted to detect fake accounts and bots on Twitter. Classification algorithms of machine learning have been utilized to choose real or fake target accounts, and those techniques were support vector machine (SVM) and neural network (NN). They proposed new algorithms support vector machine and neural networks for successful detection of fake accounts. Both approaches adopt techniques of machine learning which highlight collection and methodologies of data reduction. It was also noted that the correlation collection records quality is dynamic among the other optimization algorithms, as redundancy is eliminated. The new algorithm classified 98% of the account of training dataset using fewer features [4].

3.2 Mudasir Ahmad wania, Nancy Agarwala & Suraiya Jabinb Syed Zeeshan Hussainb

In 2018, main focus is on fake profile detection using sentiments. The study is done on the post of real account user and fake profile user and similar emotions they use. The experiment is done on Facebook user profile post. In this paper, the author mainly focuses on fake profile detection using sentiments. The study is done on the post of real account user and fake profile user and similar emotions they use. The experiment is done on Facebook user profile post. Data are trained for 12 emotions, including 8 basic emotions, positive and negativeness, by the use of machine training techniques consequently, outliers are removed using noise removal technique. To train the detection model, many learning algorithms have been used including support vector machine (SVM), multilayer perceptron, J Rip, and random forest. The result shows that in the posts of unverified accounts three types of feeling, anxiety, shock, and faith are found least. For all three measures, precision, estimation, and AUROC, random forest provides the best result [5].

3.3 *Estée van der Walt and Jan Eloff*

Described the detection of fake identities of humans vs bots using machine learning models. Numerous fake accounts are enhanced with features used to detect bot accounts, and the collection of features has been extended to different supervised learning models. This paper focuses on the detection of fake identities of humans vs bots using machine learning models. Numerous fake accounts are enhanced with features used to detect bot accounts, and the collection of features has been extended to different supervised learning models. The highlights of human and machine accounts are indistinguishable. For occasion: The title it illustrates that the traits utilized to recognize programmer accounts fizzled to recognize human account points of interest appropriately. The effects of qualified computer models are predictive 49.75% of the best F1 performance. This is due to the fact that human beings are distinct from both in terms of behavior and characteristic, which cannot be modeled in the same way [6].

3.4 *Gayathri A, Radhika S & Mrs. Jayalakshmi S. L.*

In 2018, identification of fake accounts in media application by using support vector machines and neural networks is explained. Problem definition: identification of fake accounts in media application by using support vector machines and neural networks. In this report, they reflect a profound learning pipeline for identifying fake accounts rather than utilizing presumptions. It classifies the Sybil account cluster whether it is made by the same individual. The method starts by selecting a profile, at that point extricating the fitting characteristics and passing them to a proficient classificatory that categorizes the account as untrue or veritable alongside the input. Future work: utilizing more complex calculations. The other work line is to mimic multimodels utilizing the chosen highlights of other malware-based methods [7].

Author—Year	Objective	Techniques used	Accuracy
Naman Singh, Tusshar Sharma,Abha Thakral, Tanupriya Choudhury—2018	Detection of fake accounts	Support vector machine–neural networks	93%
Gayathri A, Radhika S, Mrs. Jayalakshmi	Detecting fake accounts in media application	Support vector machine & deep neural networks	–
Mudasir Ahmad wania, Nancv Agarwala, Suiaiya Jabinb, SyedZeeshan Hussain—2019	Analysis of real and fake users in Facebook based on emotions	Naive Bayes, J Rip, random forest	Random forest

(continued)

(continued)

Author—Year	Objective	Techniques used	Accuracy
Mehmet Şimşek, Oğuzhan Yılmaz, AsenaHazal Kahriman, Levent Sabah—2015	Detecting fake Twitter accounts	Artificial neural networks	–
Oscar S. Siordia, Daniela Moctezuma—2016	Features combination for the detection of malicious Twitter accounts	Feature extraction classification. Random forest	94%
Dr. Vijay Tiwar—2017	Analysis and detection of fake profile	Honest region. Network nodes. Network edge. Benign nodes	–
Aditi Gupta and Rishabh Kaushal	Detecting fake user accounts in Facebook	Data mining	79%

4 Techniques Used in Literature Survey

4.1 Support Vector Machine–Neural Networks

SVM-NN is applied to maximize classification accuracy because it achieves maximum accuracy using a reduced number of features and is implemented on the provided dataset by performing feature reduction by splitting of data testing and training data using eight cross-folds. Neural networks are created by developing neurons and forming a model which are trained and used to predict results. The prediction accuracy is counted separately by using formula.

$$\text{Accuracy} = \text{number of detected accounts/total number of accounts} * 100$$

4.2 Random Forest

Random forest is one of a classification algorithms that is unsupervised in nature. The fundamental concept is a selection of random samples from the provided dataset to create a decision tree followed by result prediction from each tree through voting. Finally, select the most voted results as a final prediction result. It is an ensemble method that achieves the highest accuracy as it reduced the overfitted data in sentiment analysis to identify true and Sybil accounts [5].

4.3 Artificial Neural Networks

Artificial neural networks system framework is a computational processing system that incorporates various interconnected computational nodes that work in a distributed manner to accumulate data from the input to optimize the final output. In implementations of ANN, a connection link is an actual number and the output is determined through a nonlinear input function in each neuron. The objective of the ANN is solving problems like a human brain, for example, in image recognition, pictures containing dogs can be detected by examining pictures manually marked as "dogs," "no dogs" or by using the results to recognize dog in other pictures. This occurs without prior knowledge that dogs have hair, ears, or dog-like heads, for example, characteristics are created by examples they identified automatically. ANN is used for classification, pattern recognition, clustering, regression analysis, prediction, social networks, and even in activities that earlier only humans can do like a painting [8].

4.4 Feature Detection

A convolutional neural network has a special architecture in which complex data characteristics are detected feature is referred to as an "interesting" portion of an in general, image is processed as the first-pixel operation has been performed and each pixel is examined to determine whether a function exists on that pixel. When this belongs to a larger algorithm, the algorithm typically only scans the image in the function field. As a prerequisite for integrated function detection, the Gaussian kernel usually smooths the image input to the size display and calculates one or more feature images that are often expressed concerning the local image derivative.

4.5 Feature Extraction

It is a process of reducing random variables from high-dimensional datasets (data with more than 10 dimensions). It is broken down into two more parts that are feature selection which is a simple approach to find subsets of input variables or attributes and feature extraction. Dimensionality reduction along with feature extraction is applied as a preprocessing step using techniques linear discriminant analysis and K-nearest neighbor to reduce. Feature reduction is required to store time and storage [9].

4.6 Classification

Classification is a complex phenomenon introducing the definition of classes according to the characteristics of the image following the selection of features such as color, texture, and multitemporal data. The feature dataset obtained is trained with supervised or unsupervised learning algorithms. Then, various classification techniques like extraction and segmentation are applied to the trained dataset to get appropriate results. At last, the classification technique is applied to all pixels by suing pixel classification or per-field classification. Image classification covers all unique features of an image and an important factor in the digital environment [3].

5 Proposed Methodology

The proposed algorithm utilizes the convolutional neural network for face recognition [10] and has an age classifier [12]. The features detected from CNN are classified and compared with existing data in the data warehouse for the data comparison.

Convolutional neural networks (CNNs) are a category of deep neural networks, most commonly used for visual imaging processing. They consist of neurons that optimize themselves through learning. CNN is also called as space-invariant artificial neural network (SIANN). CNN is used for image recognition, video analysis, image classification, time series forecasting, recommendation systems, natural language processing, and medical. An input image can be taken by each neuron and operated based on numerous ANN. The only significant distinction between current ANNs and CNNs is that they are primarily used inside objects in the field of pattern detection. CNN comprises of three types of layers. The layers are convolutional layers, pooling layers, and fully connected layers. CNN's fundamental functionality can be explained as:

- The pixel values for the image act as the input and form the input layer.
- The convolutive layer determines the output of the neurons and the linear system is intended to apply CNN's "elemental" activation function to the activation output produced by the previous.
- The pooling layer just downsamples the contribution along with the spatial dimensionality and decreases the number of cases, hence reducing computational times.
- The fully connected layers then perform the indistinguishable tasks as generic ANNs and attempt to generate category results from classifying activations
- CNN makes developing network architecture simpler.

5.1 Proposed Algorithm

The approach this system takes for fake profile detection is too limiting each user, to one and only one account on a particular platform. The system utilized facial recognition and age classification using CNN to create a unique facial print id of the account creator to identify him/her. This helps us to uniquely identify the customer by binding the facial print to the account, eliminating the possibility for a user to create a new fake account. The facial data is utilized to identify two different data, i.e., facial features and then the age prediction. This eliminates the possibility of creating new fake accounts by any user.

The input data after face detection and preprocessing are supplied to the two CNNs for face recognition and age classification. The data are processed by the convolutional network by passing through various layers of convolution, ReLU, pooling/sampling, and finally classified through a fully connected layer.

The process to detect a fake profile involves the following steps: data collection, optimization, face detection, face recognition, age classification, profile association, and profile detection. These steps are explained further in the section.

5.2 Data Collection

The data are collected from the user. This is collected from the social media platform and then passed to the system. This step collects two types of data firstly the profile details like name, age, and facial data from the sensors for face recognition. These data are used to process and identify whether the user who is trying to create an account is genuine or not.

5.3 Optimization

The raw data collected in the collection phase are optimized into the format required. The optimization is one of the important steps before processing of the data as it prepares the data and enhances the data so that during the processing of data the algorithm can produce better results.

5.4 Face Detection

Detection involves the detection of the facial data from the collected data and identifies the points required to be processed and specific to the face rather than the whole picture. The detection phase involves facial detection by identifying the points that are of use to us and eliminating the rest unnecessary points. This is done through template matching function. It defines a standard template for all the faces and where the different features can be identified independently like eyes, nose, mouth, contour, etc.

5.5 Face Recognition

Face recognition [11] is performed using the convolutional neural network. The data are preprocessed before feeding to the convolutional network. CNN utilizes various hidden layers of convolution, ReLU, and pooling. These layers are arranged in some fashion repeatedly to form the network. After passing through the various hidden layers, the output is put to the fully connected layer of the classifier for classification. The data from the output layer are put to comparison by the dataset in the data warehouse for profile detection.

5.6 Age Classification

Age classification involves identifying the age of the user using CNN to get near about the age of the user. A different CNN is used for the age classifier. The age predicted by the classifier is given as an estimated range. If the input of the user lies between the range, the profile is allowed for creation. This is an extra parameter just to verify the data input by the user and the verification facial data match the data input by the user at the time of creating the profile [13].

5.7 Profile Detection

This involves the main comparison of the data from the data warehouse that utilizes snowflake schema to store the data. The data after the process of classification are passed to this phase. The data are compared with the data in feature data, in the data warehouse. If a match is found, then the profile is flagged as fake and the account is not created. If the face is unique then the age detection data are utilized to compare the age detected by the algorithm with the data entered by the user. If the age matches near about with a minor difference, then the profile is created else it is detected as a fake verification. In the end, if both the tests are accepted then the profile is created and the facial data are moved to the data warehouse for mapping.

5.8 Profile Association

The data are associated with the user profile once it is created. The data are added to the data warehouse creating a unique identification of every use and also the data to which all platforms the user is signed up to are maintained. The data warehouse utilizes are snowflake schema to store the data by identifying the user by its facial patterns. This eliminates the possibility of creating a fake profile of the user on a platform he/she is not utilizing.

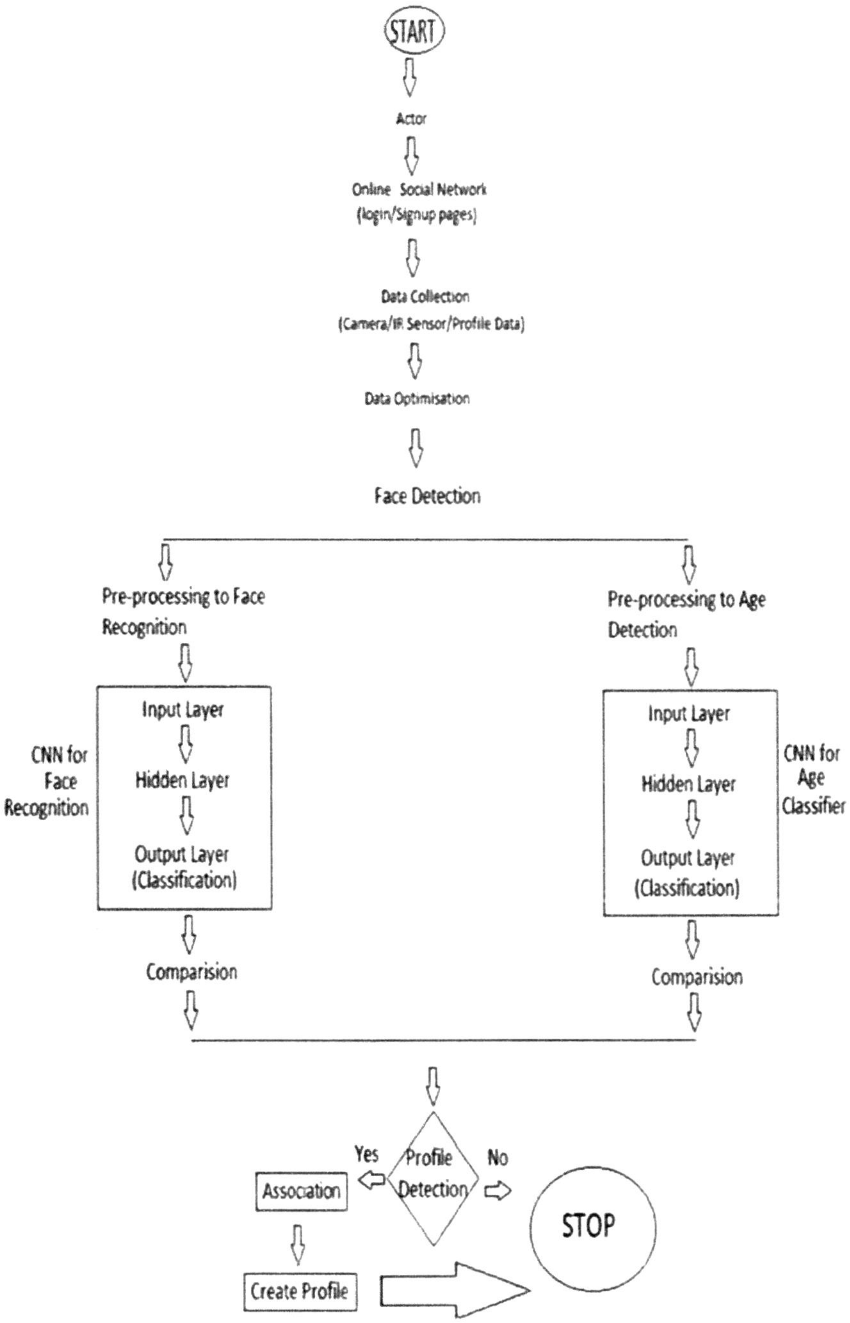
START
Actor
Online Social Network
(login/Signup pages)
Data Collection
(Camera/IR Sensor/Profile Data)
Data Optimisation
Face Detection
Pre-processing to Face
Recognition
Pre-processing to Age
Detection
Input Layer
Hidden Layer
Output Layer
(Classification)
CNN for
Face
Recognition
Input Layer
Hidden Layer
Output Layer
(Classification)
CNN for
Age
Classifier
Comparision
Comparision
Yes
Profile
Detection
No
Association
Create Profile
STOP

6 Future Scope

This paper proposes a method that utilizes CNN for face recognition and age classification to eliminate profile. Further implementation can be done using CNN and the accuracy and success of the method can be identified. The paper only proposes the idea of limiting each person to have only one profile on a social media platform to prevent the creation of a fake profile by mapping the creator's facial signature and profile data verification of age.

7 Conclusion

This paper tried to solve the identified problem of fake accounts on the online social media platform by the usage of neural networks and user profile. The paper proposes a system that is expected to prevent the creation of fake profiles as compared to the previous system that utilized machine-neural networks that eliminate fake profiles. Our knowledge does not have any such system that has taken this approach for fake profile elimination, as proposed in the system. The accuracy is not acclaimed as it has to experiment in the future implementation of this system.

References

1. Gayathri, A., Radhika, S., Jayalakshmi, S.L.: Detecting fake accounts in media application using machine learning. Special Issue Published in Int. J. Adv. Netw. Appl. (IJANA)
2. Khaled, S., El-Tazi, N., Mokhtar, H.M.: Detecting fake accounts on social media. In: IEEE International Conference on Big Data, vol. 6, pp. 101–110 (2018)
3. Hudson, B., Voter, B.R.: Profile characteristics of fake twitter accounts. Big Data & Society (2016)
4. Yang, Z., et al.: Uncovering social network sybils in the wild. Trans. Knowl. Discovery Data (TKDD) **8**(1) (2014)
5. Mudasir Ahmad wania, B., Agarwala, N., Jabinb, S., Hussainb, S.Z.: Analyzing real and fake users in Facebook network based on emotions. In: 11th International Conference of Communication System & Networks (2018)
6. Van der Walt, E., Eloff, J.: Using machine learning to detect fake identities: bots vs humans. IEEE Access (2018)
7. Gupta, A., Kaushal, R.: Towards detecting fake user accounts in Facebook. Indira Gandhi Delhi Technical University for Women, Delhi, India
8. Tiwari, V.: Analysis and detection of fake profile over social network. In: International Conference on Computing, Communication and Automation (2017)
9. Şimşek, M., Yilmaz, O., Kahriman, A.H., Sabah, L.: Detecting fake Twitter accounts with using artificial neural networks. In: 2018 Artificial Intelligence Studies (2018); David, I., Siordia, O.S., Moctezuma, D.: Features combination for the detection of malicious Twitter accounts. In: IEEE International Autumn Meeting on Power, Electronics and Computing (2016)
10. Oloyede, M.O., Hancke, G.P., Myburgh, H.C.: Improving face recognition systems using a new image enhancement technique, hybrid features and the convolutional neural network. IEEE Access **6**, 75181 – 75191. https://doi.org/10.1109/ACCESS.2018.2883748

11. Yin, X., Liu, X.: Multi-task convolutional neural network for pose-invariant face recognition. IEEE Trans. Image Process. **27**(2), 964–975. https://doi.org/10.1109/TIP.2017.2765830
12. Rafique, I., Hamid, A., Naseer, S., Asad, M., Awais, M.: Yasir, T.: Age and gender prediction using deep convolutional neural networks. In: 2019 International Conference on Innovative Computing (ICIC). https://doi.org/10.1109/ICIC48496.2019.8966704
13. Chen, S., Zhang, C., Dong, M., Le, J., Rao, M.: Using ranking-CNN for age estimation. in: 2017 IEEE Conference on Computer Vision and Pattern Recognition (CVPR). https://doi.org/10.1109/CVPR.2017.86

Error Correction Technique Using Convolution Encoder with Viterbi Decoder

K. B. Sowmya, D. N. Rahul Raj, and Sandesh Krishna Shetty

Abstract Communication method conveys information from a transmitter to receiver over an intermediate medium. The competence of received information is determined by medium and clatter. Therefore, medium and noise directly influence the efficiency of the data received. There is a great need for robust error correction techniques with ever accumulative usage of wireless expedients such as portable electronic items and broadband modems. Modern wireless communication schemes depend on advanced error rectification techniques for their suitable functioning. Hence, transferring and receiving information with less or no error, while utilizing the accessible bandwidth is a foremost proposal. As a result, there is a need for efficient error correction technique for present digital wireless communication systems comprising of high throughput, low power consumption, and less area. Although several block coding techniques have already been in use for error correction, and they are not suitable when dealing received information with random errors. Since convolution coding enables soft information decoding and soft information output which is available at communication channels helps in significantly improving error correction capability, they are preferred over conventional block codes. Convolution codes are best suited as forwarding error correction technique but as the constraint length increases the design decoders becomes very complicated.

Keywords Convolution encoder · Hard decision · Path metric unit · Viterbi algorithm

K. B. Sowmya (✉) · D. N. Rahul Raj · S. K. Shetty
Department of Electronics and Communication Engineering, RV College of Engineering, Bengaluru, India
e-mail: kb.sowmya@gmail.com

D. N. Rahul Raj
e-mail: rahulrajdn.ec17@rvce.edu.in

S. K. Shetty
e-mail: sandeshks.ec18@rvce.edu.in

P. Karuppusamy et al. (eds.), *Sustainable Communication Networks and Application*, Lecture Notes on Data Engineering and Communications Technologies 55,
https://doi.org/10.1007/978-981-15-8677-4_20

1 Introduction

Error correction codes are essential to make sure that information content of the message if altered could be corrected without having to add much of overhead to the system. Therefore, an efficient error-correcting technique which is simpler in terms of implementation is of main concern. Error correction techniques can be classified into block codes and convolution codes. The former involves an operation on chunks of data of predefined size which called as a packet in networking jargon. The most popular block coding is the hamming code which is a linear block coding technique. They are also referred to as algebraic codes because they involve generation of polynomials.

Whereas the convolution coding technique involves working on bit by bit basis, i.e., they operate on a stream of bits of some arbitrary length. While performing encoding using block coding, each packet or frame is considered to be independent of each other, but convolution codes operate on a continuous stream of data on a real-time basis. It then performs encoding operation to form a larger codeword which will be the actual data word that carries information in an encrypted pattern. Convolutional encoding is often used to correct random errors and is suited for communication channels with a limited capacity [1–4].

For decoding a convolution encoded data, maximum likelihood decoding is looked up which is accomplished here through the Viterbi algorithm. This is regarded as a hard decision for the codewords of considerable size, while there for soft decision techniques that find applications in more complex situations. The Viterbi decoder uses the trellis diagram for performing maximum likelihood decoding and this makes the computation much simpler. The Viterbi algorithm has become the most fundamental algorithm in decoding techniques and finds applications in CDMA, GSM, satellite systems, speech recognition, enhancement, and synthesis of speech and in many other technologies that employ data decoding. Also, a Viterbi decoding method can be employed to a codeword of larger size by segmenting the codewords into the words of some compatible size [4–7].

The main objective here is to perform functional analysis of error control coding technique, i.e., convolution encoding in terms of device utilization and power. The functional verification is performed, simulation using Verilog HDL in Vivado simulator.

2 Convolutional Encoder

Convolution encoding basically involves producing output codeword for any arbitrary input message bits. The encoding logic depends on the generator polynomial as specified by the designer. The size of the generator polynomial and the number of outputs produced per bit depend on the constraint length of the encoder. The encoder employs basic shift registers and the binary adders. For an encoder with a constraint

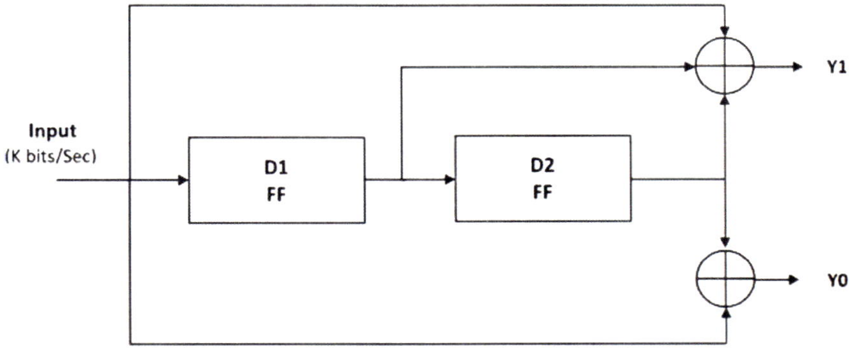

Fig. 1 Convolution encoder

length of 3, a total of two shift registers and two binary adders are required. As discussed, two entities define the performance of the encoder; they are constraint length(K) and coding rate (k/n). Here, k/n denotes that n bits are produced as output for k input bits.

The convolution encoder with a constraint length of 3 and the code rate of (1/2) is depicted in Fig. 1. Constraint length indicates how the output is dependent on the memory in the encoder. A code rate of (1/2) means that for everyone input bit, two bits are produced at the output. In fact, code rate and constraint length directly influence the efficiency of the encoding technique. Lower the code rate, better is the performance. Therefore, for high-end applications, the low code rate is preferred. And increasing constraint length directly influences the error correction capability but up to a certain limit after which code rate has to be decreased.

The generator polynomial for the given encoder is g0 = [101] and g1 = [111] as shown. The values of the registers at each instant denotes a state; these register states and the corresponding outputs for all possible input are replicated in the form of state diagram shown in Fig. 2.

Initially, it is assumed that the state is at s0 and for each input bit, the transition to the next state and a 2-bit output is produced in accordance with the state diagram depicted above.

For example, for the input bit sequence, input = {1,0,1,1,0,0,1,0} (begin from the all zero state) the transition to next states happens in accordance to each individual bit and resulting state transition sequence will be states = {10,01,10,11,00,00,10,01} and the generated encoded output sequence is outputs = {11,10,00,01,00,00,11,10}.

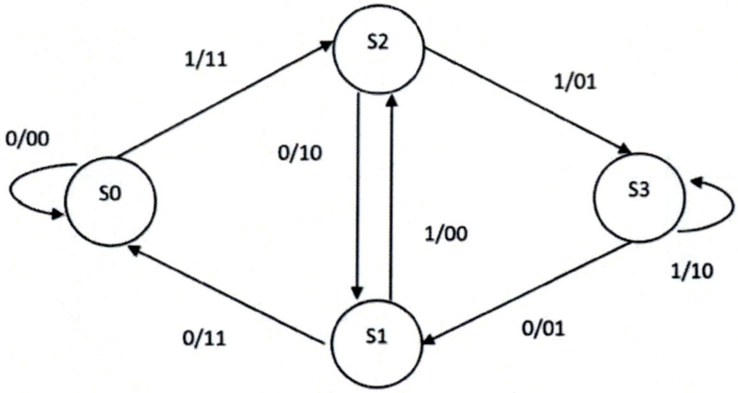

Fig. 2 State diagram of convolution encoder

3 Viterbi Algorithm

The encoded data produced by the predefined logic will be transmitted through a communication channel through a suitable line coding/digital modulation scheme. The communication channel is not ideal and hence noise interference leads to corruption in bits. Therefore, the received information sequence will no longer be the same as that originally transmitted sequence. The Viterbi decoding algorithm is regarded as the maximum likelihood algorithm that can correct any random error that occurs in the information sequence. The state diagram of the convolution encoder is now transformed in the form of a trellis diagram. This is essential as it makes the decoding simpler.

To draw the trellis diagram for this code, one has to draw four dots corresponding to each state for each clock cycle and then connect them according to various transitions that can take place between states. The trellis diagram for this code is shown in Fig. 4. From Fig. 4, the time axis corresponds to clock cycles, the four states are denoted by black dots, and the transitions between states are indicated by branches connecting these dots. On each branch connecting two states, two binary symbols indicate the encoder output corresponding to that transition. Also note that one always starts from the all-0 state (state 00), move through the trellis following the branches corresponding to the given input sequence, and return to the all-0 state. Therefore, codewords of a convolution code correspond to paths through the corresponding trellis, starting at the all-0 state and returning to the all-0 state. The number of states in the trellis increases exponentially with the constraint length of the convolution code.

The operations performed in the decoder is bifurcated into blocks. The operations performed in each block is described below. Block diagram of the Viterbi decoder is shown in Fig. 3.

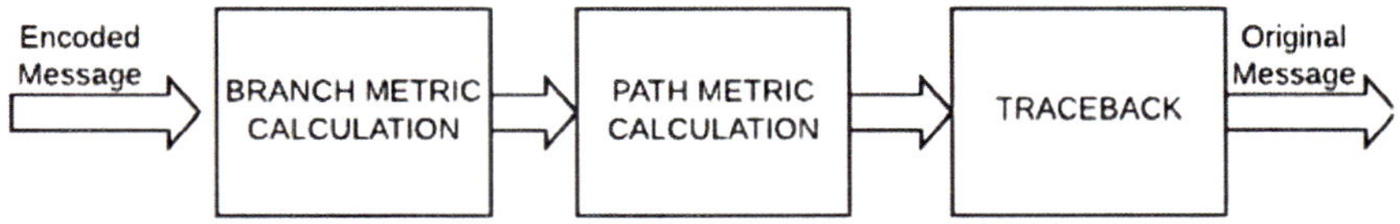

Fig. 3 Block diagram of the Viterbi decoder

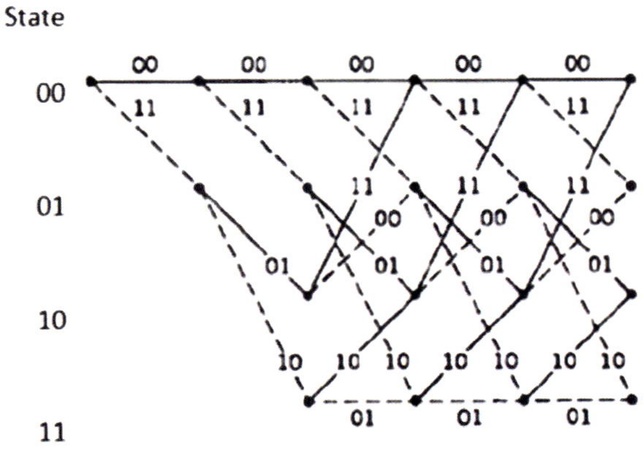

Fig. 4 Trellis diagram for the Viterbi algorithm

3.1 Branch Metric Unit

Since it is a maximum likelihood of hard decision, decoding the calculation of hamming distance is essential in this procedure. The received sequence is fragmented into blocks of two and each of these blocks is compared with the branch output value. The number of differing bits is noted at each of the nodes.

3.2 Path Metric Unit

The hamming distance is calculated at each of the nodes [8–10]. The metric is considered cumulatively that means at each node, the present node hamming distance is added with that of the previous node linked to it in trellis diagram. In case when two paths are leading to a particular node, the one which leads to minimum metric value will be considered.

3.3 Survivor Memory Unit

This unit depends on the results of the path metric unit [11, 12]. Starting from the initial node tracing the path will give us the decoded sequence. Therefore, tracing the path with minimum metric value leads to a decoded sequence which can be easily interpreted as the lines based on input bit 0 or 1 can be distinguished.

The flowchart of the Viterbi decoder with its components and their working logic is depicted in Fig. 5 (Fig. 4).

Verilog code for convolution encoder has been developed based on the state diagram shown in Fig. 1. Verilog code for Viterbi decoding is also developed. The code is written such that any arbitrary input bit sequence after encoding goes through AWGN channel where a random 1-bit error is introduced. Additive white Gaussian noise (AWGN) channel is a noisy channel used to model most of the communication

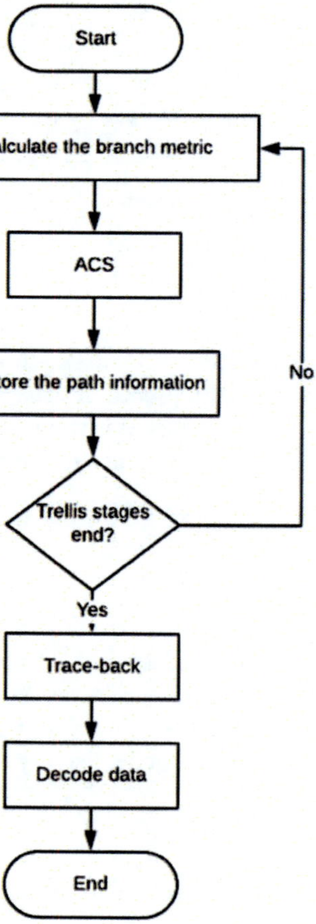

Fig. 5 Viterbi decoder flowchart

processes. It produces the same effect of any random processes that might occur in nature. It is called an additive because it adds to the noise inherently present in the system. The erroneous codeword is decoded in the Viterbi decoder, where the original bit sequence can be retrieved even though there is a one-bit error. That is according to Viterbi algorithm, the path with a minimum metric is selected which represents the original data.

4 Implementation Results and Analysis

The functional verification of the convolution encoder and Viterbi decoder was done using Vivado simulator for an 8-bit input data. For the 16-bit transmitted data with the 1-bit random error introduced in the channel, correction is done using the aforementioned Viterbi algorithm.

Vivado simulator is a component of Vivado Design Suite. It is a compiled-language simulator that supports mixed-language (VHDL, Verilog, etc.), TCL scripts, encrypted IP ,and enhanced verification. It supports behavioral and timing simulation. Vivado includes the in-built logic simulator ISIM that is used for high-level logic synthesis.

Synthesis for both the encoder and Viterbi decoder was done using Vivado tool and the device utilization and power consumption is tabulated in Table 1. Simulation results for both the designs written in Verilog HDL is shown below in Figs. 6 and 7. The synthesized design for encoder and decoder are shown in Figs. 9 and 10 respectively.

Table 1 Device utilization and power consumption

	Convolution technique	
	Encoder	Decoder
Slice LUT (20,800)	9	32
Slice Registers (41,600)	25	–
F7 MUX (16,300)	–	16
F8 MUX (8150)	–	8
Bonded IOB (106)	19	24
BUFGCTRL (32)	1	–
POWER	11.022 W	4.564 W

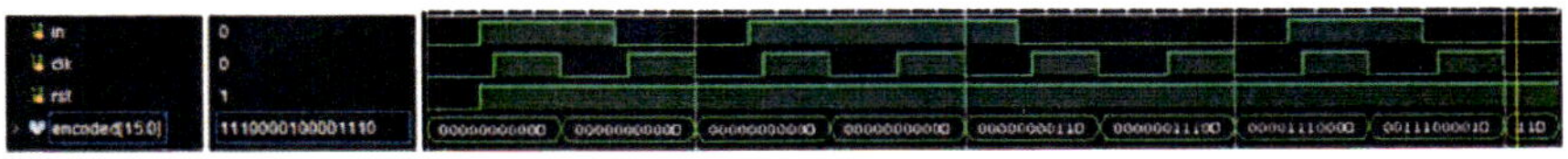

Fig. 6 Convolution encoder output

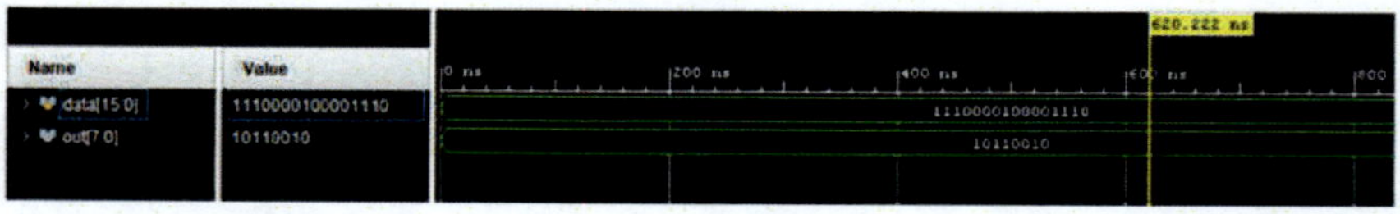

Fig. 7 Viterbi decoder output

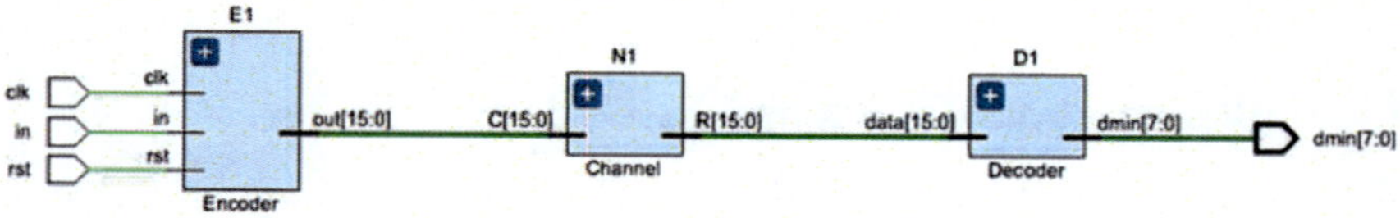

Fig. 8 RTL schematic of convolution coding technique

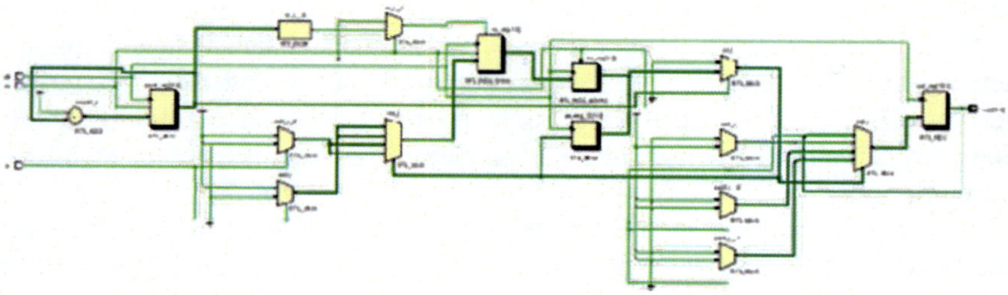

Fig. 9 RTL schematic of convolutional encoder

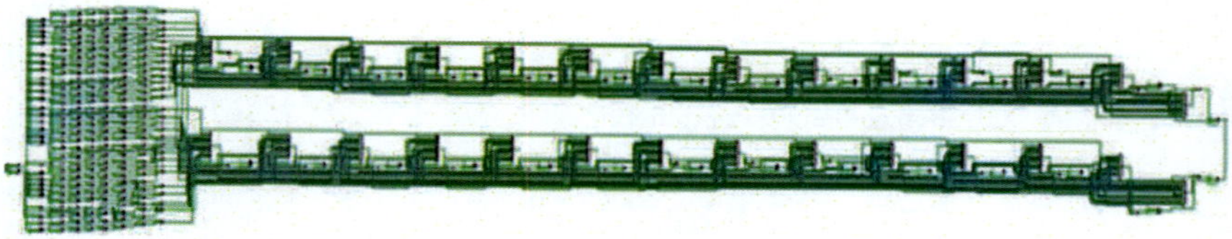

Fig. 10 RTL schematic of Viterbi decoder

Input message X = [10110010];
Encoded word Y = [1110000100001110];
Erroneous received word R = [1110001100001110].
Decoded sequence Z = [10110010].

Separate test benches were written for both encoder and decoder and the behavioral simulation output is shown in the above figures. The time resolution was fixed at 1 ps while simulation.

For convolution encoder, elapsed time for launching the simulation was found to be 33 s. Memory used: 693.75 MB.

For Viterbi decoder, elapsed time for launching the simulation was found to be 28 s. Memory used: 749.60 MB (Figs. 8, 9 and 10).

5 Device Utilization and Power Consumption

Table 1 shows the distribution of the components in the synthesis of encoder and decoder. It can be seen that the decoder is much more complex and involves more LUTs.

6 Conclusion

Convolution encoding is found to be a suitable error correction technique for code-words which might be prone to random errors. Here, an encoder and decoder have been designed for a code rate of 1/2 and constraint length of 3; further, an encoding technique can be used with different code rates while utilizing the channel band-width efficiently. The synthesis for the whole technique has been performed using Vivado. The Viterbi algorithm as discussed earlier finds the maximum likelihood path through the trellis. From the implementation point of view, whenever there is an ambiguity due to the same metric value while decoding, one path is selected arbi-trarily. This may or may not lead to the expected result. Therefore, when using in high-end applications such as deep space communications, the code rate is kept low so that decoding leads to a unique path.

Future work involves developing an area and power-efficient architectures for error correction. Nowadays, a power-efficient error correction technique which can correct even multi-bit error is of major concern. Although there are some modern techniques to correct burst errors, they do not qualify for power efficiency.

References

1. Mathur, N., Sharma, S.: Simulation of convolution encoder and Viterbi decoder using Verilog HDL. Int. J. Eng. Res. Gener. Sci. **4**(1) (2016). ISSN 2091-2730 539
2. Rambabu, M.K.: Implementation of convolution encoder and Viterbi decoder using Verilog. Int. J. Electron. Commun. Eng. **11**(1), 13–21 (2018). ISSN 0974-2166. © International Research Publication House
3. Pachlegaonkar, P.R., Patki, S.S.: Implementation of reconfigurable convolutional encoder and optimum adaptive Viterbi Decoder with multi booting and error detection on FPGA. Int. J. Eng. Res. Technol. (IJERT) **3**(10) (2014). ISSN: 2278-0181.
4. Habib, I., Paker, Ö., Sawitzki, S.: Design Space exploration of hard-decision Viterbi decoding: algorithm and VLSI implementation. IEEE Trans. Very Large Scale Integr. (VLSI) Syst. **18**, 794–807 (2010)

5. Tomas, V.: Decoding of convolutional codes over the erasure channel. IEEE Trans. Inf. Theory **58**(1), 90–108 (2012)
6. Jabeen, M., Khan, S.: Design of convolution encoder and reconfigurable Viterbi decoder. Int. J. Eng. Sci. 1(3), 15–21 (2012). ISSN: 2278-4721
7. Wong, Y., Jian, W., HuiChong, O., Kyun, C., Noordi, N.: Implementation of convolutional encoder and Viterbi decoder using VHDL. In: Proceedings of IEEE International Conference on Research and Development Malaysia, November 2009
8. Shaker, S., Elramly. S., Shehata, K.: FPGA Implementation of a reconfigurable Viterbi Decoder for WiMax Receiver. In: IEEE International Conference on Microelectronics, pp. 246–267 (2009)
9. Hema.S, Suresh Babu, V., Ramesh, P.: FPGA implementation of Viterbi decoder. In: Proceedings of the 6th WSEAS ICEHWOC (2007)
10. Azim, C., Monir, M.: Implementation of Viterbi decoder for WCDMA system. In: Proceedings of IEEE International Conference (NMIC 2005), pp. 1–3 (2005)
11. Viterbi, A.J.: Error bounds for convolutional coding and an asymptotically optimum decoding algorithm. IEEE Trans. Inform. Theory **2**, 260–269 (1967)
12. Fettweis, G., Meyr, H.: Feed forward architecture for parallel VITERBI decoding. J. VLSI Signal Process. **3**, 105–119 (1991)

Rice Grain Quality Determination Using Probabilistic Neural Networks

Kavita V. Horadi, Kshithij R. Kikkeri, Shravya S. Madhusudan, and R. M. Harshith

Abstract Food remains an eternal need for our survival. More so, rice is the staple diet of most south Asian countries. Rice quality is often degraded when impurities like broken or damaged seeds are present. It is a herculean task to determine the quality of grains using computer vision. In this paper, we have come up with a system that ascertains the class of rice grains. Rice grain images are acquired and preprocessed, and probabilistic neural network (PNN) algorithm is applied to the images. The classification has been done in accordance with chalkiness, HoG, and GLCM features. The system outputs good, average, or bad quality rice grains using PNN classifier. Our proposed model can be implemented in agriculture-based industries for grain evaluation purposes which will simplify the grading of rice grains for the consumers.

Keywords Computer vision · Rice grain quality · Probabilistic neural networks

1 Introduction

Rice is a source of vital minerals and vitamins and is nutritionally highly enriching and healthy. Rice is mostly consumed after it is boiled, or in some cases, it is milled into flour and consumed. It is a staple meal for 80% of the world's population [1]. It is energy-rich and nutrient-rich and has a low glycaemia score. India stands among the top two global producers of rice. Agriculturists conduct the study of grain size, gradation, and quality attributes manually. Such approaches are vulnerable to many issues, such as being extremely subjective, affected by human factors and working environments leading to incoherent performance. Also, the rate of salvage cleanup and recovery is low. A significant element in the agriculture industry is the harmless assessment of the food products to grade the product quality.

Non-destructive food product grading determines the quality, taking into consideration the form, color, internal deformations, etc., without actually breaking the

K. V. Horadi · K. R. Kikkeri (✉) · S. S. Madhusudan · R. M. Harshith
Department of Computer Science, BNM Institute of Technology, Bengaluru, India
e-mail: kshithij@ymail.com

P. Karuppusamy et al. (eds.), *Sustainable Communication Networks and Application*, Lecture Notes on Data Engineering and Communications Technologies 55,
https://doi.org/10.1007/978-981-15-8677-4_21

food products into its constituent parts. It is measured using advanced computer vision techniques without disturbing the internal somatic constituent particles of these items. Rapid developments in digital image processing hardware and software have stimulated hardcore research on the development of cutting-edge image processing systems for evaluating the class of various food items. Progression in computer technology is an invaluable contribution to the food processing domain such as ranking, sorting, and quality inspection. The accuracy of the system can be improved by employing features of texture. Accuracies in classification are very strong when specific features of the checked varieties are used. There are many benefits of artificial neural networks and probabilistic neural networks over fuzzy classifiers and statistical classifiers. The obvious and viable option for the classification of food products is neural network-based machine learning algorithms. The following parts of this paper are organized as follows. Section 2 presents a review of the existing literature. Section 3 gives a detailed description of PNN. The methodology is discussed in Sect. 4. Results are presented in Sect. 5. The paper is concluded with future enhancements in the final section.

2 Review of the State of the Art Techniques

Extensive research on rice grain identification based on various factors are being carried out till date. Of them all, computer vision and neural networks are apparent and natural choices for researchers for classification purposes. Siddagangappa et al. have developed an automated system using the probabilistic neural network to identify and assess the quality of three Indian rice varieties into their respective grades according to the quality. Six characteristics only, i.e., average RGB colors and three geometric elements, achieve reasonable accuracy in the ranking. The average identification success rate is 98%, average grade determination is 90%, and rice grading success rate is 92% [1]. Zahida Parveen et al. proposed an image processing algorithm to ascertain the quality of rice based on length, width, area, chalkiness, and color [2]. Another method used to grade rice using image processing was proposed by Wanputri et al. Previous researchers have applied various approaches to classifying rice grain images by employing particular characteristics such as rice form, length, chalkiness, color, and grain deformations. The authors have taken a total of 285 images and obtained an accuracy of 46.6% [3]. Harpreet Singh et al. proposed an image processing technique for grain classification. The techniques which are used to extract these features are kernels separation by sieving method, grain analysis by various image processing methods, seed count, image acquisition-based analysis of rice grains, and an inspection of kernel quality by automatic/semi-automatic grain analysis [4]. Deepika Sharma et al. proposed a neural network classifier-based technique to ascertain the grade of wheat and rice samples. The processing and the consequent classification have been done by using a neural network classifier. Grains are graded into good, medium, and bad quality [5]. Lilik et al. have come up with a physical tool to recognize rice varieties using LVQ neural network algorithm.

They have obtained an average accuracy of 70.3% but have obtained an accuracy of 90% for training dataset [6]. Pabamalie et al. have used concepts of backpropagation neural network with two hidden layers are employed for classification purposes. GLCM and color features are used for classification which yielded an accuracy of 68% [7]. Yet, another interesting application using probabilistic neural networks has been developed by Arun et al., which focuses on the classification of rice grains based on their aroma. Effective usage of sensor arrays has been the foundation of the startling research work on the electronic nose. Test results of unknown samples have been reported to give accuracy above 80% [8]. In a similar such work, Lili wu et al. have proposed another aromatic gas sensor-based application combined with probabilistic neural networks and principal component analysis to give an average identification accuracy of 91.67% [9]. Cutting-edge research has been carried out using advanced methods like mask region-based convolutional neural networks in the work of Kittinun et al. to obtain an accuracy of 67.25% [10]. Innovative methods like spatio-spectral deep convolutional neural networks have been employed in the work of Itthi et al. for non-destructive rice grain classification. They have employed hyperspectral image acquisition techniques to parallelly obtain the spatial and spectral features of grain images. The classification accuracy which they have obtained is 91.09% [11].

3 Probabilistic Neural Networks: A Discussion

Donald Specht's pioneering research on PNN had a profound impact on classic neural network classification-based applications. PNN is a highly complex structure and a feed-forward neural network. It contains input, pattern, summation, and output layers as depicted in the adjacent picture. Albeit PNN being a highly complex network, it has only one smoothing training parameter. Hence, there is not much of an effort required to train the database. Also, it is interesting to note that PNN works very efficiently even when the database consists of a small sample of data.

The structure of the PNN for the proposed system is depicted in the above Fig. 1. All the symbols and notations have been adopted as specified in the original research by Donald F Specht. MATLAB neural network toolbox too uses the same interpretation in the inbuilt function newpnn().

Input Layer: In Fig. 1, p represents the input vector, which is represented as the vertical bar. The dimension of the input layer is $R \times 1$. R have been considered to be 3.

Radial Basis Layer: In this layer, a computation of the dot product of the weight vector of each row of weight matrix W and the given input vector p is done. $Q \times R$ is the dimension of W. The vector distance between the ith row of W and the input vector p gives the ith element of the distance vector $||W - p||$. The dimension of this is $Q \times 1$. The symbol "−" represents the distance between the vectors. Now, the calculated $||W - p||$ is fused with the bias vector b as shown in Fig. 1 in a component-by-component

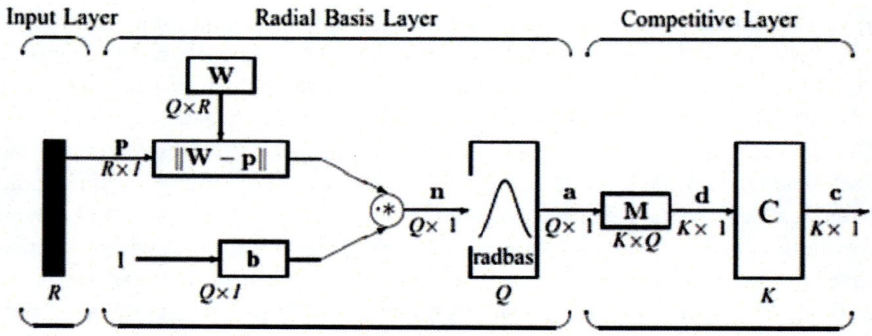

Fig. 1 PNN network for $R = 3$, $Q = 1500$, $K = 3$

form which is depicted as "*" in the figure. The outcome of the calculation is shown as $n = \|W - p\| \cdot * \, p$.

Competitive Layer: The concept of bias is absent in the competitive layer, unlike the radial basis layer. Here in this layer, a product of the output of radial basis function, a, and the weight matrix M is considered. This results in another output vector named d. This layer produces 1 correlating to the highest component of the vector d and produces zeroes to all the others. Competitive function results in another output depicted as c in Fig. 1. All values equating to 1 in c represent the quantity of rice grains which the proposed system classifies rightly. It can be used as the index to look for the scientific name of this plant. The output vector's dimension (K) is 3 as 3 types of rice grains have been considered.

4 Methodology

The proposed system could be used in the food and agriculture industries extensively for grading purposes. Figure 2 depicts the flow of events in our system. The initiation of the process is done with image acquisition. The proposed system considers rice grain image samples which are captured by a mobile camera of 16 megapixels and

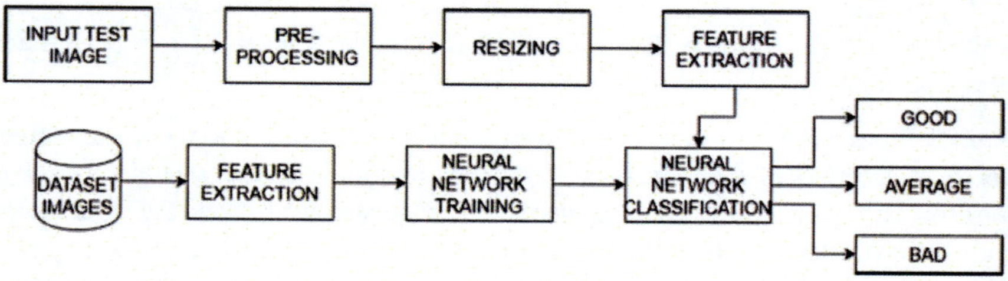

Fig. 2 Proposed architecture

then fed as an input to the image processing unit. MATLAB R2018a is employed to process the data under consideration. The classification result is then showcased on the MATLAB console in the form of bad or average or good quality rice grains based on the test input. A total of 1500 images have been considered and 500 images each of bad, average, and good quality rice grain images for our work, out of which 1200 images are used for training and the other 300 images are used for testing purposes. The images are primarily labeled according to their quality. For simplification purposes, only Indian basmati rice has been taken to be of good quality. All other slender, long, and short rice grains are considered to be of average quality. Broken rice grains, grains with deformations, are considered to be of bad quality. The images are preprocessed using standard preprocessing techniques like filtering, thinning, erosion, and dilation. Features which include chalkiness, histogram of oriented gradients, and gray level co-occurrence matrix are considered. Geometric features are not considered due to the nature of the database images. The features extracted are stored in feature vectors. All the images are then trained. Sample rice grain images acquired for testing purposes are then tested for their accuracies which are labeled as good, average, and bad quality rice grains. The percentage of chalkiness is determined using standard techniques. The flow of the system is depicted as below.

1. **Image Acquisition**: Grain images are taken with the help of a camera. Preferably a high-resolution camera may be used so that the accuracy does not vary.
2. **Preprocessing**: Various preprocessing techniques are applied to increase the classification accuracy. The techniques include filtering, thinning, erosion, and dilation.
3. **Image Dataset**: Dataset image is taken into consideration
4. **Feature extraction**: A variety of features are taken into consideration. These features are:

   a. Chalkiness
   b. Histogram of oriented gradients
   c. Gray level co-occurrence matrix

5. **PNN Training and classification**: PNN is applied to train all the images. Further, the features which were extracted earlier are employed to classify the rice grain images into three classes, viz bad, average, and good quality rice grains.

5 Results

See Figs 3, 4, 5, 6, 7, 8, 9, and 10.

The results have been tabulated as below (Tables 1 and 2):

Of the 300 images taken as a testing dataset, 50 images in each grade are considered for accuracy calculation to get an average accuracy of 97.89%.

Fig. 3 Good quality grain image is taken from Dataset

Fig. 4 Average quality grain image is taken from Dataset

6 Conclusion and Future Work

The proposed system classifies rice grains based on chalkiness, histogram of oriented gradients, and gray level co-occurrence matrix features into good, average, and bad classes using probabilistic neural networks (Table 3).

The comparison reveals that the proposed algorithm outperforms other methods. Routinely, PNN is known to outpace other algorithms in a machine vision application. The proposed system is completely automated and proves to be of great use to the agriculture-based industries and has found out that the results for rice grain grading using probabilistic neural networks to be highly lucrative and obtained high levels of accuracy.

Fig. 5 Bad quality grain image is taken from Dataset

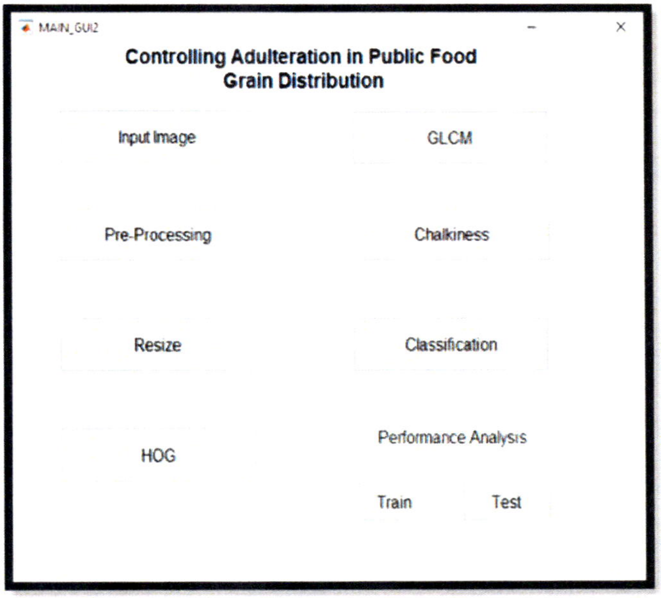

Fig. 6 GUI of the proposed model

Images have been acquired without any advanced technology-enabled camera in research conditions. Acquiring images could have been done under different lighting conditions. The proposed method could be extended to a wide variety of other species of rice grains. Further, our method can be extended to other food grains. The major research extension would be to optimize the neural network to increase the rate of accuracy.

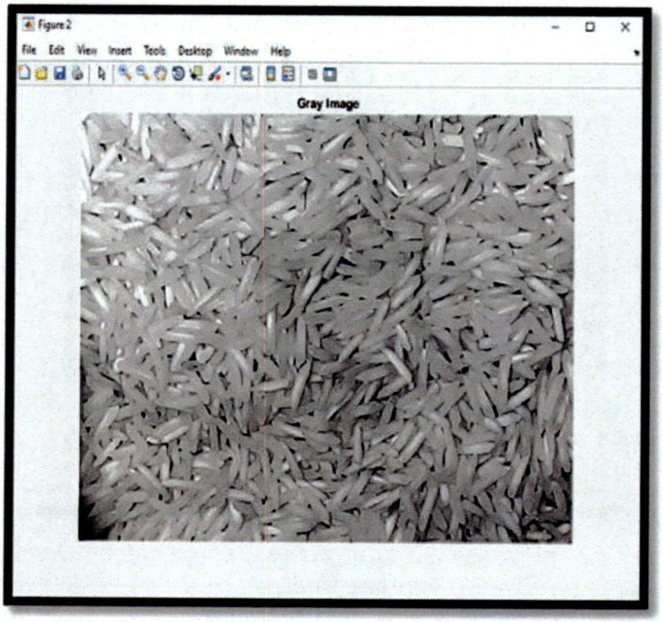

Fig. 7 Grayscale image

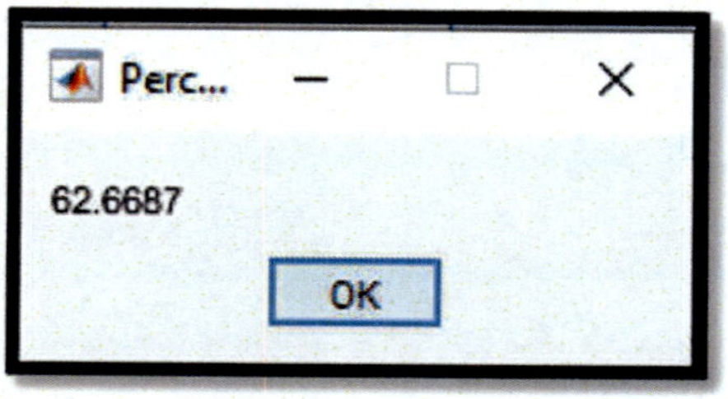

Fig. 8 Percentage of chalkiness

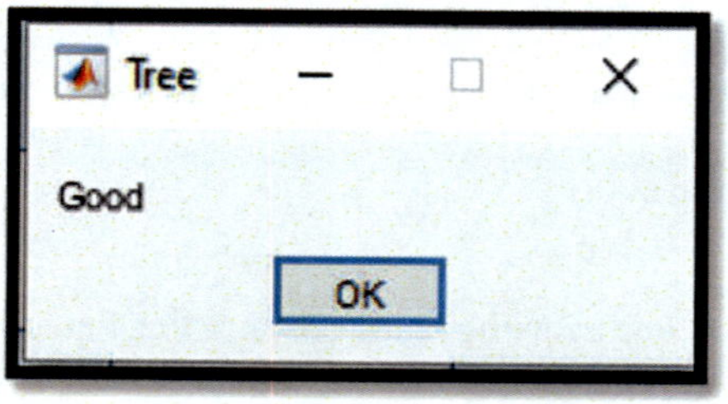

Fig. 9 Classification result

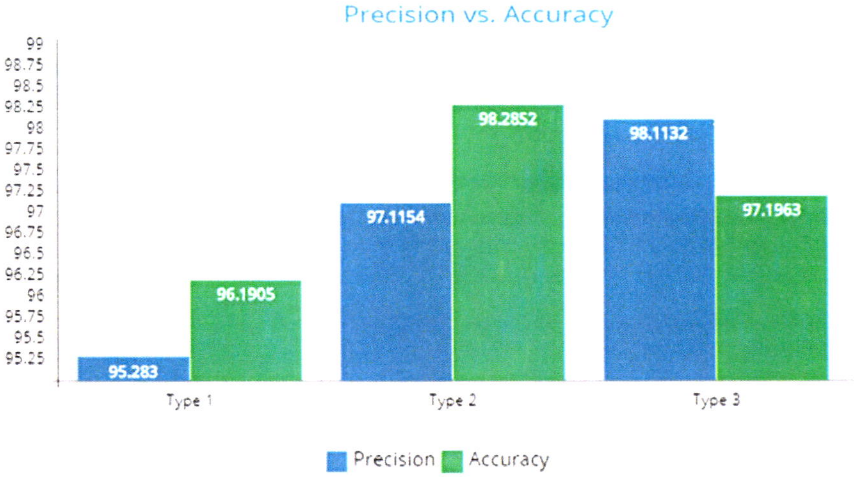

Fig. 10 Precision versus accuracy for each grain type

Table 1 Recognition rates for various types of grains

Type of rice grain	Recognition rate (%)
Good	100
Average	100
Bad	100

Table 2 Results based on various metrics

Metric	Value (%)
Accuracy (Train)	100
Accuracy (Test)	97.89
Precision	96.83
Recall	97.33

Table 3 Accuracy comparison

Scheme	Accuracy
Deep CNN [6]	95.5
LVQ neural network [7]	70.3
BPNN [8]	80.5
PNN [9]	93
PNN [10]	87.5
Region proposals based CNN [11]	67.25
Spatio-spectral deep CNN [12]	91.09
Proposed in [13]	80.64
Our approach	97.89

References

1. Siddagangappa, M.R., Kulkarni A.P.A.: Classification and quality analysis of food grains. IOSR J. Comput. Eng. 16
2. Neelam Jyoti Gupta: Identification and classification of rice varieties using neural network by computer vision. Int. J. Adv. Res. Comput. Sci. Soft. Eng. **5**(4), 992–997 (2015)
3. Singh, H., Rawat, C.S.: Image processing techniques for analysing food grains. In: International Conference on Computing Methodologies Communication (ICCMC) 978-1-4244-7164-5. IEEE (2019).
4. Sharma, D., Sawant, S.D.: Grain quality detection by using image processing for public distribution. In: International Conference on Intelligent Computing and Control Systems 978-1-5386-2745-7. IEEE (2017)
5. Lin, P., Li, X.L., Chen, Y.M., He, Y.: A deep convolutional neural network architecture for boosting image discrimination accuracy of rice species. Springer Science+Business Media, LLC, part of Springer Nature (2018)
6. Sumaryanti, L., Musdholifah, A., Hartati, S.: Digital image based identification of rice variety using image processing and neural network. TELKOMNIKA Indonesian J. Electr. Eng. **16**(1), 182–190 (2015)
7. Pabamalie, L.A.I., Premaratne, H.L.: A Grain Quality Classification System. IEEE (2010)
8. Jana, A., et al.: Classification of aromatic and non-aromatic rice using electronic nose and artificial neural network. In: 2011 IEEE recent advances in intelligent computational systems
9. Wu, L., Yuan, C., Lin, A., Zheng, B.: Identification of early moldy rice samples by PCA and PNN. In: ICCIP 2012, Part I, CCIS 288, pp. 506–514 (2012). © Springer-Verlag Berlin Heidelberg 2012
10. Aukkapinyo, K., Sawangwong, S., Pooyoi, P., Kusakunniran, W.: Localization and Classification of Rice-grain Images Using Region Proposals-based Convolutional Neural Network. © Institute of Automation, Chinese Academy of Sciences and Springer-Verlag GmbH Germany, part of Springer Nature 2019
11. Chatnuntawech, I., Tantisantisom, K., Khanchaitit, P., Boonkoon, T., Bilgic, B., Chuangsuwanich, E.: Rice classification using spatio-spectral deep convolutional neural network arXiv: 1805.11491v3
12. Qin, S., Liu, C., Huang, S.: Identification rice varieties based on k-means clustering algorithm and BP neural network. In: 2017 2nd International Conference on Advanced Materials Science and Environment Engineering (AMSEE 2017) ISBN: 978-1-60595-475-2
13. Sinha, A.: Dimension analysis and gradation of rice grain using image processing technique. J. Mod. Trends Sci. Technol. **6**(10), 69-73 (2020)

Maintenance of Automobiles by Predicting System Fault Severity Using Machine Learning

S. Shivakarthik, Krishnanjan Bhattacharjee, M. Swathi Mithran, Swati Mehta, Ajai Kumar, Lulua Rakla, Soham Aserkar, Shruti Shah, and Rajkumar Komati

Abstract The automobile and transport industries have always tried to reduce downtime by the way of preventive maintenance. In the past decade, electromechanical sensors have become more accurate along with novel innovations in IoT and machine learning, and automobiles have leveraged this. In this paper, an end-to-end open-source predictive maintenance solution is presented to predict the severity of faults in a car using onboard historical and real-time sensor data using IoT and machine learning. Sensor data is collected from a Suzuki Swift VXi Model, and classifiers like logistic regression, random forest, and gradient boosting trees are used to train the data with imputed faults. F1 score and AUC are used as evaluation metrics. An end-to-end onboard diagnostics (OBD) data to the user dashboard pipeline is proposed with final predicted faults visible on a real-time dashboard.

S. Shivakarthik · K. Bhattacharjee · M. S. Mithran · S. Mehta · A. Kumar
Centre for Development of Advanced Computing, Pune 411008, India
e-mail: shivakarthiks@cdac.in

K. Bhattacharjee
e-mail: krishnanjanb@cdac.in

M. S. Mithran
e-mail: swathimithran@cdac.in

S. Mehta
e-mail: swatim@cdac.in

A. Kumar
e-mail: ajai@cdac.in

L. Rakla · S. Aserkar (✉) · S. Shah · R. Komati
MIT College of Engineering, Pune 411038, India
e-mail: soham339@gmail.com

L. Rakla
e-mail: raklalulua@gmail.com

S. Shah
e-mail: shahshrutiketan@gmail.com

R. Komati
e-mail: rajkumar.komati@mitwpu.edu.in

P. Karuppusamy et al. (eds.), *Sustainable Communication Networks and Application*, Lecture Notes on Data Engineering and Communications Technologies 55,
https://doi.org/10.1007/978-981-15-8677-4_22

Keywords Internet of things · Predictive maintenance · Machine learning

1 Introduction

Maintenance, since the invention of machines, has relied upon human specialists in the field of machines to diagnose the faults and problems in the structural, electrical, and software subsystems that make up the machine. With advancing technology and cheaper manufacturing, more machines are produced at a higher rate than they can be accommodated for repair in case a need for the same arises. At times, the wrong intuition of technicians often costs the customer unnecessary repair in their machines while other times cause significant downtime leading to losses. The former was an example of preventive maintenance and the latter, of run-to-failure. This approach has high chances of vehicle downtime and unexpected failures. Predictive maintenance involves the use of data-driven analytics to predict faults before they occur by using relevant machine learning algorithms as the backbone of the system, which can reduce the downtime of the vehicles and the cost of maintenance while increasing consumer safety. Predictive maintenance is highly cost-effective, saving roughly 8–12% over preventive maintenance, and up to 40% over reactive maintenance (according to the U.S. Department of Energy) [1]. A report by Deloitte Insights, predictive maintenance (PdM) promises a 20 to 50% reduction in time required to plan maintenance, a 10–20% increase in equipment uptime and availability, and a 5–10% reduction in overall maintenance cost [2–4].

2 Literature Survey

Machine learning techniques have spread throughout both the manufacturing and service industries to improve productivity, quality, and efficiency of entity maintenance. Toyota's Lexus has extensive systems that alert dealerships and owners when problems occur. General Motors can even predict failures of individual components. The predictive part of maintenance alerts is still in its infancy. However, as data on components grow and machine learning improves, more drivers will get notices about impending failures rather than alerts about current ones [5]. Arindam Chaudhuri proposed predictive maintenance for Industrial IoT of vehicle fleets using the hierarchical modified fuzzy support vector machine [6]. A hydraulic motor fault detection system based on a grid-based classifier developed by Alzghoul [7] was able to attain 89% classification accuracy when using predicted data [8, 9]. PdM is not just limited to cars but is also used in all forms of vehicles such as buses, trucks, trains, and tanks [10–13]. Air-pressure system faults in Scania trucks have been predicted by using random forest and data mining [10]. Corazza et al. relied on a maintenance software to analyze data coming from sensors assessing the engine oil quality, therefore detecting potential breakdowns and replacing spare parts in advance [11]. ARIMA model has been used for predictive maintenance of railways in [12]. A paper from the Netherlands Defense Academy mentions the use of predictive maintenance for their military vehicles using physical failure models [13]. Support vector machines-based cuckoo search algorithm was implemented by Dr. Shakya to predict faults in wind turbines for efficient windmill performance [14]. Dr. Manoharan proposed a way to identify leaks in the air conditioning of automobiles using image detection [15].

After the conclusion of the literature survey, it was found that major predictive maintenance approaches use proprietary data that is generated by the vehicle. This paper proposes using the ELM327 OBD-II sensor in combination with a micro-controller NodeMCU and Bluetooth module HC-05, and an end-to-end open-source solution which can be taken up by all the cars complying with the OBD-II port for connectivity and run machine learning algorithms using sensor data for fault prediction for the various parts of the vehicle.

3 Proposed Real-Time System

The electronic control unit (ECU) is an embedded system in automotive electronics that controls the electrical systems or subsystems in a vehicle. It stores all the fault codes related to the sensor parameters and governs each subsystem to ensure the smooth and efficient functioning of the vehicle and gives different warnings to the vehicle dashboard. Controller area network (CAN) protocol is used for communicating with the ECU. ECU has an outlet through which the maintenance professionals can communicate using the onboard diagnostics (OBD) port to receive all the sensor values. OBD-II scanner, with the help of its on board controller, converts the hex values of sensor parameter IDs to decimal values. The fields are filled in a comma-separated values (.csv) file and exported from the controller to the Bluetooth transmitter.

Hence, this paper suggests an independent remote monitoring system consisting of a Bluetooth module to receive the data from the ELM327 OBD-II scanner, which is controlled by a NodeMCU acting as the controller to send and receive packets of data as soon as the OBD-II dispatches a specific amount of information. NodeMCU, having received the data in a comma-separated value (.csv) format from the Bluetooth module, sends the same to a Web server. The Web server hosts the trained machine learning model as a pickle file and the script, which converts the output of the sensor readings to a JavaScript Object Notation (.JSON) format. This JSON formatted data can be visualized on the real-time dashboard along with the fault prediction of the model. Figure 1 shows the architecture of the proposed real-time system.

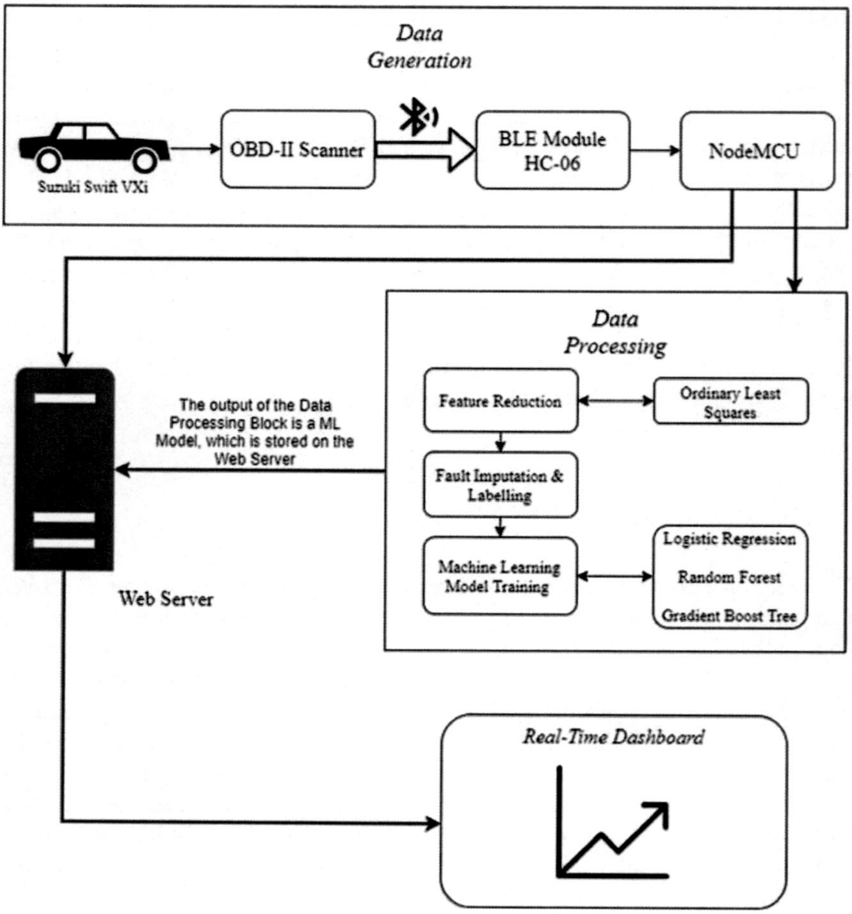

Fig. 1 Architecture of the proposed real-time system

4 Data Generation

OBD-II scanner is plugged into the OBD-II port of the vehicle, present under the steering wheel of the car. OBD-II scanner is a microcontroller circuit that communicates with the ECU of the vehicle, converts the encoded data stream sent by the ECU from hexadecimal format to the decimal format for more comfortable viewing. The scanner is having Bluetooth connectivity and can be connected to any device having Bluetooth.

The OBD-II scanner comes with an open-source app for both Android and iOS, which connects to the phone to deliver real-time sensor values generated from the car. It also features a logging capability of these values, which is being utilized in our project. The logging was enabled in two conditions—driving on a highway and in heavy traffic. The logs are exported as CSVs. The various parameters available are GPS coordinates, acceleration, intake air temperature, intake manifold pressure, barometric pressure, bearing, torque, engine power, engine load, and so on.

From all the available parameters, firstly, those parameters which are not relevant to predicting a fault were dropped like GPS coordinates, device time, and altitude, among others. The dataset had 2630 samples.

5 Feature Selection

To reduce the number of features, the ordinary least squares method was used [16].

OLS uses the following equation-

$$m = \frac{\sum(x_i - \overline{x})((y_i - \overline{y})}{\sum(x_i - \overline{x})^2} \tag{1}$$

where
$x=$ independent variables.
$\overline{x}$ = average of independent variables.
y = dependent variables.
$\overline{y}$ = average of dependent variables

$$Y = \beta_0 + \beta_1 X_1 + \beta_2 X_2 + \cdots + \beta_k X_k + \epsilon \tag{2}$$

where ϵ is the error term, β are the true parameters of the regression, and X are the independent variables.

From the method of least squares regression, if the p-value for a variable is less than the significant value (0.01), then it implies that changes in independent variables are associated with changes in the target-dependent variable with 99% confidence.

Running OLS on the features, it was found how the dependent variables like engine load, engine RPM, torque, etc., are related to the independent variables like engine coolant temperature, barometric pressure, intake manifold temperature, etc. Eight independent features were isolated which contained sensor readings of various subsystems like ignition, exhaust, and fuel. These comprise of intake air pressure, intake air temperature, CO_2, O_2, engine coolant temperature, barometric pressure, fuel flow rate, and throttle position.

6 Fault Imputations and Modeling

By analyzing the central tendencies of the raw data, and the manufacturer's reference manual, the normal range of sensors was observed. Next, a fault range (above typical

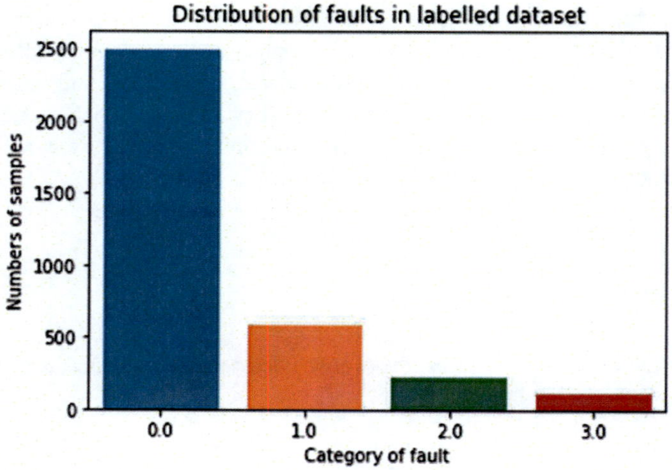

Fig. 2 Distribution of dataset by fault

values) for our features is defined. The fault thresholds of every sensor were further split into three parts to signify minimum fault, medium fault, and maximum fault. The output was labeled '1' if maximum faults were in minimum range, '2' if maximum faults were in medium-range, and '3' if maximum faults were in maximum range.

Around 800 samples of fault were generated randomly, which were then randomly combined with the original data of 2630 samples.

The main characteristics of the dataset are that the output is multiclass with continuous input feature variables.

Figure 2 shows the distribution of the dataset based on the category/severity of the fault on the *X*-axis and the number of samples on the *Y*-axis. It is seen that the dataset is imbalanced. Hence, the classifiers used to train the data are logistic regression, random forest, and gradient boost tree.

6.1 Multiclass Logistic Regression [17] is used as a base classifier due to its simplicity.

$$f(k, i) = \beta_k \cdot x_i \tag{3}$$

A linear predictor function $f(k, i)$ is used to predict the probability of the observation i which has outcome k, where β_k is the set of regression coefficients associated with outcome k and x_i (a row vector) is the set of features associated with observation i.

6.2 Random Forest [18], an ensemble classifier, operates by constructing a multitude of decision trees at training time and outputting the class that is the mode (maximum) of the classes of individual trees. It was selected as it does not overfit and is robust to outliers. After training, predictions for unseen samples x' can be made by taking the majority vote in the classification trees.

$$\hat{f} = \frac{1}{B}\sum_{b=1}^{B} f_b(x') \quad (4)$$

RF is applied using Gini (GDI) Diversity Index (a measure of node impurity) for splitting criteria calculated by

$$(\text{gdi}) = 1 - \sum_i p^2(i) \quad (5)$$

One hundred estimators were trained using the Gini Index.

6.3 Gradient Boosting Tree (GBT), another ensemble model was used. The gradient boosting method assumes a real-valued y and seeks an approximation $\widehat{\boldsymbol{F}}(\boldsymbol{x})$ in the form of a weighted sum of functions $\boldsymbol{h}_i(\boldsymbol{x})$ from some class H, called base (or weak) learners. In GBT, after calculating the loss, to perform the gradient descent procedure, a tree is added to the model that reduces the loss.

$$\hat{F}(x) = \sum_{i=1}^{M} \gamma_i h_i(x) + \text{constant} \quad (6)$$

As GBT yielded the best results, the team tuned it further with the best parameters obtained from grid search. Those were *100 estimators* and *0.8 subsampling*.

7 Results

The evaluation metrics used are accuracy, F1 score, and AUC (Area under receiver operating curve).

- Accuracy is the percentage of the total number of predictions that were found correct. Since the data is imbalanced, it is not entirely reliable.
- F1 score is accuracy, which is measured using Precision and Recall, where Precision is true positive accuracy (TPA), and Recall is the true positive rate (TPR). It is calculated from the confusion matrix [19–20].

$$\text{Precision} = \frac{\text{TP}}{\text{TP} + \text{FP}}$$

$$\text{Recall} = \frac{\text{TP}}{\text{TP} + \text{FN}}$$

$$\text{F1score} = 2 * \frac{\text{Precision} * \text{Recall}}{\text{Precision} + \text{Recall}}$$

8 Confusion Matrices

Figures 3, 4, 5, and 6 showcase the confusion matrices of different algorithms which describe the performance of a model based on a set of data for which true values are known. Also, Table 1 lists evaluation metrics: AUC score, F1 score, and accuracy of different classifiers.

Figure 7 visualizes the evaluation metrics of different classifiers listed in Table 1. Based on these metrics, the best performing model was selected.

The best performing model—*tuned gradient boost tree*—was then exported as a serialized object—a pickle file that contained the final trained parameters of the model. A service postman is used to simulate real-time data sending to make fault predictions. An open-source service freeboard.io is used, which takes the data from a flask server (having the model) and visualizes it on a dashboard. The dashboard consists of nine gauges—one for each sensor reading and one for fault prediction.

The results seen on the dashboard are shown in Fig. 8.

Figure 8 shows a case of a medium fault, which is represented by the fault level 2.

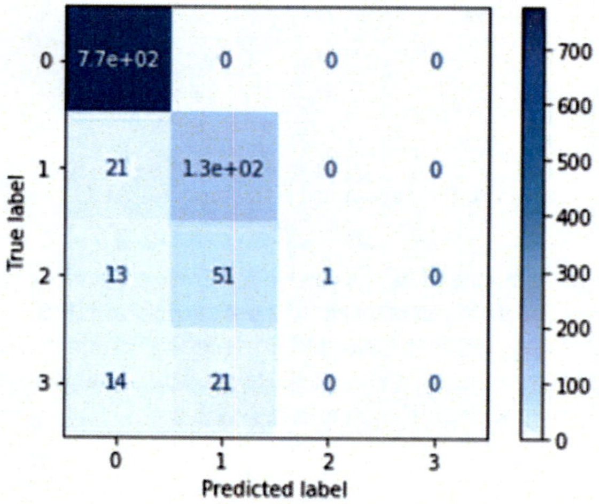

Fig. 3 Logistic regression

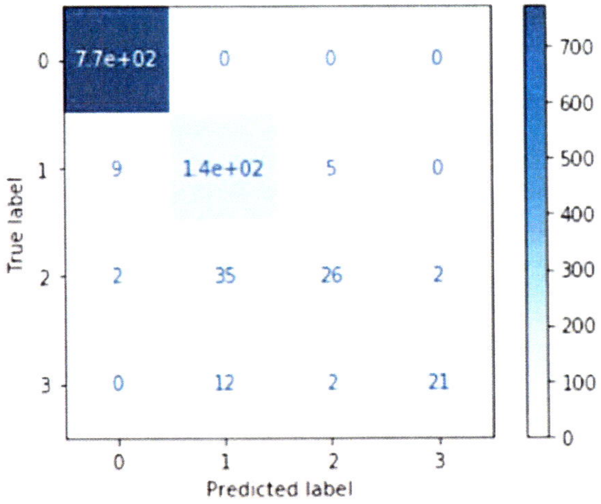

Fig. 4 Random forest

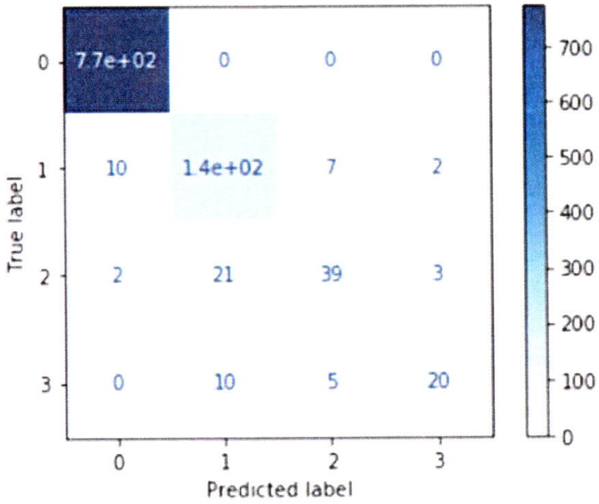

Fig. 5 Gradient boost tree

9 Conclusion

As more vehicles become "smart" and machine learning models can be trained on "Big Data," companies are increasingly leveraging these resources to make their vehicles have less downtime and lower costs of failures. A way to monitor vehicles and predict faults is presented in this paper. Three algorithms, including logistic regression, random forest classifier, and gradient boost tree, have been used for fault prediction. The main objective is to reduce the fault frequency of systems in vehicles.

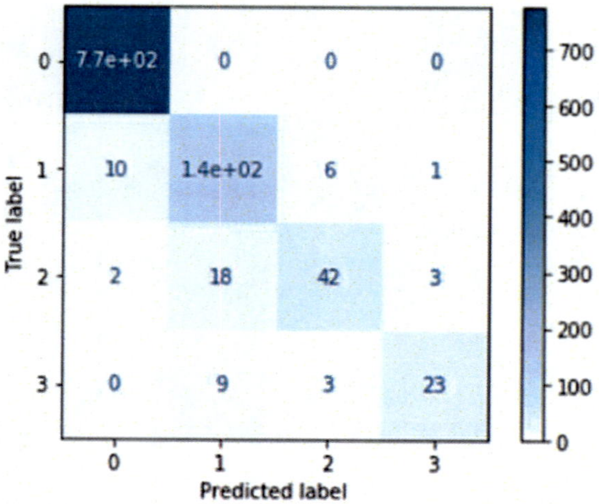

Fig. 6 Gradient boost tree (tuned parameters)

Table 1 Evaluation metrics of different classifiers

Model name	F1 score	AUC	Accuracy
Multiclass logistic regression	0.84	0.96	0.88
Random forest	0.92	0.95	0.96
Gradient boost tree	0.93	0.99	0.94
Gradient boost tree (with hyperparameter tuning)	0.947	0.996	0.95

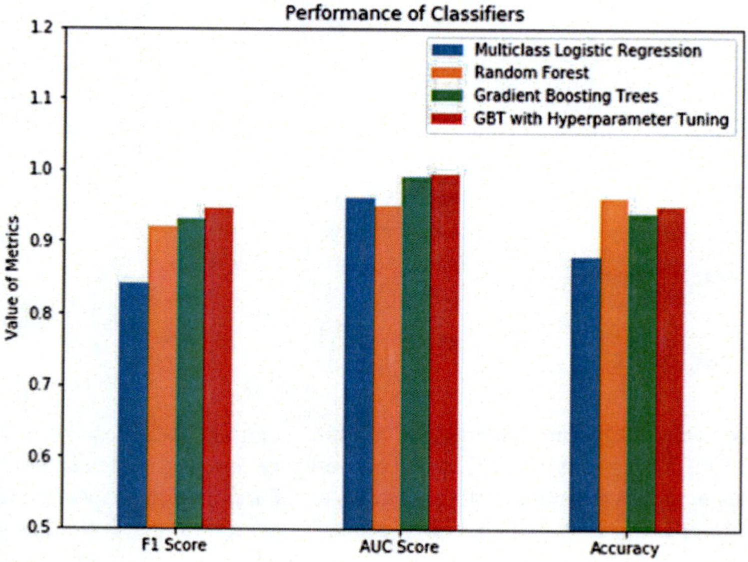

Fig. 7 Evaluation metrics visualized

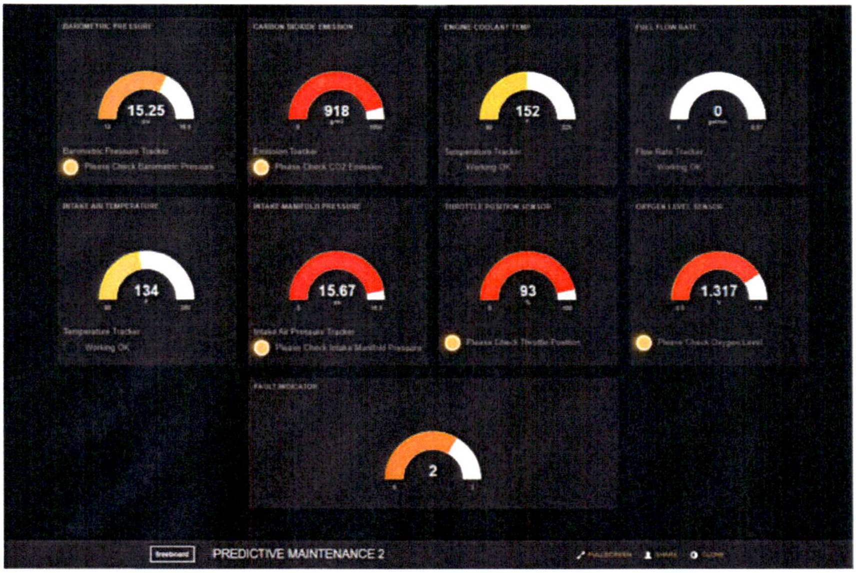

Fig. 8 Fault is medium, i.e., 2

Our contributions to this paper include a proposed open-source end-to-end automobile predictive maintenance system. Real-time car sensor data is collected using OBD-II sensor. The car sensors that are majorly responsible for the maintenance and breakdown of the system have been selected for training. Next, the fault is induced in our data using the threshold values and uses a novel methodology to label our data with three categories of fault. This labeled data is used to train classifiers. Lastly, the trained model is hosted on a dashboard, which takes in real-time sensor readings and can predict faults.

10 Future Scope

This research can be extended by turning the algorithms to maximize efficiency and adding more data. Also, future work can be made to focus on receiving inputs from the N-number of vehicles in real-time, and each user can have a unique dashboard link. The system can be made customized for different models of a vehicle or even types of vehicles (EV, hybrid, buses, etc.). The users can have a unique login ID and password provided by the manufacturers to access the data and view the real-time fault predictions made using the trained model relevant to their vehicle, which adds a security layer for the user and makes the system robust.

References

1. CSS Electronics: OBD2 Explained—A Simple Intro (2020)
2. Making maintenance smarter: (n.d.). https://www2.deloitte.com/us/en/insights/focus/industry-4-0/using-predictive-technologies-for-asset-maintenance.html
3. Automotive Predictive Maintenance Market: (2019). https://www.transparencymarketresearch.com/automotive-predictive-maintenance-market.html
4. Automotive Predictive Maintenance Market Projected for Healthy Growth Rate During the Forecast Period. (2020, April 13). https://techresearch.over-blog.com/2020/04/automotive-predictive-maintenance-market-projected-for-healthy-growth-rate-during-the-forecast-period.html
5. Kwanten, A.: Big data lets OEMs, dealers predict when vehicles will need service
6. https://www.autonews.com/fixed-ops-journal/big-data-lets-oems-dealers-predict-when-vehicles-will-need-service
7. Arindam, C.: Predictive maintenance for industrial IoT of vehicle fleets using hierarchical modified fuzzy support vector machine (2018)
8. Alzghoul, A., Löfstrand, M., Backe, B.: Datastream forecasting for system fault prediction. Comput. Ind. Eng. **62**, 972–978 (2012). https://doi.org/10.1016/j.cie.2011.12.023
9. Yoon Bo-ram(2018) Hyundai Kia Motors develops new technology to detect and report engine abnormalities
10. https://www.yna.co.kr/view/AKR20181018111200003?input=openapi
11. Hee-Nam, N.: Hyundai Motor develops technology to detect car engine vibration and inform vehicle anomalies. https://www.businespost.co.kr/BP?command=article_view&num=99304 (2018)
12. Gondek, C,. Hafner, D., Sampson, O.: Prediction of failures in the air pressure system of Scania trucks using a random forest and feature engineering (2016). https://doi.org/10.1007/978-3-319-46349-0_36.
13. Corazza, M.V., Mangialardo, S., Petracci, E., Tozzi, M., Vasari, D., Verdalle, E.: Research and innovation in predictive management for bus fleets: the Ravenna case study (2018)
14. Francis, F., Mohan, M.: ARIMA Model-based real-time trend analysis for predictive maintenance, 735–739 (2019). https://doi.org/10.1109/ICECA.2019.8822191.
15. Tinga, T.: Predictive maintenance of military systems based on physical failure models. Chem. Eng. Trans. **33**, 295–300 (2013). https://doi.org/10.3303/CET1333050
16. Durbin, J., Watson, G.S.: Testing for serial correlation in the least squares regression. I. In: Kotz, S., Johnson, N.L. (eds.) Breakthroughs in Statistics. Springer Series in Statistics (Perspectives in Statistics). Springer, New York, NY (1992)
17. Manoharan, S.: Image detection, classification, and recognition for leak detection in automobiles. J. Innovative Image Proc. (JIIP) **1**(02), 61–70 (2019)
18. Shakya, S.: Performance analysis of wind turbine monitoring mechanism using integrated classification and optimization techniques. J. Artific. Intell. **2**(1), 31–41 (2020)
19. Cox, D.R.: The regression analysis of binary sequences. J. R. Stat. Soc. Ser. B **20**(2), 215–242 (1958)
20. Ho, T.K.: Random decision forests. In: *Proceedings of the 3rd IJDAR*, pp. 278–282 (1995)
21. Liu, Q., et al.: Non-destructive monitoring of netted muskmelon quality based on its external phenotype using random forest. PLoS One **14**(8), e0221259 (2019) (Public Library of Science)
22. Hand, D.J., Till, R.J.: A simple generalization of the area under the ROC curve for multiple class classification problems. Mach. Learn. **45**, 171–186 (2001)
23. Metrics and scoring: quantifying the quality of https://scikit-learn.org/stable/modules/model_evaluation.html
24. Suzuki Swift Service Manual

Endoscopic Wireless Capsule Compressor: A Review of the Existing Image and Video Compression Algorithms

B. Sushma

Abstract Wireless Capsule Endoscopy (WCE) is considered as one of the non-intrusive diagnostic methods to explore diseases in the complete digestive tract without any discomfort or need for anodyne. To enhance the quality of diagnosis, some essential specifications of WCE such as video/image resolution, frame rate and transmission speed need to be refined. However these features demands more power. Existing capsule endoscopes are powered by button cell batteries and most of the power is utilized by video transmission which restricts obtaining quality video with better frame rate and better resolution. In order to overcome these issues many researchers proposed design of a low-complexity and low power video/image compression algorithms that produces higher frame rate with better quality, lower bandwidth and transmission power. This paper summarizes the existing image and video compression methods for WCE, with a significant focus on understanding their design and architecture in-terms of complexity, memory, power and diagnostic yield.

Keywords Wireless capsule endoscopy · Compression · Reconstruction quality · JPEG · DPCM · Discrete cosine transfor · Distributed video coding · Wiener-Ziv coding

1 Introduction

The gastrointestinal (GI) or digestive system endoscopy is a paramount procedure to spotting GI disorders [1, 2] such as GI cancer, tumours and bleeding. Endoscopy enables the direct view of human GI including the oesophagus, large bowel, colon and parts of the small intestine [3, 4]. Many other technical methods have been evolved to identify GI tract diseases, such as X-Radiography, angiography, ultrasonography

B. Sushma (✉)
CMR Institute of Technology, Bengaluru, India
e-mail: sushma.b@cmrit.ac.in
URL: https://www.cmrit.ac.in/

P. Karuppusamy et al. (eds.), *Sustainable Communication Networks and Application*, Lecture Notes on Data Engineering and Communications Technologies 55,
https://doi.org/10.1007/978-981-15-8677-4_23

and scinitigraphy. Unfortunately all these are revealed to have very poor diagnostic yields even for bleeding region identification [5].

The eminent way to detect GI diseases is direct observing of the GI tract. The development of wired endoscopy enabled to view all parts of the GI tract except lower part of the intestine. Because of the capability of allowing physicians to directly observe the GI tract, diagnosing of GI abnormalities by using endoscopy has become a common method. However, restricted by physical reasons the wired endoscopy cannot examine the whole GI tract, leaving the lower small intestine as an unsighted region due to its complex and curvy structure. Wired endoscopy is very inconvenient and causes intense pain for patients and dispirits the patient to go through the procedure.

In order to provide comfortable and painless diagnosis to the patient, WCE is used in recent days. It is capable enough to get into some critical parts of the GI tract [6, 7] which is not possible by wired capsule. WCE is a very important technology in clinic based on its assets. In WCE, after eight hours of fast, the patient engulf a capsule which moves through the GI tract by peristaltic motion. While in motion through the GI tract, the capsule captures images and transfer image data through RF transmission to a mobile data recorder unit attached to the patients abdomen as shown in Fig. 1. During this course of action full monitoring of the specialist is not required and patients are allowed to do other activities which saves their time.

Even with lot of primacy of WCE, this system is still reviewed as an undeveloped and lot of features needs an improvement. Diagnostic yield (DI) [8] remnants as the major case and to achieve a better DI, important factors such as quality and resolution of the image, frame rate and capsule life need to be improved. In terms of DI, existing WCE capsules are sub-standard when compared to wired endoscopy. Most of the commercial WCE capsules operate at 320×320 pixels frame resolution and 2–12 fps frame rate with a battery life of 7–12 hours. Low image resolution and frame rate leads to wrong diagnosis due to failure of detecting sensitive regions.

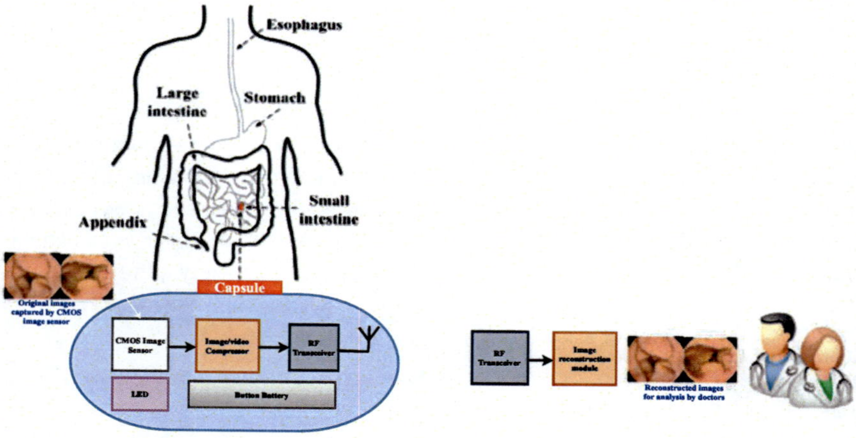

Fig. 1 Scanning of GI tract using WCE capsule and RF transmission of captured image data

Image resolution is an important criterion for the quality diagnosis because images may be distorted while physician enlarges images for detailed diagnosis. Most of the battery power is consumed by RF transmission of the image data and Image resolution enhancement causes increase in power consumption for processing and transmission. Also, capsule battery should afford sufficient power to the capsule to be in life for more than 16 hours [9] for the complete examination of GI tract. More power utilization leads to fast draining of the battery which leads to incomplete examination of GI tract.

The power requirement of WCE increases with the quality and resolution of the captured image. The most basic commercial wireless capsule essentially needs around 30 mW of battery power [10]. When the condensed dimensions and latest battery technologies are considered, the amount of power yielded by a typical WCE battery is practically close to around 25 mW [11]. Employing high resolution endoscopy capsules to improve the image quality utilizes high resolution image/video devices, high light intensity and more bandwidth of the RF transmitter. All these features results in increase in the power requirement that ranges upto 200 mW. To overcome all these serious issues, wireless power transmission (WPT) system is recommended as WCE future.

WPT system refers to a method that transfers electrical power using wireless mode from a transmitter to receiver in the form of electromagnetic radiation. Focusing on working of WPT system is beyond the scope of this paper but lists the drawbacks of WPT system used in WCE to emphasize the importance of video/image compression. The WPT is a proven technology which is useful to powerup health care devices, especially in biomedical implants. However, it is found to be useful in biomedical implantable devices in pacemakers and retinal implants. In WCE the approach of transferring power, fails due to (i) Remarkable distance between the transmitting coil and receiving coil; (ii) Unstability in the amount of received power due to unpredictable capsule motion and movement ; (iii) Size of the capsule increases due to additional receiving coil and its power harvesting circuits; (iv) Adverse health effects in the patient due to current density induced in human tissues; and (v) Damage caused inside the GI tract due to overheating of the receiving coil.

Due to all these limitations of WPT, capsule endoscope still depends on the cell battery for the power requirements. As mentioned previously, to increase the battery life and to reduce the power of the image data transmission which saves the communication bandwidth, the image data need to be compressed first. So, the challenge is how to transmit images of high resolution with low power consumption. There have been lot of works done in designing low complexity and low power consumption image compression algorithms with the goal of improvising quality of the image. This review paper imparts technical details of the existing image and video compression algorithms designed for WCE with a focus on understanding their low power architectures interms of complexity, compression rate and reconstructed image quality.

2 Wireless Capsule Endoscope System

The typical WCE system consists of three main components. (a) A swallowable battery powered disposable electronic capsule endoscope (CE). (b) Data recorder unit which consists sensing system with sensing pads and a battery pack. (c) A workstation computer with application software to process the video/image. In WCE procedure [12], the patient swallows the capsule which transits through GI tract by peristaltic motion. While in moving across the GI tract, the capsule captures images and transmits the image data by a wireless RF transmitter to a image data recorder. CMOS based image sensor with an LED and lens to capture images while passing along the GI tract, the ASIC based transceiver with an antenna transmits image data to the receiving unit. Power is supplied for all these operations by two miniature sized batteries. The image recording subsystem consists eight sensing pads clamped to the chest and abdomen. This helps in recording image data and in detection of the position of the capsule. The recorded image data is processed by a image processing software in the workstation computer for the doctor's review. Since the development of the endoscopic capsule, three major companies have introduced the WCE capsules approved by Food and Drug administration (FDA) in different names for use in medical diagnosis of complete GI tract. Table 1 lists the commercially available wireless endoscopic capsule specifications [13–15]. Given imaging is the leading contributor in the field of capsular endoscopy and manufactured different capsules for different parts of GI. Pillcam SB/SB2/SB3 for a complete small bowel imaging, Pillcam ESO/ESO2 for imaging oesophagus and colon imaging by Pillcam COLON/COLON2. All the commercially available capsules uses RF for wireless data transmission except Microcam which uses human body as conductive medium for data transmission. In most of the commercial WCE systems the image compressors are placed in the capsule hardware. The detailed information about image compression techniques used in these systems is not available in literature. But some vague information about the compression algorithms is available by patents published by some inventors. Table 2. lists compression methods used in different capsules along with the name of the inventor. For the good quality diagnosis the image should have minimum 512×512 frame resolution and around 10 fps [16]. Though pill-cam could achieve the required frame rate, it fails in achieving the required resolution. Among the existing capsules Olympus endocapsule achieves good frame resolution compare to other capsules. But its frame rate is too less (2 fps). To achieve better frame rate with better frame resolution, compression rate (CR) needs to be improved at the cost of low power consumption.

3 Compression Principles Suitable for WCE

The main objective of using WCE is for diagnosing the human GI tract for any diseases or abnormalities. Image quality interms of frame resolution and frame rate is very important for the quality diagnosis by the physician. Since the endoscopic

Table 1 Specifications of commercially available capsules

Company	Capsule	Size L × D in mm	Weight in gms	Battery life in hours	Frame resolution	Frames per second	Field of view in degrees	Transmission type	FDA approval
Given imaging	Pillcam SB	26 × 11	3.7	6–8	256 × 256	2	140	RF	Yes
	Pillcam SB2	26 × 11	3.7	8	256 × 256	2	156	RF	Yes
	Pillcam SB3	26.2 × 11.4	3	8	340 × 340	2–6	156	RF	Yes
	Pillcam ESO	26 × 11	3.7	20 min	256 × 256	14	140	RF	Yes
	Pillcam ESO2	26 × 11	3.7	20 min	256 × 256	18	169	RF	Yes
	Pillcam colon	31 × 11	2.9	10	256 × 256	4	156	RF	No
	Pillcam colon2	31.5 × 11.6	–	10	256 × 256	4–35	172	RF	No
Olympus America	Endocapsule	26 × 11	3.5	8–10	1920 × 1080	2	145	RF	Yes
Intro-medic	Microcam	24.8 × 10.8	3.25–4.7	10–12	320 × 320	3	170	Human body	Yes
Jianshan	OMOM	27.9 × 13	6	6–8	640 × 480	2	140	RF	Yes

L Length, *D* Diameter, *RF* Radio frequency, *FDA* Food and drug administration

Table 2 Compression in commercial Endoscope capsules

Company	Inventors	Compression technique
Given imaging	Glukhovsky et al. [17]	JPEG and MPEG Compression algorithms
	Zinaty et al. [18]	Image data in mosaic form is color space transformed into (Y, Cb, Cr, Gdiff) and compressed explicitly by a cost effective and fast lossless image compression algorithm
	Avni et al. [19]	The residue obtained by taking the difference between the previous and present frames is transmitted instead of the entire frame
	Horn et al. [20]	Images are captured with different frame rate in different parts of GI. Compression is applied only in the case of higher frame rate
Olympus America	Shigemori et al. [21]	Uses the conventional compression algorithm, where some pixels are removed from the raw image before compression
	Bandy et al. [22]	A new concept is used where frame rate and resolution varies based on the speed of the capsule. If the speed is less, images are captured at low resolution and data is compressed by combining group of blocks from multiple frames

capsule depend on button batteries which supplies 3 V at 55 mAh for approximately 8 h which implies an average power delivery of around 20 mW. This amount of power is hardly sufficient to transmit even low-resolution images at low frame rates. Keeping high image quality with reduction in the transmission power and bandwidth is a major challenge and paves the way for developing the high quality compression algorithms. Though, there exists some traditional compression algorithms, they are unsuitable for WCE application due to their high memory requirements and more power consumption.

Design of a right compression algorithms will increase the lifetime of the capsule by reducing data size and power consumption. A particular data compressor in WCE has to satisfy the following compression principles.

- Large scale internal image processing cannot be performed due to low power supply. Each image element can consume very few low complexity operations. Other, conventional compression algorithms are not suitable, because they need more than 100 mW to perform a sufficient compression [23].
- An experimental study [24] recommends more frames per second which can detect more lesions and decreases the missing lesions rate. A minimal frame rate of 10 fps is highly needed for flawless endoscopic diagnosis. when the capsule is traveling with high speed, better frame rate is expected. At least a frame rate of 25 fps is required to accomplish the smoothness quality of conventional wired endoscopy.
- For diagnostic accuracy, the reconstructed image quality which is measured by peak signal to noise ratio (PSNR) should be high. High quality images consist high PSNR. According to a study [25] minimum medical image quality in PSNR required is around 35 dB.
- Memory size should be less for compression. Large silicon area is consumed by memory. Also accessing memory consumes more power. Therefore saving intermediate data on chip should be avoided and it should enable the data process in one pass without using feedback.

The next section covers the image compression techniques involved in WCE related research.

4 WCE Image Compression Techniques

WCE Image compression algorithms can be divided into lossy, near Lossless and lossless compression techniques. In lossless compression, redundancy is removed without any information loss and this process is reversible. Certain degree of distortion is introduced in near-lossless compression with out loss of any valuable information. On the other way, high compression ratio of 50:1 is achieved using lossy compression techniques preserving visual image quality of the reconstructed image. Lossy compression is more beneficial in case of power constrained applications.

4.1 Lossless Image Compression Techniques

Lossless image compression techniques mainly depends on decorrelation and entropy encoding. Spatial redundancy between pixels can be removed by using prediction based techniques. In predictive coding, pixels are decorrelated by a good predictor and data is converted to a pattern, so that there is no dependency between adjacent image pixels. Mainly, these includes delta modulation, Differential pulse code modulation (DPCM) or JPEG compression based techniques. However, most of the predictive endoscopic compression techniques consider DPCM and simple estimation. All these approaches are mainly lossless and near-lossless with one or two exceptions. The following section discusses various predictive based techniques and their architectures.

Prediction Based Image Compression Architectures Using DPCM Most of the prediction based compression architectures are usually lossless and consists of reversible colour transformer, DPCM and Golomb Rice encoder. Khan et al. [26] proposed lossless image compressor which consumes less power. This type of compressor encodes and outputs a 32-bit vector. A parallel-to-serial converter is used to connect to commercial transceiver which accepts data serially using serial peripheral interface protocol. Most of the CMOS sensors scans the image in the raster order so that commercial CMOS image sensors can be directly interfaced with the proposed compressor. Fante et al. [27] proposed another DPCM based compression technique. Here the pixels are quantized before DPCM coded. This technique proposes a new modified Golomb Rice encoder with Golomb Rice parameter calculated adaptively using a single context. khan et al. [28] proposed another DPCM based predictive encoder. The difference between two luminance component is encoded in Golomb-Rice code where the chrominance difference is encoded in unary code. Malathkar et al. [29] proposed an hybrid DPCM technique to improve the compression efficiency by taking advantage of WCE image properties. Another DPCM method proposed by Malathkar et al. [30] works in near-lossless mode and uses fixed threshold. It gives better compression ratio compared to other near-lossless methods.

Low Complexity JPEG Based Image Compression Architectures Compression performance of the JPEG based low complexity compression architectures is the highest compare to other lossless compression algorithms. Controlling the quality and extraction of region of interest (ROI) features are added into the low pass filter (LPF) stage, which improves the compression efficiency by compressing ROI in lossless mode while remaining regions are compressed in lossy mode. Many JPEG based compression algorithms have the capability to incorporate quality controlling in the LPF. But detection of ROI region algorithms suitable for WCE systems is not used in any state of the art compression schemes.

Xie et al. [31] proposed a different low power hardware architecture mainly comprises the internal circuitry of WCE system used for diagnoses and real-time endoscopic image scanning. Additionally, a low complexity micro-current generator is deployed to stimulate the muscles in the GI tract to contract and push the capsule forward when a physician wants to ignore some regions. Liu et al. [23] proposed a modified form of JPEG-LS compression algorithm where each pixel consumes very few operations and processed in single pass. This method acquires a context template pixel by causal interpolation method. Median predictor is replaced by gradient predictor, which improves pixel prediction accuracy. Finally, the context index is obtained by quantization of gradient context. The hardware architecture consists three parallel pipelined data paths which operates in normal mode, interruption regular mode. To avoid computation overhead, based on mode determination a single pipeline is activated. Chen et al. [32] proposed context free compression algorithm to reduce the memory usage and introduced prediction module to increase the compression performance. This architecture of the proposed image compressor comprises a register bank, predictor, restoration unit of pixel, a predictor, a run mode module, and a modified Golomb rice encoder to improve compression performance. Wireless capsule endoscope captures around 50,000–80,000 images during its journey in GI tract and all the images does not carry any information. Very few frames are only considered for diagnosis. Extracting only ROI based parameters and high temporal correlation between consecutive frame is considered which improves the compression performance.

4.2 *Lossy Image Compression Techniques*

DCT Based Image Compression Architectures DCT is one of the most significant and primary transforms in the area of image and video data compression and widely used in many applications. DCT based compression algorithms are lossy and gives very good compression performance due its energy compaction property. Computation of DCT is highly complex and consume lot of power and requires complex hardware. Lot of research has been done to propose computationally simple algorithms which require very less number of operations. Various multiplier based DCT designs have been proposed in the aim of decreasing multiplier count and complexity. These multiplier based DCTs have complicated structure due to heavy wiring which occupies more silicon area and more power is consumed at the encoder. To optimize DCT computation some research works suggests to minimize DCT coefficient computations, while some approaches aim to reduce the processing of DCT coefficients. In [32, 33] two different methods of adaptive JPEG compression based on pruning also called as region selection have been presented. Only a selected region (in [32] triangular and in [33] squared portion) is processed by dividing into blocks. DCT coefficients are computed for each block in the selected region. Method for region size also called as zonal mask is not given in any of the references. Also selection of zonal mask depends on image properties such as brightness, texture and image quality assessment. For the accurate selection of the zonal mask size, algorithms based on local and global characteristics are proposed. Optimum size is considered for entire DCT block of image in global based approach [34]. In case of local approach, the main image is partitioned into sub images. Each sub-image is again processed using global based approach. Sub image size selection is based on the visual quality. When more quality is required, sub image size has the larger size.

In WCE the compression algorithm typically consists Reversible color transformation, two dimensional forward transformation of typical image block of size 4×4 or 8×8, and quantization followed by an entropy encoding. Lin et al. [35] proposed WCE image compressor which processes the image colour components in the order of R, G1, G2 and B. The image sensor outputs the image data in raster order. Therefore, block based image compressor requires intermediate memory units to store each block of data. In order to synchronize the computational operations between quantizer and an entropy encoder, an 8-by-8-memory array is used.

Lin et al. [36] proposed sub-sample based image compression algorithm which samples G1 and G2 components at 2:1 and B component at 4:1 ratio. Turcza et al [37] proposed an hardware architecture for WCE image compressor which requires only two clock cycles for processing a single pixel. Other than the image compressor unit, the proposed work consists an embedded camera and First in-First-out queue to store the entropy encoded data. A feature for forward error correction at the encoder is added to protect the transmitted data from random and burst errors.

Shabani et al. [38] proposed a compressor based on CORDIC-DCT suitable for WCE application. The computation of DCT is done by using look ahead concept and it consumes low power compare to other DCT architectures. A novel architecture is proposed called as look ahead CORDIC (LA-CORDIC) which exploits the pipeline property of LA-CORDIC. Comparing to other DCT based WCE compression architecture, this work consumes less power and uses much low complexity hardware with the favourable reconstruction quality and compression. WCE image processing speed is slightly reduced, but it is negligible because of low image resolution and it slightly reduces the speed of the proposed structure. The speed reduction is negligible due to the low rate of frames and almost low image resolution in WCE application. Turcza et al. [39] proposed DCT based low-complexity efficient WCE image compressor which is an optimal combination of DCT and predictive coding. It is a compact and ultra low power compression system very much suitable for integrating with RF transmitter for WCE system.

DWT Based Image Compression Architectures Wavelet based image compression schemes leads to an higher CR with reduced imaging artifacts. In DCT based compression systems reconstruction loss is more due to blocking artifacts, though it results in higher CR. This limitation can overcome by DWT in which an image is processed by an analysis filter bank with decimation. Haar wavelet is most suitable for WCE application because it offers simple and low computation implementation. Wavelet transform consists a low pass and high pass filters which divides a block of images into low and high frequency components. Low frequency components are successively sub divided into subbands called as sub-band coding. Thone et al. [40] proposed a Haar wavelet based compression algorithm. Because of sparse colour information in the endoscopic image, only luma component is transformed using Haar wavelet followed by Zig-Zag scanning and entropy coding. Better PSNR and CR is achieved compared to DCT based compression algorithms.

Abdelkrima [41] proposed compression task which is based on a hybrid 2D-DWT-DCT architecture that allows the application of the DWT and the DCT techniques sequentially. This method is also called as DWT-Gall 5/3 that is constituted by three main steps: split, predict and update. 2D-DWT is constituted line computation unit which computes LH coefficients, the transposition unit performs the management of the existent coefficients of the line computation unit in order to facilitate the calculation of the details and the approximation coefficients by the line computation unit. The DCT architecture part is composed of 2D-DCT modules addressing the pixels and generating the columns quantization matrix in parallel. This architecture is merged with a quantization block in parallel manner in order to allow the quantification of the coefficients. The merged DCT and the quantification modules allow simultaneous quantification of 8 coefficients per clock cycle, thus a column is calculated during one clock cycle. The gated clock technique is included to reduce the switching activity, thus reduces the power consumption, based on a finite state machine controller.

H.264 Based Video Compression Architectures As per previous discussion, an image coder removes spatial redundancy among pixels to compress the images. In contrast, a video coder achieves the compression by removing temporal redundancy between frames. Image compression algorithms proposed and developed till now provide a low frame rate and low resolution. To overcome this, video coding algorithms exploits temporal correlation between frames to remove the redundancy. The main issue in achieving this is the use of high computational hardware, large silicon area and consumption of more power which is not affordable by a capsule. Hence, modified algorithms are developed for video coding which consumes low power and provide low complexity hardware. Dung et al. [42] proposed H.264 intra-frame prediction technique to reduce computational complexity and lower battery power consumption. The developed video compressor works on higher resolution images of 512 by 512 images at 2 fps with CR 82%. The GI Cam Intra-frame Encoder processes the image in the order of R, G1, G2 and B. Because image sensor streams data in the raster scan order, intraframe video encoder uses a buffer for block based data processing.

Distributed Video Coding Compression Algorithms Distributed video coding (DVC) is an alternative for conventional video coding algorithms which provide low complexity encoding and very much suitable for low power compression schemes. Based on the two famous theorems proposed by Slepian and Wolf [43] and its extension Wyner-Ziv (WZ) [44], DVC allows independent encoding and joint decoding of video frames. DVC provides low complexity encoder which enables interframe temporal correlation exploitation by shifting the high computation load of motion estimation and compensation from encoder to decoder side. This confirms a simple encoder system with low power. DVC is more suitable for an application such as an endoscopic capsule where power and area are the main constraints and also demands high compression efficiency. DVC has been recognized as a potential strategic component for a wide range of lightweight video encoding applications, including WCE and it needs more exploration as an application to WCE. Nikos Deligiannis et al. [45] have proposed a Wyner-Ziv video coding based compression for WCE. The method provides low encoding complexity and accelerates quality, temporal scalability and provides high compression performance by using a novel hash-driven motion estimation technique. Djamel Eddine Boudechiche et al. [46] have presented a distributed video coding (DVC) architecture for wireless capsule endoscopy. This method uses an adapted vector quantization (VQ) with a searching complexity. VQ allows for the creation of side information (SI). Thus, SI was created from a codebook (CB) instead of motion compensated prediction. The method reduces the complexity of the encoder efficiently. But does not improve the performance of the system interms of frame rate. To overcome the limitations of these architectures Sushma et al. [47] proposed a DVC architecture consisting of texture conditioned keyframe encoder. Low frequency components of WZ frame are intra coded and parity bits of high frequency bands are obtained by low density parity check codes. Side information of the WZ frame is generated at the decoder side by using block motion estimation. This scheme exploits interframe temporal correlation and performs better interms of

CR and PSNR compare to other DVC architectures. Though implementation is not done on the hardware, DVC scheme proves to be one of the best method to achieve greater compression by using low power and low computational complexity.

Compression in Wireless Capsule Endoscopic System with Multiple Cameras Multiple cameras can be embedded into the capsule to minimize missing spots caused due to single camera. Multiple cameras provide wide field view as the total angle view covered proportionally increases. The commercial capsule endoscope Pillcam by Given Imaging embedded with two cameras. One camera is placed at front end and another at back end of the endoscope capsule. It has largest field of view of approximately around 150°. To achieve 360° coverage of view of the GI tract more cameras needs to be placed. But using more cameras make the capsule bulky and also increases the complexity of the system. This in turns increases power consumption and processing power. Capsule endoscope with four cameras is the optimal option to take up this problem. Gu et al. [48] proposed a design to reduce the frame miss rate by using multiple cameras in endoscopic capsule which also provides wider visual field view. More power is consumed due to multiple cameras, processing and RF transmission of captured image data. Wireless transmission of image data is avoided and instead captured data is directly stored in flash memory incorporated inside the endoscopic capsule. Advantage of this system is patient need not wear data recorder and missing significant lesion information is reduced. Disadvantage of this system is once the capsule is expelled from the patient, data should be collected from the storage which is more unhygienic. The design strategy proposed does not consist RF transmitter and saves 50% of the power consumption involved in image processing and transmission. The compression architecture consists 4×4 block integer transform comprising two 1-D integer transform units and an intermediate transpose buffer. This architecture also consists configuration module to alter the parameters of quantizer and Huffman entropy encoder.

4.3 *Comprehensive View of the Existing Compression Algorithms*

Table 3 provides the summary of existing compression schemes and their performances in-terms of hardware utilization, compression and quality. Though prediction based compression algorithms provide lossless compression with high quality, the compression ratio is very low. JPEG-LS based algorithms consume more power due to memory usage and complex hardware. High compression is possible with transform based compression schemes but they are lossy. Some researchers have proposed compression schemes which provide acceptable quality images with high compression ratio. A block based compression algorithms consisting DCT requires intermediate buffer to store the pixels scanned in raster order. This increases the complexity of hardware implementation and makes less suitable for low complexity encoder. FIFO buffers with large outputs are used to minimize the buffer requirement.

Table 3 Summary of compression algorithms in WCE

Work and compression scheme	Features	Hardware implementation	Die and core dimension in mm^2	Gate count	Memory in Bytes	Latency per pixel	DCLK frequency	Core power consumption	CR (%)	PSNR	FR	BL in hours
Khan et al. DPCM	Supports both NBI and WLI	0.18 μm CMOS	1.6 × 1.6, 1 × 1	24 K	0 B	1 clock cycle	42 MHz	1.72 mW	80.2	48.2	2	–
Fante et al. DPCM	Supports both lossless and lossy, Modified Golomb Rice encoder	ASIC UMC 130 nm CMOS	0.018 × 0.018, NA	2.3 K	8 B	1 clock cycle	393.21 KHz	32 μW	90.35(L), 73.62(LL)	40.66, ∞	2	–
Khan et al. DPCM	Supports WLI and NBI	FPGA 65 nm	11 × 28	2112 LUT	0 B	1 clock cycle	405 MHz	1.63 mW	84	∞	2	8.5
Xie et al. JPEG LS	ROI based Lossless compression	UMC 0.18 μm CMOS 6-layer process	3 × 4.2	–	103 k bits	–	40 MHz	8.2 mW	56.24	46.89	15	8
Li et al. JPEG LS	Fully pipelined hardware with clock management scheme	FPGA 0.18 μm CMOS	–	17.6 K	18 k bits	28 clock cycles	40 Mhz	7.1 mW	54.13	46.43	–	–

(continued)

Table 3 (continued)

Work and compression scheme	Features	Hardware implementation	Die and core dimension in mm^2	Gate count	Memory in Bytes	Latency per pixel	DCLK frequency	Core power consumption	CR (%)	PSNR	FR	BL in hours
Chen et al. JPEG LS	on-chip oscillator	0.18 µm CMOS process	3.4 × 3.3	23.4 K	764 k bits	–	20–24 MHz	1.3 mW	54.13	46.43	2	8
chen et al. JPEG LS	Prediction, Run mode, Modified GR encoder	90 nm TSMC CMOS	30.625 × 30.625	12.4 K	10.2 k bits	240 MHz	3.1 mW	49.75	46.43	30	–	–
Liu et al. JPEG LS	No LPF, Causal Interpolation, gradient predictor	0.18 µm CMOS process	5 × 5,11 × 26	–	–	40 MHz	–	56.89	46.31	–	–	–
Liu et al. JPEG LS	Interpolation without LPF	ASIC 0.18 µm CMOS	–, 11 × 25	–	–	–	–	11.7 mW	72.89	46.2	24	WPT
Lin et al. DCT	LZ entropy coding, No color transform	TSMC 0.18 µm 1P6M CMOS	1.25 × 1.25,–	–	288 Bytes	–	12.58 MHz	14.92 mW	79.65	32.51	2	–
Lin et al. DCT	Subsampling of green and Blue components	TSMC 0.18 µm 1P6M CMOS	1 × 1,–	318 k	–	–	7.96 MHz	9.17 mW	83	40.73	2	9
Turcza et al. DCT	DPCM along with variable length huffman codes	FPGA 65nm chip	–,–	–	204 k bits	2 clock cycles	12 MHz	5.9 mW	90.8	35.36	12	–

(continued)

Table 3 (continued)

Work and compression scheme	Features	Hardware implementation	Die and core dimension in mm^2	Gate count	Memory in Bytes	Latency per pixel	DCLK frequency	Core power consumption	CR (%)	PSNR	FR	BL in hours
Wahid et al. DCT	DCT based on algebraic integer quantization	0.18 µm CMOS process	–,–	325 k	–	–	150 MHz	10 mW	87.13	32.95	–	12
Turcza et al. DCT	Transform based coding with predictive coding	Verilog UMC 180 nm CMOS	0.96 × 0.54,–	1.8 k	7.68 k bits	–	5.25 MHz	0.106 mW	74.35	46.5	2	–
Shabani et al. DCT	Low-power Lookahead (LPLA) CORDIC based DCT	TSMC 0.13 m technology	–	133 k	–	–	–	–	–	–	–	
Thone et al. DWT	Subsampling of G	B, Haar transform of luma	Silicon Blue 65L04 low power FPGA	–	–	–	–	–	96	36	–	–
Dung et al. H.264	Only Intraframe prediction of video	TSMC 0.18 µm 1P6M CMOS	–	60 k	–	–4 MHz	–0.91 mW	82	36.24	2	–	–

FR Frame rate, *CR* Compression rate, *PSNR* Peak signal to noise ratio, *WPT* Wireless power transmission, *BL* Battery Life

Next generations of WCE video capsules are expected to transmit more frames per second with high video quality and reduced battery power consumption. Therefore, efficient and low complexity video compression system at the least feasible computational cost needs to be designed. Distributed video coding based on WZ coding is the best possible solution for this. Many research works proposed this method as the most suitable for WCE system as it adopts removal of temporal redundancy to provide high frame rate at the best possible low encoder computational complexity.

In order get wide angle view of the endoscopic images multiple cameras are used. Transmitting the images captured by multiple cameras consume more bandwidth as well as more power. In order to reduce the transmission power distributed video coding for multiview imaging can be used. Some researchers have proposed schemes which avoid transmission and instead store the compressed images in flash memory. Here suitable algorithm need to be used to compress the images. Distributed video coding without feedback is an optimum compression scheme because the reconstruction is offline.

5 Future Scope and Conclusion

In WCE system power management is the major challenge. Reducing the frame rate is the key solution for this issue, but it increases the rate of missing lesions. More batteries embedding into the system increases the volume of capsule and making it difficult to swallow. Controlling the frame rate according to the speed of the capsule is the best solution. For this, speed controlling system needs to be incorporated at the recording unit. Also to provide wider view of the GI tract, multi view endoscopic capsule with multiple cameras should be developed. Suitable low complexity compression algorithms for compressing multi-view images need to be developed. Also by using GI tract semantic features between consecutive frames, transmission of chroma components of some frames in between can be avoided. This will save processing and transmission power required for chroma. Distributed video coding is also a promising technology to compress the images captured by multiple cameras.

WCE allows painless and comfortable scanning of the entire GI tract. It is a major breakthrough in the field of endoscopy. Still WCE has drawbacks compare to wired endoscopy in terms of DI and real time detection of abnormalities. In order to overcome all these hindrances the imaging unit plays an important role to produce images with high quality at reduced bandwidth and minimal power consumption. This paper reviews the existing compression algorithms and their pros and cons. Suitable compression algorithm can be adopted based on the physician's specifications and whether the reconstruction is real time or offline. More research is required to design an efficient compression algorithm that exploits textures, colour semantics and temporal redundancy. Also more exploration is needed to find the technique for blocking non-informative frames at the encoder which saves the power required for the transmission.

References

1. Luis, M., Tavares, A., Carvalho, L.S., Lara-Santos, L., Araújo, A., de Mello, R.A.: Personalizing therapies for gastric cancer: molecular mechanisms and novel targeted therapies. World J. Gastroenterol.: WJG **19**(38), 6383 (2013)
2. Myer, P.A., Mannalithara, A., Singh, G., Singh, G., Pasricha, P.J., Ladabaum, U.: Clinical and economic burden of emergency department visits due to gastrointestinal diseases in the united states. Am. J. Gastroenterol. **108**(9), 1496–1507 (2013)
3. Ciuti, G., Menciassi, A., Dario, P.: Capsule endoscopy: from current achievements to open challenges. IEEE Rev. Biomed. Eng. **4**, 59–72 (2011)
4. Reavis, K.M., Melvin, W.S.: Advanced endoscopic technologies. Surg. Endosc. **22**(6), 1533–1546 (2008)
5. Bonnington, S.N., Rutter, M.D.: Surveillance of colonic polyps: are we getting it right? World J. Gastroenterol. **22**(6), 1925 (2016)
6. Moglia, A., Menciassi, A., Dario, P., Cuschieri, A.: Capsule endoscopy: progress update and challenges ahead. Nat. Rev. Gastroenterol. Hepatol. **6**(6), 353 (2009)
7. Pan, G., Wang, L.: Swallowable wireless capsule endoscopy: progress and technical challenges. Gastroenterol. Rese. Pract. **2012**, (2012)
8. Mylonaki, M., Fritscher-Ravens, A., Swain, P.: Wireless capsule endoscopy: a comparison with push enteroscopy in patients with gastroscopy and colonoscopy negative gastrointestinal bleeding. Gut **52**(8), 1122–1126 (2003)
9. Wahid, K., Ko, S.B., Teng, D.: Efficient hardware implementation of an image compressor for wireless capsule endoscopy applications. In: 2008 IEEE International Joint Conference on Neural Networks (IEEE World Congress on Computational Intelligence). pp. 2761–2765. IEEE (2008)
10. Pan, G., Xin, W., Yan, G., Chen, J.: A video wireless capsule endoscopy system powered wirelessly: design, analysis and experiment. Meas. Sci. Technol. **22**(6), 065802 (2011)
11. Carta, R., Tortora, G., Thoné, J., Lenaerts, B., Valdastri, P., Menciassi, A., Dario, P., Puers, R.: Wireless powering for a self-propelled and steerable endoscopic capsule for stomach inspection. Biosens. Bioelectron. **25**(4), 845–851 (2009)
12. Secretariat, M.A.: Wireless capsule endoscopy: an evidence-based analysis. Ontario Health Technol. Assess. Ser. **3**(2), 1 (2003)
13. Given imaging company. http://www.givenimaging.com
14. Basar, M.R., Malek, F., Juni, K.M., Idris, M.S., Saleh, M.I.M.: Ingestible wireless capsule technology: a review of development and future indication. Int. J. Antennas Propagation (2012)
15. Li, C.Y, Zhang, B.l., Chen, C.x., Li, Y.M.: Omom capsule endoscopy in diagnosis of small bowel disease. J. Zhejiang Univ. Sci. B **9**(11), 857–862 (2008)
16. Menciassi, A., Ciuti, G., Cavallotti, C.: Future developments of video capsule endoscopy: hardware. In: Video Capsule Endoscopy, pp. 543–556. Springer (2014)
17. Glukhovsky, A., Avni, D., Meron, G.: Diagnostic device using data compression (2003), uS Patent App. 10/202,626
18. Zinaty, O., Horn, E., Bettesh, I.: In-vivo imaging device providing data compression (2015), uS Patent 9,113,846
19. Avni, D., Meron, G., Horn, E., Zinaty, O., Glukhovsky, A.: Diagnostic device, system and method for reduced data transmission (2010), uS Patent 7,664,174
20. Horn, E.: Device, system, and method for reducing image data captured in-vivo (2008), uS Patent 7,336,833
21. Shigemori, T., Matsui, A.: Body-cavity image observation apparatus (2011), uS Patent 8,038,608
22. Bandy, W.R., Jamieson, B.G., Powell, K.J., Salsman, K.E., Schober, R.C., Weitzner, J., Arneson, M.R.: Ingestible endoscopic optical scanning device (2013), uS Patent 8,529,441
23. Liu, G., Yan, G., Zhao, S., Kuang, S.: A complexity-efficient and one-pass image compression algorithm for wireless capsule endoscopy. Technol. Health Care **23**(s2), S239–S247 (2015)

24. Fernandez-Urien, I., Carretero, C., Borobio, E., Borda, A., Estevez, E., Galter, S., Gonzalez-Suarez, B., Gonzalez, B., Lujan, M., Martinez, J.L., et al.: Capsule endoscopy capture rate: has 4 frames-per-second any impact over 2 frames-per-second? World J. Gastroenterol. WJG **20**(39), 14472 (2014)
25. Alam, M.W., Hasan, M.M., Mohammed, S.K., Deeba, F., Wahid, K.A.: Are current advances of compression algorithms for capsule endoscopy enough? a technical review. IEEE Rev. Biomed. Eng. **10**, 26–43 (2017)
26. Khan, T.H., Wahid, K.A.: Lossless and low-power image compressor for wireless capsule endoscopy. VLSI design **2011** (2011)
27. Fante, K.A., Bhaumik, B., Chatterjee, S.: Design and implementation of computationally efficient image compressor for wireless capsule endoscopy. Circ. Syst. Sig. Proc. **35**(5), 1677–1703 (2016)
28. Khan, T.H., Wahid, K.A.: Design of a lossless image compression system for video capsule endoscopy and its performance in in-vivo trials. Sensors **14**(11), 20779–20799 (2014)
29. Malathkar, N.V., Soni, S.K.: Low complexity image compression algorithm based on hybrid dpcm for wireless capsule endoscopy. Biomed. Sig. Proc. Control **48**, 197–204 (2019)
30. Malathkar, N.V., Soni, S.K.: A near lossless and low complexity image compression algorithm based on fixed threshold dpcm for capsule endoscopy. Multimedia Tools and Appl. 1–16 (2020)
31. Xie, X., Li, G., Chen, X., Li, X., Wang, Z.: A low-power digital ic design inside the wireless endoscopic capsule. IEEE J. Solid-State Circ. **41**(11), 2390–2400 (2006)
32. Chen, S.L., Liu, T.Y., Shen, C.W., Tuan, M.C.: Vlsi implementation of a cost-efficient near-lossless cfa image compressor for wireless capsule endoscopy. IEEE Access **4**, 10235–10245 (2016)
33. Mammeri, A., Khoumsi, A., Ziou, D., Hadjou, B.: Modeling and adapting jpeg to the energy requirements of vsn. In: 2008 Proceedings of 17th International Conference on Computer Communications and Networks. pp. 1–6. IEEE (2008)
34. Kaddachi, M.L., Soudani, A., Lecuire, V., Makkaoui, L., Moureaux, J.M., Torki, K.: Design and performance analysis of a zonal dct-based image encoder for wireless camera sensor networks. Microelectron. J. **43**(11), 809–817 (2012)
35. Lin, M.C., Dung, L.R., Weng, P.K.: An ultra-low-power image compressor for capsule endoscope. Biomed. Eng. Online **5**(1), 14 (2006)
36. Lin, M.C., Dung, L.R., Weng, P.K.: A cardinal image compressor for capsule endoscope. pp. 146–149. IEEE (2006)
37. Turcza, P., Duplaga, M.: Low power fpga-based image processing core for wireless capsule endoscopy. Sens. Actuators A: Phys. **172**(2), 552–560 (2011)
38. Shabani, A., Timarchi, S.: Low-power dct-based compressor for wireless capsule endoscopy. Sig. Proc.: Image Commun. **59**, 83–95 (2017)
39. Turcza, P., Duplaga, M.: Near-lossless energy-efficient image compression algorithm for wireless capsule endoscopy. Biomed. Sig. Proc. Control **38**, 1–8 (2017)
40. Thoné, J., Verlinden, J., Puers, R.: An efficient hardware-optimized compression algorithm for wireless capsule endoscopy image transmission. Procedia Eng. **5**, 208–211 (2010)
41. Abdelkrim, Z., Ashwag, A., Majdi, E.: Low power design of wireless endoscopy compression/communication architecture. J. Electr. Syst. Inf. Technol. **5**(1), 35–47 (2018)
42. Dung, L.R., Wu, Y.Y., Lai, H.C., Weng, P.K.: A modified h. 264 intra-frame video encoder for capsule endoscope. In: 2008 IEEE Biomedical Circuits and Systems Conference, pp. 61–64. IEEE (2008)
43. Slepian, D., Wolf, J.: Noiseless coding of correlated information sources. IEEE Trans. Inf. Theory **19**(4), 471–480 (1973)
44. Wyner, A., Ziv, J.: The rate-distortion function for source coding with side information at the decoder. IEEE Trans. Inf. Theory **22**(1), 1–10 (1976)
45. Deligiannis, N., Verbist, F., Barbarien, J., Slowack, J., Van de Walle, R., Schelkens, P., Munteanu, A.: Distributed coding of endoscopic video. In: 2011 18th IEEE International Conference on Image Processing, pp. 1813–1816. IEEE (2011)

46. Boudechiche, D.E., Benierbah, S., Khamadja, M.: Distributed video coding based on vector quantization: application to capsule endoscopy. J. Vis. Commun. Image Represent. **49**, 14–26 (2017)
47. Sushma, B., Aparna, P.: Distributed video coding based on classification of frequency bands with block texture conditioned key frame encoder for wireless capsule endoscopy. Biomed. Sig. Proc. Control **60**, 101940 (2020)
48. Gu, Y., Xie, X., Li, G., Sun, T., Wang, D., Yin, Z., Zhang, P., Wang, Z.: Design of endoscopic capsule with multiple cameras. IEEE Trans. Biomed. Circ. Syst. **9**(4), 590–602 (2014)

Design of Retrodirective Arrays Using Hybrid Couplers for Autonomous Car

P. Mahalakshmi and R. Vallikannu

Abstract To overcome the traffic access issue in cloud services over 4G networks, mm waves has been proposed. In radiofrequency spectrum, 30 GHz to 300 GHz is allocated for millimeter-wave communications. It has been used for many applications including industrial and academic purposes. To achieve high data rate communication and low latency, 5G has been designed. The proposed work is to design retrodirective antenna modes to work for vehicular communication (self-driving cars), which comes under mission-critical communication in 5G. The retrodirective mode is obtained using hybrid (rat-race) couplers simulated in ANSYS, which operates at 2.4 Ghz.

Keywords Rat-race couplers · Retro-directive antenna · Mm-wave communications · Return loss · ANSYS · FR4 substrate

1 Introduction

In recent years, interest in autonomous automation is increasing due to the development of vehicle communication technology. In the 1920s, the first radio-controlled electric was designed. This self-autonomous vehicle is powered by embedded circuits on the roads. By the 1960s, electronic guide systems were integrated inside the autonomous cars. From 1980s, vision-guided autonomous vehicles came into the market, which was a milestone in automation technology. Due to the increasing number of vehicles and the human population, road safety is a goal that can be never achieved [1]. The features such as automatic braking, lane maintenance, and integration with vision-guided features and extended network systems have been

P. Mahalakshmi (✉) · R. Vallikannu
Hindustan Institute of Technology and Science, Chennai, India
e-mail: mahapearl57@gmail.com

R. Vallikannu
e-mail: vallikannu@hindustanuniv.ac.in

P. Karuppusamy et al. (eds.), *Sustainable Communication Networks and Application*, Lecture Notes on Data Engineering and Communications Technologies 55,
https://doi.org/10.1007/978-981-15-8677-4_24

designed for safe transportation. The retrodirective antenna has simpler construction than another conventional antenna, it is widely used in vehicle communication where high directivity is important. Because of high speed, high directivity and self-tracking processes make this method used for millimeter communications [2], military communications, radars, and civilian applications. Retrodirective array operates by responding toward a source upon incident interrogation with a directivity base on array theory. Van Atta array and heterodyne mixing method are widely used to achieve high directivity. Since the heterodyne mixing method [3] uses mixers and local oscillators which act as power-hungry devices, so to reduce power consumption passive devices like Van Atta array is used. To design for large arrays, the design of the interweaving feed lines can become intricate, couplers have been used.

2 Hybrid Couplers as Retrodirective Array

The block diagram for the proposed work is shown in Fig. 1. If any signal is detected from obstacle sensor in an autonomous car which is sent to rat-race with antenna unit, where port 3 is excited and port 4 is terminated, then Retrodirective mode with an out-of-phase signal is produced, and braking operation occurs. If no signal is detected from the obstacle sensor, then port 3 is terminated, port 4 is excited, and the retrodirective mode with an in-phase signal is obtained which ensures safe driving function in an autonomous car.

Hybrid couplers or rat-race couplers act as retrodirective antenna if the port 1 and port 2 are connected to the antenna. The antenna acts as transmitter and receiver. While port 3 and port 4 are terminated with reflection cofficient, the scope of using this method is to compare traditional retrodirective methods such as Van Atta array [4] and heterodyne method for its compatibility and low power consumption. These couplers have a wide bandwidth than other couplers and can be used for high-frequency application, and signal degradation problem can be overcome. Instead of active retrodirective antenna arrays, passive retrodirective array is used to overcome the propagation losses and requires no power to scan multiple angles. The phase reversal property can be derived from the S-matrix.

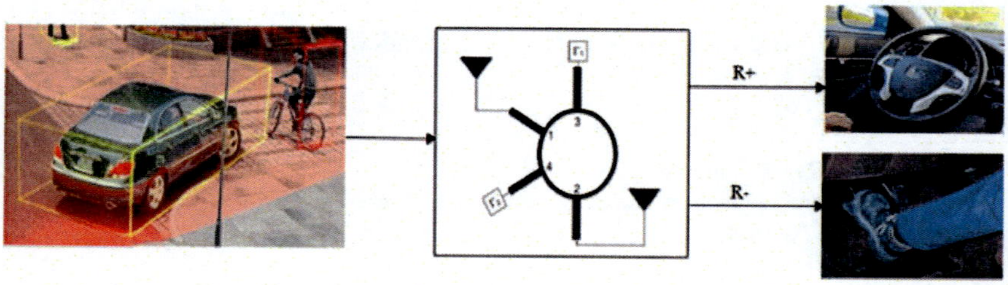

Fig. 1 Block diagram for proposed work

A method of what Hsieh and Chu [5] is used to deriving the retrodirectivity conditions in a rat-race coupler. It is obtained from the scattering matrix, as

$$[S] = \frac{1}{\sqrt{2}}\begin{bmatrix} 0 & 0 & e^{-j\frac{\pi}{2}} & e^{-j\frac{\pi}{2}} \\ 0 & 0 & e^{-j\frac{3\pi}{2}} & e^{-j\frac{\pi}{2}} \\ e^{-j\frac{\pi}{2}} & e^{-j\frac{\pi}{2}} & 0 & 0 \\ e^{-j\frac{\pi}{2}} & e^{-j\frac{\pi}{2}} & 0 & 0 \end{bmatrix}$$

The output waves (b_3 and b_4) at port 3 and port 4 are the input waves (a_1 and a_2) at port 1 and port 2 are

$$b_3 = \frac{1}{\sqrt{2}}\left(a_1 e^{-\frac{j\pi}{2}} + a_2 e^{-\frac{j3\pi}{2}}\right) \tag{1}$$

$$b_4 = \frac{1}{\sqrt{2}}\left(a_1 e^{-\frac{j\pi}{2}} + a_2 e^{-\frac{j\pi}{2}}\right) \tag{2}$$

$$a_1 = \Gamma_1 b_3 \tag{3}$$

$$a_2 = \Gamma_2 b_4 \tag{4}$$

By setting the condition $\Gamma_2 = -\Gamma_{1,}$, the retrodirective condition in rat-race couplers is obtained. By comparing the phase difference of the input waves to that of the output waves, the phase of the output is negative concerning input. Thus, the rat-race couplers reverse the incoming phase and act as retrodirective. When port 3 is terminated as an open circuit, and port 4 is terminated as short circuit, thus coupler acts as retrodirective mode of operation. When port 3 is excited and port 4 is terminated, it gets equally split into two ports, traveling clockwise and anti-clockwise direction. However, port 4 acts as an isolated port since the path difference when the signals traveling from port 3 to port 4 is λ/2 where no output appears at this port [6]. The distance traveled by the signal from port 3 to port1 is λ/4, and the distance traveled from port 3 to port 2 is 3λ/4. The output at port 1 and port 2 in out-of-phase retrodirective mode of operation is obtained (Fig. 2).

When port 3 is terminated as short circuit and port 4 is terminated as open circuit, thus coupler acts as a retrodirective mode of operation. However, port 3 acts as an isolated port since the path difference when the signals traveling from port 4 to port 3 is λ/2, where no output appears at this port [7]. The distance traveled by the signal from port 4 to port1 is λ/4, and the distance traveled from port 4 to port 2 is λ/4. The output at port 1 and port 2 in-phase retrodirective mode of operation is obtained.

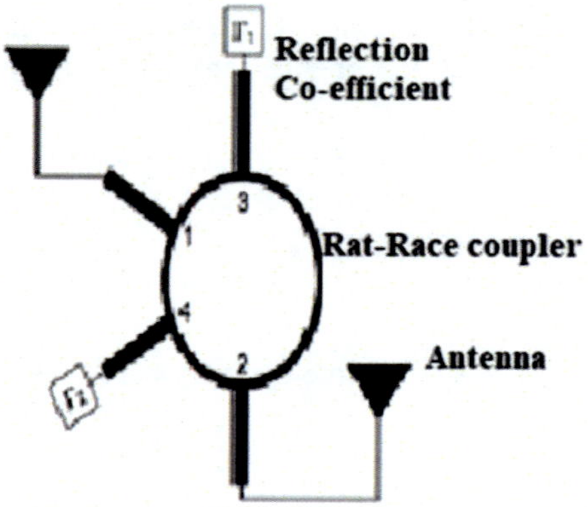

Fig. 2 Hybrid coupler in which port 1 and port 2 are connected to antennas and port 3 and port 4 is terminated with loads of specific reflection coefficient

3 Design and Evaluation

Proposed work has been carried out using the following steps shown in the flowchart. To realize in planar structure, microstrip patch antennas are widely used [8]. The rectangular patch is separated from the ground plane with FR4 epoxy substrate with relative permittivity of $\varepsilon_r = 4.4$, and the thickness of the substrate is 1.58 mm (Fig. 3).

The design procedure is carried out in three steps. Design a microstrip patch antenna at an operating frequency of 2.45 GHz. The simulation parameters used to design patch antenna are length and width which are 40 and 60 mm, and the thickness of the substrate is 1.58 mm. The second step is to design a hybrid coupler with inner and outer radius which are 17.56 and 16.46 mm, and length and width of the feedline are 3 and 9.36 mm. Now, combine the designed microstrip patch antenna with rat-race couplers. Determination of the impedance matching between the antenna and rat-race couplers is obtained, and simulation is performed using ANSYS-EM simulation software, and return loss and the radiation pattern are obtained.

3.1 *Coupler Design*

This proposed model presents a comprehensive description to model rat-race couplers. The dielectric material is glass epoxy substrate fr4 (ε_r=4.4), and the height is $h = 1.58$ mm for the operating frequency of 2.4 GHz [9].

- The inner and outer radius of the rat-race coupler is calculated using

$$R = \frac{3\lambda g}{4\pi} \tag{5}$$

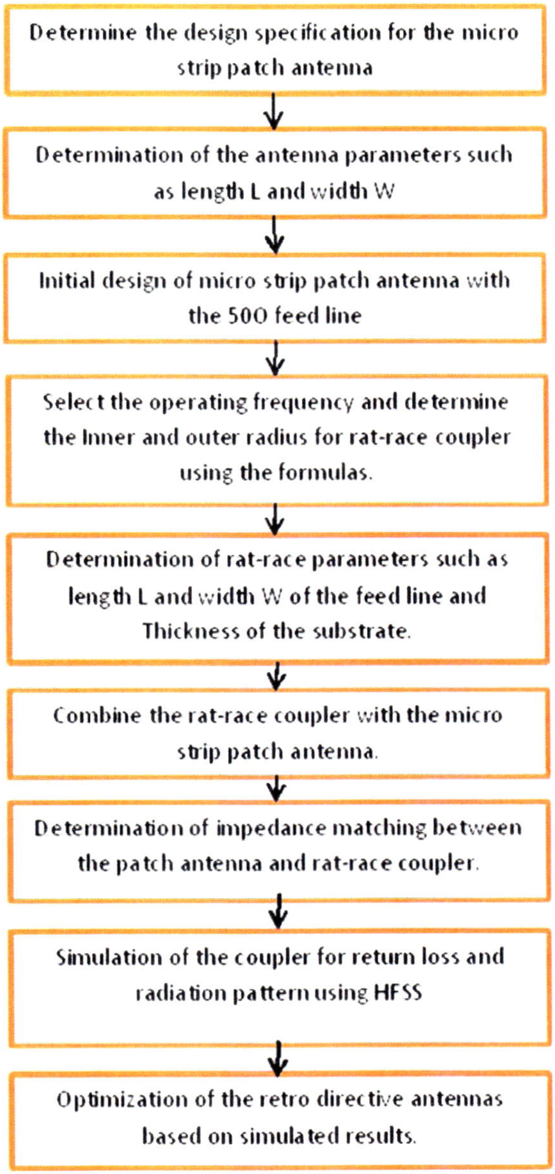

Fig. 3 Flow diagram for proposed work

Table 1 Design parameters for hybrid couplers

Parameters	Value (mm)
Inner radius (r1)	17.56
Outer radius (r2)	16.46
Length of the feed line	3
Width of the feed line	9.36
Thickness of the substrate	1.58

$$\text{Inner radius of the ring } r1 = \left(r - \frac{w}{2}\right) \quad (6)$$

$$\text{outer radius of the ring } r2 = \left(r + \frac{w}{2}\right) \quad (7)$$

- The wavelength of the coupler is calculated with the velocity of light $C = 3 \times 10^8$ m/s and the operating frequency $f = 2.4$ Ghz.

$$\lambda 0 = \frac{c}{f} \quad (8)$$

$$\lambda g = \frac{\lambda 0}{\sqrt{\varepsilon \text{eff2}}} \quad (9)$$

- The effective dielectric constant $\varepsilon_{\text{eff1}}$ and $\varepsilon_{\text{eff2}}$ is calculated using

$$\varepsilon_{\text{eff}} = \frac{\varepsilon_r + 1}{2} + \frac{\varepsilon_r - 1}{2}\left(\frac{1}{\sqrt{1 + 12\frac{h}{w}}}\right) \quad (10)$$

- The width for the port input and width for the branches are calculated using the formula

$$\frac{w}{h} = \frac{8e^A}{e^{2A} - 2} \text{ for } \frac{w}{d} < 2 \quad (11)$$

where,

$$A =; \frac{z}{60}\sqrt{\frac{\varepsilon_r + 1}{2}} + \frac{\varepsilon_r - 1}{\varepsilon_r + 1}\left(0.23 + \frac{0.11}{\varepsilon_r}\right) \quad (12)$$

where λ_g represents the guided wavelength, r and w represents radius and width, ε_r represents the dielectric permittivity, and h denotes the thickness of the substrate.

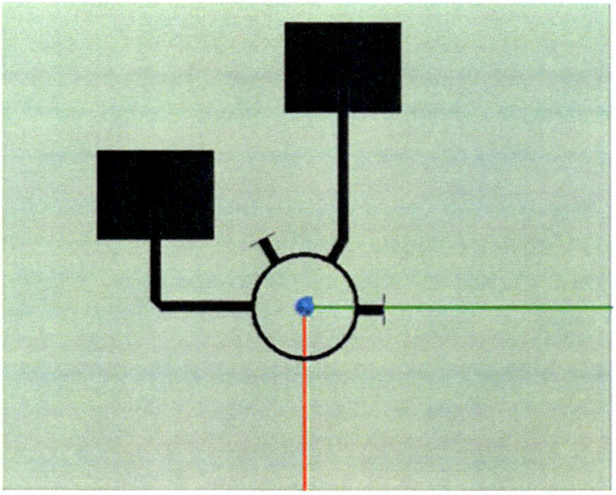

Fig. 4 Conventional diagram of rat-race coupler and antenna

The simulation parameters utilized in the look of hybrid couplers are illustrated in Table 1. The coupler operates at center frequency 2.45 Ghz that has been simulated using HFSS EM simulation software (Fig. 4).

4 Results and Discussion

4.1 Return Loss

The optimization is done for this coupler dimension, to evaluate the results of scattering parameters. The S11 parameter is the return loss, which is known as the ratio of power reflected from the port to the power communicated into the port. The couplers operate at a center resonant frequency of 2.45 GHz, and very low return losses are observed, nearly as −21.129 dB (Fig. 5).

4.2 Radiation Pattern

The radiation pattern is obtained for hybrid couplers [10], which is going to act in two modes of operation retrodirective with in-phase mode and retrodirective with out-of-phase mode.

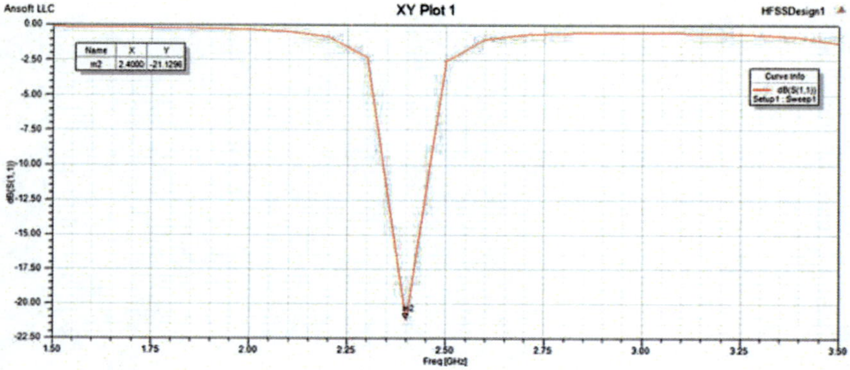

Fig. 5 Scattering parameter versus frequency

4.2.1 Retrodirective with In-Phase Mode

If port 1 and port 2 are connected to an antenna, while port 3 and port 4 are terminated with reflection coefficient for the coupler to act as retrodirective in-phase mode, port 3 is terminated as short circuit and port 4 is terminated as open circuit. When port 4 is excited, it gets equally split into two ports, traveling clockwise and anti-clockwise direction. Port 3 act as an isolated port. From the simulated radiation pattern, the two signals are in-phase, where copolarization exist and positive gain value is obtained (Fig. 6).

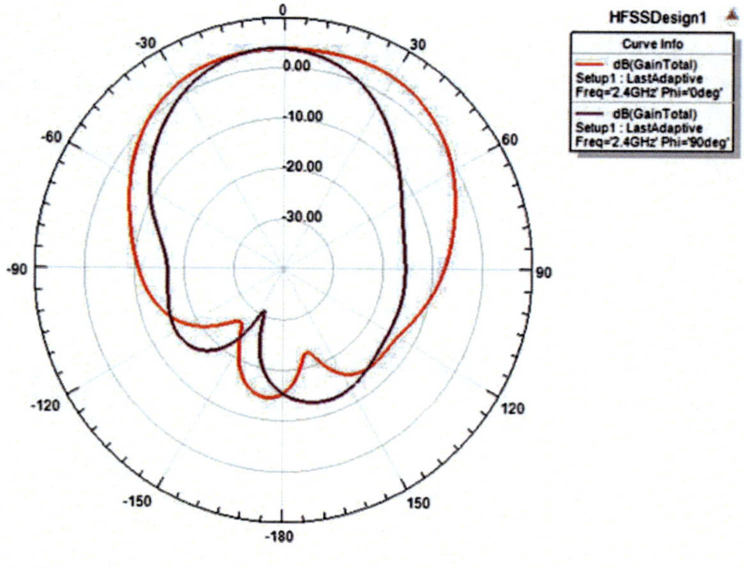

Fig. 6 Radiation pattern when port 3 is terminated and port 4 is excited

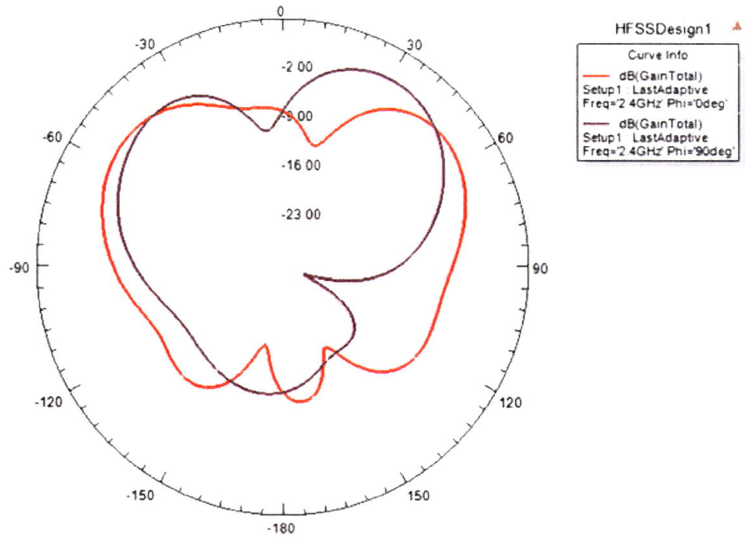

Fig. 7 Radiation pattern when port 3 is excited and port 4 is terminated

4.2.2 Retrodirective with Out-of-Phase Mode

If port 1 and port 2 are connected to an antenna, while port 3 and port 4 are terminated with reflection coefficient for the coupler to act as retrodirective out-of-phase mode,, port 3 is terminated as open circuit and port 4 is terminated as short circuit [11]. When port 3 is excited, it gets equally split into two ports, traveling clockwise and anti-clockwise direction. Port 4 act as an isolated port. From the simulated radiation pattern, the two signals are out-of-phase, where cross-polarization exist and negative gain value is obtained (Fig. 7).

5 Conclusion

The retrodirective modes of in-phase and out-of-phase have been achieved using rat-race couplers. The rat-race couplers interconnected with antenna are designed and evaluated using ANSYS simulation software. From the simulated results return loss, radiation pattern was obtained. The processor controlled sensor unit can be replaced into couplers with antenna, the coverage of the antenna will be high compared to existing technique [12], and it is also very light when compared to bulky processor unit. Multiple usage of couplers with antenna can be used to replace the processor unit. As an extension of the current work, this hybrid coupler retrodirective antenna is connected in array format and will act as feeding network. To improve the gain of the antenna, series fed array is interconnected along with each antenna in an array for vehicle communication used for self-driving autonomous car, where each sensor unit is going to act as individual retrodirective transceiver.

References

1. Zheng, L.Z., Tseng, J.D., Wu, B.S., Li, J.M.: Hybrid type rat-race coupler designs. In: APMEC (2015)
2. Dobias, F., Grabow, W.: Adaptive arrays antennas for 5.8 GHz vehicle to roadside communications. In: IEEE 44th Vehicular Technology Conference. Stockholm, Sweden (1994)
3. Williams, D.A.: Millimeter wave radars for automotive applications. In: IEEE MIT-S International Microwave Symposium Digest. Albuquerque, NM (1992)
4. Nielsen, E.D.: Square Van Atta reflector with conducting mounting plane. IEEE Trans. Antennas Propag. **AP-18** (1970)
5. Chiou, Y.C., Kuo, J.T., Chan, C.H.: New miniature dual band rat-race coupler with microwave C-sections. In: 2009 IEEE MTT-S International Microwave Symposium Digest, pp. 701–704 (2009)
6. Pon, C.Y.: Hybrid-ring directional couplers for arbitrary power divisions. IRE. Trans. Microw. Theory Techn **9**(6), 529–535 (1961)
7. Pon, C.Y.: Retrodirective array using the heterodyne technique. In: IEEE Transmission Antennas Propagation, vol. 12 (1964)
8. Meinel, H.H.: Commercial applications of millimeter waves: history, present status and future trends. IEEE Trans. Microwave Theory Tech., **43** (1995)
9. Muthumani. S.E., Vallikannu, R.: Compact slot loaded Kotch fractal microstrip patch antenna. In: IEEE Applied Electromagnetics Conference, AMEC (2013)
10. David, M.: Microwave Engineering, 3rd ed. Willey (2005)
11. Lee, K.F., Dahele, J.S.: Characteristics of microstrip patch antennas and some methods of improving frequency agility and bandwidth, Handbook of Microstrip antennas (1989)
12. Manoharan, S.: An improved safety algorithm for artificial intelligence enabled processors in self driving cars. J. Artif. Intell. 1(02) (2019)

Enhanced Analysis of Brain MR Images for Detection of Abnormal Tissues Using Deep Learning

Check for updates

Jyotindra Dharwa and Shivang Patel

Abstract Due to the self-learning ability and probability to get more precise outcomes within less time span in the detection of abnormal tissues from a large amount of brain MR images, the automated segmentation approaches using various deep learning techniques like CNN, AlexNet, ResNet, Inception, etc. are proven more efficient nowadays in accurate diagnosis, risk assessment and treatment planning by medical practitioners. Here, multiple approaches used for the same domain will be reviewed with their diagnostic efficiency. Also, an approach will be proposed based on the recent study of deep learning, use it for analyzing the problem and compare results with different other methods so it will help make more truthful decisions to diagnose the actual problem. T1/T2 weighted and FLAIR MR images have been included for experimental study from different imaging centers, as they provide rich and structural information about the anatomy of brain tissues compared to other imaging modalities like PET, CT, etc. Various classified results are compared with manual segmentation results by radiologists and they were found accurate in terms of detection and segmentation of abnormal tissues. This will lead to improved treatment planning, monitoring and clinical acceptance of proposed methods depending upon the simplicity of computation and degree of involvement of the user in the entire process. Also, here hybrid semantic segmentation approach—RRDNN with U-Net (Recurrent Residual Deep Neural Network with U-Net) is introduced. To train a model in detail, the residual unit plays an important role, while recurrent residual layers with U-Net promises effective segmentation representation. Using this approach, higher classification accuracy up to 94% can be attained compared to existing approaches on the same platform in abnormal tissue identification from brain MR images.

Keywords Brain MR images · Image segmentation · Image classification · Deep learning techniques · Recurrent neural networks

J. Dharwa (✉)
AMPICS, Ganpat University, Kherva, Mehsana, Gujarat, India
e-mail: jyotindra.dharwa@ganpatuniversity.ac.in

S. Patel
CPICA, Gujarat University, Ahmedabad, Gujarat, India
e-mail: professorsmpatel@gmail.com

P. Karuppusamy et al. (eds.), *Sustainable Communication Networks and Application*, Lecture Notes on Data Engineering and Communications Technologies 55,
https://doi.org/10.1007/978-981-15-8677-4_25

1 Introduction

In today's era of automation, the medical field does also not remain untouched. Machine learning and robotics play a great role in the healthcare sector. Artificial Intelligence algorithms acquire features from bulky clinical data. Before AI systems, it is required to train data based on specific features and after comparison, the interested region can be extracted. Also, automated processing will reduce human error, time and cost. Recently, many deep learning-based techniques and clinical applications are introduced and used for early and robust identification of various brain tissues related defects such as strokes, disorders, tumor, infection, swelling, hemorrhage, etc. Deep learning-based methods have replaced traditional model-based methods and with the remarkable improvements in almost all image processing stages for analyzing a large amount of digital images. With optimizations in algorithms and improved computational devices, it is also possible to extract and analyze required features with improved quality from a large amount of imaging data [1]. In image analysis, data are divided into different phases for learning like captioning, training, validation, testing. For validating data, plenty of algorithms are introduced in the literature. Generally, for testing, ground truth statistics are considered using manual delineations. Automated analysis of brain images is a challenging task due to factors like the variation in intensity, contrast, noise, etc. Specifically, for brain MR images, few preprocessing steps have to follow like registration, noise reduction, intensity normalization, skull stripping, filtration, bias field correction, etc. for accurate diagnosis and improvement in result findings [2]. Still, few questions have to be reviewed like—is there any standalone technique to be present by applying can get accurate results, can more precise results is achieved, i.e. accurate classification of abnormal tissues within the stipulated time period, is there any common method which can be applied in general, etc. [3]. The main objective of this research survey is to analyze the loop falls in the field of medical treatment where various segmentation techniques were used to detect and classify ROI and also to propose an approach which will be useful in gaining notable accuracy in less number of iterations.

2 Segmentation Methods—A Survey

Image segmentation performs a vital role in image classification, as it divides an image into semantically meaningful regions having similar attributes like intensity, color, texture, etc. such that it becomes easier to identify sections by labelling. Segmentation and classification are interrelated, as classifier subtly segments the regions of an image [4]. Brain MR image analysis consists of detection, segmentation and classification tasks for differentiating between normal and abnormal tissues. A significant experimental study has been accompanied by many researchers in the field of brain MRI over the past decades. Many researchers have presented various methodologies using available approaches [5]. This study represents a quick review of a few existing segmentation methods with traditional machine learning approach also. In literature, various segmentation methods are elaborated such as edge and region-based, clustering techniques like k-means, fuzzy-based, expectation–maximization, etc., artificial neural network, feedforward, backpropagation, multilayer perceptron, classifiers like SVM, k-NN, CNN, probabilistic neural network, etc. [6–8]. Let us discuss few machine learning as well as few deep learning-based recognition, segmentation and classification of brain MR images from the literature. El-Dahshan et al. [9] presented PCA to reduce features to be detected. He used ANN with 3 layers in which the input layer comprises of 7 nodes, a hidden layer comprises of 4 neurons and the output layer holds 1 neuron. He used two classifiers: ANN and KNN to study 60 abnormal and 10 normal images. The performance measure for this research study has been found 90% in terms of specificity. Saritha et al. [10] has suggested a technique to organize brain MR images based on normality and abnormality using wavelet entropy. From 75 images, she used 52 images to train model and 23 images to assess the model. Yang et al. [11] proposed three-level 2D-DWT and SVM classifier on 90 T2 256 $\times$ 256 pixels images. The features are reduced at a remarkable level. Xuan et al. [12] recommended a method to detect a tumor. He extracted 3 types of features from input images based on intensity, texture and symmetry. He has used T2 images having 24 slices volume for training and testing. Machhale et al. [13] has designed a system to detect cancer tissues. He has used median filter and skull masking in the preprocessing phase. He has used a hybrid approach SVM and KNN in the detection of abnormal tissues from the dataset of 96 brain MR images. Goswami et al. [14] used a hybrid approach of neural network and fuzzy logic. In this approach, he used 100 brain MR images for extracting features. He has used in his study a gray level occurrence matrix. Nathan et al. given an approach by combining the geometric model with statistical parametric mapping and expectation maximization. Sachdeva et al. [15] has developed a model for detection and segmenting various tissues using the content-based model with active contour. For feature extraction, various functions like Laplacian of Gaussian, intensity-based, directional Gabor texture feature was applied. Havaei et al. [16] mentioned a semiautomatic model centered on KNN methodology with conditional random field generation. Li et al. [17] has combined an approach of Markov Random Field with sparse representation to develop a fully automated segmentation model for T1 and T2 as well as FLAIR brain MR images. Ellah et al. [18] used CNN for feature extraction. For error correction, he has used SVM and achieved an accuracy of 99%. Havaei et al. [19] presented an approach based on deep neural networks with notable 0.88 dice score by reducing segmentation time with the help of GPU. Zhao et al. [20] used a combined approach of CNN and CRF to train network in 3 stages of image slices and patches and attained 0.87 dice score. Xiao et al. [21] projected deep learning-based segmentation method with stacked autoencoder network and morphological filter for classification. Abd-Ellah et al. [22] has given a kernel SVM approach for tumor classification. His dataset consists of 120 MRI images for testing and training purposes. Pan et al. [23] compared the performance of CNN and back-propagation neural network and calculate accurate specificity and sensitivity in the outcome by training network with 60 k images of 195 patients. Let's summarize different approaches included in the survey with research gaps (Table 1).

3 Approaches Towards Deep Learning

Deep learning—a branch of ML, aids computers to discern complex patterns from large data sets by computer vision models with deeper neural networks. These techniques grow into the standard to solve many research problems in different domains like image processing, speech recognition, natural language processing, signal processing, etc. with accurate and improved outcomes compared to traditional methods [24]. The main focus of deep learning techniques is feature learning. Also, activation functions resolve the training deficiencies as well as the dropout ratio regulates the network parameters in this approach. Various literature is found for deep learning techniques for different purposes and applications. Let us discuss few methods and their strengths and weaknesses in general as well as in the medical field especially for brain MR images along with possible solutions in the proposed technique.

3.1 Convolutional Neural Networks

In the early 90s, CNN was applied for the analysis of medical images but the success ratio was not so remarkable. It represents a form of artificial neural network. The key aim is to preserve spatial relationships in data by connecting different layers. The main drawback is the necessity of a large amount of labelled data for classification during the learning phase. Here, each layer is operating upon the previous layer. CNN is a fully connected network with multiple layers—convolutions, activations and pooling layers [25]. Due to its robust architecture, CNN is used in the classification task by encoding with activation functions. Feature map is created in the next phase as input through layers. As the number of feature map increases, the number of network parameters is also increased. To get target classes, computation is done at the end of the network [26]. Multiple layers allow the machine to learn multiple features of imaging data. These layers are trained using a specific filter like gradient descent and methods like backpropagation or any other ANN. For a fully connected network, layers have a large number of parameters to study as all are associated with some weight and bias. In convolutional layers, the activation is convolved from previous layers with some filters. Several layers generate a hierarchy of features. With the help of all convolutional layers, one can generate a feature map by following the steps until maximum pooling. It makes it possible to feed a feature map using Non-linear activation functions—ReLU. Feature map is now pooled to pooling layers [27]. Let us discuss few variations in CNN architecture based on training approach follows [28]: Patch-based architecture: Here, the model is trained based on extracted patches

Table 1 Deep learning approaches—survey, research findings and its lacunae

Approach	Suggested by	Findings	Drawback
PCA, ANN, KNN	El-Dahshan et al. [9]	Reduced features to be detected	Different specificity results
Wavelet entropy	Saritha et al. [10]	Classification based normality and abnormality	More time-consuming training phase
three levels 2D-DWT and SVM classifier	Yang et al. [11]	Reduced features at a remarkable level	Problems in the initialization of total iterations
Statistical structure analysis	Xuan et al. [12]	Based on block size, lengthy process	Depends on block size so sometimes not detecting few characteristics
Hybrid approach-SVM, KNN	Machhale et al. [13]	Detect brain cancer cell	Preprocessing requires
Neural network fuzzy logic	Goswami et al. [14]	Benefits of both approaches	Difficulty in choosing fuzzy membership
PCA-ANN	Sachdeva et al. [15]	Multiclass tumor classification	The lengthy process to label dataset
KNN with random field	Havaei et al. [16]	Generalized approach	Non-trivial and time-consuming
MRF with sparse representation	Li et al. [17]	Probabilistic model	Voxel-wise labelling problem
CNN, SVM	Ellah et al. [18]	Two-phase multi-model system for localization	İncrease complexity while increasing layers
DNN	Havaei et al. [19]	Works on global and local data	Cascade requires multi-phase training
CNN, CRF	Zhao et al. [20]	Training based on image patches and slices	Pre and post-processing requires
DNN with autoencoder	Xiao et al. [21]	Correlation between normal and pathological structures	Deformation process takes time and requires notable effort
Kernel SVM	Abd-Ellah et al. [22]	Two-stage classification	Difficulty to choose a kernel
CNN and backpropagation neural network	Pan et al. [23]	Grading processing for classification	Requires attention to include kernels at various layers

from a given image and labels are given to classes. This will help in reducing class imbalance in multiple regions. Based on these classes abnormal and normal tissues from brain MR images are classified. Patch size may differ according to requirements while training the model. Semantic-based architecture: This approach is similar to

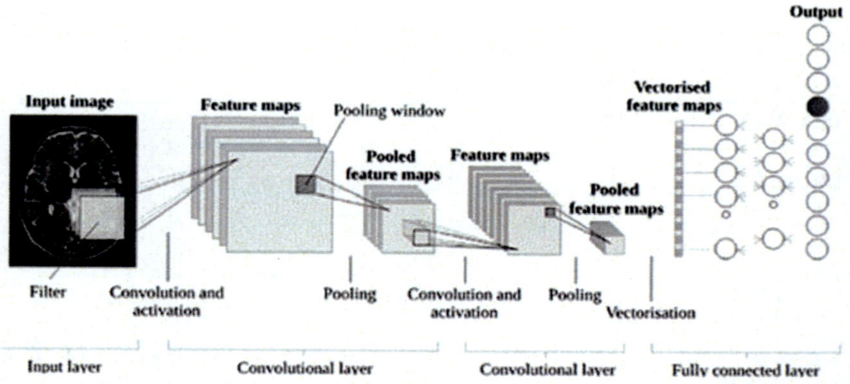

Fig. 1 Basic building blocks of a CNN [27]

auto-encoder in which higher and lower level features are combined to classify pixels. Here, backgrounds are assigned with labels and foreground regions are assigned with a target class (Fig. 1).

Can be referred to it as a local pathway by means of small receptive fields and global pathway with large fields. Here, mapping is done in such a way that it minimizes the loss function. This will help in identifying the desired contour. Cascade based architecture: Here, the chain approach is used in which the output of one CNN classification result is used as an input for another CNN to get accurate class labels.

3.2 *Deep Belief Networks*

This model extracts the representation of training dataset deeply. It comprises of the addition of many superimposed Restricted Boltzmann Machine. This model contains two layers: visual and hidden. Both are interconnected. This network model is trained sequentially by both unsupervised learning methods and supervised learning methods with backpropagation by labelled dataset. There are basically two types—preprocessing and fine-tuning—training phases [29]. In preprocessing each RBM is trained individually using a greedy approach layer-wise. The dataset for the hidden layer is acquired using conditional probability. After some computation hidden layer became a visual layer for further processing. Both layers are related to the energy function. In the fine-tuning phase, the trained network is refined to classify the required ROI. For this binary classification—normal and abnormal—of tissues, few parameters have been set in the network using one of the approaches like stochastic gradient descent, root mean square, adaptive gradient, etc. From literature, observed that error rate for the hidden layer is low, i.e. 0.1 for 100 neurons. The fundamental hurdle is the optimization of parameters in the dataset.

3.3 Deep Neural Networks

It is nothing but a deep convolutional neural network to outshine the learning hierarchy of specific features. This technique solves the problem faced in many machine learning methods during pixel classification by neglecting local dependencies of labels by forming cascading architecture. In this method, there are three cascade architectures—input concentration, local pathway concentration and pre-output concentration [30]. In input concentration, the outcome of the 1st convolutional network is provided to 2nd convolutional network directly so they are considered as additional image channels in the patch. In the local pathway, the first hidden layer is concentrated. In pre-output, concentration is done just before the output layer. Training is also into multiple phases. This method was applied on T1, T2 and Flair brain MR images and trained for segmentation labels like infected tissues, tumor, swelling, etc. on clinical image data sets. The remarkable results found by the input cascade method in terms of Dice measure up to 84% to classify abnormal tissues [31]. The main drawback of this approach is that the training is not simple as the errors are backpropagated to previous layers and also the learning phase is very slow.

4 Research Findings

According to the survey, DL-based approaches are applied universally applied on almost all medical image modalities like X-ray, PET, CT-scan, MRI, etc. In medical image processing and especially in brain MRI domain, finding out present and absence of abnormal tissues is a quite challenging task. The number of algorithms and models are proposed in the literature with the study of different parameters and the results may vary in most of the cases. Most of the approaches follow semiautomatic and automatic methods for detection, classification and feature extraction. There is also an issue of data scarcity and unbalancing in classification [32]. If the classification is compared with segmentation, then classification requires less network parameters than segmentation approach. So, it is demanding to design an efficient CNN approach (recurrent and residual) which uses less number of network parameters. The selection of the best approach is a difficult task. Also, many techniques can be hybrid to achieve a notable amount of accuracy by removing the limitations of individual approach. Few very commonly used techniques are fuzzy methods, SOM, ANN, MRF, SVM, KSVM, CNN, etc. along with specific filters. So it is tough to link all the approaches due to lacking regular datasets, ambiguity in feature selection and undefined standard framework to implement experiments on different platforms. Also many models based on deep learning was proposed like GoogleNet [33], Dense Net [34], AlexNet [35], ResidualNet [36], etc. DL-based approaches to detect abnormal tissues from brain MR images can be arranged to succeed in effective training by registration process to segment precise grouping and performance. DL-based models entail a large number of datasets for training to achieve notable accuracy in the outcome. The major issue in deep learning-based approaches is that it is somehow time-consuming to train huge datasets. Also, these techniques require GPUs to promptness the training process [37].

5 Materials and Methods

The research design of this experimental study can be represented by the below framework. The whole structure is distributed into four different phases: the first phase is image acquisition, the second phase is image enhancement (pre-processing), the third phase is image segmentation, and the fourth one is image representation (post-processing). At the beginning phase, input images are transformed into a gray scale using rgb2gray function after reduction/removal of noise by filters like Poisson or Gaussian. It is advisable to remove backgrounds like skull or scalp or eyes or any other non-interested regions from input image dataset using surface extractor algorithms. Some automated brain surface extractor algorithms based on thresholding intensity morphology are already available as they are useful to extract or classify brain tissues with non-brain tissues which are unwanted for an experimental research study. In the next phase, required features are extracted using a hybrid clustering approach like k and fuzzy c-means to take advantage of saving time for the next phase by applying the required number of iterations. Using these features, the model will be trained and labelled to extract the required tissues. Here deep learning approach is used to train the network model with few parameters like the number of iterations, layers, etc. and confusion matrix. After training the model, it will be used in the classification phase. In this phase, deep learning-based hybrid classifier—RRDNN with U-Net is used for sensing brain MR abnormal tissues. The result is analyzed with various statistical measures with comparison as shown in the next section (Fig. 2).

Below images depict few intermediate outcomes of the experimental framework (Figs. 3, 4 and 5).

6 Experiments and Results

With remarkable numbers in accuracy in the medical imaging domain, by implementing the approach of deep learning, its different variants are experimented and applied to achieve better diagnostic results with cost and time optimization, as well as to reduce human error in medical information processing by efficient automated processing. One of the deep learning-based segmentation methods—U-Net became most popular in medical image classification and detection. A hybrid deep learning approach is applied, centered on Recurrent Residual Deep Neural Networks with U-Net models in our experimental study on different image data sets of brain MR images. The main objective to apply this hybrid approach is to achieve remarkable performance through training and validation accuracies and to reduce computing cost

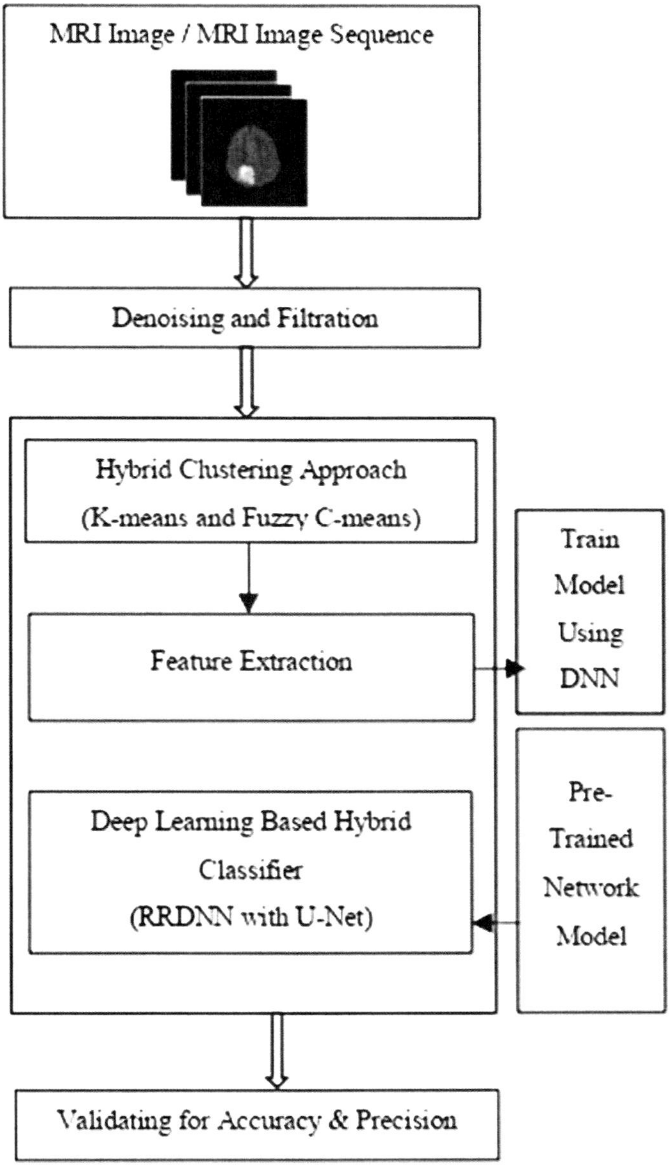

Fig. 2 A proposed framework for a hybrid approach of segmentation

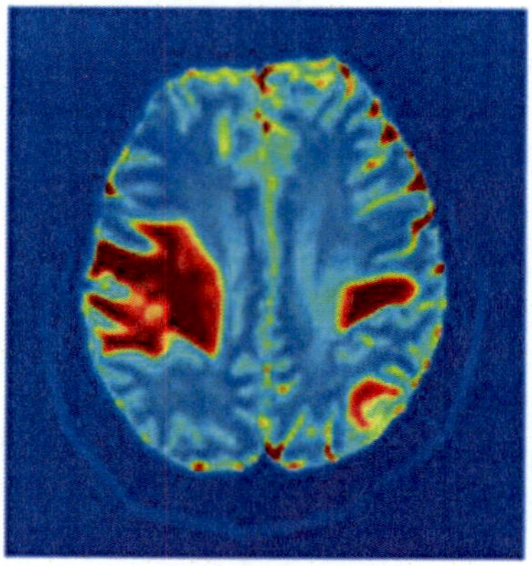

Fig. 3 Abnormal brain MR image

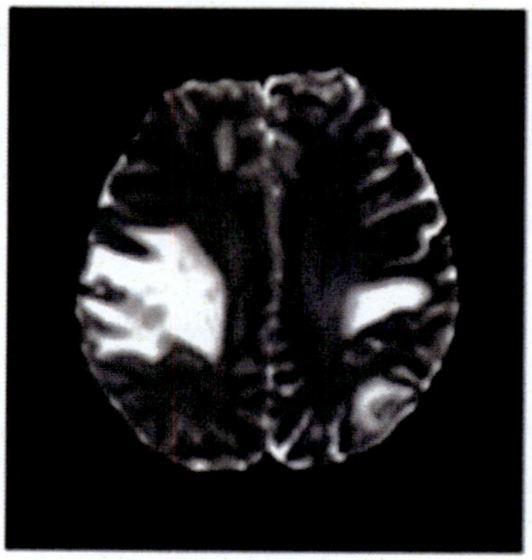

Fig. 4 Gray scale abnormal brain MR image

and time in the entire process of detection and classification [38]. Using this hybrid approach, more features at each subsequent layer can be extracted, i.e. output feature map of one is taken as input to other which are ultimately useful in efficient analysis. U-Net consists of convolutional encoding and decoding units. Also, it preserves most of the context of input images. Here, activation function ReLU is used in both units. For down sampling 2×2 max-pool operations are executed in encoding unit. In decoding unit, de-convolutional operations are executed by feature maps. Proposed approach concentrates on feature maps from encoding to decoding unit. The experimental image-datasets and the resultant segmentation maps are cast-off to learn the model. For minimizing the overhead and optimizing the GPU memory, it is advisable to use large input tiles by reducing the batch size to a single image. The energy function in this approach is calculated by soft-max with cross-entropy function. For this experimental study, 72 brain MR images of size 256×256 pixels have been

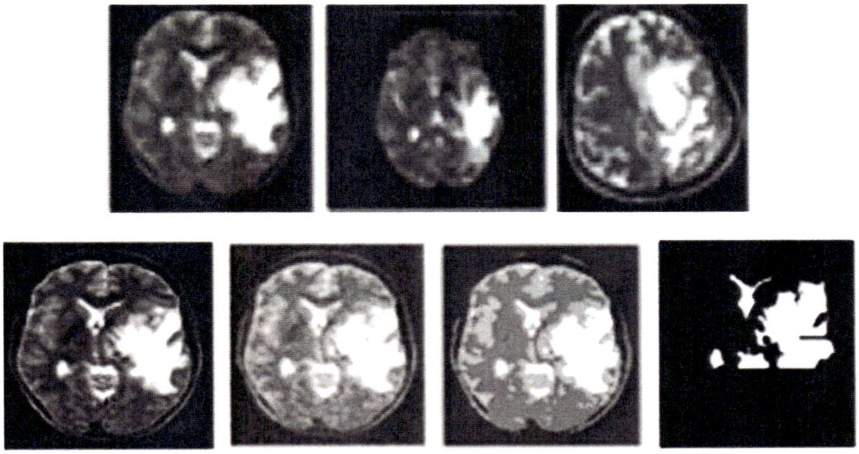

Fig. 5 Various phases during classification

used [39]. The hardware used here is configured as: 8 GB RAM, intel core i5 7th generation processor with 1 GB GPU built-in. For analyzing results, performance parameters like TP, TN, FP, FN have been approached to calculate accuracy, specificity, sensitivity, precision and dice coefficient [40]. To do this calculation, standard formulae are used after getting image matrices by implementing this hybrid approach on MATLAB. The implementation of this algorithm is going to plan on frameworks like Keras and TensorFlow using Python for making this study more transparent and smoother. In this experimental study, maximum training accuracy has been tried to achieve. To evaluate this research study, the following performance parameters have been considered: finding sensitivity, specificity, accuracy by dice similarity index [40].

$$\text{sensitivity} = \text{TP}/(\text{TP} + \text{FN}) * 100\% \tag{1}$$

$$\text{specificity} = \text{TN}/(\text{TN} + \text{FP}) * 100\% \tag{2}$$

$$\text{accuracy} = (\text{TP} + \text{TN})/(\text{TP} + \text{TN} + \text{FP} + \text{FN}) * 100\% \tag{3}$$

$$\begin{aligned}\text{dice similarity index} = {} & (\text{segmented area (i.e. number of pixels) by proposed} \\ & \text{approach} * \text{actual area in ground truth image})/ \\ & ((\text{segmented area by proposed approach} + \text{actual area in ground truth image})/2)\end{aligned} \tag{4}$$

The following table depicts a comparison of various readings after applying the proposed approach and few existing approaches also including deep learning on the specified dataset of brain MR images. Here, in this experiment, gained 0.9433 dice similarity index for proposed RRDNN with U-Net with $t = 3$. From the actual results of ROI, ground truth images data are compared and found this approach more reliable and fruitful than other segmentation approaches with the same experimental parameters (Table 2).

Table 2 Analytical comparison of accuracy in abnormal tissue detection from MR images of brain through different deep learning approaches

Methods	Performance matrix							
	TP	TN	FP	FN	AC	PR	SE	SP
K-NN ($k = 3$)	28	18	17	9	63.89	62.22	75.68	51.43
SVM	33	15	13	11	66.67	71.74	75.00	53.57
K-NN & SVM	36	14	12	10	69.44	75.00	78.26	53.85
ANN	39	17	9	7	77.78	81.25	84.78	65.38
CNN	44	16	6	6	83.33	88.00	88.00	72.73
G-CNN	46	14	7	5	83.33	86.79	90.20	66.67
DBN	47	18	5	2	90.28	90.38	95.92	78.26
Proposed RRDNN with U-Net	53	14	3	2	93.06	94.64	96.36	82.35

TP TruePositive, *FP* FalsePositive, *TN* TrueNegative, *FN* FalseNegative, *SE* Sensitivity/Recall, *SP* Specificity, *AC* Accuracy, *PR* Precision (Fig. 6).

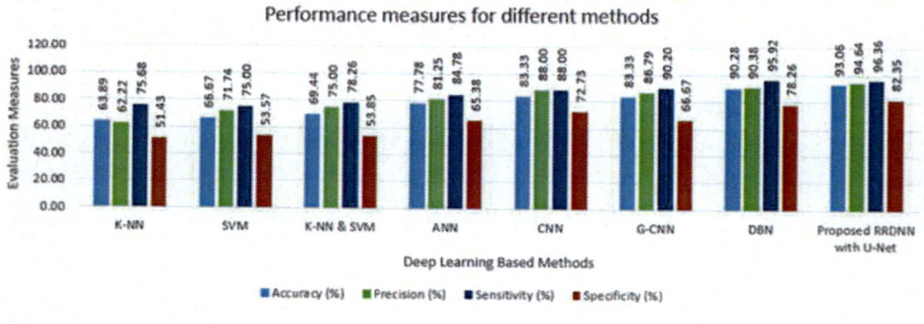

Fig. 6 Comparative chart of performance measures of various deep learning-based methods for detection of anomalous tissues from brain MR images

7 Conclusions

In this research study, various segmentation approaches used for detecting abnormalities have been discussed from brain MR images and highlighted a few problems and difficulties in existing methods. Also, a hybrid segmentation approach RRDNN with U-Net is suggested for detecting abnormalities of brain tissues from MR images. The experimental study was conducted on various brain MR datasets from different imaging centers. Different criteria and parameters are considered and applied multiple approaches to the same datasets for detection and classification. The result was compared with ground truth image dataset and tested all required characteristics under experimental study. The experimental study and the outcomes demonstrate improved performance (94%) in segmentation of abnormal tissues throughout testing phases. The proposed method has been proven more trustworthy in the medical research domain compared to similar studies in the same environment. Also, it is time-efficient to achieve required outcomes. By applying this hybrid approach, medical practitioners will get more precise results within time constraints at every stage of diagnosis.

References

1. Lundervold, A.: An overview of deep learning in medical imaging focusing on MRI. Z Med. Phys. (2018)
2. LeCun, Y., Hinron, G.: Deep learning. Nature **521**, 436 (2015)
3. El-Baz, S. Charya, R., Suri, J.S. Multimodality State-of-the-Art Medical İmage Segmentation and Registration Methodologies. SpringerLink, vol. II ISBN: 978-1-4419-8203-2
4. Deshmukh, R.D., Jadhav, C.: Study of different brain tumor MRI image segmentation techniques. İnt. J. Comput. Sci. Eng. Technol. **4**(4), 133–136. ISSN: 2231:0711
5. Dass, R., Priyanka, Devi, S.: Image segmentation techniques. İnt. J. Electron. Commun. Tech. **3**(I), (2012) ISSN 2230–7109 (online), ISSN 2230-9543 (print)
6. Thilangamani, S., Shanthi, N.: A survey on image segmentation through clustering. İnt. J. Res. Rev. Inf. Sci. **1** (2011)
7. Litjens, G., Kooi, T., et al.: A survey on deep learning in medical image analysis. Med. Image Anal. **42**, 60e58 (2017)
8. Despotović, I., Goossens, B.: MRI segmentation of the human brain: challenges, methods and applications. Hindawi publishing corporation. Comput. Math. Methods med **2015**, Article ID 450341
9. El-Dahshan, E., Mohsen, H.: Computer-aided diagnosis of human brain tumor through MRI: a survey and a new algorithm. Expert Syst. Appl. **41**(11), 5526–5545 (2014)
10. Saritha, M., Joseph, K.P., Mathew, A.T.: Classification of MRI brain images using combined wavlet entropy based spider web plots and probabilistic neural network. Pattern Recognizaiton (2013). ISSN 3416 2151-6
11. Zhang, Y.G., et al.: Automated classification of brain images using wavelet-energy and biogeography-based optimization **26**, 1–17 (2015)
12. Xuan, X., Liao, Q.: Statistical structure analysis in MRI brain tumor segmentation. In: International Conference on Image and Graphics—ICIF-2007, pp. 421–426 (2007)
13. Machhale, K., Nandpura, H.B.: MRI brain cancer classification using hybrid classifier (SVM-KNN). In: International Conference on İndustrial İnstrumentation and Control (ICIC), pp. 60–65 (2015)

14. Goswami, S., et al. A hybridneuro-fuzzyapproach for brain abnormalitydetection using GLCM based featureextraction. In: İnternational Cnference on Emerging trends in Communication, Control, Signal PRocessing and Computing Applicatiions (C2SPCA), pp. 1–7 (2013)
15. Sachdeva, J., Kumar, V.: Segmentation, featureextraction and multiclass braintumor classifiaction. J. Digit İmaging **26**(6), 1141–50 (2013)
16. Havaei, M., Jodoin, P.M.: Effective interative braintumor segmentation as within brain kNNclassification. In: İnternational Conference of Pattern Recognization, pp. 556–561 (2014
17. Li, Y., Jia, F., Qin, J.: Brain tumor segmentation from multimodal magnatic resonance images via sparse representation. Artif. Intell. Med. **73**, 1–13 (2016)
18. Abd-Ellah, M.K, Awad, A.I., et al.: Two-phase multimodel automatic brain tumor diagnosis system from magnetic resonance images using convolutional neural networks **97**(1), 1–10 (2018)
19. Havaei, M., Davy, A., et. al.: Brain tumor segmentation with deep neural networks. Med. İmage Anal. **35**, 18–31 (2017)
20. Zhao, X., Wu. Y., et al.: Brain tumor segmentation using a fully convolutional neural network with conditional random fields, pp. 75–87 (2016)
21. Xiao K., Liang A., et al.: Extraction and application of deformation based feature in medical images. Neurocomputing **120**, 177–184 (2013)
22. Abd Ellah, M.K., Hamed, H.F.A., et al.: Design and implementation of a computer aided diagnosis system for brain tumor classification. In: International Conference on Microelectronics, pp. 73–76
23. Pan, Y., Huang, W., et al.: Brain tumor grading based on neural networks and convolutional neural networks. İn: İnternational Conference of IEEE Engineering in Medicine and Biology Society (EMBC), pp.699–702 (2015)
24. Maier, A., Syben, C., et al.: A gentle introduction to deep learning in medical processing 2019, zemedi-10780
25. Krizhevsky, A., Hinton, G.: Imagnet classificaton with deep convolutional neural networks. Adv. Neural İnf. Proc. Syst. 1097–1105 (2012)
26. Ravi, D., Wong, C., et al.: Deep learning for health informatics. IEEE J. Biomed Health İnf. **21**(1), 4e21 (2017)
27. Ahmed, K.B., Liu, R., et al.: Fine-tuningconvolutionaldeepfeatures for MRI basedbraintumor classification. In: Medical İmaging Computer AİdedDiagnosis, İnternational Society for Optics and Photonics, vol. 10134 (2017)
28. Alexander, S., Arvind, L.: An overview of deep learning in medical imaging focusing on MRI. Zemedi-10775 (2018)
29. Hinton, G.E.: Deep belief networks. Scholarpedia **4**(5), 5947 (2009)
30. Srivastava, N., Salakhutdinov, R.R.: Multimodel learning with deep boltzmann machines. Adv Neural İnf. Proc. Syst. 2222–2230
31. Anuse, A., Vyas, V.: A novel training algorithm for convolutional neural network. Contr. Inell. Syst. **2**(3), 221e34
32. Ism, A., Sah, M.: Review of MRI-based brain tumor image segmentation using deep learning methods. Sciencedirect, Elsevier Inc. (2016)
33. Szegedy, C., Reed, S., et al.: Going deeper with convolutions. In: Proceedings of IEEE Conference of Computer Vision and pattern recognition, pp.1–9 (2014)
34. Huang, G., Liu, Z., et al.: Densely connected convolutinal networks. CVPR **1** (2016)
35. Krizhevsky, A., Surskever, I.: ImageNet classification with deep convolutional neural networks. In: Advances in Neural İnformation Processing System Curran Associates inc., pp. 1097–1105 (2012)
36. He, K., Zhang, X., et al.: Depp residual learning for image recognition. İn: Proceeding of IEEE Conference on Computer Vision and pattern recognition, pp. 770–778
37. Ronneberger, O., Brox, T.: U-Net: convolutional networks for biomedical image segmentation, arXiv:1505.04597v1 (2015)
38. Fischer, P., Brox, T. U-Net: convolutional networks for biomedical image segmentation. In: İnternational Conference on Medical İmage Computing and Computer-assisted İntervention, Springer, pp. 234–241 (2015)

39. Kofler, A., Dewey, M., et al.: A UNETs cascade for sparse view computed tomography. In: İnternational Workshop on Machine Learning for Medical İmage Reconstruction, Springer (2018)
40. Havaei, M., Davy, A., et al.: Brain Tumor Segmentation with Deep Neural Networks. Elsevier B. V., pp. 1361–8415 (2016)

A Comprehensive Study Toward Women Safety Using Machine Learning Along with Android App Development

Karthik Hariharan, Rishi Raj Jain, Anant Prasad, Mridhul Sharma, Prashant Yadav, S. S. Poorna, and K. Anuraj

Abstract In this modern era, women have shown excellence in all walks of life and stand equally with men in all sectors. So has increased violence against women due to this exposure. Hence, women security during travel is one of the critical issues faced by society, from the past and even today. The paper aims at providing a technological solution to this problem. For this, a system for forewarning and safeguarding the female cab users using machine learning is developed. An Android application is developed to collect data, and a virtual panic environment was created to collect the anomalies. The accelerometer data along the three coordinates was captured corresponding to normal and anomalies. Three machine learning models viz support vector machines, logistic regression, and Naïve Bayes are trained on data collected, and their test performance is compared in terms of accuracy. This trained model is further integrated with an Android application for real-time data collection and feedback.

Keywords Women safety · SVM · LR · NB · Android · Machine learning · Cloud database

1 Introduction

In the past few years, cab services have become a very important part of the Indian transport system, especially in metropolitan cities where Ola, Uber, mytaxi, and many more have become a daily commute option with affordable rates and wide network. On one hand having all the pros of using such services, on the other hand, women especially face a lot of uncomfortable encounters with the cab drivers and which leads to molestation, harassment, and rape. Nearly, 23% of women surveyed had to report an uncomfortable encounter with a cab driver, 10% of women were hit on, and 8% of the women had to get the police involved [1, 2]. Despite the intense debates about women's safety and promises of improvement, not much has changed

K. Hariharan · R. R. Jain · A. Prasad · M. Sharma · P. Yadav · S. S. Poorna (✉) · K. Anuraj
Department of Electronics and Communication Engineering, Amrita Viswa Vidyapeetham, Amritapuri, India
e-mail: poorna.surendran@gmail.com

P. Karuppusamy et al. (eds.), *Sustainable Communication Networks and Application*, Lecture Notes on Data Engineering and Communications Technologies 55,
https://doi.org/10.1007/978-981-15-8677-4_26

for women on roads. While the administration did launch apps, the efforts have not borne the expected fruit. Further, only a part of the population made use of such applications. The statistics are shocking, and the painful truth is that women in our country are not safe.

With technology and digitalization, this problem could be tackled to an extent. In this work, a few existing applications are studied and a new application with enhanced features is introduced, to ensure better safety of women. The key features include real-time live location tracking, detection of wrong routes, and anomaly detection using machine learning algorithms [3–5] viz support vector machines, logistic regression, and Naïve Bayes. Some of the related researches in this area are as follows.

2 Literature Review

Many apps have been developed for the safety of women. Some applications that offer similar services for Android and other platforms are discussed here. In [6, 7], a mobile application, Nirbhaya, was developed, which sends a message with the user's GPS coordinates to a list of emergency contacts when a button on the app screen is touched. The coordinates are updated and reset with every 300m change in location. iMace mobile application described in [7] produces a high-pitched alarm upon shaking of the phone and alerts friends and law enforcement in the location, a warning of the attack.

Another application, Secureme beta, mentioned in [8] helps to raise alert and get help in case of life-threatening emergencies. Initially, a PIN has to be entered for security purposes and then emergency contacts must be registered in the app. It notifies the contacts with location coordinates, by pressing on the secure button. In [8] another Android application, Abhaya had an added advantage of continuously alerting the emergency contact by sending messages, every 5min until the person gets notified. It also helps to monitor the location of the user continuously. Some of the other related apps include Raksha, Vanitha Alert, Street safe, and Fightback [8]. Among these, Vanitha Alert and Street safe will automatically update the victim's Facebook account with their recent location.

A model to track and monitor car for ensuring safety while traveling was developed in [9]. The system is capable of sending SMS and MMS, along with an added advantage of including the intruder's picture being sent to the police. With the help of a database, this system is capable of accessing the owner's information along with GPS data, which in turn helps the police to track the vehicle fast. A smartphone application developed in [10], HearMe provides access to it even if the phone is locked. Further, it can provide a panic alarm on both the victim and receiver sides. It is capable of turning the device camera to a spy cam mode, without shutter sound and flashlight. It also gives information about the nearest law assistance.

The main feature of this developed application is the use of a machine learning trained model for automatically identifying women in a panic situation while traveling. The following sections include the methodology adopted training the aforementioned model and the corresponding app development for women safety.

3 Proposed Model

The detailed methodology is explained with the help of a block diagram given in Fig. 1, which explains the application's backend and frontend. The prime focus of this paper is in machine learning-based backend development, especially the database, authorization, and trained models (highlighted in the diagram below in red). In the block diagram, the frontend application is connected to Firebase which is a mobile and Web application development platform plays as the backbone of the system. Real-time information collected from users is stored in the Firebase cloud. Storing data in the cloud gives better security and backup in addition to giving easy access to the administrator. The data that is stored in the clouds include the details of the user and their family members in a structured form. It uses messages as a service of the Firebase to send a notification to users and their family when there is a trigger. Each user is uniquely identified with an IMEI number of their phones in the database. The app authorization deals with users receiving an OTP in their Android device to register in the app and hence accepting the terms and conditions. The final trained machine learning models are also saved in the cloud for integration with the app. The database also plays a role in storing the live location sent by the app when in distress, so that the last active location could be tracked before the phone connection is lost. Features of this app also include wrong route tracking and live location tracking, which is explained in detail in Sect. 5.2.

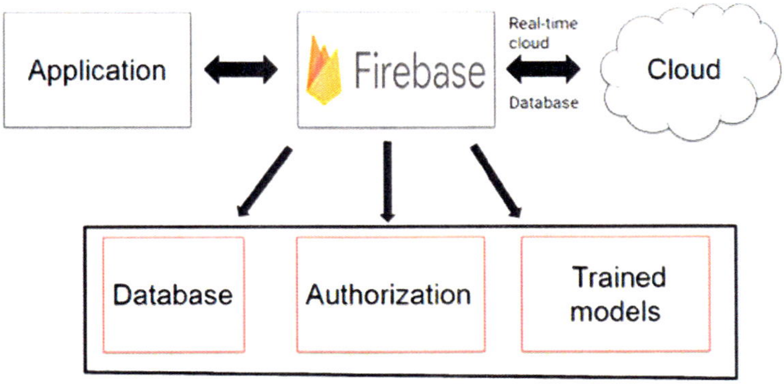

Fig. 1 Architecture diagram

4 Methodology Adopted for Machine Learning-Based Backend

The machine learning backend could be modeled as a "binary classification." Suppose the person faces a sudden impact due to unforeseen circumstances or her vehicle crashes, the output could be categorized as "anomaly" else "normal." The application pre-trained on seed normal and anomaly data, if classifies the real-time data as an anomaly, will immediately send an emergency message to the nearest police station and close contacts. Naïve Bayes (NB), logistic regression (LR), and support vector machine (SVM) are the supervised learning techniques adopted in this study.

4.1 Collection of Seed Data

A separate application was made to collect the data from the users. For this, eight volunteers with host mobile phones were selected. The application used the accelerometer in the mobile phones to read the change in velocity and log it into the cloud database viz Firebase. To be specific, it measures the change in velocity along with the three-dimensional Cartesian coordinate system in the form of (X, Y, Z). This, in turn, is used by the machine learning algorithms. The data is then cleansed and modified according to our application. It was observed that when volunteers were in a normal state, the accelerometer readings were very minimal ranging from negative axis to positive 20 along the three axes but on the other hand, when the subjects created forced impact situations such as suddenly dropping their phones or sudden jerks, the accelerometer readings were very high in the range of 30–100. Further, this data was logged in to Firebase, cloud database. Data was then cleaned by writing a script to exclude unwanted values and labeled.

4.2 Selection of Machine Learning Algorithm

After cleaning the data, a machine learning algorithm needs to be chosen whether the person is either in a dangerous situation or not. Various machine learning algorithms were tested based on our understanding of the data and the algorithms used which are discussed here: Logistic regression uses a logistical function to model a binary dependent variable. Here, the logarithm of the likelihood of the value labeled "1" is a linear combination of one or more independent variables, called predictors [11]. In Naïve Bayes, it uses a collection of probabilistic classifiers which are based on the mathematical Bayes' theorem [12]. A support vector machine (SVM) model represents the examples as points in space, mapped in such a manner that the examples of the separate categories are divided by a hyperplane which maximizes the associated margin [13].

Once the training is finished, next the model has to be exported to the Android platform. To create a "frozen" version of the trained model, Android studio IDE needs TensorFlow clone. Also, a file consisting of "labels" need to be included.

5 Architecture and App Development

The women safety app is written in Java. The frontend uses XML format to display data and services. The backend part is used to access the data that needs to be displayed. This is a very common architecture which is followed to keep up with the usual convention and the usual flow. The backend of the app will be integrated with our trained machine learning model for detecting jerks. For the app development first, a mock-up of the user interface (UI) was created to get a more comprehensive idea about the final UI. Then, it was implemented in the XML layout using the libraries and modules provided by Android. The backend was implemented using Firebase. These two components were then fused seamlessly. Firebase platform has been used as a cloud database, and access to the database is provided as a service. All data collected is stored in Firebase.

5.1 Implementation of App

For the frontend of app, UI was designed for the alpha version for developers and testers to login and add necessary data to our database, and to test and train our machine learning model. Once the data was collected, the design of the app was created separately for each page and was implemented using the XML layout with the help of libraries and modules provided by Android. Firebase is used as a backend for the app. This was chosen for easy integration and scalability. Firebase is a cloud service provided by Google for decreasing the lag or delay between elements in receiving data. One of the reasons for choosing firebase is that it provides a real-time database. Also, the structure for our database is a NoSQL type, and hence, the data is stored in tree form, so while fetching data that is required for that particular call can be quickly fetched with going through the whole chunk. The data stored in the cloud include the details of the user and their family members. It uses SMS as a service of the firebase to send a notification to the users and their family when there is a trigger. Each user is uniquely identified with an IMEI number of their phones in the database. The final trained model is saved for integration with the app. The database also stores the live location sent by the app, when in distress, so that the last active location could be tracked before the phone connection is lost.

5.2 *Features of Android Application*

Salient features of this app include wrong route tracking and live location tracking. The Google Maps application program interface (API) is used to integrate Maps into the system. The Direction API is used to obtain all the routes, given a source and a destination in the Maps. The routes are obtained from the API, parsed and stored in an array. Further, the current route is checked with all other routes stored in the array. If there is no route which matches the array, an alert message is sent and the user is also notified that he or she is deviating from the actual route. Location data in Android app is used in live location tracking. The user's location ID is displayed on Google Maps, which gives reverse geocode coordinates into a street address and the details regarding combining location data with the Places API. This location data only ever exists locally on the user's device. For remote access, of this data, it has to be linked with the Firebase real-time database. An app that essentially acts as a GPS tracker may sound presumptuous; this functionality forms the basis of lots of different applications. The ability to view a device's live location remotely will be very useful, as the user can see his current position on the map.

6 Results and Discussion

An Android app for the safety of women while traveling was implemented with the advantages of wrong route detection and live location tracking. Further, a study on different machine learning models was carried out to aid anomaly/jerk detection. The tools used include Android Studio for developing the app, Firebase for real-time cloud database, Google Collaboratory for testing machine learning algorithms, and TensorFlow lite to convert and deploy machine learning models to Android app.

UI was written in XML for the frontend of the app. The final UI for the alpha version is shown in Fig. 2a. The app offers the additional feature of showing all possible routes, and it enables the emergency contacts with remotely monitoring the users live locations, which is one of the highlights required for a safety application. Since NoSQL type database is used by Firebase, fetching and retrieval of data becomes quick and hence advantageous for real-time applications. Screenshots of SOS messages sent to the emergency contact numbers are shown in Figs. 2b and 3. show all possible routes from the maps. After live location tracing, mockup of live location, sign-up, and emergency contacts screen obtained is shown in Fig. 4.

For the study based on machine learning, using logistic regression, NB classifier and SVM, the testing and training accuracies are as given in Table 1. The seed data for this analysis is collected from a phone accelerometer for detecting jerks. Of the three methods analyzed, SVM gave the best test accuracy, followed by NB for the data under study. Due to this, SVM was chosen to be the frozen model, which was deployed in the app.

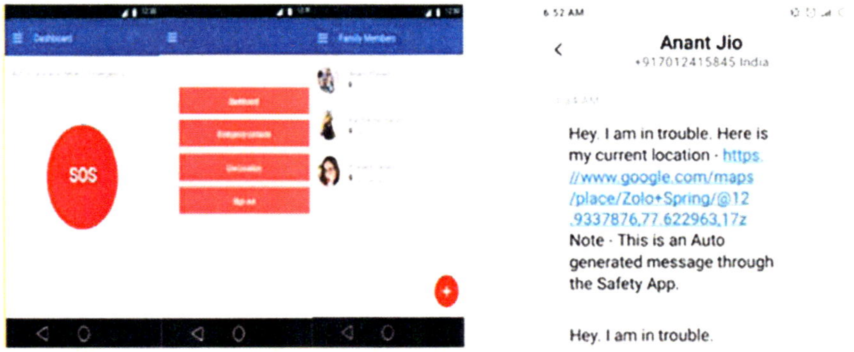

(a) (b)

Fig. 2 **a** App main screen, **b** Screen shots of SOS messages sent to the emergency contact numbers

Fig. 3 Showing all possible routes

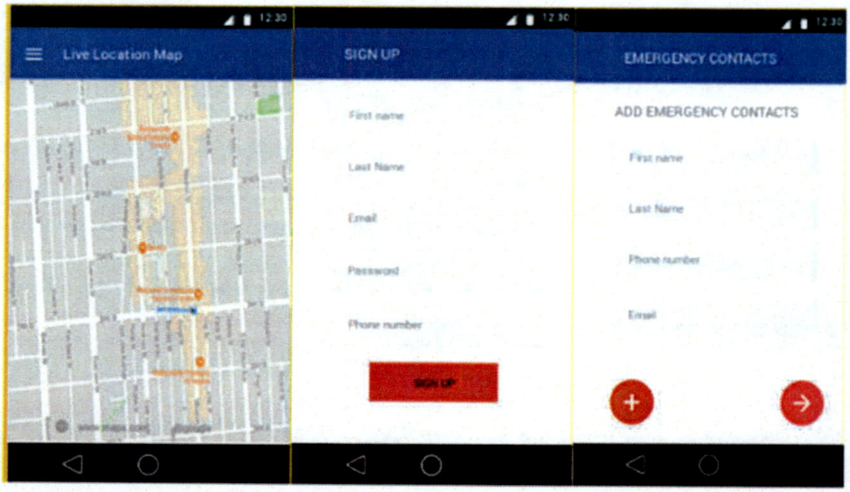

Fig. 4 Mockup of live location, sign-up, and emergency contacts screen

Table 1 Training and testing accuracies of the machine learning algorithms

	NB	LR	SVM
# Training samples	10,592	10,592	10,592
# Test samples	2648	2648	2648
Training accuracy (%)	87.73	78.37	98.27
Test accuracy (%)	87.34	77.56	89.5

The app can be simply downloaded from Play Store and can be used in any Android devices. Support across all versions is also provided. Its simple UI design makes it easy for users to handle the app. Women while traveling can use this app, so that all the information and notification could be traced by any of their family members. The app can be used when the user senses that there is a danger in some region while traveling and can be switched on in the background to keep the emergency contacts notified. The app asks for audio or camera access before using it, in order not to breach your privacy. Apart from that being potentially a women safety app, it can be used even by men or children to track and alert their emergency contacts when in danger. The most important feature of this app is that it automates the alert process during the danger, so that makes it very useful during danger or emergencies where the user does not have time to do things manually.

7 Conclusion and Scope for Future Work

Architecture for implementing safety application, with ML backend is given in this paper. The app offers additional features such as live location tracking and prediction of wrong routes. Further, three machine learning models have been tested, out of which SVM classifier works best giving an accuracy of 89.5% with our dataset. Furthermore, the app is capable of fetching all possible routes from the user's source to the destination where he/she wants to go and if the driver of the cab takes any of the other routes and tries to deviate, it sends a warning, after which an SOS message will be sent. Other features include marking all safe zones in and around the victim, sending MMS during emergencies. The merit of this application would be that the app can function in auto mode during distress situations. Also, the live location of the person in danger is shared with the person's emergency contacts including the police. Also since a cloud database is being used, the number of users exceeding will never be an issue. As technology emerges, it is possible to upgrade the system and make it even more robust and adaptable to the user. Furthermore, it can be developed in the IOS platform as well. Thus, this app can help society in emergencies.

References

1. Is Uber Safe? How Safe are Uber and and Lyft for Women, January 14. https://www.alarms.org/uber-lyft-womens-safety-report/ (2020)
2. Only 9% Indian Women Feel Public Transport is Safe, But They Still Use It: Report, NEWS18. March 7. https://www.news18.com/news/buzz/indian-women-feel-public-transport-is-unsafe-but-they-still-use-it-report-2059233.html (2019)
3. Suma, V.: Computer vision for human-machine interaction-review. J. Trends Comput. Sci. Smart Technol. (TCSST) **1**(02), 131–139 (2019)
4. Smys, S.: DDOS attack detection in telecommunication network using machine learning. J. Ubiquit. Comput. Commun. Technol. (UCCT) **1**(01), 33–44 (2019)
5. Manoharan, S.: An improved safety algorithm for artificial intelligence enabled processors in self driving cars. J. Artif. Intell. **1**(02), 95–104 (2019)
6. Nirbhaya: Be Fearless. https://digitalindia.gov.in/content/nirbhaya-app
7. Pandey, S., Jain, N., Bhardwaj, A., Kaur, G., Kumar, V.: Reach360: A comprehensive safety solution. In: 2017 Tenth International Conference on Contemporary Computing (IC3) 2017 Aug 10, pp. 1–3. IEEE (2017)
8. Yarrabothu, R.S., Thota, B.: Abhaya: An Android App for the safety of women. In: 2015 Annual IEEE India Conference (INDICON) 2015 Dec 17, pp. 1–4. IEEE (2015)
9. Hameed, S.A., Abdulla, S.: Effective car monitoring and tracking model. Australian. J. Basic Appl. Sci. **6**(1), 1–8 (2012)
10. Akash, S.A., Al-Zihad, M., Adhikary, T., Razzaque, M.A., Sharmin, A.: Hearme: A smart mobile application for mitigating women harassment. In: 2016 IEEE International WIE Conference on Electrical and Computer Engineering (WIECON-ECE) 2016 Dec 19, pp. 87–90. IEEE (2016)
11. Shah, K., Patel, H., Sanghvi, D., Shah, M.A.: Comparative analysis of logistic regression, random forest and knn models for the text classification. Augmented Hum. Res. **5**(1), 1–6 (2020)

12. Dey, S., Wasif, S., Tonmoy, D.S., Sultana, S., Sarkar, J., Dey, M.A.: Comparative study of support vector machine and naive bayes classifier for sentiment analysis on amazon product reviews. In: 2020 International Conference on Contemporary Computing and Applications (IC3A) Feb 5, pp. 217–220. IEEE (2020)
13. Poorna, S.S., Reddy,M.R.K., Akhil,N., Kamath,S., Mohan, L., Anuraj K.: Computer vision aided study for melanoma detection: a deep learning versus conventional supervised learning approach. In: Pati, B., Panigrahi, C., Buyya, R., Li, K.C. (eds.) Advanced Computing and Intelligent Engineering. Advances in Intelligent Systems and Computing, vol. 1082, Springer, Singapore (2020)

Automated Plant Disease Identification and Detection with Multi-features

Sumathi Ganesan

Abstract The identification of the diseases from the leaf images is an important feature of the agriculture field. The preprocessing contains the boundary identification and separating the images are the common features of identification purpose. The kernel function with filter values is used to reshape the boundary for better classification purpose. The proposed multi-feature identification algorithm is constructed to provide better accuracy and faster run speed. The comparison of disease was analysed by support vector machine algorithm and segmented by k-means clustering with the severity values were calculated. It provides significantly accurate results in a short duration of time by identifying up to 95% of the pests and in some cases 97% of the pests upon comparison with a neural network-based approach.

Keywords Plant disease identification · Classifier · Multi-feature · Extraction

1 Introduction

Pest deduction is used to secure crops and ensure food quality. In this paper, a novel and fast methodology is developed to deduct and enumerate the pests present in an image using rapid feature deduction algorithm [1]. The amount of pesticides being used for agriculture pollutes the environment. The implementation of the algorithm can make machines to utilize pesticides effectively by using image processing [2]. Human labour is used for the manual deduction of pests which is not highly accurate [3]. These types of algorithms can be used for targeted pesticide usage in crops to help reduce soil pollution [4]. Studies on plants have been made for flowers, leaves, seeds besides fruits. Though there exist millions of various kinds of plant, many sub-types are still unidentified and they will invariably die and become non-existent before they become known [5]. Thus, there is a need for detecting and classifying the plant automatically for speeding up the action of learning the individual plant types. To

S. Ganesan (✉)
Department of Computer Engineering, Government Polytechnic College, Aranthangi, Tamilnadu, India
e-mail: sumi.ganesan@yahoo.com

P. Karuppusamy et al. (eds.), *Sustainable Communication Networks and Application*, Lecture Notes on Data Engineering and Communications Technologies 55,
https://doi.org/10.1007/978-981-15-8677-4_27

suggest innovative methods to identify plant types, biologists and computer experts have been constantly playing their roles [6]. Computerized vision methodology has transformed the task of classifying the plant automatically to find ideal. For reducing the huge data set to a lesser subset, the algorithms of feature extraction become essential with daily gradual development [7].

2 Related Work

A researcher incorporated contour and strain besides colour and quality attributes for classifying a leaf and the classifying agent used probabilistic neural network (PNN) [8]. The outcome of the investigation revealed that the procedure for classifying yielded normal precision of 93.75% on testing with Flavia data set consisting of 32 types of plant leaves, concluding that the procedure used yielded superior performance in comparison with the original task. While classifying plants, the shape of the leaves plays a prominent role. The greatest prominent part has been incorporated for deciding besides processing of data is the recognition of shape [9]. A researcher used feed-forward neural systems to mechanize for recognizing leave for classifying the plants [10]. A new procedure has implemented to classify pictures of leaves by taking advantage of the concept of gaining data and discovered the effectiveness of education algorithms of MLP to classify plant leaf. The outcome will be an enhancement in data gain procedure for MLP with batch reverse transmission algorithm-based knowledge acquiring computing efficiency by enhancing the precision of classification. The suggested algorithms outperformed MLP with additional orientation besides getting educated at Levenberg–Marquardt level to classify plant leaf on testing with nine species [11]. An optimum methodology to extract feature and choosing algorithm leaf's classification based on genetic was suggested wherein selecting optimum attribute's subset besides the classification became a significant procedure in classifying leaves. The causative attribute arrangement is taken off the leaf pictures applying GA procedure, and these extracted attributes were put in place for training the support vector machine [12].

3 Proposed Work

Figure 1 determines the input image for producing the leaf classification of healthy brinjal leaf and the affected leaf images. The proposed method is constructed to identify the disease from the input image for the leaf classification process. The proposed technique is constructed to detect the infection of leaves at the initial stage based on its features. It is a step-by-step process. In each step, a particular task is accomplished. Therefore, it is divided into different modules, and each module has specific output. The final output can be obtained through these steps in Fig. 2.

Fig. 1 Healthy Brinjal leaf and affected area

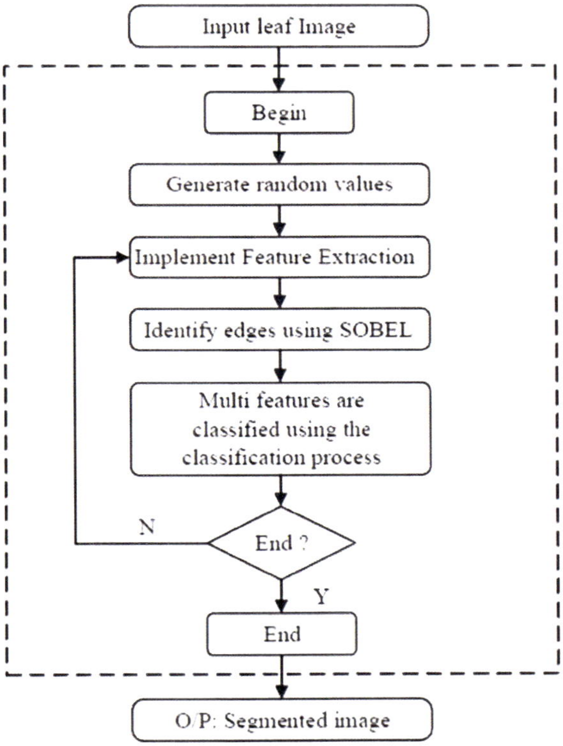

Fig. 2 Flow chart for the execution process

The feature extraction process is initiated when the initial object is submitted to a specific set of affine formal alterations or an indiscriminate amalgamation of the same, and the shape or appearance of the object in a feature should remain unaltered. A 2D shape describing device should not be sensitive to translation, scale changes (uniform in both the X and Y-coordinates) besides rotations. This principle for describing shape denotes that the describing device is capable of carrying out regularization for different looks the object may occur in. The real representation of the object must be able to map a similar model of the pattern.

3.1 SOBEL Edge Detector

The SOBEL operator extracts the appropriate derivative function from the input image that the gradient value from the image boundary from both directions has been computed to evaluate the kernel matrix as 3 × 3, and it is illustrated in Fig. 3. The gradient value for the centre pixels is computed through the middle pixels in both directions.

The gradient matrix with every pixel from the input image has computed in kernel convolution parameter, and the edge within the columns is illustrated in Fig. 4. The computation of the pixel values from the image has the initial value, and the edges are classified through the gradient matrix.

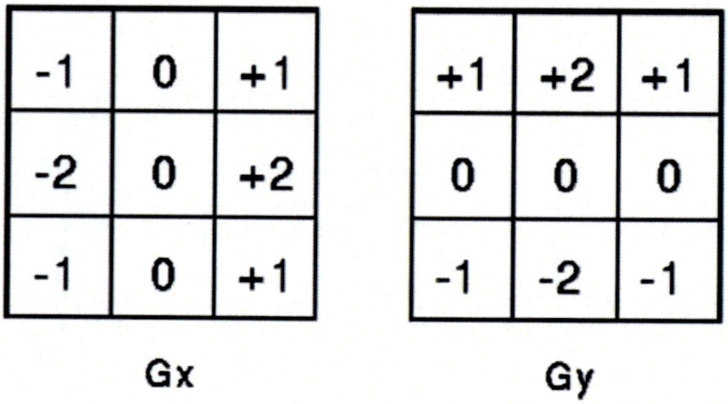

Fig. 3 Kernel matrix formation

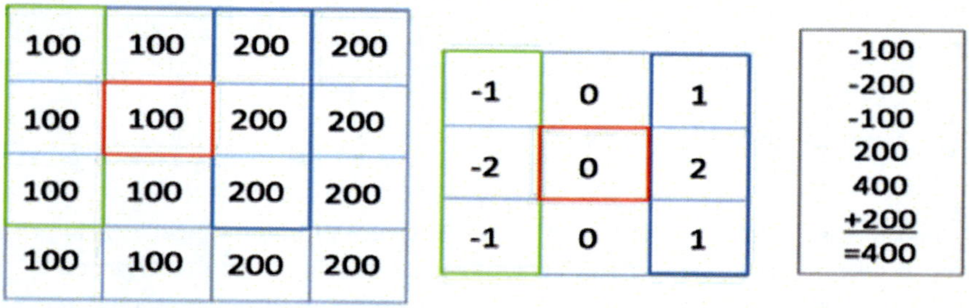

Fig. 4 Edge between the columns

3.2 Algorithm—Multi-feature Identification

Input: Plant leaf image.

Output: Disease identification.

Step 1: The initial process is to capture the leaf image in high quality.

Step 2: Feature extraction is performed to eliminate the distortion from the input image using the filter technique.

Step 3: The threshold value is calculated to identify the components of the pixel values.

Step 4: The infected area is identified using the SOBEL edge detector.

Step 5: Multi-feature identification method is utilized to generate the output segments to classify the leaf diseases.

Step 6: The classification is computed to identify the similarity values.

The classification values $(\delta_i(x))$ are generated by computing the mean of every pixel $(\alpha_i(x))$ in Eq. (1)

$$\delta_i(x) = \sum_{i=1}^{n} (\alpha_i(x)) \tag{1}$$

The fitness function (Fit_i) is computed in Eqs. (2) and (3).

$$\text{Fit} = \sum_{i=1}^{n} \text{Fit}_i \tag{2}$$

$$\text{Fit}_i = \sum_{i=1}^{n} |\alpha_i(x) - \delta_i(x)| \tag{3}$$

4 Performance Evaluation

Figure 5 illustrates the edge detection process from the input image, and the input image is converted into the resized image for producing the grey image. The grey image is transferred into the SOBEL gradient image that the borders are identified and marked, and finally, the edge detection is used to identify the plant disease identification. To optimize the features of colour and boundary sequences, the filter technique was used to enhance the total overview by the action based on the matching precision.

Fig. 5 Edge detection from the input image

The SOBEL gradient was applied to obtain the false positive and negative attributes. The outcome is applied to the edge detection for choosing attribute subset using the classifying agent-established computing efficiency besides improving the precision in comparison with the proposed technique for leaf pattern classifications. The assessment demonstrated that gaining data helps in handpicked attributes resulting in specific MLP enhancements with batch reverse transmission algorithm classifying action having 94.81% accuracy. Applying the attributes choosing procedure, a strong enhancement has been implemented in the normal foretelling arrangement precision action outcomes, simultaneously, requiring time to calculate the foretelling precision from the reduced total data set due to the existence of a smaller number of handling inconsistencies is remaining in the data set. The accuracy level for identifying the category of leaf diseases is illustrated in Fig. 6 for the proposed technique.

5 Conclusion

The proposed multi-features plant disease identification from the brinjal leaves has been successfully implemented. The detection of the specified objects from the complex environment is by joining the image acquisition and the standard disease identification system. The SOBEL edge detection technique is used to identify the borders after converting the input image into the grey images. The accuracy of the plant disease identification is measured and identified as more than 95% of accuracy for plant disease identification process.

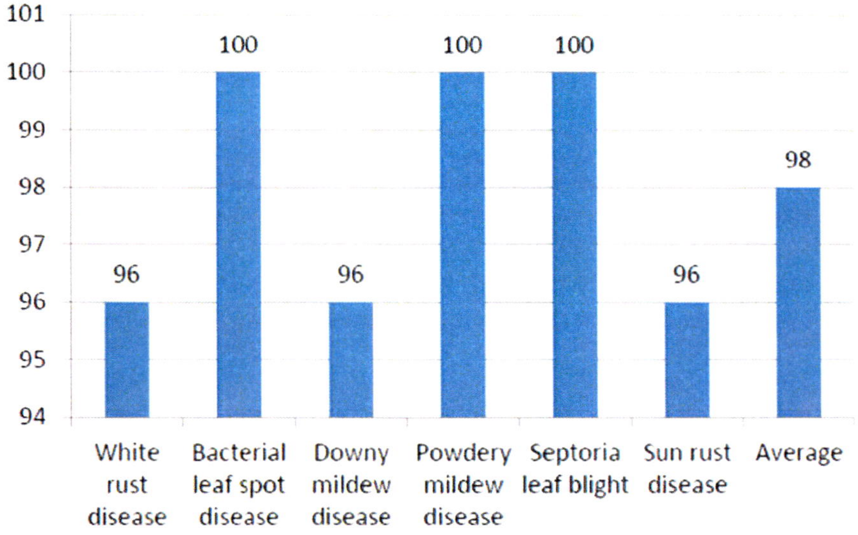

Fig. 6 Accuracy

References

1. Jadoun, V.K., Gupta, N., Niazi, K.R., Swarnkar, A.: Dynamically controlled particle swarm optimization for large scale non-convex economic dispatch problems. In: The International Transactions on Electrical Energy Systems. Wiley (2014)
2. Al-Bashish, D., Braik, M., Bani-Ahmad, S.: Detection and classification of leaf diseases using Kmeans-based segmentation and neural-networks-based classification. Inform. Technol. J. **10**, 267–275 (2011)
3. Kulkarni, A.H., Patil, A.: Applying image processing technique to detect plant diseases. Int. J. Mod. Eng. Res. **2**(5), 3661–3664 (2012)
4. Singh, V., Misra, A.K.: Detection of plant leaf diseases using ımage segmentation and soft computing techniques. Inf. Process. Agric. **4**(1), 41–49 (2016). https://doi.org/10.1016/j.inpa.2016.10.005
5. Sowmya, G.M. et al.: Disease detection in pomegranate leaf using ımage processing technique. Int. J. Sci. Eng. Technol. Res. (IJSETR) **6**(3) (2017)
6. Sannakki, S.S., Rajpurohit, V.S., Nargund, V.B., Arun Kumar, R., Yallur, P.S.: A hybrid ıntelligent system for automated pomgranate disease detection and grading. Int. J. Mach. Intell. **3**(2), 36–44 (2011). ISSN: 0975-2927
7. Beyyala, A.K., Beyyala, S.P.: Application for diagnosis of diseases In cropsusing ımage processing. Int. J. Life Sci. Bt. Pharm. Res. **1**(2), 171–176 (2012). ISSN 2250-3137. www.ijlbpr.com
8. Nazki, H., Yoon, S., Fuentes, A., Park, D.S.: Unsupervised image translation using adversarial networks for improved plant disease recognition. Comput. Electron. Agric. **168**, 105117 (2020). ISSN 0168-1699.
9. Sweetwilliams, F.O., Matthews, V.O., Adetiba, E., Babalola, D.T. and Akande, V.: Detection of Sigatoka disease in plantain using IoT and Machine learning techniques. J. Phys. Conf. Ser. **1378**(2) (2019)
10. Devi, R.D., Nandhini, S.A., Hemalatha, R., Radha, S.: IoT enabled efficient detection and classification of plant diseases for agricultural applications. In: 2019 International Conference

on Wireless Communications Signal Processing and Networking (WiSPNET), Chennai, India, pp. 447–451 (2019)
11. Liao, W., Ochoa, D., Zhao, Y., Rugel, G.M.V., Philips, W.: Banana Disease Detection By Fusion Of Close Range Hyperspectral Image And High-Resolution RGB Image. IEEE (2018)
12. Aruraj, A., Alex, A., Subathra, M.S.P., Sairamya, N.J., George, S.T., Ewards, S.E.V.: Detection and classification of diseases of banana plant using local binary pattern and support vector machine. In: 2019 2nd International Conference on Signal Processing and Communication (ICSPC), Coimbatore, India, pp. 231–235 (2019)

Deep Learning Techniques for Optical Character Recognition

Naragudem Sarika and NageswaraRao Sirisala

Abstract Handwritten character recognition (HWCR) is an area of pattern recognition and optical character recognition (OCR) to identify and recognize the characters by the system or any electronic device. Optical character recognition is the process that converts handwritten text into machine-encoded text. In this paper, OCR techniques such as digitization, preprocessing, segmentation, feature extraction, and recognition are discussed. HWCR using different machine learning algorithms like Bayesian classifier, Bayesian decision theory, and SVM are discussed. Recent techniques of HWCR are analyzed for native language and compared their functionality and features. The comparative study of these recent techniques is summarized in table format.

Keywords HWCR · OCR · Bayesian classifier · Zoning features · SVM

1 Introduction

Optical character recognition (OCR) converts handwritten or printed text into machine format. The fundamental procedure of optical character recognition associates in checking the text of a document and converts the characters into code that can be used for data processing. OCR sometimes mentioned as text recognition. The process of OCR is commonly used to convert hard copy or documents into PDFs [1]. OCR can be used for automatic plate recognition, extraction business cards information from the contact list, book scanning, Google books, and convenient technology for a blind user. Character recognition consists of two types: offline and online [2]. Offline is divided into handwritten and printed.

N. Sarika · N. Sirisala (✉)
Vardhaman College of Engineering, Hyderabad, India
e-mail: nagsirisala@gmail.com

N. Sarika
e-mail: sarikanaragudem@gmail.com

P. Karuppusamy et al. (eds.), *Sustainable Communication Networks and Application*, Lecture Notes on Data Engineering and Communications Technologies 55,
https://doi.org/10.1007/978-981-15-8677-4_28

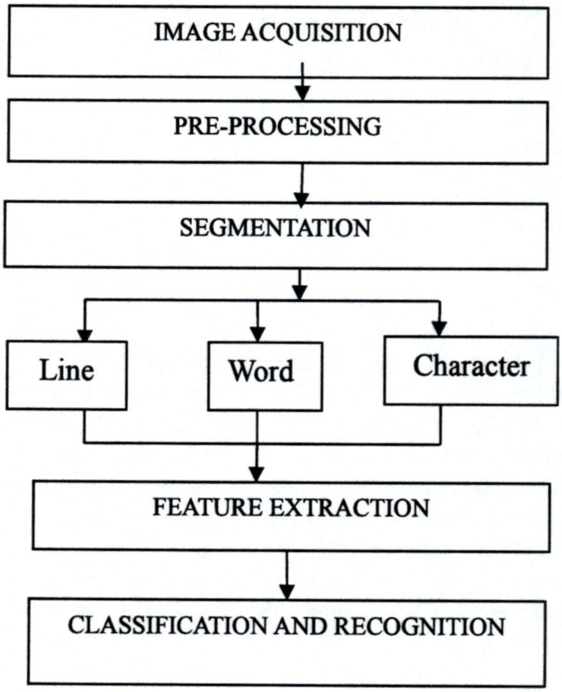

Fig. 1 Phases of OCR

Optical character recognition categorized into many phases such as image acquisition, preprocessing, segmentation, feature extraction, classification, and recognition (Fig. 1).

1.1 Digitization

Digitization is the process of converting a handwritten document into an electronic document [3]. In this, information is categorized into different units of data and it is mentioned separately [4]. The scanner captures images and converts it to an image file. Digitizing information is easy to protect, save, and share. Digitization uses a different scanner. The process of translating a multilevel image into a bi-level image of the binary image is called thresholding. In OCR, thresholding is used to recover memory and thresholding is performed on a scanner. A gray-level image is below the threshold value, then it indicates black and the level of the image is above the threshold value, then it indicates white [5]. To reduce the threshold value problem, the best process is contrast and brightness. These methods are used in the multilevel scanning of the image.

1.2 Preproceesing

Preprocessing is a sequence of operations accomplished on the scanned input image. After completion of the scanning process, a specific amount of noise is obtained. The application of thresholding technique leads to the character are broken or shattered is proposed in [3]. In the preprocessing, smoothing and thinning methods are applied to remove the noise and clear the image. In some cases, the normalization method is also applied [4]. Normalization is applied for the result of consistent size and shape. The next process is segmentation.

1.3 Segmentation

Segmentation is a process of dividing a digital image into multiple segments and it is also used to simplify the representation of an image. Handwritten character segmentation is more complex than the printed copies or documents because of lines and characters of a word [6].

1.4 Feature Extraction

In this, individual character features are extracted. The main purpose of feature extraction to apprehend the important characteristics of the symbols or characters [3]. It is the major problem in handwritten character recognition.

1.5 Classification

Classification is a process used to classify an image into visual content. In the OCR system, classification is used to convert an image into a digital image. Character recognition is obtained with the process of segmentation by dividing characters into an individual image. Classification and recognition are steps for postprocessing in the OCR system [7, 8].

2 Different Algorithms

The different algorithms like Bayesian classifier, SVM, Bayesian decision theory, and zoning features are discussed.

2.1 HWCR Using Bayesian Classifier

Offline Telugu HWCR is a complex task to recognize Telugu characters of human writing. Optical character recognition made offline after the characters have been printed. For training and testing the image, the OCR techniques are used. In OCR techniques, the first technique is digitization done with an optical scanner, second is preprocessing which is used for binarization and smoothing. It is used to improve the scanned image. Binarization is used to handle the threshold data in the binary image. It is used to authorize the removal of noise. The third is segmentation. Fourth is feature extraction and last is classification. The steps of OCR phases are firstly read and scan the image. Next, resize the image, and converting image into gray and then into binary and next applying filtering. Finally, testing and training of Bayesian classifier. In this, feature extraction uses the method called histogram of oriented gradients (HOG) [9]. This is a feature selection method that gives the parameters. HOG uses for finding the edges of a character and standardized them into gray scale image. The gradient is calculated for the intensity of characters.

Bayesian classifier algorithm is used in the Telugu handwritten character recognition to identify a single character. The Bayesian classifier is depended upon Bayes theorem. Bayes theorem is classified into two types: posterior probability and prior probability.

$$\text{Posterior Probability} = P(E/X)$$
$$\text{Prior Probability} = P(E)$$

where E is the hypothesis and X is the data.

According to Bayes theorem [9]

$$P(E/X) = P(X/E)P(E)\big/P(X)$$

Bayesian classifier is used to predict the tuple and class of probability. The merits of Bayesian classifier are: It is used to handle the noisy data which is obtained in the image. Further, Nearest neighborhood method is also applicable for Telugu handwritten character recognition. The standard deviation of Bayesian classifier accuracy is less than a standard deviation of nearest neighborhood accuracy. The character recognition accuracy depends upon the number of positives and negatives of the training and testing the data sets.

2.2 HWCR Using Zoning Features

HWCR is a process of organizing handwritten characters based upon the feature extraction from each character image. Handwritten character recognition is categorized into different phases like preprocessing, segmentation, feature extraction, and postprocessing. These phases are used for training and testing the image. In Telugu HWCR, Euclidean distance is used for identification and classification. In the zoning feature method, a character is separated into a certain size. Binarization is the first step in the Telugu HWCR. Binarization is used to scan the input image. In this process, the threshold technique is used to obtain an optimum value [10]. Handwritten character is a different size and resized to 50×50 pixels. In any region, combining of all pixels is obtained in the form of the feature vector of that particular region. After calculating feature vectors, the values are sequentially displayed one after another. In the zoning method, the database image is standardized to specific pixels. Every image is partitioned into a hundred zones. The portion of each zone should be five rows and five columns. The determination of feature vectors obtained by adding pixel intensities feature extraction plays a major role to recognize an image differently. In the zoning method, the result is obtained through a feature vector by partitioning into different zones.

2.3 HWCR Bayesian Decision Theory

Handwritten character recognition involves the translation of document or text into an image. For handwritten characters, image processing techniques are used for conversion of binary image to extract feature vectors. For character recognition, there are several techniques: Preprocessing, segmentation, feature extraction, and postprocessing. Preprocessing technique is used to noise reduction and normalization of the data. Segmentation is classified into different types like line, word, and character segmentation. Feature extraction is used to release the feature vectors to identify and classify an image. In postprocessing, Bayesian decision theory is used [6].

Bayesian decision theory is used to reduce the error in the classification process. It is a basic procedure for the pattern classification problem. Bayesian decision theory is based upon the Bayes theorem. Bayes theorem uses posterior probability and prior probability. Bayes theorem is used to reduce the error in the classification problem. Handwritten character recognition without errors during classifying a binary image, Bayesian decision theory is used. Hidden Markova model is also used to classify an image. Bayesian decision theory is used to classify missing data and reduce the error compared to the Hidden Markova model (Fig. 2).

2.4 Online HWCR Using Support Vector Machines

Handwritten character recognition is used to interpret the documents into a text document by scanning an image. Offline character recognition deals with the scanned digital image whereas online character recognition deals with the identification of printed characters. The automatic character recognition is used to process a large number of handwritten characters. A handwritten character varies in the text size

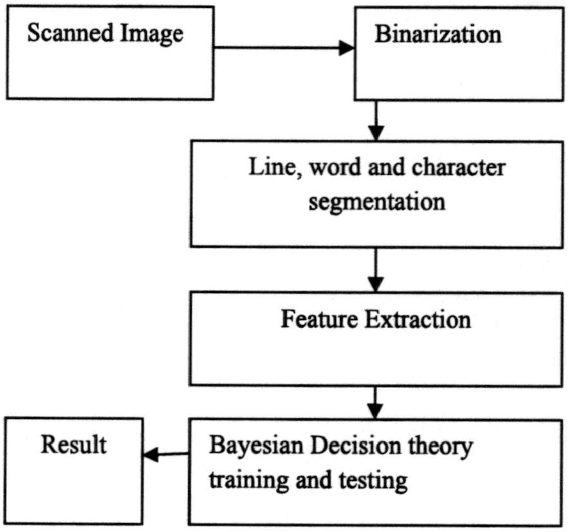

Fig. 2 Bayesian decision theory process

and shape of the character. The text size and shape display a problem in character recognition. Based on this problem, there are few techniques in the HWCR.

2.5 *Preprocessing*

The preprocessing involves in the binarization method [11]. An image is converted into a digital image, and then these operations are applied to remove the noise and clean an image. Binarization is an important operation in the preprocessing technique. In binarization, Otsu's method is commonly used. Noise is obtained after scanning an image. Smoothing is used to reduce the noise in an image. Smoothing and noise removal are the processes of filtering (Fig. 3).

Thinning is a processing operation used to remove unnecessary information and maintain the characteristic features of the image. Thinning plays a major role in the reduction of unwanted data in Telugu character recognition. Skew detection is obtained due to document scanning and document copying. There are different methods for the correction of document images: Projection profile, Hough transforms, Fourier method, nearest neighborhood method, and correlation. The skew detection method is used to detect copied images and documents. This method is used in the document image analysis system. Slant correction is used to segment the corrected word into characters. For better word recognition, slant correction is used in handwritten character recognition.

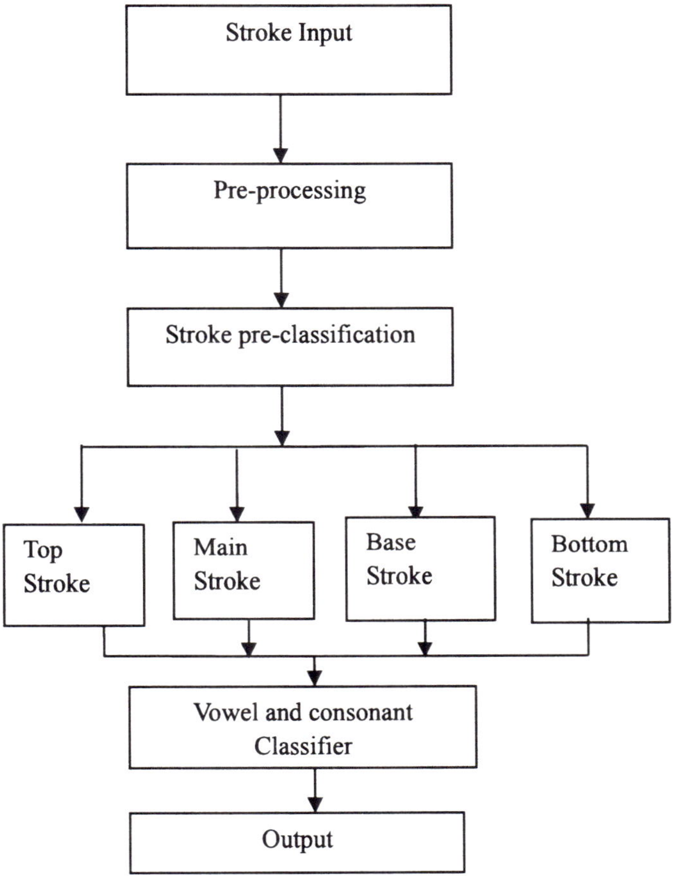

Fig. 3 Stroke recognition

2.6 Stroke Pre-classification

Telugu characters are divided into three level strokes: First-level stroke, middle-level stroke, and last-level stroke [11, 12]. Each character has a base stroke and there is a size variation in the strokes. There are two methods to arrange three-level strokes. Base stroke is used to recognize the stroke graphically and SVM is used to arrange strokes. First-level stroke is preclassified and the feature vector is obtained with the primary stroke. Strokes are preclassified and best output for preclassification is obtained by different transformation methods and statistics [13]. In feature extraction, individual characters are removed (Fig. 4).

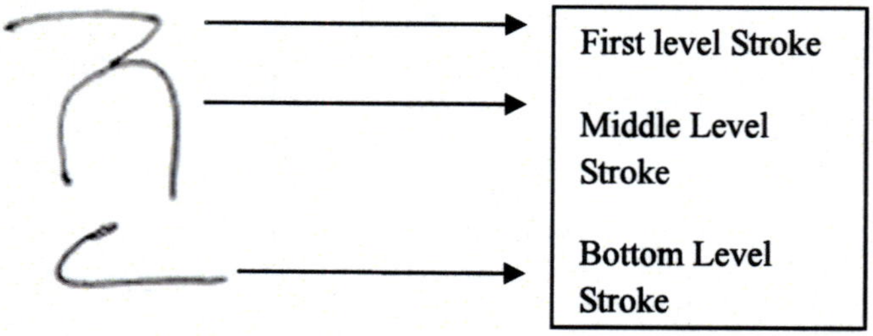

Fig. 4 Stroke Pre-classification

2.7 *Stroke Recognition Based On SVM*

SVM is used to identify the particular stroke form the first-level, middle-level, and last-level stroke preclasses which features are extracted [11]. Classifying the primary stroke, three SVM's are used [11]. Primary classifier, first-level classifier, middle-level classifier, and last-level classifier consist of vowel classifier (CV) and consonant classifier (C2) [11]. This character classifiers are used to classify the recognized characters from the primary stroke.

2.8 *Character Recognition Based on SVM*

A consonant classifier is consistently occurring in the last-level stroke [11]. Preclassification is obtained through the SVM method. Support vector machine is used for creating a stroke recognition engine.

2.9 *Comparison between different algorithms*

S. No.	Algorithm	Technique	Merits	Demerits
1	Bayesian classifier [9]	Bayes theorem	Bayesian classifier generates high Character recognition rates using nearest neighborhood method	Bayesian classifier model is always depended upon the standard deviation

(continued)

(continued)

S. No.	Algorithm	Technique	Merits	Demerits
2	Adaptive and static zoning method [14]	Genetic algorithm	Static zoning method implements with the KNN algorithm to obtain a set of features	In a genetic algorithm, a large set of training samples needed
3	Bayesian decision theory [6]	Probability	This is used to classify missing data to reduce error	It is time taken process to identify missing data
4	Zoning features [10]	Zonal-based feature extraction	This method is used to identify an image into different zones to get good output	There is no standard algorithm for this process
5	Support vector machine [11]	Preclassification	SVM is used for stroke recognition and also used for preclassification to obtain a recognized character	Character-level recognition is higher than stroke-level recognition
6	Hybrid HMM or ANN [15]	Multilayer perception	This method is used for recognizing unconstrained Telugu lines	Better trained ANN's are not implemented
7	Raspberry Pi [16]	KNN and DTW	It is used to recognize the letters for blind and deaf people	Required more number of trained samples
8	HMM [17]	HMM classifier	Telugu character recognition by using time-domain features and frequency domain features	It does not recognize the Telugu characters using line segmentation
9	CNN model [18]	Four-layer architecture	CNN model is a sturdy character recognition technique compared to other techniques	This method is complicated in the preprocessing step
10	Radial sector coding algorithm [19]	Center of mass feature	This method is used to detect the inconsistent text position in an image	It does not detect the more complex images

3 Conclusion

In this paper, different phases of Optical Character Recognition(OCR) are discussed. Several recent machine learning algorithms for handwritten character recognition (HWCR) are analyzed. The procedures for OCR techniques are discussed in this paper. Character recognition using Bayesian classifier, zoning features, adaptive and static zoning method and support vector machine are surveyed in this paper. Comparisons for algorithms by using its techniques, merits, and demerits are explored.

References

1. Sujatha, P., Lalitha Bhaskari, D.: Telugu and Hindi Script recognition using deep learning techniques. Int. J. Innov. Technol. Explor. Eng. **8**, 2278–3075 (2019)
2. Chakradhar, C.V., Rajesh, B., Raghavendra Reddy, M.: A study on online handwritten Telugu character recognition. Int. J. Sci. Eng. Technol. Res. **5**(7) (2016)
3. Modi, H., Parikh, M.C.: A review on optical character recognition techniques. Int. J. Comput. Appl. **160**(6) (2017)
4. Avinash, A., Sita Mahalakshmi, T.: Hand written Telugu character recognition-review. Int. J. Manag. Technol. Eng. **8**(VII) (2018)
5. Vishwanath, N.V., Manjunathachari, K., Satyaprasad, K.: Handwritten Telugu composite character recognition using morphological analysis. Int. J. Pure Appl. Math. **119**, 667–676 (2018)
6. Vijayakumar, Joseph, S.: Handwritten character recognition using Bayesian decision theory. Int. J. Comput. Sci. Emerg. Technol. **1**(2) (2010)
7. Prameela, N., Anjusha, P., Karthik, R.: Offline Telugu handwritten characters recognition using optical character recognition. In: International Conference on Electronics, Communication and Aerospace Technology. IEEE (2017)
8. Jain, A., Sharma, J.: Classification and Interpretation of Characters in Multi-Application OCR System. IEEE (2014)
9. Mohana Lakshmi, K., Venkatesh, K., Sunaina, G., Sravani, D., Dayakar, P.: Hand written Telugu character recognition using Bayesian classifier. Int. J. Eng. Technol. **9** (2007)
10. Sastry, P.N., Vijaya Lakshmi, T.R., Koteswara Rao, N.V., Rajinikanth, T.V., Wahab, A.: Telugu Handwritten Character Recognition Using Zoning Features. IEEE (2014)
11. Vijaya Kumar, K., Rajeshwara Rao, R.: Online handwritten character recognition for Telugu language using support vector machines. Int. J. Eng. Adv. Technol. **3**(2) (2013)
12. Rajkumar, J., Mariraja, K., Kanakapriya, K., Nishanthini, S., Chakravarthy, V.S.: Two schemas for online character recognition of Telugu script based on support vector machines. In: International Conference on Frontiers in Handwritten Recognition. IEEE (2012)
13. Dogra, S., Prakash, C.: Hindi handwritten character recognition system based on SVM. Int. J. Comput. Sci. Eng. **4**(5) (2012)
14. Prasad, S.D., Kanduri, Y.: Telugu Handwritten Character Recognition Using Adaptive and Static Zoning Methods. IEEE (2016)
15. Vikram, C., Shoba Bindu, C., Sasikala, C.: Handwritten character recognition for Telugu scripts using multi layer perceptrons. Int. J. Adv. Res. Comput. Eng. Technol. **2**(12) (2013)
16. Naga Deepa, C.H., Sri Divya, C., Balaji, N., Padmaja, V.: Real time implementation of Telugu character recognition using Raspberry Pi. Int. J. Sci. Res. **3**(12) (2014)
17. Espana-Boquera, S., Castro-Bleda, M.J., Gorbe-Moya, J., Martinez, F.Z.: Improving offline handwritten text recognition with hybrid HMM/ANN models (vol. 33, no. 4). IEEE (2011)

18. Marzuki, P., Syafeeza, A.R., Wong, Y.C., Hamid, N.A., Nur Alisa, A., Ibrahim, M.M.: A design of license plate recognition system using convolutional neural network. Int. J. Electr. Comput. Eng. **9**(3), 2196–2204 (2019)
19. Kaur, M., Randhawa, N.S., Garg, V.: Proposed approach for layout and handwritten character recognization in OCR. Int. Res. J. Eng. Technol. **6**(4) (2019)

Toward Effectual Group Formation Method for Collaborative Learning Environment

Neeta Sarode and J. W. Bakal

Abstract Group formation is a significant process of collaborative learning. Collaborative learning helps individuals to be more focused on the purpose, goals, and approach related to the task that they have been assigned. Groups are incontrovertibly beneficial for many of the intricate objectives of learning, related to critical thinking, decision making, problem-solving, preserving, or modifying attitudes. The success of the group undoubtedly depends on the compatibility and aptness of members of the group. This paper attempts to solve this group formation problem using optimized genetic algorithm. The proposed method combines both static and dynamic characteristics of students to achieve productive collaborative learning groups. The system is implemented and the experimental results are shown for a varied number of class size. The algorithm gives the solution to the problem in time and achieves better quality leading to a better solution for forming groups of students having heterogeneous characteristics. The performance has also competed with the k-means clustering algorithm for the same set of characteristic features.

Keywords Group formation · Collaborative learning · Genetic algorithm · Learning style · Academic achievement

1 Introduction

Collaborative learning is a joint intellectual effort of two or more students to learn something. Nowadays, learning in groups has become an influential part of the teaching and learning process. It has become a good opportunity for students to learn new things. It has become an essential skill. It helps them to learn better and retain more information. Group work and collaborative work help the students to get

N. Sarode (✉)
Thadomal Shahani Engineering College, University of Mumbai, Maharashtra, India
e-mail: neetame07@gmail.com

J. W. Bakal
Principal, S.S. Jondhale College of Engineering, University of Mumbai, Maharashtra, India

P. Karuppusamy et al. (eds.), *Sustainable Communication Networks and Application*, Lecture Notes on Data Engineering and Communications Technologies 55,
https://doi.org/10.1007/978-981-15-8677-4_29

prepared for their forthcoming perspective as most of them will be a part of multi-disciplinary and multi-professional teams when they go to industry [1]. Group work and discussions help to stimulate the thought process, foster new ideas, and transmit constructive information. In other words, a collaborative learning environment helps students together, to refurbish their theories, analyze ideas, and construct testimony of their learning. Also, the students gain from varying competence-based potentials and innovative thinking skills of a combined group to advance self abilities.

The significance of collaborative learning is simple to comprehend and hence it is the prerequisite for a passable approach in the group formation process. The groups can be formed in three different ways: self-select, random-select, and instructor selected. In self-selected groups [2], the groups are formed by the students themselves and it is the genuine tendency of selecting their friends as group members of the group. Mostly, such groups formed will be of the homogeneous type as students apt with others having similar characteristics. Such groups exhibit poor performance. In random select [3], the groups are formed randomly by a teacher or with the help of a computer. The group forming process is fastest and effortless but at the same time disorganized and inconsiderate. The members of the group can be widely incompatible in terms of behavior, readiness to group work, and level of participation. In the instructor selection group formation process, the instructor actively participates in forming groups based on some specific criteria [4] (skills, learning style, knowledge, etc.). These can be a homogenous or heterogeneous type of groups. Research shows that such groups are better for achieving particular aims in a structured manner.

This work uses a genetic algorithm for the group formation process and shows good performance in achieving the solution.

This paper is structured as follows: The first section is the present one giving the introduction to the topic. The second section gives a brief about the related work done in this area. The third section explains the genetic algorithm and its operators in short. The fourth section explains about the proposed approach followed by experimental work section. The final section concludes the work done.

2 Related Work

Automatic group formation problem was studied by numerous researchers. In most of the studies, the major goals were to reduce the amount of time required to form groups and form a heterogeneous type of groups to improve performance. Still, there is no standard methodology that defines the stages of forming effective groups.

O'Malley and Scanlon [5], in 1990, first investigated the study of collaborative learning supported by a computer system. Their research aimed to show that by integrating technology, the group learning process and knowledge sharing among members of the group become easy and fast. It was termed as computer-supported collaborative learning. This is the important development of online learning with the major advances in information and communication technologies. Many researchers have shown that group formation step is the first and foremost step in computer-supported collaborative learning for achieving the desired success [6, 7].

The work by Tanimoto [8] has devised an algorithm to raise the level of compatibility among students of the same group. The algorithm is called as 'squeaky wheel'. The compatibility was based on the scale of the willingness of a particular student to work with the other student in the class. In the work mentioned in [9], the authors have considered three main characteristics which may affect the success of the group: team roles, social networks, and psychological styles and have formed a computer-supported collaborative system. K-means clustering algorithm is proposed in [10] to form groups of learners considering their behavior and personality features. Ant colony optimization algorithm is proposed in [11], to assign each student the most appropriate group to increase the heterogeneity of the group. A mathematical prototype is proposed that considers students' fluency in the language used, personality features, and performance score in the subject. In research [12], the author provides students with a questionnaire and the answers confirm their learning style. Now the groups are formed based on their learning style. In the work mentioned in [13], the author uses a genetic algorithm to form heterogeneous groups. The students were grouped based on their educational background such as GPA, prior classes, performance, and personality test score. The results of the experiment lead to better solutions when compared to self- select technique. Constraint optimization is used to form groups based on Jigsaw CLFP in [14]. A binary integer programming method is employed to give an optimal solution. A semantic group formation algorithm and application are implemented in [15]. It assists teachers to manage knowledge diversity in the classrooms.

The work presented here is a method for group formation for a collaborative learning environment. Every student exhibits different characteristics and with a substantial amount of distinct combinations of such characteristics, and it is required to implement a suitable method to find a probable optimal solution. The genetic algorithmic approach proposes an optimization technique which assists to reach an agreeable and appropriate solution when it is tedious and difficult to test each and every probable possibility. Genetic algorithm is an evolutionary technique which searches for global solutions over local solutions. Here, the groups are characteristically heterogeneous with special consideration to the skills that are needed for achievement in the task and genetic algorithm displays higher proficiency to deal with questions which shows such high combinatorial characteristics.

The group size can be kept small. Ideally, there is no specific group size defined. It is normally the instructor's choice, but in [16], they advise a group size of 3, 5, and 7. For a larger group, the interpersonal communication may shrink and the involvement of some group members may start to deteriorate. Larger the group size, lesser is the productivity of the group members.

3 Genetic Algorithm

Genetic algorithm (GA) is said to be the evolutionary optimization technique. It is based on Charles Darwin's theory of evolution and states the survival-of-the-fittest concept. A logical utilization of a random search is defined within a local search space to find a solution to a problem. Only the fittest will survive and reproduce and procreate and successive generation will become better and better compared to previous generations [17]. The genetic algorithm simulates the process of evolution. Simulating or imitating the process of evolution, the key concept is evolution is an optimizing process. Means, each generation is like iteration in numerical methods and each of the iterations brings progressive improvement of the objective function or the fitness function. It is a successive iteration of design variables. Higher the fitness of an individual, greater is the probability of reproduction [20].

Here, the possible solution is represented as a chromosome in the form of array data structure. Each individual is seen as a possible solution. The three genetic operators are performed in each generation: selection, crossover, and mutation. Selection is executed by implementing the roulette wheel method. The bigger part of the wheel has more probability to get selected during reproduction and hence represents the higher fitness value. Crossover is used to inter-mix the genetic representation of the parent population to create a better fitness score of the resulting offspring [21]. Mutation introduces random, slight change that adds value to the genetic algorithm. It helps in overcoming local minima in the search assessment.

The general structure of the genetic algorithm is as follows:

1. Initialize the population
2. Calculate the fitness of each individual
3. Generation = Generation + 1
4. Do selection (evaluate the fittest individuals)
5. Do crossover
6. Do mutation under random probability
7. Add the fittest offspring to the population
8. Calculate new fitness value
9. If the population gets an individual with maximum fitness or have reached maximum generation, end the algorithm. Otherwise, go to step 3.

This work presents an approach to form balanced groups of learners, such that the groups (intra) are as heterogeneous as possible and groups (inter) are as homogeneous as possible. The group formation method considers both static as well as dynamic behavior for forming features of learners.

4 Proposed Approach

To implement the genetic algorithm as stated above, it is mandatory to devise a process to produce random initial population, to decide genetic operators for reproduction phase, survival measures to be met for individuals to reach the next iteration, coding representation for a solution, an objective function to relate different individuals, and set of parameters for testing phase.

In a class of n, the number of students, shown as $S = (s_1, s_2, \ldots s_n)$, each student has its characteristics defined in a vector $(v_1, v_2, \ldots v_k)$. This work considers the student characteristics as: gender, CGPA/academic achievement, grade in a prior similar course, logical reasoning test score, and students' learning style. Every student's characteristics have a set of distinct possible values which makes it hard to solve. The set of student's characteristics is not easy to evaluate but still can lead to an effective group formation process. The first feature is the gender of the student; boys and girls to be equally distributed among the groups. The second feature is the cumulative grade point average or the academic examination score. The third feature is the grade achieved in a prior similar course. For example, if the student has enrolled for Advanced Java Programming courses, then the grade for previous Basic Java Programming course to be considered. The fourth, logical reasoning test will be designed by the instructor and the score of the test to be considered. Finally, the fifth is the learning style test (Index of Learning Styles, ILS Learning Style Questionnaire) [18, 19] to identify whether the student has Activist/Reflector, Sensing/Intuitive, Visual/Verbal, Sequential/Global style of learning. Equally, distribute all types of learners in each group. The second to fourth features represent the student's educational skill and performance.

For n students of a class, the chromosome contains n-sized ordered sets. Every element of chromosome denotes the group to which the student belongs. These n students of the class are divided into t groups, and the ordered set can be denoted as G_m, and $0 \leq m \leq t$. For example, for a class of 12 students, if three groups are formed($t = 3$), the length of the chromosome will be 12 and the ordered set for each chromosome will be G_1, G_2, or G_3. One of the group formation solutions can be as shown in Fig. 1.

Where; $G_1 = \{s_2, s_3, s_9, s_{12}\}$; $G_2 = \{s_1, s_4, s_6, s_{11}\}$; and $G_3 = \{s_5, s7, s8, s10\}$.

Since each student has five features, a separate table is maintained to store the student's information. Table 1 shows how the attributes of each student are considered.

The student ID is the unique identification number of the student. The gender attribute will store either 0 or 1 for boy or girl, respectively. CGPA, grade of prior course, and logical reasoning test attributes have three categories as good, average, or

S1	S2	S3	S4	S5	S6	S7	S8	S9	S10	S11	S12
G2	G1	G1	G2	G3	G2	G3	G3	G1	G3	G2	G1

Fig. 1 Chromosome representation (A class of 12 students)

Table 1 Student attributes

Student ID	Gender	CGPA (out of 10)	Grade of prior course	Logical reasoning test (Max.20)	Learning style
Number	0-Boy 1-Girl	Above 8-good 5,6,7—Average 4 and below —Poor	A, A−, B+—good B, B−,C—Average D+, D, F—Poor	14 and above—good 8–14-Average Below 8—Poor	0/1

poor and so are represented as 1, 2, or 3, respectively. The learning style is represented as zero or one number for each of the learning style. Now, here, the first learning style, Active/Reflective is only considered as this learning style is related to group work. Active learners are more inclined toward group work compared to reflective learners who like to do their work alone. So if, for example, the student having ID as 11, and the subsequent attribute sequence is [0,2,2,1,0]; that means the student with ID as 11 is a boy, possessing average CGPA and grade of prior course, a good grade for logical reasoning and is an active learner. The students' details such as student ID, gender, CGPA, and grade of a prior course are to be updated as the student logs into the system. Then the student has to appear for the logical reasoning test and also the test to identify the learning style. The scores of both these test contribute to the fourth and fifth attribute of the student.

The objective function evaluation depends on two conditions in an effective group formation process. The first is, students, be equally distributed among all groups and the second is, every individual group should be identical to other rest of the groups. There should be a balanced allotment of students in each group regarding their characteristics. This is done by calculating the minimum and maximum value for each characteristic feature. For each group, the minimum and maximum value of each characteristic are evaluated, and the group score is checked. And then the average score for each group is calculated. To find the distance among two groups the absolute value of average score subtraction is performed using Eq. 1.

$$\text{Dist}_{\text{Grps}} = \left|\text{Grp}_{i\text{Avg}} - \text{Grp}_{j\text{Avg}}\right| \tag{1}$$

where $\text{Grp}_{i\text{Avg}}$ is the ith group average score and $\text{Grp}_{j\text{Avg}}$ is the *j*th group average score.

The total of these average subtractions gives the total distance of all groups using Eq. 2.

$$\text{TotDist}_{\text{Grps}} = \sum_{i=1}^{t-1} \sum_{j=i+1}^{t} \text{Dist}_{\text{Grps}} \tag{2}$$

where t is the number of groups formed.

Finally, the fitness value is evaluated as in Eq. 3.

$$\text{FitnessValue} = \text{GroupScore} + \frac{1}{(\text{TotDist}_{\text{Grps}}) + 1} \tag{3}$$

Since this is a minimization problem, the total distance is considered as the denominator. If the total distance comes to zero, the denominator will be zero, hence adding 1 to the denominator. There is a uniform distribution of all type of students in every group. Good and average performing students and poor performing students are distributed in a balanced way.

The roulette wheel method is used for selection. The value of each chromosome is stated as its fitness probability. The chromosomes having high fitness probability has higher chances of selection.

Chromosome represents the set of students. No student can be eliminated while progressing from one generation to next. So to implement crossover, the permutation encoding method is used [22]. In a single-point crossover, the crossover point is randomly generated. So, till this single point, copy the permutation from the first parent, then the second parent is examined and if the number is not yet in the offspring, it is added. Due to this, no student is lost. Following is the example to explain the process of single-point crossover using permutation encoding:

$$\overset{\text{Parent 1}}{(\mathbf{1\,2\,3\,4}\;5\;6\;7\;8\;9)} + \overset{\text{Parent 2}}{(\mathbf{3\,2\,4\,6}\;8\;9\;1\;5\;7)} = \overset{\text{Offspring}}{(1\;2\;3\;4\;6\;8\;9\;5\;7)}$$

In the mutation stage, order changing method is used where the two randomly generated numbers are selected and simply exchanged.

$$(1\;\mathbf{2}\;3\;4\;6\;8\;9\;\mathbf{5}\;7) \geq (1\;\mathbf{5}\;3\;4\;6\;8\;9\;\mathbf{2}\;7)$$

The mutation rate is calculated by considering the average fitness value on a specific generation. The current mutation rate is 0.9 times the previous generation's mutation rate if the difference between the average fitness and previous generation's average fitness is greater than some specified threshold value. Otherwise, it is 1.1 times the previous generation's mutation rate.

For generating the new population, the elitist selection strategy is implemented [17]. In this strategy, while constructing the new population, the best fittest individuals are allowed to carry over from present to the next generation unchanged. It assures that the quality of the solution remains intact and will not fall from one generation to the next.

5 Experimental Results

The proposed method for forming heterogeneous groups using genetic algorithm is tested for different configurations settings and is carried out using Java Programming construct. Experiments were carried out on graduate-level students enrolled for three different courses and the numbers of students was 21, 30, and 42 students. So it is considered: 21 students (as a small class), 30 students (as an average class), and 42 students (as a large class) which also correspond to educational realism.

A different set of values of the parameters was treated for the testing purpose for each type of class. For a small class, the population size considered was 50, 100, 150, and 200 with many generations as 150. For an average size class, the population size considered was 100, 200, 300, and 400 with the number of generations as 400. And for a large size class, the population size considered was 300, 500, 700, 900, and 1000 with the number of generations as 900. The crossover probability considered for testing is 70–100%, whereas mutation probability considered for testing is 1–6% for every type of class. An algorithm is tested for these different values of parameters and is tested for convergence for varied class sizes.

The results are as shown in Table 2. For the parameter values shown in the table, best solution or results were obtained. The column names denote the different class size that is considered and hence the analysis is developed.

The observations for different class size are summarized as follows:

Small class size: Convergence to the optimum value is reached faster, with a bigger population size compared to small population size and the execution time is just a few seconds. With a bigger population, the number of iterations required to reach the optimum value is less.

Medium class size: For population size more than 300, it required more number of generations to reach the optimum value. For population size less than 250, the optimum value was not reached.

Large class size: For the population size less than 500, the algorithm needs a large number of iterations but does not give good fitness value. With a bigger population size, it requires less number of iterations to reach the optimum value.

The running time of the algorithm is noted as shown in Table 3 for varied class sizes and varied group sizes (as 3, 5, or 7 students in a group).

It is observed that as the number of students' increases, the running time also increases. It is due to the time required to compute an increased number of students'

Table 2 Results for different parameter values

Parameters	Class size		
	21	30	42
Population size	150	300	600
Number of generations	50	300	200
Crossover (%)	100	100	80
Mutation (%)	3.5	4	3.65

Table 3 Running time of the algorithm in seconds

Class size/Number of students	Group size		
	3	5	7
21	0.78	5.86	7.15
30	0.94	23.02	50.06
42	1.53	95.46	356.10

Table 4 Running time of clustering algorithm in seconds

Class size/Number of students	Group size		
	3	5	7
21	0.80	5.50	5.15
30	1.02	21.32	45.30
42	1.60	90.07	289.14

feature vector. Also for larger group size, the optimizer algorithm takes more running time, as the number of values for a group to be calculated increases. So, notably, smaller groups should be preferred to larger groups. As it is already mentioned in [6], those students' in smaller groups work more interactively and this leads to the more effective and successful completion of the task.

Group formation process can be executed by using the k-means clustering algorithm also. For the same set of attributes of the students and with same class size, k-means clustering algorithm is experimented with to form homogeneous groups as this algorithm is best known and accepted for homogeneous grouping. After experimentation, the performance was evaluated and the accuracy of the objective function is calculated. The performance slightly decreases with more number of students but for large group sizes, the algorithm performs well. Compared to the genetic algorithm, the running time required for this algorithm is less in all cases as shown in Table 4 and graphically in Fig. 2.

The performance of the students was analyzed and students' feedback was collected. Accordingly, it was found that for the subjects like mathematics, programming skills, etc., heterogeneous grouping helps a lot as weaker students tend to learn from stronger students and as a whole, the group develops. While for language subjects and theoretical subjects, homogenous grouping is effective as sharing, understanding and interacting are among the same-level individuals and their performance according to their ability gives them more satisfaction and ease.

6 Conclusion

This work has used genetic algorithm for the group formation process. Mainly the work aimed to find how a genetic algorithm can form groups with varied characteristic features showing the static and dynamic behavior of students. The process is tested for

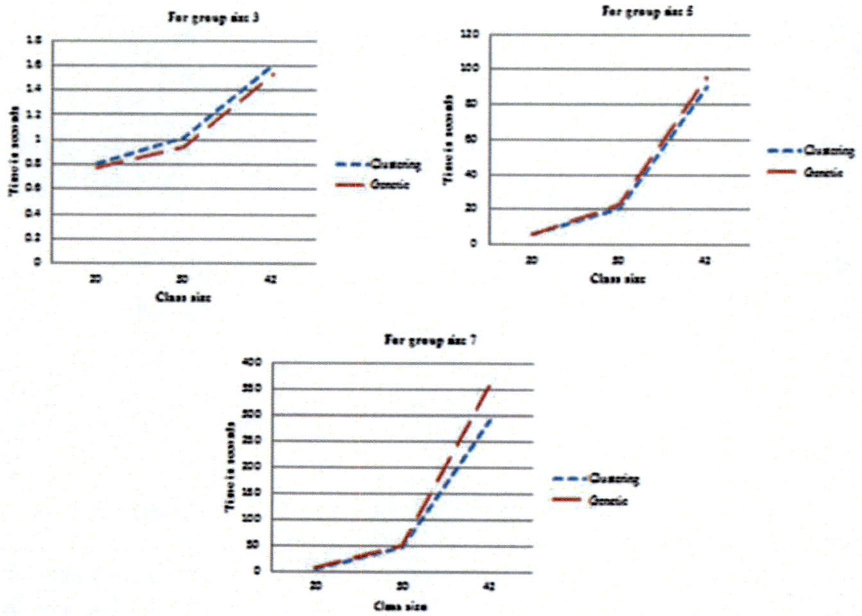

Fig. 2 Comparison of running time between genetic algorithm and clustering algorithm for the same set of attributes

various values of parameters (population size, number of generation, crossover, and mutation) and determined the values which give best results. The method proposed here is extended to be used as a Web application. Teachers can use it as an effective tool for group formation in collaborative learning. The educational and psychological experts have interacted and the proposed approach has gained positive reviews. The study can be continued by adding more specific criteria for group formation. The introduced approach can also be used for team formation for projects in the software industry considering the psychological and technical skills of team members.

References

1. Dennick, R.G., Exley, K.: Teaching and learning in groups and teams. Biochem. Educ. **26**(1998), 111–115 (1998)
2. Bacon, D.R., Stewart, K.A., Silver, W.S.: Lessons from the best and worst student team experiences: how a teacher can make a difference. J. Manag. Educ. **23**(5), 467–488 (1999)
3. Liu, Y., Liu, Q., Wu, R., Chen, E., Su, Y., Chen, Z., Hu, G.: Collaborative learning team formation: a cognitive modeling perspective. In: Navathe, S.B., Wu, W., Shekhar, S., Du, X., Wang, X.S., Xiong, H. (eds.) Database Systems for Advanced Applications, pp. 383–400. Springer International Publishing Switzerland, Dallas (2016)
4. Oxford, R.L.: Cooperative learning, collaborative learning, and interaction: three communicative strands in the language classroom. Mod. Lang. J. **81**(4), 443–456 (1997)

5. O'Malley, C., Scanlon, E.: Computer-supported collaborative learning: problem solving and distance education. Comput. Educ. **15**(1–3), 127–136 (1990)
6. Hogg, M.A., Gaffney, A.M.: Group processes and intergroup relations. In: Wixted, J.T. (ed.) Stevens' Handbook of Experimental Psychology and Cognitive Neuroscience, pp. 1–34. Wiley, New York (2018)
7. Stewart, G.L.: A meta-analytic review of relationships between team design features and team performance. J. Manag. **32**(1), 29–55 (2006)
8. Tanimoto, S.L.: The squeaky wheel algorithm: automatic grouping of students for collaborative projects. In: Proceeding of Workshop on Personalization in Learning Environments at Individual and Group Level in Conjunction with the 11th International Conference on User Modeling, pp. 79–80 (2007)
9. Balmaceda, J.M., Schiaffino, S.N., Pace, J.A.D.: Using constraint satisfaction to aid group formation in CSCL. Inteligencia Artificial, Revista Iberoamericana De Inteligencia Artificial **17**(53), 35–45 (2014)
10. Jin, D., Qinghua, Z., Jiao, D., Zhiyong, G.: A method for learner grouping based on personality clustering. in: Proceedings of the 10th International Conference on Computer Supported Cooperative Work in Design, pp. 1–6, Nanjing (2006)
11. Graf, S., Bekele, R.: Forming heterogeneous groups for intelligent collaborative learning systems with ant colony optimization. In: Ikeda, M., Ashley, K.D., Chan, T.-W. (eds.) Proceedings of the 8th International Conference on Intelligent Tutoring Systems: Lecture Notes in Computer Science, vol. 4053, pp. 217–226. Springer, Jhongli, Taiwan (2006)
12. Martín, E., Paredes, P.: Using learning Styles for dynamic group formation in adaptive collaborative hypermedia systems. In: ICWE Workshops; July 28–30, Munich, Germany (2004)
13. Sukstrienwong, A.: A genetic-algorithm approach for balancing heterogeneous group of students. In: Proceeding of 2016 International Conference on Advances in Software, Control and Mechanical Engineering, pp. 1–7 (2016)
14. Amarasinghe, I., Hernandez-Leo, D., Jonsson, A.: Intelligent group formation in computer supported collaborative learning scripts. In: IEEE 17th International Conference on Advanced Learning Technologies (2017)
15. Manske, S., Hoppe, H.U: Managing knowledge diversity: towards automatic semantic group formation, IEEE 17th International Conference on Advanced Learning Technologies. (2017).
16. Kravitz, D.A., Martin, B.: Ringelmann rediscovered: The original article (1986)
17. Goldberg, D.E.: Genetic Algorithms in Search, Optimization and Machine Learning. Addison-Wesley (1989)
18. Felder, R.M., Soloman, B.A.: Index of Learning Styles (2015). https://www.ncsu.edu/felder ublic/ILSpage.html.
19. Felder, R.M.: Reaching the second tier: learning and teaching styles in college science education. J. Coll. Sci. Teach. **23**(5), 286–290 (1993)
20. Whitley, D.: A genetic algorithm tutorial. Stat. Comput. **4**(2), 65–85 (1994). https://doi.org/10.1007/BF00175354
21. Michalewicz, Z., Fogel, D.B.: How to Solve It: Modern Heuristics. Springer (2000)
22. Umbarkar, A.J., Sheth, P.D.: Crossover operators in genetic algorithms: a review. ICTACT J. Soft Comput. **06**(01) (2015)

On Total Domination Number of Hypercube and Enhanced Hypercube Networks

S. Prabhu, S. Deepa, G. Murugan, and M. Arulperumjothi

Abstract A set $D \subseteq V(G)$ is named a dominating set if each element in $V(G) \setminus D$ is connected to some element belongs to D. Further S is a total dominating set (TDS) of G if for every member of $V(G)$ is next to some member in S. An integer representing the cardinality of this set S is named as total domination number. In this paper, we find a new bound which is exactly half of the value given by Adel P. Kazemi in the reputed journal Utilitas Mathematica 2013. We also find the TDS for enhanced hypercube networks.

Keywords Domination · Total domination · Regular graphs · Hypercube · Enhanced hypercube

1 Introduction

Domination theaters a dynamic role in numerous areas such as communication systems, facility positioning problems, social net problems, and biological models. To determine the domination number for an r-dimensional hypercube, Q^r is an essential problem in coding theory, computer science, and of course in graph theory. In coding theory, this exercise is equivalent to the determination of (Q^r) is to find the size of a minimal covering code of length r and covering radius 1. In computer science, different dissemination type of problems on interconnection networks can

S. Prabhu (✉)
Department of Mathematics, Sri Venkateswara College of Engineering, Sriperumbudur 602117, India
e-mail: drprabhu@svce.ac.in

S. Deepa
Department of Mathematics, Easwari Engineering College, Chennai 600089, India

G. Murugan
Department of Mathematics, Chennai Institute of Technology, Chennai 600069, India

M. Arulperumjothi
Department of Mathematics, Loyola College (Autonomous), Chennai 600034, India

P. Karuppusamy et al. (eds.), *Sustainable Communication Networks and Application*, Lecture Notes on Data Engineering and Communications Technologies 55,
https://doi.org/10.1007/978-981-15-8677-4_30

be demonstrated by domination invariants, where hypercube in opportunity forms a vital model for interconnection networks.

The parameter TDS of graphs was familiarized in [1] and is now fit considered in graph theory. The works on this theme have been charted and meticulously studied in the two excellent domination books by Haynes, Hedetniemi, and Slater [2, 3] who did an owing job of unifying results distributed through some 1200 domination papers. A subset S of G is named to be a dominating set (DS) if each member in $V \setminus S$ is connected to some member belongs to S. We say S a TDS if each member in V is connected to some member belongs to S (i.e., $N(S) = V$). The least integer that represents the size of a TDS is symbolized by $\gamma_t(G)$. Denote $N_1(s) = \{t \in V \mid st \in E\}$ and $N_1[s] = N_1(s) \cup \{s\}$ as an open neighborhood and closed neighborhood of a vertex in V. Generally, an open r-neighborhood of $s \in V$ is the set $N_r(s) = \{t: d(s, t) = r\}$, where $d(s, t)$ is the number of edges present in shortest $s - t$ path. Similarly, the closed r-neighborhood of a vertex $s \in V$ is $N_r[s] = N_r(s) \cup \{s\}$. The degree of a vertex s is also $d_G(s) = |N_1(s)|$. The greatest and smallest degrees of G are symbolized by Δ and δ respectively. For k-regular graph, $\Delta = \delta = k$.

Theorem 1 [4] For any G of order $n, \gamma_t(G) \geq \frac{n}{\Delta}$.

One can easily see those regular graph tights the above bound. In this paper, we classify some regular graphs for which the bound is tight.

A dominating set $D \subseteq G$ is said to be efficient if every member is in V, there is exactly one member in D dominating the member in V.

Equivalently, if D is the efficient dominating set, then for every $\{u, v\} \in D$ we have $N_1(u) \cap N_1(v) = \varphi$. The complexity of efficient domination with respect to various graph community and other locating problems was consulted in [5–10]. For efficient total dominating set, we have $N_1(s) \cap N_1(t) = \varphi$ and $N_1[s] \cap N_1[t] = \{s, t\}$.

2 Hypercube Network

The efficiency of similar computers is frequently discovered by its communication network. The interconnection network is a significant constituent of a parallel processing system. An interconnection network with less topological cost and shortest diameter is referred to as a good interconnection network in literature of computer and communication network [11]. In parallel computing, hypercube is termed as a versatile and most efficient vertex-transitive interconnection network. The dimension of hypercube with cardinality of its vertex set and diameter are respectively connected by an exponential and linear relation. Many combinatorial problems have been discussed for hypercube and its variations. Conditional resolving number of hypercube network has been discussed in [12]. For variations of hypercube, reader can refer [13, 14].

Let Q^r represent the graph of an r-dimensional hypercube, $r \geq 1$. The set $V(Q^r) = \{\{x_0x_1 \ldots x_{r-1}\} : x_i = 0 \text{ or } 1\}$ represents the vertex set. Two vertices $(x_0x_1 \ldots x_{r-1})$ and $(y_0y_1 \ldots y_{r-1})$ are joined by an edge if and only if they differ exactly in one position.

Figure 1 depicts the hypercube of dimension 1, 2, and 3. The cardinality of $V(Q^r)$ is 2^r and cardinality $E(Q^r)$ is $r2^{r-1}$. Q^r is r-regular with diameter r. It is bipartite, Hamiltonian when $r \neq 1$ and Eulerian when r is even [15].

It is clear from the definition that there are four copies of Q^{r-2} in Q^r. We denote them as Q_1^{r-2}, Q_2^{r-2}, Q_3^{r-2} and Q_4^{r-2}. Figure 2 exhibits the four copies of Q^3 in Q^5. The vertex u' is the image of u, and u'' is the image of u' and so on. The edges between the vertices and its images are not shown in the figures of higher-dimensional hypercube and its variants. See Figs. 3 and 5.

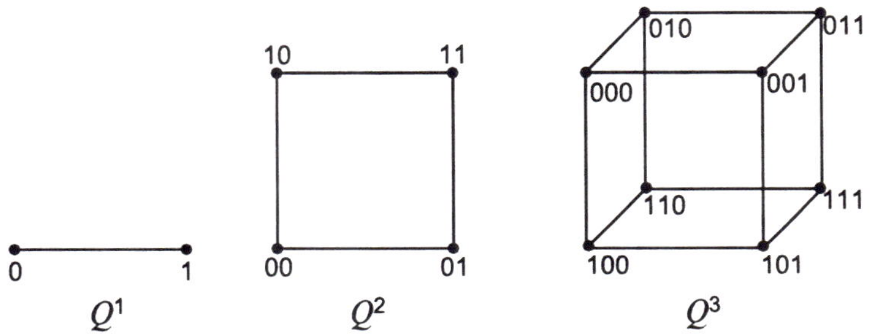

Fig. 1 Hypercube

Fig. 2 Four copies of Q^3 in Q^5

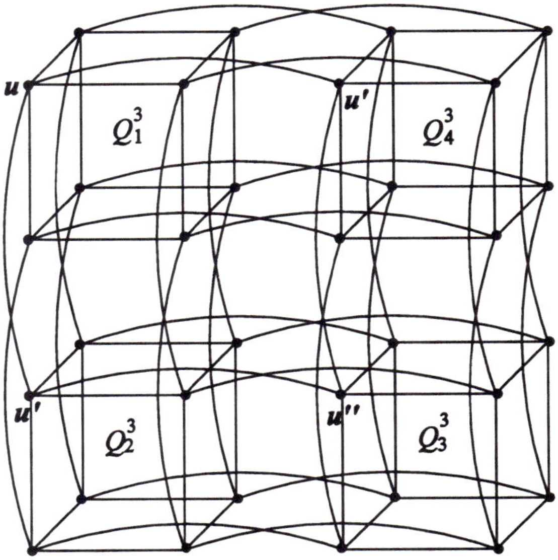

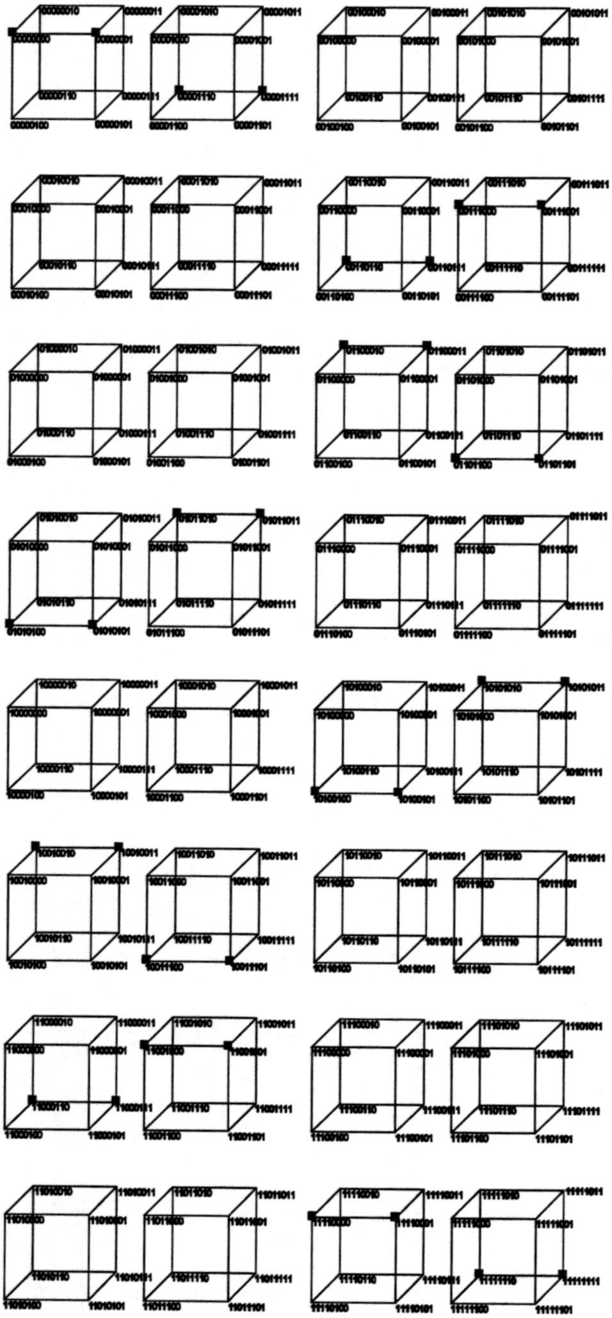

Fig. 3 Total dominating vertices in Q^8

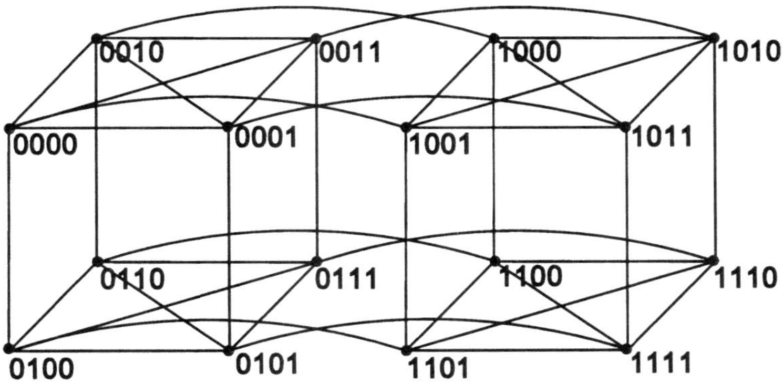

Fig. 4 Enhanced hypercube $Q^{4,2}$

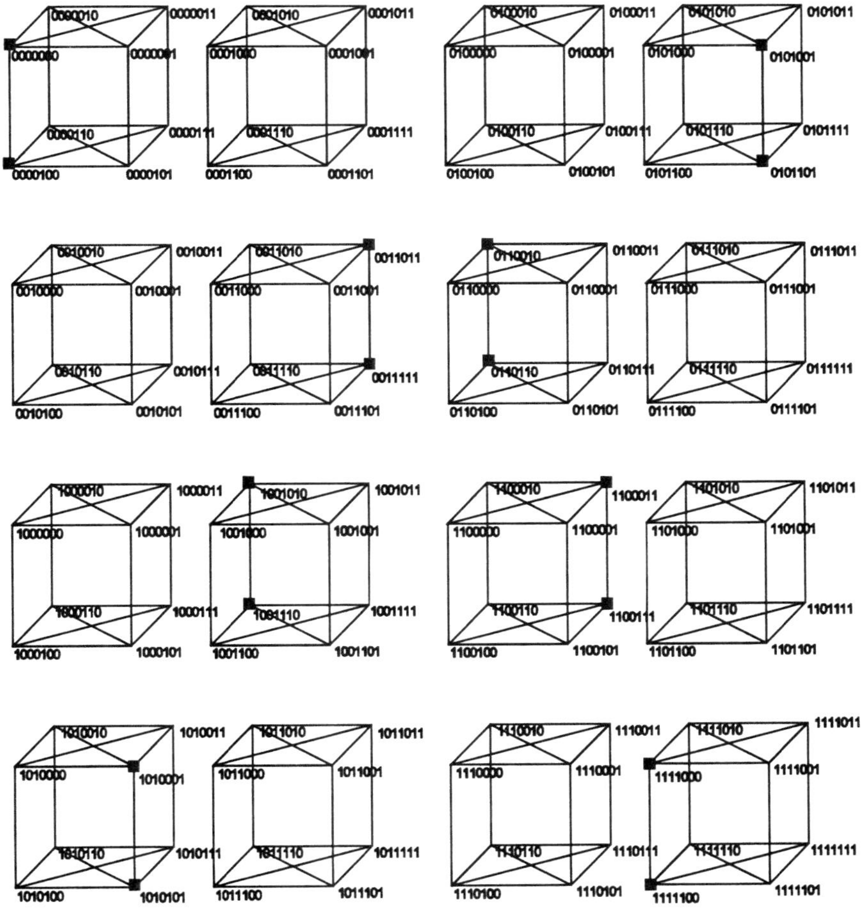

Fig. 5 Total dominating set in $Q^{7,2}$

Theorem 2 [16] Let $G \cong Q^r$. Then $\gamma_t(Q^r) = \begin{cases} 2^r, & r = 1 \\ 2^{r-1} & r = 2, 3 \\ 2^{r-2} & r = 4, 5 \end{cases}$

Theorem 3 [16] Let $G \cong Q^r$. Then $\gamma_t(Q^r) \leq 2^{r-2}, r \geq 6$.

It is also conjectured in [16] that the $\gamma_t(Q^r) = 2^{r-2}, r \geq 6$. The following theorem gives exactly half of the value given in [16].

Theorem 4 $\boldsymbol{\gamma_t(Q^8) = 2^5}$.

Proof The Q^8 has 2^8 vertices, and it is 8—regular. By Theorem 1, the $\gamma_t(Q^8) \geq \frac{2^8}{8} = 2^5$.

Clearly, the set S is the total dominating set for Q^8.
From Fig. 3, we can see that $\gamma_t(Q^8) \leq 2^5$. Hence $\gamma_t(Q^8) = 2^5$.
$S = \{(00000000), (00000001), (00001110), (00001111), (00110110), (01101101), (01011010), (01011011), (01010100), (01010101), \overline{(00000000)}, \overline{(00000001)}, \overline{(00001110)}, \overline{(00001111)}, \overline{(00110110)}, \overline{(00110111)}, \overline{(00111000)}, \overline{(00111001)}, \overline{(01100010)}, \overline{(01100011)}, \overline{(01101100)}, \overline{(01101101)}, \overline{(01011010)}, \overline{(01011011)}, \overline{(01010100)}, \overline{(01010101)}\}$.

Theorem 5 Let Q^r be an r-dimensional hypercube. Then, $\boldsymbol{\gamma_t(Q^r)} \leq 2^{r-3}, r > 8$.

Proof Divide the Q^r into $Q_1^8, Q_2^8, Q_3^8, \ldots\ldots Q_{2^{r-8}}^8$. By Theorem 4, each Q^8 is totally dominated by 2^5 vertices. Hence,

$$\gamma_t(Q^r) \leq 2^{r-8}2^5 = 2^{r-3}.$$

Conjecture: Let $G = Q^{2^k}$ be a 2^k-dimensional hypercube. Then, $\gamma_t(G) = 2^{2^k - k}$ for $k \geq 2$.

3 Enhanced Hypercube Network

The hypercube has acknowledged substantial consideration primarily due to its systematic structure, small diameter, and respectable assembly with a comparatively small vertex degree [15]. It is very popular, versatile, and vertex-transitive interconnection network [13]. Many dissimilarities of the hypercube have been recommended to advance the performance [13]. One of the variations is in the enhancement [17] of the hypercube with a same number of vertices. The enhanced hypercube is copious additional attractive than the normal one because of their potentially nice topological properties. Locating set for this network has been discussed in [18].

The new invariant enhanced hypercube $Q^{r,k}$, $0 \leq k \leq r-1$, with $V(Q^{r,k}) = V(Q^r)$ and $E(Q^{r,k}) = E(Q^r) \cup$

$x_0x_1 \ldots x_{k-2}x_{k-1}x_k \ldots x_{r-1}, x_0x_1 \ldots x_{k-2}\overline{x_{k-1}x_k \ldots x_{r-1}}$. The edges of Q^r in $Q^{r,k}$ are called the parent edges, and the other edges are called skips or complementary edges[17]. See Fig. 4.

The $Q^{r,k}, 0 \leq k \leq r-1$ is $(r+1)$-regular having 2^r vertices and $(r+1)2^{r-1}$ edges. Furthermore, it is bipartite if and only if r and k have the same parity [14,17].

Theorem 6 Let $G = Q^{r,2}$ be an r-dimensional enhanced hypercube. Then, $\gamma_t(G) = 2^{r-2}, 3 \leq r \leq 6$.

Proof For $r = 3$,

$Q^{3,2}$ has two $K_4's$. Take one vertex from each K_4 such that they are adjacent. This forms the total dominating set for the $Q^{3,2}$. Therefore, $\gamma_t(Q^{3,2}) = 2$.

For $r = 4$.

$Q^{4,2}$ is 5-regular and has 2^4 vertices. By Theorem 1, we have $\gamma_t(Q^{4,2}) \geq \frac{16}{5} = 3.1$. Since $\gamma_t(Q^{4,2})$ is an integer, we can have $\gamma_t(Q^{4,2}) \geq 4$. The other argument says $Q^{4,2}$ has two copies of $Q^{3,2}$. This argument gives us $\gamma_t(Q^{4,2}) \leq 2, \gamma_t(Q^{3,2}) = 2 \cdot 2 = 4$.

For $r = 5$.

$Q^{5,2}$ is 6-regular and has 2^5 vertices. By Theorem 1, we have $\gamma_t(Q^{5,2}) \geq \frac{32}{6} = 5.3$. Therefore, $\gamma_t(Q^{5,2}) \geq 6$.

Case1: $\gamma_t(Q^{5,2}) \neq 6$.

Assume that $\gamma_t(Q^{5,2}) = 6$. Let V_i, $1 \leq i \leq 8$ are the eight vertex partition of the vertex set of $Q^{5,2}$ such that the cardinality of each V_i is 4 and the induced graph of any V_i, $1 \leq i \leq 8$ is isomorphic to K_4. Let us construct the dominating set of cardinality 6 by selecting one vertex in each V_i, then there exists at least one V_j left without dominating vertex. By construction definition of $Q^{5,2}$, each K_4 induced by V_i, $1 \leq i \leq 8$ is adjacent to at most three K_4. This implies that the vertices of V_j are dominated by at most three vertices. Hence, there exists at least one vertex in V_j which is not dominated by any vertex in the dominating set, a contradiction.

Case 2: $\gamma_t(Q^{5,2}) \neq 7$.

The arguments of Case 2 are similar to Case 1.

Clearly, the dominating set $S = \{v_i \text{ess } V_i: 1 \leq i \leq 8\}$ is a total dominating set of $Q^{5,2}$. Therefore, $\gamma_t(Q^{5,2}) = 8$. Hence, the theorem

Theorem 7 $\gamma_t(Q^{7,2}) = 2^4$.

Proof It is clear from the definition of hypercube $Q^{7,2}$ has 2^7 vertices, and it is 8—regular. By Theorem 1, the $\gamma_t(Q^{7,2}) \geq \frac{2^7}{8} = 2^4$.

The following set

S = {(0,000,000), (0,000,100), (0,011,011), (0,011,111), (0,101,010), (0,101,110), (0,110,001), (0,110,101), (1,001,001), (1,001,101), (1,010,010), (1,010,110), (1,100,011, (1,100,111), (1,111,000), (1,111,100)} is a total dominating set for $Q^{7,2}$. See Fig. 5. This proves that $\gamma_t(Q^{7,2}) \leq 2^4$. Hence, $\gamma_t(Q^{7,2}) = 2^4$.

Theorem 8 $\gamma_t(Q^{r,2}) \leq 2^{r-3}$ for $r \geq 8$.

Proof It is clear that $Q^{r,2}$ has 2^{r-7} copies of $Q^{7,2}$.

$$\begin{aligned}\gamma_t(Q^r) &\leq 2^{r-7} \cdot \gamma_t(Q^{7,2}) \\ &= 2^{r-7} \cdot 2^4 \\ &= 2^{r-3}.\end{aligned}$$

4 Conclusion

In this paper, we find a total dominating set in a hypercube network and enhanced hypercube network. We also disprove and restate the conjecture which is proposed in [16] which is the reputed journal Utilitas Mathematica. The construction of proof for the conjectures and other domination parameters of regular graphs are under investigation. Investigation of total domination of varietal hypercube network and cube connected cycles is the proposed future work toward this direction of research. The combination of metric dimension (Resolving set) [19] and power domination [8] is yet another field of domination called resolving power domination [20]. Identification of resolving power domination of hypercube and its derived architecture is still open.

Acknowledgements This research is primarily supported by UGC-IGSGC Fellowship No. F. 8-1/2014-16/SGC-OBC-2014-23907/(SA-III), University Grants Commission, New Delhi, Govt. of India.

References

1. Fischermann, M.: Unique total domination graphs. Ars Combin. **73**, 289–297 (2004)
2. Henning, M.A., Swart, H.C.: Bounds on a generalized total domination parameter. J. Combin. Math. Combin. Comput. **6**, 143–153 (1989)
3. Kang, L., Shan, E.: Total minus domination in k-partite graphs. Discrete Math. **306**, 1771–1775 (2006)
4. Henning, M.A., Yeo, A.: Total domination in graphs. Springer Monographs in Mathematics (2013). https://doi.org/10.1007/978-1-4614-6525-6
5. Bange, D.W., Barkauskas, A.E., Host, L.H., Slater, P.J.: Generalized domination and efficient domination in graphs. Discrete Math. **159**, 1–11 (1996)
6. Biggs, N.: Perfect codes in graphs. J. Combin. Theory (B) **15**, 289–296 (1973)
7. Chang, G.J., Pandu Rangan, C., Coorg, S.R.: Weighted independent perfect domination on co-comparability graphs. Discrete Appl. Math. **63**, 215–222 (1995)
8. Prabhu, S., Arulmozhi, A.K., Arulperumjothi, M.: On Power Domination in Certain Chemical Graphs. Int. J. Pure and Appl. Math. **118**(11), 11–19 (2018)

9. Lu, C.L., Tang, C.Y.: Weighted efficient domination problem on some perfect graphs. Discrete Appl. Math. **117**, 163–182 (2002)
10. Yen, C.C., Lee, R.C.T.: The weighted perfect domination problem and its variants. Discrete Appl. Math. **66**, 147–160 (1996)
11. El-Rewini, H., Abd-El-Barr, M.: Advanced Computer Architecture and Parallel Processing. Wiley, Hoboken (2005)
12. Rajan, B., William, A., Rajasingh, I., Prabhu, S.: On Certain connected resolving parameters of hypercube networks. Appl. Math. **3**, 473–477 (2012)
13. Chodum, S.A., Sunitha, V.: Augmented cubes. Networks **40**(2), 71–84 (2002)
14. Hongmei, L.: The structural features of enhanced hypercube networks. In: Fifth International Conference on Natural Computatation, vol. 1, pp. 345–348 (2009)
15. Adel, P.K.: Total dominator coloring in product graphs. Utilitas Math. Accepted (2013)
16. Xu, J.: Structures and analysis of interconnection networks. Kluwer Academic Publishers, Dordrecht (2001)
17. Tzeng, N.F., Wei, S.: Enhanced hypercubes. IEEE Trans. Comput. **40**(3), 284–294 (1991)
18. Rajan, B., Rajasingh, I., Chris, M.M., Manuel, P.: Metric dimension of enhanced hypercube networks. J. Combin. Math. Combin. Comput. **67**, 5–15 (2008)
19. Prabhu, S., Flora, T., Arulperumjothi, M.: On independent resolving number of TiO_2 [m, n] nanotubes. J. Int. Fuzzy Sys. **35**(6), 6421–6425 (2018)
20. Sudeep, S., Bharati, R., Cyriac, G., Albert, W.: Resolving-power dominating sets. Appl. Math. Comput. **256**, 778–785 (2015)

Neural Network-Based Classification of Toxic Gases for a Sensor Array

V. V. Ragila, Ramya Madhavan, and U. Sajesh Kumar

Abstract Electronic noses or array sensors are very popular in the last decades because of their ability to avoid the cross-sensitivity issue in semiconductor metal oxide (SMO) gas sensors. The identification and discrimination of toxic gases have a significant role in industrial applications. This work encompasses the classification of carbon monoxide (CO) and methane (CH_4) toxic gases using a gas sensor array. Classification algorithm based on artificial neural network (ANN) with one hidden layer is used for identifying the gas type from the gas mixture. This metal oxide gas sensor array is built with six SMO gas sensors, which are sensitive to several types of gases. The ANN model ensures a training accuracy of 94.57% and a validation accuracy of 93.33%. For practical applications, the gas concentration is randomly assigned in the training stage. Neural network-based classification algorithm provides better performance in identifying the type of gas.

Keywords Array sensor · Artificial neural network · Classification

1 Introduction

Global demand for a fast, inexpensive, and accurate measurement system for detecting the concentration of a selective gas from the gas mixture has been increasing in these recent decades. Electronics nose (E-nose) or array sensor can be used for many applications including military, commercial, environmental, food quality checking, and for diagnosing human health problems [1–3]. Most of the current apparatus for gas sensing applications are very expensive and difficult to use for portable applications. Electronic nose incorporates a matrix of sensors. Each sensor

V. V. Ragila (✉) · R. Madhavan · U. Sajesh Kumar
Government College of Engineering Kannur, Kannur, Kerala, India
e-mail: ragilavv@gmail.com

R. Madhavan
e-mail: ramya10madhavan@gmail.com

U. Sajesh Kumar
e-mail: sajesh@gcek.ac.in

P. Karuppusamy et al. (eds.), *Sustainable Communication Networks and Application*, Lecture Notes on Data Engineering and Communications Technologies 55,
https://doi.org/10.1007/978-981-15-8677-4_31

readings needs to be analyzed to classify the analytes from the gaseous mixture [1, 3]. Metal oxide materials, namely SnO_2, In_2O_3, ZnO, and WO_3 are the common sensing material used for sensing gases, because of their high sensitivity ability [2, 4]. The array of sensors with a signal conditioning circuit produces a change in the resistance value of sensing materials for a particular gas. There is a difficulty in finding gases from the gaseous mixture, because of the weak selectivity of the SMO gas sensor [3, 4]. By considering cross-sensitivity with the pattern recognition algorithm, the sensor array provides better results [5–7].

To identify the target gas from the mixture and to measure the quantity of that analyte accurately, many researchers have developed different algorithms. The most commonly used classification techniques for quantitative analysis of E-nose are support vector machine (SVM) and principal component analysis(PCA) [6, 8]. In 2010, T. Tang et al. presented a portable E-nose, which can classify four different fruit gases [5]. It consists of eight metal oxide gas sensors with PCA and *k*-nearest neighbors' (KNN) algorithm for classification. Neural networks and decision tree learning methods are introduced by Zhiyun Chen and Yangong Zheng for the classification and prediction of the mixture of gases in an E-nose [9]. In 2019, M. Abdelkhalek et al. proposed an embedded Electronic nose for VOC and non-VOC gases classification for robot applications [10]. They used the KNN classification algorithm with an array of 4 MOS gas sensors. It has 98% accuracy for the classification of gases. Likewise, a multi-layer perception (MLP) network is an effective method for the classification of gases. Rabeb et al. used this method for the detection and classification of atmospheric pollutant gases [11].

Methane (CH_4) and carbon monoxide(CO) are toxic gases, which harmfully affect human health [3]. So the detection of these gases is very significant in industrial applications. This work aims to develop an array gas sensor system with an artificial neural network model for accurately identifying CO and CH_4 gases. The sensor array is constructed using MQ4 and MQ7 metal oxide gas sensors [12]. The ANN classification model is designed in the python language using the virtualenv tool [13]. It provides better classification performance with 94.57% training and 93.33% validation accuracy.

2 Experimental Method

2.1 Gas Sensor and Electronic Nose

There are different gas sensor types based on sensing methods of the material. Semiconductor Metal Oxide (SMO), polymers, carbon nanotubes are the examples of gas sensors which have the mechanism of sensing based on the change in electrical variations [3, 4]. SMO gas sensor consists of sensing material layer, heater, and electrode. Figure 1 shows the structure of a semiconductor metal oxide gas sensor. If gas molecules are adsorbed on the sensing material surface, the electrical resistance of

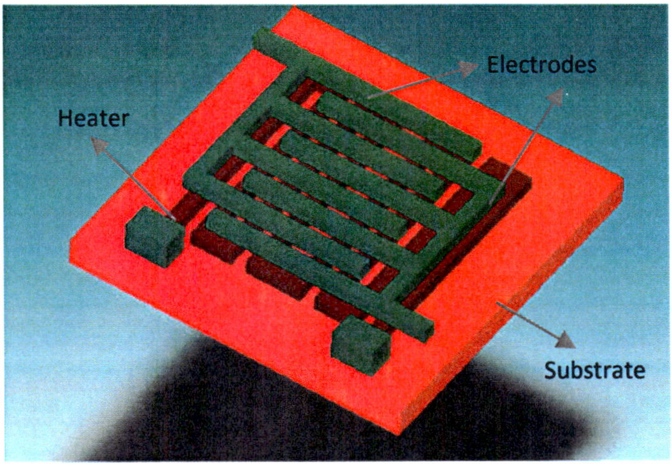

Fig. 1 Structure of a metal oxide gas sensor

the material changes [3]. Sensing material consists of several grains. In the presence of oxygen in the air, each grain loses some electrons and forms a depletion layer [1]. A molecule can react better at a higher temperature. So when heating the sensitive material, its sensitivity increases and the reaction sites also increases. Lower selectivity or cross-sensitivity is the main challenge while using the SMO gas sensor [12, 14]. So by using the array sensor, the aim is that for each gas, it should show different sensitivity.

This sensor array consists of three MQ4 gas sensors and three MQ7 gas sensors. Also, they can sense different kinds of gases. The sensitivity of the sensor array increases with an increase in humidity and temperature of the environment [1, 4]. So a temperature–humidity sensor is also employed with the sensor array to measure the temperature and humidity of the environment. Because of their chemical composition, MQ4 is highly sensitive to CH4 gas, whereas MQ7 shows more sensitivity toward CO gas. Also, they can sense different kinds of gases. From the datasheet of the MQ4 gas sensor, it can sense LPG, H_2, CO, alcohol, smoke other than CH_4. Likewise, the MQ7 sensor senses CO, H_2 LPG, CH_4, and alcohol gases. Both of the sensors can sense CH_4 and CO gases, so the presence of a mixture of these gases produces a cross-sensitivity issue. Two main functions are happening in an E-nose. The first processing part is the sensing mechanism in the sensor array [15].

Each sensor has a different R_0 (resistance of the sensor in a clean environment) and R_g (sensor resistance to a particular gas) [10, 12]. Hence, R_0 was determined using the sensors in a clean atmospheric condition. The sensor response is calculated as follows:

$$S \text{ reducing gas} = R_a/R_g \quad (1)$$

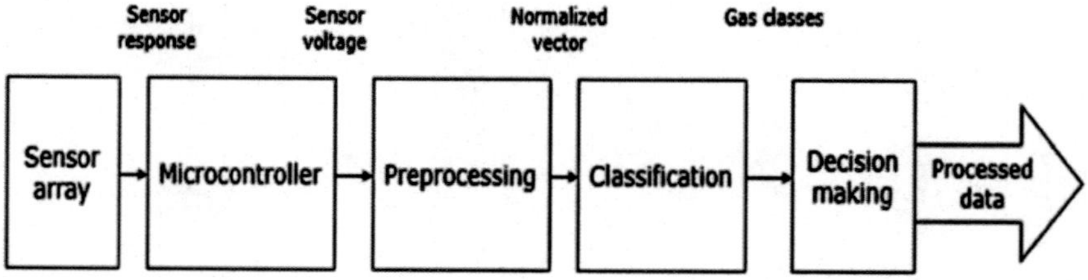

Fig. 2 Steps used for signal processing in a sensor array

$$S \text{ oxidising gas} = R_g/R_a \tag{2}$$

The response of the sensor (*R*) changes according to the gas concentration, *C* as the following:

$$R = 1 + xC^n \tag{3}$$

where x and n are constants depending on the sensitive material. The sensor response has a nonlinear relationship with the gas concentration.

Figure 2 shows the steps used for signal processing in an array sensor or E-nose. The first section is the hardware part. According to the change in concentration of the gases, each sensor in the sensor array produces corresponding analog readings [11]. These readings from this sensor array passed to the microcontroller unit, which converts these analog readings to sensor voltages. The pre-processing stage normalizes these sensor voltages for the ease of training purpose. The classification stage uses this normalized vectors from this pre-processing stage.

2.2 *Preprocessing*

Each sensor in the sensor array produces different voltages. Hence, the output data from sensor array has a different mean and offset [11]. Initially, in a clean environment condition, each sensor is placed and calculated the R_0 value. Since our institute situated in an industrial area, the air itself shows these toxic gas presence. Also, the data samples prepared using vehicle exhaust, burning situations, etc… The classification process requires large datasets, so some dummy values corresponding to each gas included from the datasheet information of the sensors. A total of 500 data samples is arranged for the dataset preparation. 70% of this dataset is transferred to the training phase, 15% to the validation stage, and 15% to the testing stage. Let X_k be the feature matrix from the received sensor signals for 'k' set of samples. There are 24 sets of methane gas samples and 20 sets of carbon monoxide samples. Here, 'i' = 1, 2, …, 6 denotes the number of sensors.

$$k = \begin{cases} 1, 2, \ldots, 24 \text{ ; for methane gas} \\ 1, 2, \ldots, 20 \text{ ; for carbon monoxide gas} \end{cases} \tag{4}$$

$$X_k = \begin{bmatrix} x_{11} & x_{12} & \cdots & x_{1i} \\ x_{22} & x_{21} & \cdots & x_{2i} \\ \vdots & \vdots & \ddots & \vdots \\ x_{k1} & x_{k2} & \cdots & x_{ki} \end{bmatrix} \tag{5}$$

The dataset is normalized to make zero mean. Normalization is a necessary step in the pre-processing stage since it makes the dataset suitable for the classification and training process. Also, normalization reduces the chance of occurring error during prediction. The data output from the array sensor, X_{ki} is normalized to get X_{ni} using the formula:

$$X_{ni} = (X_{ki} - M)/\text{Sd} \tag{6}$$

where M and Sd are mean and standard deviation. The output normalized feature matrix is,

$$X_n = \begin{bmatrix} x_{11} & x_{12} & \cdots & x_{1i} \\ x_{22} & x_{21} & \cdots & x_{2i} \\ \vdots & \vdots & \ddots & \vdots \\ x_{n1} & x_{n2} & \cdots & x_{ni} \end{bmatrix} \tag{7}$$

2.3 ANN Model for Classification

The normalized vector from the pre-processing stage is forwarded to the artificial neural network (ANN)-based classification model [9]. ANN is a bio-inspired neural network system [11]. Interconnected nodes construct a network in the ANN model. The signal path inside this network is called a synapse. This ANN classification model consists of one input layer, one hidden layer, and one output layer shown in Fig. 3. These layers calculate the output data from the given data using nonlinear functions. The input layer consists of 6 neurons, which represent the normalized sensor readings. The hidden layer includes 12 neurons [12]. The output layer has only one neuron representing the output of classification. There are different types of neural network models, for example, convolution neural network (CNN), recurrent neural network (RNN), generative adversarial networks (GAN) [16, 17]. Usually, convolution neural network-based classification models are used for image classification purposes, also it needs a large amount of dataset for increasing the accuracy of the model [17]. In this classification model, artificial neural network architecture provides high accuracy for

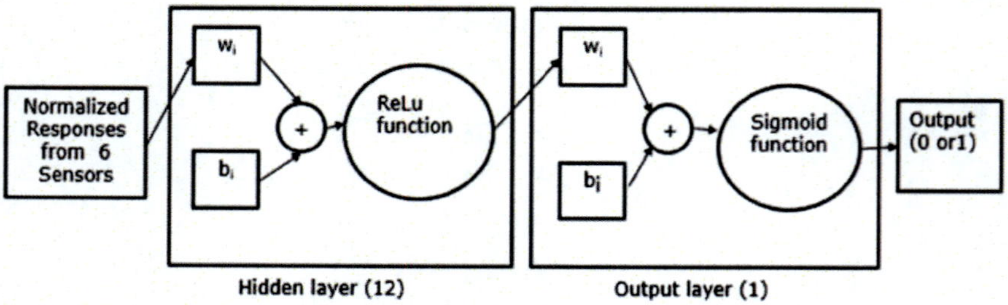

Fig. 3 Classification model based on ANN [9]

the given set of data. Also, the training time needed for this model is very less. This is a feed-forwarded ANN with backpropagation for adjusting the weight and bias. To increase the accuracy of the model, Adam optimization algorithm [18] is used.

Two datasets are fed into the input layer, corresponding to each gas. This whole dataset is divided into three portions and passed to the training, validation, and testing process. The hidden layer activation function [18, 19] is,

$$f(x) = \max(0, x) \tag{8}$$

Activation function of output layer is,

$$f(x) = 1/\left(1 + e^{x}\right) \tag{9}$$

The normalized data is weighted by the factors w_i and b_i. It is given to a ReLu function $g(X_n)$, which is calculated as [19]:

$$g(X_n) = \sum_{i=1}^{6} f(w_i(X_{ni}) + b_i) \tag{10}$$

The model iteratively trained to obtain good accuracy and to reduce the loss [18]. The cost function is calculated using the formula:

$$C = (12N)\sum_{1}^{N}(T\log(g(X_n)) - (1 - T)\log(1 - g(X_n))) \tag{11}$$

where $g(X_n)$ is the predicted gas type output and T is the desired gas output. For getting a good accuracy, the weight factor, wi is updated as per weight updation rule [18]:

$$w_i + 1 = w_i - a\frac{\partial C}{\partial w_i} \tag{12}$$

where 'a' is the learning rate.

3 Simulation Results

Figure 4 shows the epoch versus accuracy graph of the model. Here, x-axis indicates the number of epochs and the y-axis shows the accuracy of the ANN model. The results of training and validation indicate that after each iteration accuracy and performance of this ANN model improves. Initially, the accuracy of the model is very low. But with the increases in epochs, the accuracy of the model also increases. Here, the blue curve shows training set accuracy, and the orange curve indicates the validation set accuracy.

Improvement in the performance of the ANN model minimizes the cost function, which reduces the loss happening in the model. Figure 5 shows the epoch versus loss graph. The blue and orange curves indicate the training and validation loss, respectively. Initially, the loss of the model is high. As the number of epoch increases, validation and the training set loss decrease. Finally, the validation loss and training set loss become stable.

The model accurately identifies the gas type within 50 epochs. That assures the quality of the ANN classifier. Figure 6 shows the testing and validation, which confirms a training set accuracy of 94.57% and validation set accuracy of 93.33%.

Five samples are randomly selected from the test data set and used them for the prediction. Figure 7 shows the results of testing, which accurately predicts the gas type.

Figure 8 shows a comparison of the test samples results with the actual data type. Most of the samples accurately predict the gas type. 15% of the dataset is used for testing the model. Figure 9 shows the results of test data prediction. From this test samples, 98.67% results show a correct prediction of gas type.

Figure 10 shows the confusion matrix obtained by the ANN model. This results in a classification accuracy of 96%. Table 1 shows the evaluation measures calculated from the confusion matrix.

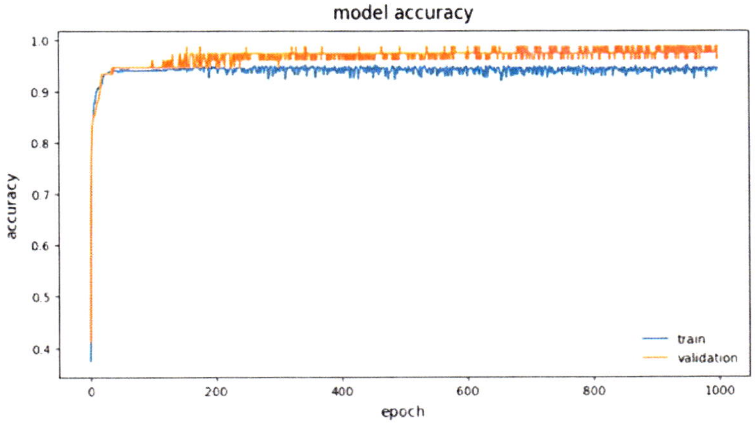

Fig. 4 Epoch versus accuracy graph

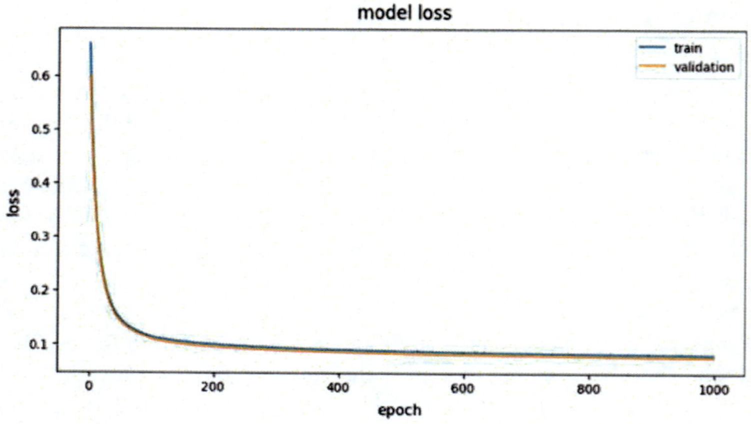

Fig. 5 Epoch versus loss graph

```
11/11 [==============================] - 0s 5ms/step - loss: 0.1730 - accuracy: 0.9429 - val_loss: 0.1524 - val_accuracy: 0.9333
Epoch 44/50
11/11 [==============================] - 0s 4ms/step - loss: 0.1714 - accuracy: 0.9429 - val_loss: 0.1504 - val_accuracy: 0.9333
Epoch 45/50
11/11 [==============================] - 0s 5ms/step - loss: 0.1699 - accuracy: 0.9457 - val_loss: 0.1486 - val_accuracy: 0.9333
Epoch 46/50
11/11 [==============================] - 0s 5ms/step - loss: 0.1684 - accuracy: 0.9457 - val_loss: 0.1475 - val_accuracy: 0.9333
Epoch 47/50
11/11 [==============================] - 0s 4ms/step - loss: 0.1671 - accuracy: 0.9457 - val_loss: 0.1459 - val_accuracy: 0.9333
Epoch 48/50
11/11 [==============================] - 0s 4ms/step - loss: 0.1659 - accuracy: 0.9457 - val_loss: 0.1450 - val_accuracy: 0.9333
Epoch 49/50
11/11 [==============================] - 0s 4ms/step - loss: 0.1643 - accuracy: 0.9457 - val_loss: 0.1433 - val_accuracy: 0.9333
Epoch 50/50
11/11 [==============================] - 0s 4ms/step - loss: 0.1632 - accuracy: 0.9457 - val_loss: 0.1423 - val_accuracy: 0.9333
```

Fig. 6 Training and validation

test_data

	Sample	Input gas concentration	MQ7 voltage1	MQ7 voltage2	MQ7 voltage3	MQ4 voltage1	MQ4 voltage2	MQ4 voltage3	humidity	temperature	class	gas_type	predicted_gas
53	3	80	0.658006	0.419164	0.511352	0.128922	0.151492	0.138843	48.1	34.2	0	CO	CO
12	1	110	1.214196	0.778452	1.040020	0.131838	0.162300	0.144690	45.3	33.4	0	CO	CO
187	8	200	0.055631	0.047669	0.052791	0.296639	0.329676	0.304735	50.9	34.5	1	CH4	CH4
49	3	40	0.099926	0.069684	0.077659	0.128445	0.150715	0.137929	48.1	34.2	0	CO	CO
371	15	300	0.060625	0.050579	0.057674	0.224622	0.289104	0.242578	53.3	35.6	1	CH4	CH4

Fig. 7 Testing using randomly selected samples

4 Conclusion

In this work, an array of sensors composed of semiconductor metal oxide (SMO) gas sensors is used to identify the toxic gases methane and carbon monoxide. Classification of this gaseous mixture is performed using an artificial neural network. Using neural network-based classification, it is evident that the training accuracy is 94.57% and validation accuracy is 93.33%. And also the variation in the responses

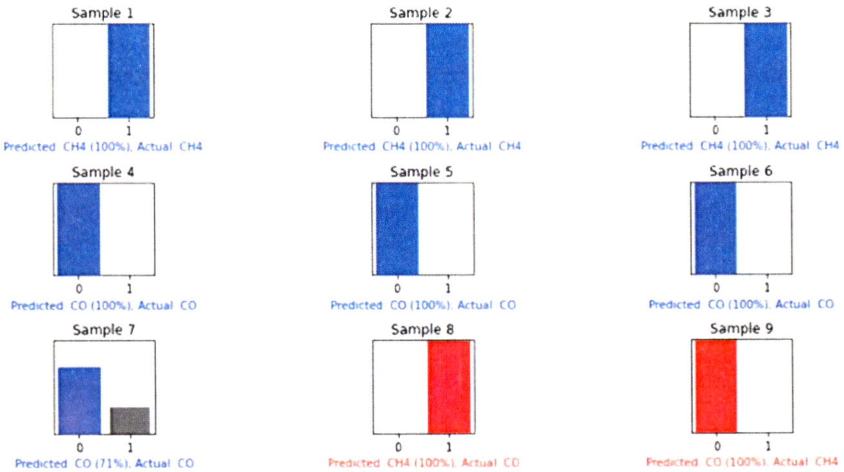

Fig. 8 Test results of samples

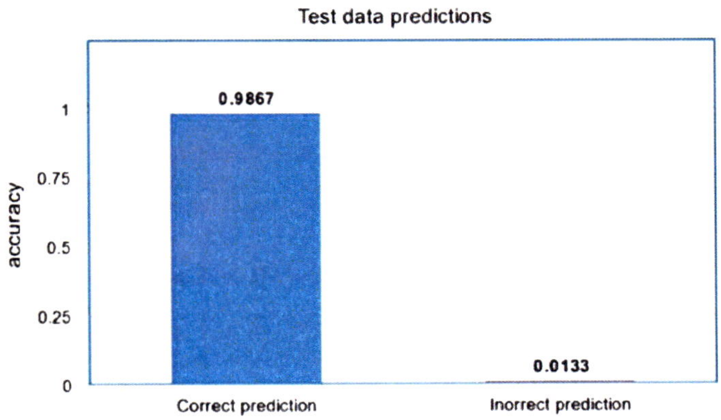

Fig. 9 Prediction using test data samples

of sensors and environmental condition significantly affects the performance of the gas sensor array. This ANN method can be used for the classification in the future work on E-nose practical applications with less implementation cost.

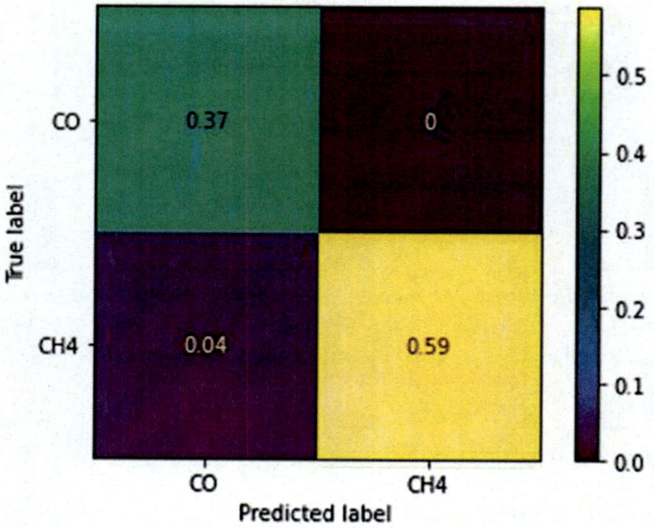

Fig. 10 Confusion matrix from the results of classification

Table 1 Evaluation measures from a confusion matrix

Measure	Calculated value
Accuracy	0.96
Sensitivity (recall)	0.902
Specificity	1
Precision	1

References

1. Dey, A.: Semiconductor metal oxide gas sensors: a review. Mater. Sci. Eng. B **229**, 206–217 (2018)
2. Mahajan S., Jagtap, S.: Metal-oxide semiconductors for carbon monoxide (CO) gas sensing: a review. Appl. Mater. Today (2019)
3. Kant, R., Battacharya, S.: Sensors for air monitoring. In: book: Environmental, Chemical and Medical Sensors, pp. 9–30, 2017.
4. Bochenkov, V.E., Sergeev, G.B.: Metal Oxide Nanostructures and Their Applications, vol. 3, pp. 31–52. American Scientific Publication (2010)
5. Tang, K.T., Chiu, S.W., Pan, C.H., Hsieh, H.Y., Liang, Y.S., Liu, S.C.: Development of a portable electronic nose system for the detection and classification of fruity odors. IEEE Sens. **10**, 9179–9193 (2010)
6. Ansari, A.Q., Khusro, A., Ansari, M.R.: Performance evaluation of classifier techniques to discriminate odors with an E-Nose. In: 2015 Annual IEEE India Conference (INDICON), 2016.
7. Guo, D., Zhang, D., Li, N., Zhang, L., Yang, J.: A novel breath analysis system based on electronic olfaction. IEEE Trans. Biomed. Eng. **57**, 2753–2763 (2010)

8. Akbar, M.A., Ali, A.A.S., Amira, A., Bensaali, F., Benammar, M., Hassan, M., Bermak, A.: An empirical study for PCA- and LDA-based feature reduction for gas identification. IEEE Sens. J. **16** (2016).
9. Chen, Z., Zheng, Y., Li, H., Chen, K., Jian, J.: Concentration estimator for mixed VOC gases using sensor array with neural networks and decision tree learning. IEEE Sens. J. **17**(6) (2017).
10. Abdelkhalek, M., Alfayad, S., Benouezdou, F., Fayek, M.B., Chassagne, L.: Compact and embedded electronic nose for volatile and non-volatile odor classification for robot applications. IEEE Access **7**, 98267–98276 (2019)
11. Rabeb, F., Souhir, B., Abdermaceur, K., Mounir, S.: An electronic nose for detection pollutant odorant and olfaction classification using neural network. In: IEEE International Conference on Sciences and Techniques of Automatic Control & Computer Engineering, 2014.
12. Shahid, A., Choi, J., Rana, A., Kim, H.: Least squares neural network-based wireless E-nose system using an SNO_2 sensor array. Sensors (Basel). https://doi.org/10.3390/s18051446.
13. https://docs.python.org/3/tutorial/venv.html.
14. Prajapati, C.S., Soman, R., Rudraswamy, S.B., Nayak, M., Bhat, N.: Single chip gas sensor array for air quality monitoring. IEEE J. Microelectromech. Syst. **26**(2), 433–439 (2017)
15. Yang, Y., Huixiang Liu, YuGu.: A model transfer learning framework with back-propagation neural network for wine and Chinese liquor detection by electronic nose. IEEE Access **8**, 105278–105285 (2020)
16. Zhang, G.P.: Neural networks for classification: a survey. IEEE Trans. Syst., Man, Cybern.—Part C: Appl. Rev. **30** (2000)
17. Xu, X., Ge, H., Li, S.: An improvement on recurrent neural network by combining convolution neural network and a simple initialization of the weights. In: IEEE International Conference of Online Analysis and Computing Science (ICOACS), May 2016.
18. Baxt, W.G.: Improving the accuracy of an artificial neural network using multiple differently trained networks. Neural Comput. **4**, 772–780 (1992)
19. Baxt, W.G.: Use of an artificial neural network for data analysis in clinical decision-making: The diagnosis of acute coronary occlusion. Neural Comput. **2**, 480–489 (1990)

Effect of Negative Capacitance MOSFET Devices on Circuit Applications

K. P. Krishna Priya and U. Sajesh Kumar

Abstract The conventional complementary metal-oxide-semiconductor (CMOS) technology is coming to an end due to its large power dissipation and self-heating. This can be solved by utilizing the 'negative capacitance' (NC) effect. To observe this negative capacitance, a ferroelectric (FE) material is needed to be used instead of an insulator in transistors. In this work initially, a normal Verilog A modeling of the transistor has been used to study its characteristics and then gradually employed a ferroelectric layer in the gate stack by modifying and inserting Landau Khaltnikov (L-K) model parameters in the model. This modification allows us to observe a reduction in voltage across the ferroelectrics, i.e., a negative capacitance and subthreshold swing less than 60 mv/decade. The analysis of NC transistor is performed to find the role of ferroelectric layer thickness and three ferroelectric coefficients α, β and γ. By using these concise models, the performance of various digital circuits has been evaluated with NC transistors.

Keywords Negative capacitance · Boltzmann tyranny · Subthreshold swing · Ferroelectrics

1 Introduction

The important part of technological growth in recent times is the metal-oxide-semiconductor field-effect transistors (MOSFET). The factor that limits the supply voltage scaling is Boltzmann tyranny. According to Boltzmann tyranny, the minimum value of subthreshold swing in MOSFETs is 60 mv/decade. For normal transistors, the subthreshold swing is around 80 mv/decade. This is a large value for nano-electronic devices. Various works are done in this direction toward overcoming the

K. P. Krishna Priya (✉) · U. Sajesh Kumar
Department of Electronics and Communication Engineering, Government College of Engineering Kannur, Kannur, India
e-mail: kpkrishnapriya30597@gmail.com

U. Sajesh Kumar
e-mail: sajesh@gcek.ac.in

P. Karuppusamy et al. (eds.), *Sustainable Communication Networks and Application*, Lecture Notes on Data Engineering and Communications Technologies 55,
https://doi.org/10.1007/978-981-15-8677-4_32

Boltzmann limit. From these, the negative capacitance is promising and has achieved much attention these days.

Subthreshold swing means the amount of voltage that is required to raise the current by 1 decade [1],

i.e., subthreshold swing,

$$SS = \left(1 + \frac{C_s}{C_{ins}}\right) 60\ \text{mv/decade [1]} \tag{1}$$

where C_s is the capacitance of semiconductor and C_{ins} is the capacitance of the insulator layer. Its ideal value at room temperature is about 60 mV/decade. Practically, its value is 70 mV/decade. In this condition, the operating voltage is around 200mv (For achieving a $\frac{I_{on}}{I_{off}}$ of the ratio of at least 1000). By scaling of device size, the operating voltage cannot be decresed because of the above-said condition. The reduction of voltage is possible only by reducing the subthreshold swing to a lower value. Operating voltage reduction ensures minimal power consumption and less size [2].

By utilizing the ferroelectrics voluntary polarization, it is viable to detect negative capacitance [3]. In such a material, P-V graph contains a negative slope region. This region is not stable. So it is necessary to stabilize this negative capacitance effect. For that, a normal positive dielectric capacitor in series can be used with this negative capacitance [4]. Then the net capacitance becomes positive, i.e.,

$$C_{Total} = \frac{1}{C_{ins}} + \frac{1}{C_{Normal}}\ [4] \tag{2}$$

This is the systematic method to make negative capacitance stable [4]. Figure 1 shows a negative capacitance NPN transistor. In this work, the analysis of NPN transistor is performed, because in NPN transistor, electrons are the majority charge

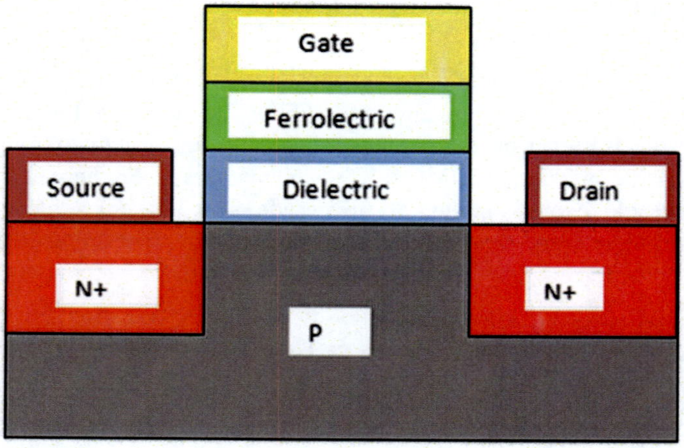

Fig. 1 Negative capacitance field-effect transistor (NCFET) [5]

carriers. While in PNP, majority charge carriers are holes. Electrons are faster than holes. So NPN transistor is faster. That is why analysis of NPN transistor is perfered in this work.

Many research works are progressing in the field of negative capacitance. As per Michel Hoffman and Milan pesic's work, the best method to overcome power dissipation due to scaling is by using ferroelectrics [6]. Polymer and perovskite ferroelectrics which are used to observe NC are inconsistent with semiconductor manufacturing. Recently, HfO_2 based ferroelectrics are discovered, and due to its enhanced scalability, it is suitable for several applications. They connected a capacitor with 18 and 27 nm thickness Gd:HfO_2 layer in series with an external resistor. Also to analyze the transient behavior of polycrystalline ferroelectrics, a multigrain L-K model was built. Victoria Li chin chen and his team called the ferroelectric substituted MOSFET as 'ferroFET' [7]. According to them, ferroFETs are more suitable over normal FETs in low power applications. Asif Islam Khan and Korok Chatterjee found that when an input voltage is applied, the voltage across FE material shows a rapid decay with time [8] and this 'inductance' like appearance can lead to novel applications. Yu Jin Kim comments that FE capacitor is an important part of microelectronic devices. The NC state of FE material corresponds to its highest energy state. In this work, they mentioned a similar reduction in voltage from an epitaxial $BaTiO_3$, a ferroelectric thin film of thickness 150 nm [9]. But their interpretation for this kind of voltage drop is as, due to the flow and consumption of charge developed through reverse domain formation in the ferroelectric capacitor.

Atanu K Saha and Sumeet K Gupta developed a preisach-based circuit compatible model for FE/AFE(Anti-ferroelectric) [10]. They proposed miller model and Landau Khaltnikov (L-K) model for ferroelectrics. A series combination of resistor and capacitor is used in both the models. But the parameters used are different. A deep study of the negative capacitance under the variation of input, resistances and capacitances is performed. They mentioned a general condition to observe negative capacitance in a resistor-ferroelectric stack and ferroelectric-dielectric stack. The same results are verified in the initial sections of this work. Subthreshold swing changes in accordance with temperature. It can be different at different values of temperature. Such a study on the effect of temperature on negative capacitance and subthreshold swing was the topic of Jaesung Jo and Changhwan shin work. They found that the increase in temperature is not good for subthreshold swing. When temperature increases, the subthreshold swing also increases due to the reduction of internal voltage gain with temperature [11].

An application-based work was conducted by Tapas dutta and Girish pahwa. They implemented this NC-based fin field-effect transistor (FinFET) in static random-access memory (SRAM). According to them as the ferroelectric layer thickness (T_{fe}) increases, the readability and write ability of SRAM increases except for very high T_{fe}[12]. At low supply voltage, the noise margin increases with FE thickness. However, at high voltages, the variation of noise margin is just the opposite. This is the major shortcoming of this paper. The improvement in the performance of analog circuits with ferroelectric FET is explained in this paper by Thomas A. Phillips, Todd C. Macleoda and Fat D. Ho. They implemented a differential amplifier circuit with two ferroelectric FETs. Then due to the effect of negative capacitance, the amplifier has four modes of operation [13]. As a result, they can able to seen improvement in the performance of analog circuits due to the negative capacitance property of ferroelectrics. Yang Li and team in their work mentioned that how the negative capacitance transistor's qualities improve the performance of analog circuits such as a differential amplifier, analog switch, resistive voltage divider, current mirrors. They implemented the negative capacitor transistor by substituting gate oxide layer by a ferroelectric material and called this transistor as FeFET [14]. According to them, high current and transconductance of negative capacitance are very suitable for analog applications. Even though the negative capacitance has so many advantages, they point out that there is no improvement in the unity gain of FeFETs compared to MOSFETs. Qianqian Huang [15], presented that NC is occurred in low frequencies, due to the phenomenon of dynamic polarization switching. Then the FE is added at the gate of tunnel FETs(TFET)and they called the structure as NC-TFETs which is suitable for applications demanding low frequency and low power. However, the major limitation of their proposed structure is that it is not suitable for high-speed applications. That means its delay is more.

The upcoming sections of this paper describe the modeling of negative capacitance by using Landau Khaltnikov (L-K) equations. Then the results obtained are discussed after analyzing the negative capacitance field-effect transistor (NCFET) to evaluate the influence of thickness, α, β and γ of ferroelectrics on the subthreshold swing. Finally, various digital circuits are built by implementing those NCFETs and verified the improvement in performance.

2 Modeling of Negative Capacitance

The properties of ferroelectrics are needed to be observed and analyzed to study the effect of negative capacitance in sub-20 nm CMOS technology. These properties include the ferroelectric capacitance, charge accumulation, ferroelectric voltage, current through the ferroelectric capacitor, thickness of the ferroelectric, area, polarization behavior with respect to the electric field, etc. The inter-relation and the possible range of the above-mentioned parameters can be characterized mainly using Landau Khalatnikov (L-K) model. Landau theory was developed in the 1930s as a model to describe the equilibrium behavior of a system near a phase transition using an analysis based on symmetry. It effectively provides a link between the microscopic models and what is observed macroscopically. This theory is particularly appropriate for homogeneous and bulk ferroelectric materials because it relies on a special averaging of the local fluctuations. The basic technique to model the performance of NCFETs has been including an extra insulating layer driven by Landau Khaltnikov (L-K) equation [7],

$$G = \alpha P^2 + \beta P^4 + \gamma P^6 - \text{EP [1]} \tag{3}$$

where E is the electric field applied across ferroelectric material, G is the energy of ferroelectric material, P is the polarization. α, β and γ are the three ferroelectric coefficients. Generally, α and β can be either +ve or −ve, but γ is always greater than zero to ensure stability. Dynamics of G is given by, $\delta \frac{dP}{dt} = -\frac{\partial G}{\partial P}$ [1]. The polarization damping factor is denoted by δ. In the steady state, $\frac{dP}{dt} = 0$. Hence,

$$E = 2\alpha P + 4\beta P^3 + 6\gamma P^5 = \frac{V_{fe}}{T_{fe}} \text{ [1]} \tag{4}$$

where V_{fe} is the voltage across ferroelectrics and T_{fe} is the ferroelectric thickness.
Therefore,

$$V_{fe} = E * T_{fe} \tag{5}$$

The standard Verilog A model of the transistor has been proposed in [16]. It is implemented by using Cadence tool. In this work, at first, the normal transistor characteristics are analyzed. Then the mentioned model is modified by employing an additional ferroelectric layer in the gate stack of a transistor. This is possible by using an additional term V_{fe} as in Eq. (5). Then the analysis of this ferroelectric substituted transistor, i.e., the negative capacitance transistor is performed.

2.1 Experimental Settings: Cadence Tool

Modeling of the transistor having a gate length of 45 nm is done in Cadence tool using Verilog A language. Cadence is a design platform and can be used for simulation. The simulation tool is started directly from the schematic editor and all the necessary netlists describing the design will be created. The simulation also requires a test bench, which is also a schematic. All the signal sources and power supply used for the simulation are defined in this test bench. The simulation results obtained are plotted using Virtuoso visualization and analysis XL tool. It is a tool for visualizing analog signals and mixed signals. Verilog A is a hardware description language. The signals obtained in the virtuoso visualization and analysis XL tool are then fed to a calculator, and by using this calculator, analysis of the obtained results can be performed. Mainly, DC analysis and transient analysis of the transistor are performed. For DC analysis, a voltage ranging from 0 to 1 V with a step of 0.01 V is applied at the gate terminal of the transistor, and for transient analysis, a pulse having delay of 1 ns, rise time 10 ns, fall time 10 ns, width 500 ns and period 1000 ns are applied across the gate.

3 Analysis of NCFET

Normal FETs consist of an insulator as a gate oxide. Whereas in NCFET, insulator is substituted with a ferroelectric material. A Verilog model of field-effect transistor, having a gate length of 45 nm, is built in Cadence tool and performed a various analysis of transistor with and without ferroelectric layer.

The transfer characteristics of normal transistor and transistor with ferroelectrics are compared in Fig. 2. From the figure, it can be seen that the slope of transfer characteristics of the transistor with ferroelectrics is high. This means that the subthreshold swing (inverse of the slope of transfer characteristics), being $S = \frac{dV_{GS}}{d(\log_{10} I_{DS})}$ (Where V_{GS} = Gate-to-source voltage and I_{DS} = Drain-to-source current) is low for this transistor which makes use of ferroelectric material. From the analysis conducted, it can be observed that the value of subthreshold swing is around 39.6 mv/decade which is less than 60 mv/decade. For any device, the off current (I_{off}) should be minimum and on current (I_{on}) should be maximum. Off current means, the current when the input voltage is zero and on current is current when there is an input voltage.

In the figure, for a normal transistor, off current is not a negligible value. It is considerably large compared to the ferroelectric-based transistors. Off current of normal FET is nearly 100 times more as that of NCFET. When the applied voltage becomes maximum, the I_{DS} value of normal transistor is less when compared with FE-based transistors. By attaining a small off current and large on current, the ratio $\frac{I_{on}}{I_{off}}$ can be maintained at a high enough value. So by using NCFET, one of the biggest challenges in new technologies can be resolved.

3.1 *Effect of Ferroelectric Thickness*

In the case of capacitor, capacitance, $C = \frac{A\epsilon}{d}$ (Where A = Area, ε = Permittivity, d = Insulator thickness), i.e., when the thickness of insulator increases, capacitance decreases. By doing transient analysis of transistors with different thickness of the ferroelectric layer (T_{fe}), the voltage drop across FE, i.e., $|V_{fe}|$ increases with an increase in T_{fe}. Then the net voltage across the gate, V_{gs} becomes more than that of input voltage, V_{App}, i.e., $V_{gs} = V_{App} + |V_{fe}|$. Transfer characteristics also become steeper with an increase of FE thickness. So a low value of off current can be achieved. The variation of off current I_{off} and the voltage drop across ferroelectrics with ferroelectric thickness, $|V_{fe}|$ is shown in Figs. 3 and 4. That means it is possible to achieve a better $\frac{I_{on}}{I_{off}}$ ratio with an increase of FE thickness. Here. the ratio obtained is maximum when T_{fe} = 12 nm. By increasing thickness beyond this, the graph becomes distorted.

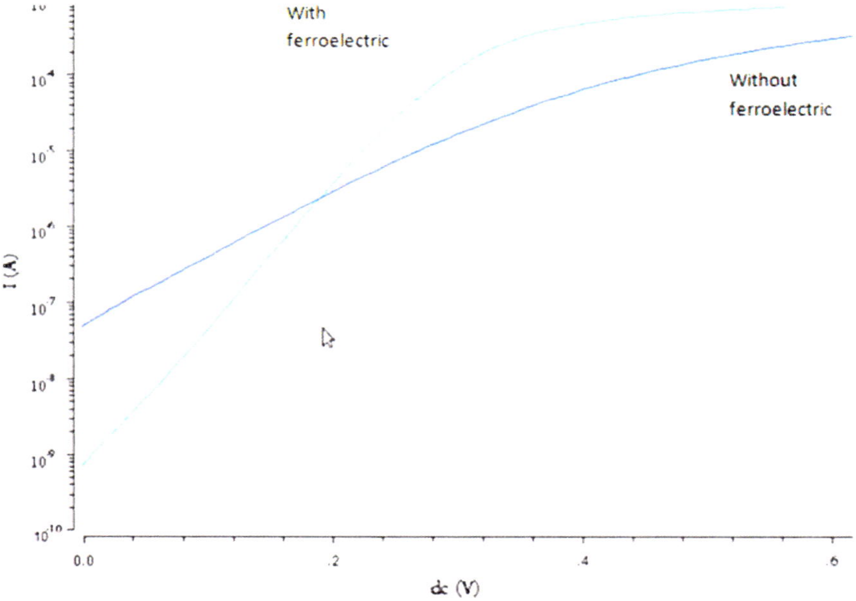

Fig. 2 Transfer characteristics of FET with and without ferroelectric (log I_{ds} v/s V_{gs} curve)

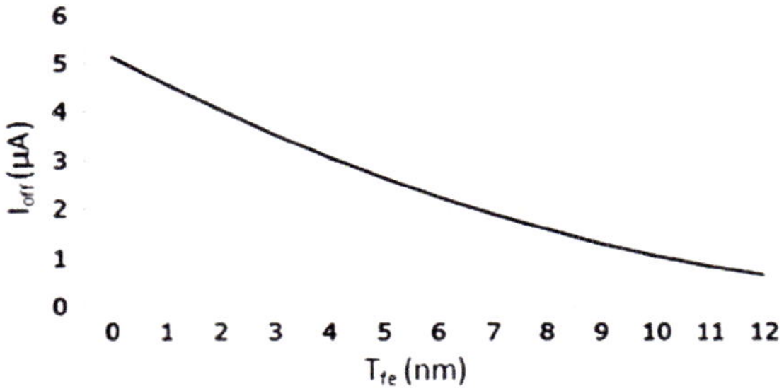

Fig. 3 Variation of off current, I_{off} with respect to FE thickness

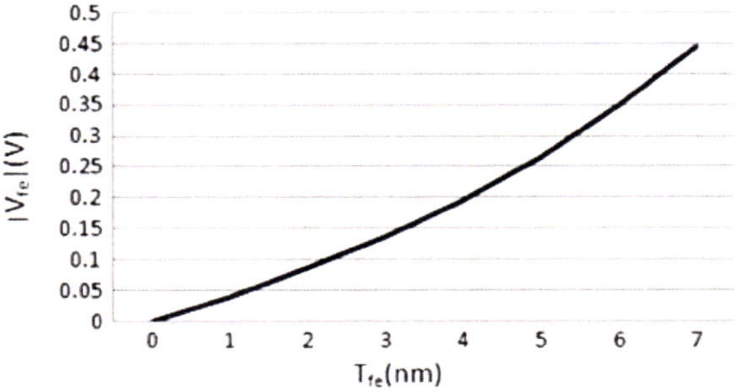

Fig. 4 Variation of $|V_{fe}|$ with respect to FE thickness

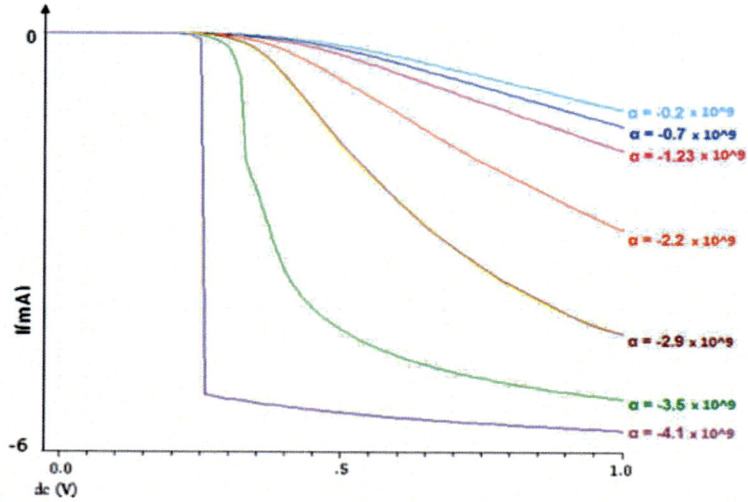

Fig. 5 Comparison of transfer characteristics with different values of α

3.2 *Effect of α*

It is an important and primary parameter while selecting a ferroelectric material. When α is positive, it is impossible to observe negative capacitance. As per Fig. 5, it can be seen that the slope of transfer characteristics becomes steeper with large negative values of α. Thereby a low value of off current and high value of on current can be achieved. The transfer characteristic curves approach the desired response with a decrease in α. When $\alpha = -4.1 \times 10^9$, the transfer characteristics with maximum slope are obtained. Also during transient analysis, the voltage drop across ferroelectrics $|V_{fe}|$ increases with a decrease in α.

3.3 *Effect of β*

Negative capacitance can be seen with positive values of β. As that of α, β is not such an important parameter during the selection of a ferroelectric material. From the analysis conducted, it is observed that β can take a maximum value of the order 10^{14}. As per the Fig. 6, all the curves obtained for values of β less than 10^{10} were found to overlap with themselves, which means at low values of β, there is no specific change in the slope of transfer characteristics. A slight variation in curves obtained only by varying β to a large extent. Here, a transfer characteristics with maximum slope can be attained when $\beta = 3.8 \times 10^1$.

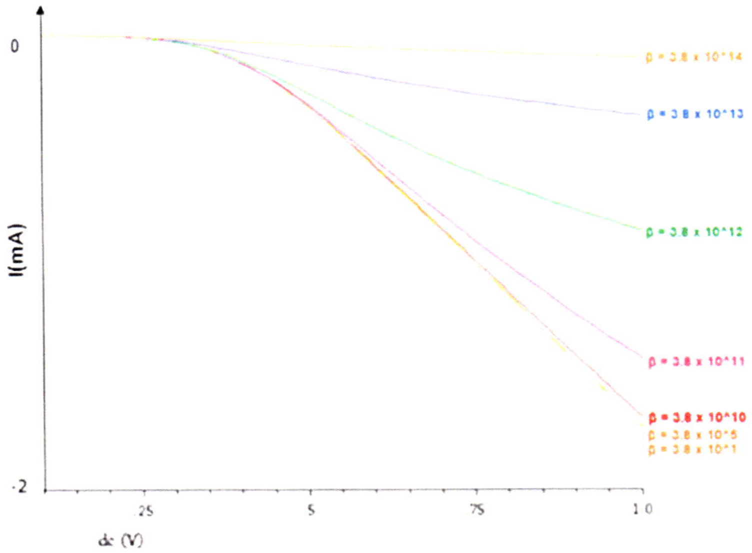

Fig. 6 Comparison of transfer characteristics with different values of β

3.4 *Effect of γ*

Compared to α and β, γ does not affect in determining properties of ferroelectrics. It can take any values, but not less than zero due to stability reasons.

4 Implementation of NCFET in Circuits

From the analysis, it can be seen that when $T_{fe} = 12$ nm, $\alpha = -4.1 \times 10^9$ and $\beta = 3.8 \times 10^1$, the slope of transfer characteristics is maximum (subthreshold swing is minimum). By using these parameters, a negative capacitance transistor (NCFET) is designed (both PMOS and NMOS) and this NCFET is used for constructing different digital circuits. A CMOS inverter is designed by cascading NMOS and PMOS together. Figure 7 compares the transfer characteristics of NCFET-based inverters and normal transistor-based inverters. The output curve of the inverter with ferroelectric transistors exhibits a steeper transition when compared to the one without ferroelectrics. Thus, faster switching is ensured when ferroelectric is used as the insulator. This indicates that the inverters which consist of transistors with ferroelectric respond faster to the input. At the same time, it gives a high current amplification. The successful implementation of the inverter circuit with ferroelectric MOSFET provides substantial support to the fact that it is possible to implement

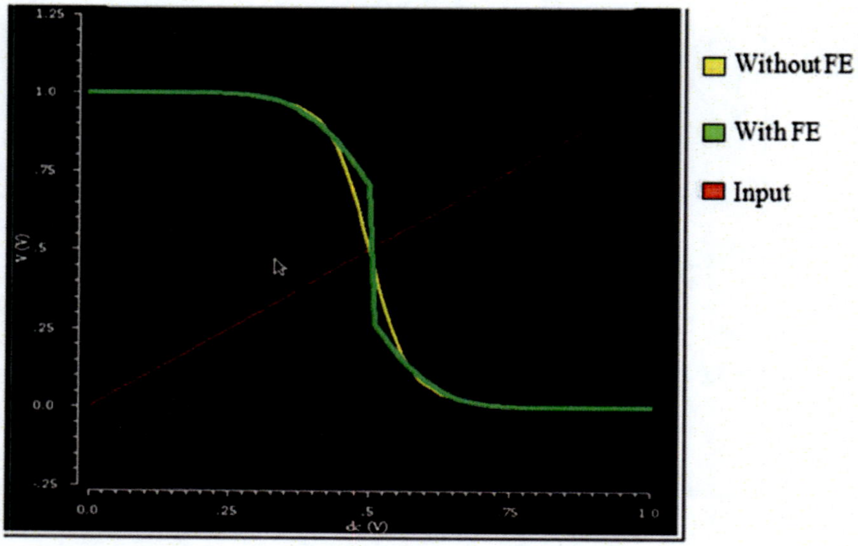

Fig. 7 Inverter (DC analysis): with and without ferroelectric

all circuits under negative capacitance effects, which provides an internal voltage gain and current amplification and hence the reduction in the operating voltage.

Using CMOS technology, digital circuits such as NAND, NOR, XOR, XNOR gates, and full adder are designed and compared its delay and dissipated power. Power and delay calculation are done by, passing the obtained voltage and current plots to the calculator in the schematic window and by doing the necessary arithmetic operations. Table 1 shows the delay of circuits with and without NC transistors. By using NC transistors, circuits can achieve faster switching (less delay).

Here, the power consumed is calculated by circuits using the calculator. It is calculated as by multiplying current and voltage plots and then integrating the product over a time period. Table 2 shows the total power consumption of circuits with and without the use of negative capacitance transistors and the static and dynamic power

Table 1 Delay comparison in circuits with and without NC transistors

Circuits	With NC (ns)	Without NC (ns)
Inverter	1.729	2.186
NAND Gate	1.68	2.2478
NOR Gate	1.213	2.285
XOR Gate	0.7	1.51
XNOR Gate	1.699	3.115
Full adder	1.787	4.74

Table 2 Power comparison in circuits with and without NC transistors

Circuits	With NC (Total power) (W)	Without NC (Total power) (W)	Static power of circuits with NC (W)	Dynamic power of circuits with NC (W)
Inverter	0.247 e−6	10.28 e−6	9.8 e−12	0.242 e−6
NAND Gate	0.894 e−6	10.25 e−6	7.52 e−12	0.891 e−6
NOR Gate	1.762 e−6	20.54 e−6	9.32 e−12	1.72 e−6
XOR Gate	3.221 e−6	16.04 e−6	13.41 e−12	3.20 e−6
XNOR Gate	1.794 e−6	16.06 e−6	9.52 e−12	1.786 e−6
Full adder	6.124 e−6	46.8 e−6	24.92 e−12	6.108 e−6

components of negative capacitance transistor-based circuits. Here, the mentioned power is the total power which is the sum of static and dynamic power components.

Static power is the consumed power when the output is a DC value or constant. Dynamic power consumption is the amount of power when there is a transition in the output. For any circuit, dynamic power is larger compared to static power. Here, by employing negative capacitance, static power in the range of pW (as in Table 2) can be achieved and the dynamic power in μW range. These values are very less compared to normal transistor-based circuits.

5 Conclusion

To overcome Boltzmann limit, and to evolve low power and fast electronic devices, steep-slope transistors are innovated and attained more recognition around the globe over the last decade. Out of such devices, negative capacitance field-effect transistors (NCFETs) has a minimum subthreshold swing (SS = 39.6 mV/decade at room temperature) along with the large value of on current and low off current. To detect the negative capacitance effect, ferroelectric materials are used due to their unique properties. The slope of transfer characteristic curve and thereby subthreshold swing depends upon the thickness of the ferroelectric layer (T_{fe}) and the ferroelectric coefficients α and β (When $T_{fe} = 12$ nm, $\alpha = -4.1 \times 10^9$ and $\beta = 3.8 \times 10^1$ maximum slope is obtained). Among the three coefficients of ferroelectrics α, β and γ, selection of α and β is important. Circuits with negative capacitance transistors exhibit superior performance over normal circuits. The switching speed of circuits which use negative capacitance transistor is about double when compared with circuits with normal transistors. The power consumed by normal circuits is 20 or 30 times more compared with NC transistor-based circuits. That means 'negative capacitance' is an ideal solution to resolve the major problems faced by CMOS technology.

References

1. Sayeef Salahuddin and Supriyo Datta.: Use of negative capacitance to provide voltage amplification for low power nanoscale devices. In: Nano letters, Vol.8,No.2, 405-410, (2008)
2. Abdellatif Bellaouar, Mohamed Elmasry.: Low-voltage low–power CMOS vlsi circuit design. In: Low Power Digital Vlsi Design: Circuits and Systems, Springer US, 978-1-4615-2355-0, (1995)
3. Changyong Oh, Amit Tewari, Kyungkwan Kim, Ulayil Sajesh Kumar, Changhwan Shin, Minho Ahn and Sanghun Jeon.: Comprehensive study of high pressure annealing on the ferroelectric properties of $Hf_{0.5}Zr_{0.5}O_2$ thin films. In: Nanotechnology, Volume 30, Number 50, 1 October (2019)
4. Asif I. Khan, Michael Hoffmann, Korok Chatterjee, Zhongyuan Lu, Ruijuan Xu, Claudy Serrao, Samuel Smith, Lane W. Martin, Chenming Hu, Ramamoorthy Ramesh and Sayeef Salahuddin.: Differential voltage amplification fron ferroelectric negative capacitance. In: Applıed Physıcs Letters 111, 253501 (2017)
5. Muhammed A. Alam, Mengwei Si, and Peide D. Ye.: Acritical review of recent progress on negative capacitance field-effect transistors. In: Appl. Phys. Lett., volume 114, 090401, (2019)
6. M.Hoffmann, M. Pesic, K. Chatterjee, A. I. Khan, S. Salahuddin, S. Slesazeck, U. Schroeder, and T. Mikolajick.: Direct observation of negative capacitance in polycrystalline ferroelectric HfO_2. In: Advanced Functional Materials, vol. 26, no. 47, pp. 8643–8649, (2016)
7. V. Chen.: Ferroelectric Field Effect Transistors: Negative Capacitance for Efficient Switching. In: PhD thesis, Pennsylvania State University (2016)
8. Asif Islam Khan, Korok Chatterjee, Brian Wang, Steven Drapcho, Long You, Claudy Serrao, aidur Rahman Bakaul, Ramamoorthy Ramesh, Sayeef Salahuddin.: Negative Capacitance in a ferroelectric Capacitor. In: Nature Materials volume 14, pages182–186, (2015)
9. Tu Jin Kim, Hyeon Woo Park, Seung Dam Hyun, Han Joon Kim, Keum Do Kim, Young Hwan Lee, Taehwan Moon, Yong Bin Lee, Min Hyuk Park, and Cheol Seong Hwang.: Voltage Drop in a ferroelectric Single Layer Capacitor by Retarded Domain Nucleation. In Nano Letters, 17, 77967802, (2017)
10. Atanu K. Saha, Suman Datta, and Sumeet K. Gupta.: Negative capacitance in resistor ferroelectric and ferroelectric-dielectric networks: Apparent or intrinsic?. In: Journal of Applied Physics, 105102,, Volume 123, Issue 10, (2018)
11. Jaesung Jo and Changhwan Shin.: Impact of temperature on negative capacitance field effect transistor. In: Electronics letters 8th January, Vol. 51 No. 1 pp. 106–108, (2015)
12. T. Dutta, G. Pahwa, A. R. Trivedi, S. Sinha, A. Agarwal, and Y. S. Chauhan.: Performance Evaluation of 7 nm Node Negative Capacitance FinFET based SRAM. In: IEEE Electron Device Letters, Volume: 38, Issue: 8, Aug., (2017)
13. Thomas A. Phillips, Todd C. Macleoda and Fat D. Ho.: Ferroelectric Field-Effect Transistor Differential Amplifier Circuit Analysis. In: Integrated Ferroelectrics: An International Journal, 105:1, 107–117, (2010)
14. Yang Li, Yuye Kang, and Xiao Gong.: Evaluation of Negative Capacitance Ferroelectric MOSFET for Analog Circuit Applications. In: IEEE Transactions On Electron Devices, Volume: 64, Issue: 10, Oct. (2017)
15. Qianqian Huang, Huimin Wang, Yang Zhao, Mengxuan Yang, Zhongxin Liang, Kunkun Zhu and Ru Huang.: New Understandıng Of Negatıve Capacıtance Devıces For Low-Power Logıc Applıcatıons. In: 2019 China Semiconductor Technology International Conference (CSTIC), 04 July (2019)
16. Dimitri A. Antoniadis, Massachusetts Institute of Technology (MIT).: VerilogA for virtual-source (VS) based self-consistent transport/capacitance model for Si MOSFET. May 27 (2013)

A Verification of Pattern-Oriented Healthcare System Using CPN Tool

U. Prabu, R. Sriram, P. Ravisasthiri, and N. Malarvizhi

Abstract The Pervasive computing otherwise called Ubiquitous computing is one of the fixed technologies which engage computing with the transmission ability and is flawlessly incorporated by means of the client. The model of pervasive computing consists of Artificial Intelligence, Context-aware Computing, and Human Computing Interface that forms computing towards the development of overall places. The healthcare plus pervasive computing is collectively well-known as Pervasive Healthcare. It concentrates on these technologies and additional ideas that combine healthcare further flawlessly daily life. A framework is proposed for diagnosing the patient and to claim the medical insurance policy for the patient through the pervasive system using architectural patterns. Various architectural patterns are also proposed such as Diagnosis Pattern, Decision Pattern and Reimbursement Pattern. These patterns are compactable; it can be used in other services by modifying according to the use. The structure is then evaluated and analyzed towards the existent-life framework through the Colored Petri net tool.

Keywords Pattern · Reimbursement pattern · Pervasive healthcare · CPN

U. Prabu (✉)
Department of Computer Science and Engineering, Velagapudi Ramakrishna Siddhartha Engineering College, Vijayawada, India
e-mail: uprabu28@gmail.com

R. Sriram
Department of Medical Informatics, Mahatma Gandhi Medical College and Research Institute, Puducherry, India
e-mail: mail4ramz@gmail.com

P. Ravisasthiri
Department of Information Technology, Christ College of Engineering and Technology, Puducherry, India
e-mail: ravisasthiri.p@gmail.com

N. Malarvizhi
Department of Computer Science and Engineering, IFET College of Engineering, Villupuram, India
e-mail: nmalarvizhi16@gmail.com

P. Karuppusamy et al. (eds.), *Sustainable Communication Networks and Application*, Lecture Notes on Data Engineering and Communications Technologies 55,
https://doi.org/10.1007/978-981-15-8677-4_33

1 Introduction

The perception of pervasive computing is in software engineering along with computer science in which computing is ready to emerge in all places [1]. The pervasive computing can take place with-in several devices, location and several formats. The idea behind pervasive computing is availability and invisibility. The initial thoughts on Pervasive computing were considered well ahead of its time given that the lack of hardware, software as well as network to support this vision. Pervasive computing is, therefore, ubiquitous [2], enabling everyday objects to become smarter and interactive such as refrigerators that can create grocery lists or buildings that can adjust temperature and lighting according to the weather and number of people in the room. Pervasive computing will, therefore, revolutionize the way humans interact with the world around them.

The vision of Ubiquitous computing is to facilitate computer-based services to be accessible ubiquitously [3–5], to sustain a spontaneous human practice. But until now, this computing is invisible to the customer. This recent concept is furthermore described as pervasive computing, every-ware, or else ambient intelligence.

The mobile computing purpose of any time, anywhere connectedness is extensive towards all time and everywhere via the integrated Pervasive holdup technologies like interoperability, invisibility, scalability, and elegance. Computers mainly as machines facilitate users to run programs in an effective environment that the user performs a job and depart when users come to an end. Pervasive computing [6] believe in the whole diverse visualization. A machine can exist like an entry into an application-data space, not a storehouse of custom software that a customer has to run.

The skill is moving further than the personal computer to daily de-vice through fixed technology and connectedness because computing devices develop into gradually portable, new influential and exist universally. Pervasive Computing is increasing quickly to the embedding technologies in daily things so that the communication became more easily with the equipment's that are fully linked and persistently available. Context-aware computing [7] is part of pervasive computing. Context-aware computing can sense their atmosphere and respond according to their behaviour. Numerous Research is completed and continuing in a variety of fields that are extra smart and attractive. Many research was going on in healthcare using pervasive computing so it combined and called as pervasive healthcare, even if the healthcare facility is accessible all over the place. It concentrates on these technologies and additional ideas that combine health care further flawlessly daily life.

2 Related Work

According to the threat in computer systems studies posed with the promising field of pervasive computing as well as a primary study of this pervasive computing to its predecessor fields [1]. Next, it identifies four novel research fields: efficient use of smart spaces, masking uneven conditioning, localized scalability, invisibility and masking uneven conditioning. It then provides two real pervasive computing fields to identify a solution, which was missing in many other fields.

The challenges of the pervasive computing which has been recognized are specified as follows. They are

- Client-objective
- Fake Foraging
- Adaptation plan
- Advanced Energy managing
- Client depth
- Context Awareness
- Paired Proactivity and simplicity
- Confidentiality and Trust.

The majority of the key application fields intended for pervasive technologies are mainly in healthcare, as well as hold up for autonomous life also wellness plus disease managing. Pervasive healthcare [8] can be defined as of two views (1) applying pervasive computing or ubiquitous computing, intelligence technologies for healthcare, proactive computing, healthiness, and wellness board (2) availability of health care all over, all-time pervasively.

Ubiquitous message passing [9] placed on mobile phone networks, wireless networks such as (LAN) local area networks, and former wireless technologies make feasible everywhere, all-time transmits and contact with all sorts of data as well as health-based information, measurement of data otherwise medical information.

The architecture of context-aware is the level that includes a sensor, every equipment and software to facilitate perspective information. The structure in healthcare services gives the circumstance of a patient like electronic patient verification, monitoring and lab reports stored in the database. The context layer provider [10] takes concern of place, time, location and other detailed framework information. All these data are added towards the semantic reasoning layer, in knowledge base every important context data into a proper context model is acknowledged as ontologies [11].

Context-Aware Real-Time Assistance System (CARA) [12] is planned to collect information with given context-aware data and identify the conclusion within the healthcare facility. To attain the context-aware, hybrid reasoning structure is given via case-based reasoning along with fuzzy rule-based reasoning. The information approaching from the sensors is processed furthermore, included among the context information, and then produces the context designed for structure case quires as well as fuzzy sets. Case-based reasoning will get back reuse, reuse and maintain information to execute abnormality recognition and home computerization. The fuzzy rule [13] is used in general, higher-level context, such as the medical state and towards identifying the existing condition of the customer. The fuzzy output is used to dynamically weight the case retrieval. The case retrieval is completed via the retrieving parallel case along with fuzzy output. In a novel case it activates the alarm, collects information, and stores the data of the case.

3 Research Contribution

The pervasive healthcare patterns are identified and proposed in the framework. The identified patterns are modelled and verified using CPN tool. The architectural pattern [14] such as diagnosis pattern, decision support pattern and health insurance pattern is proposed, along with it there are certain other patterns such as register patter, notification pattern, check-up pattern, decision pattern, admission pattern, request pattern, verification pattern, payment pattern, and reimbursement pattern which are discussed. These patterns [15] provide simple ways to get register and diagnosis from the doctor. It helps the doctor to be accurate in caring for the patient by taking the correct decision. It also gives an efficient way to claim the health insurance policy by registering the policy id when a patient gets admitted. The claiming is done easily for both network and non-network hospitals so that the patient can have all the benefits of the policy anywhere.

4 System Design

The proposed system gives the quality and accuracy of caring by the diagnosis and decision support pattern [15]. This also gives an efficient way to claim the health insurance policy by the insurance pattern proposed. Along with these, the registered pattern is used to register the patient details effectively and these are used to notify doctor and insurance issuer with simple procedures. The healthcare environment structure was authenticated and analyzed by CPN tools. The scalability and feasibility were achieved in this framework using this architectural design.

The architecture diagram [16] in Fig. 1, gives the flow of the framework and the details of the pattern proposed. The framework represents three modules or patterns and it is explained below.

The work consists of three modules or patterns and they are,

- Diagnosis Pattern
- Decision Support Pattern
- Health Insurance Pattern.

4.1 Diagnosis Pattern

The diagnosis pattern [17] consists of four patterns with it and they are,

- Register Pattern
- Authentication Pattern
- Notification Pattern and
- Check-up Pattern.

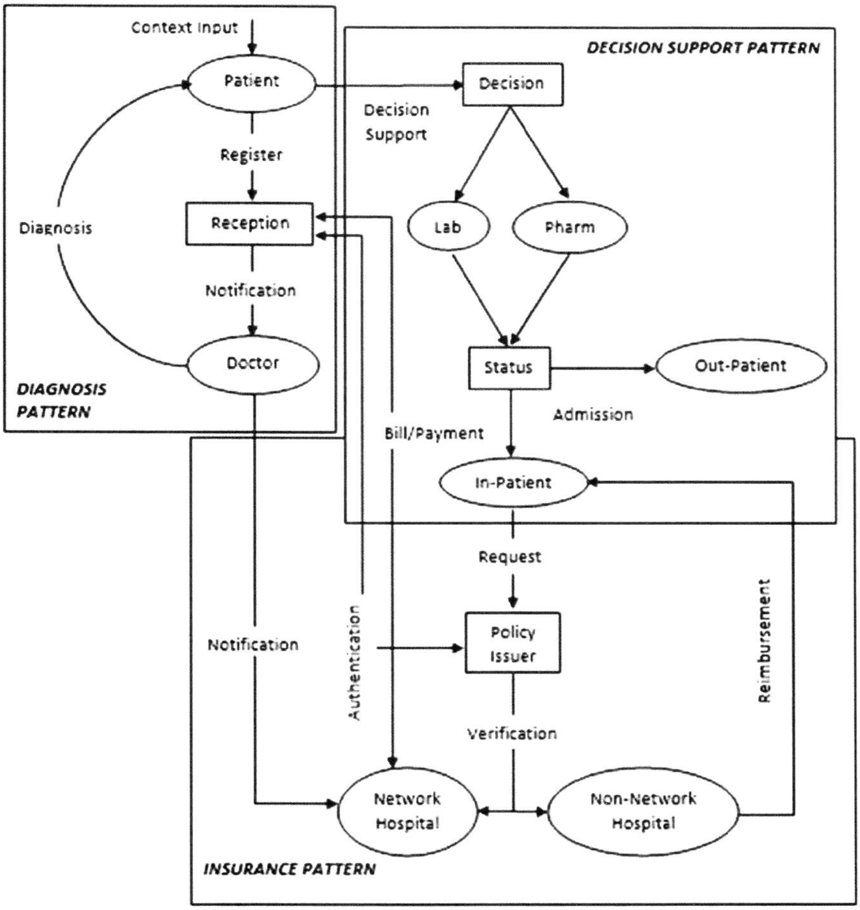

Fig. 1 System design for pattern-oriented healthcare system

The **Register Pattern** registers the context of the patient to the reception and where it is stored. This information's are filled or provided when the patient enters or request for the laboratory testing or to get the medication from the pharmacy or requesting to the health insurance. The **Authentication Pattern** authenticates the patient who gives the policy id with the relevant policy issuer. If the policy id is valid, then the patient will be provided with the details to claim the health insurance directly through the hospital and make the claiming process simple for the patient. If it is not valid the patient is asked to enter the valid id or the patient who does not have policy id, they cannot able to enjoy the benefits of the policy.

The **Notification Pattern** gives the details of the patient to the doctor. The doctor will get notified about all the treatment given to the patient through this pattern. This pattern helps the doctor to keep track of the patient's complete report and provides necessary details to the policy issuer for claiming the policy if the patient provides the policy id while registration. The **Check-up Pattern** provides the details of all problems and symptoms. It also gives the precaution and how to care and cure. In this pattern [18] the doctor diagnosis the patient by getting the details such as symptoms and problems from the patient and doctor gives the necessary treatments and takes the decision through the test from laboratory or prescribes medication to the patient collected from the pharmacy.

The context of the patient is registered to the reception. The patient's contexts such as patient name, address, phone number, policy id, etc., are registered to the reception with the help of the registered pattern. This information is stored and authenticated with the policy issuer, if the patient provides the policy id, through the authentication pattern. The reception notifies the patient details to the doctor for a further check-up, through the notification pattern. After getting notified the doctor diagnosis patient by the context. The context given while a diagnosis is the symptoms and problems faced by the patient, these are handled by the check-up pattern.

4.2 Decision Support Pattern

The decision support pattern consists of two patterns with it and they are,

- Decision Pattern and
- Admission Pattern.

The **Decision Pattern** gets the details from the diagnosis pattern and helps the doctor to decide the disease and seriousness of the patient. The decision is taken with the help of the tests in the laboratory, with the results, the medication is prescribed or if the doctor can identify the problem without the help of the test, then he prescribes medication directly to the patient. The **Admission Pattern** gives the condition to make the patient admit or not with the status of the patient given by the doctor with the help of a laboratory test. If the patient has a moderate level of the problem which can be cured by self or with medication, the pattern chooses the patient to be out-patient. If the patient suffers from a serious problem, then the pattern decides to make the patient admit to the hospital for further treatment.

In this pattern [19], the diagnosis report of the patient provided by the doctor is given to take the decision or to confirm the problem of the patient. The decision is taken by running some necessary tests from the laboratory or prescribing the medication to the patient with the help of a decision pattern. The status of the patient is noted by taking tests from the laboratory and medications from the pharmacy. If the patient is suffering a moderate level of the problem, then he/she will be the out-patient. Otherwise, the patient will be admitted and be an in-patient in the hospital, due to a serious problem. The admission pattern provides the conditions and it helps to decide to make the patient in/out with the help of decision patterns.

4.3 Health Insurance Pattern

The health insurance pattern consists of three patterns with it and they are

- Request Pattern
- Verification Pattern
- Bill/Payment Pattern and
- Reimbursement Pattern.

The **Request Pattern** helps the admitted patient to request claiming the insurance policy. The policy issuers will be noted when the patient is admitted to the hospital to claim the policy which is subscribed by the patient, to get the benefits. This pattern helps the patient to give the request to policy issuer in an efficient way. The **Verification Pattern** verifies the hospital where the policyholder is admitted and requested for the policy benefits. The pattern verifies whether the hospital is the network or non-network hospital. The hospital which has tie-up with the policy company is known as Network hospital and the hospital which does not have to tie up is known to be Non-Network hospital. If it is a network hospital, the hospital provides all the necessary details to the policy issuer to make the policy benefits for the in-patient or policyholder. If it is a non-network hospital the hospital will provide the necessary details to both in-patient and policy issuer, so the in-patient can pay the treatment bill from the pocket and claim it afterwards as reimbursement.

The **Bill/Payment Pattern** gives the details of all treatment and expense of the patient from the date of admission. The reception has the record of all caring done to the patient and provides the bill at the time of discharge. If the patient has the policy and the hospital in which he is admitted is a network hospital, then the policy issuer will help with the payment. Otherwise, the patient will pay it from the pocket and can claim it from policy issuer as reimbursement. The **Reimbursement Pattern** gives the in-patient policy benefits after discharging from the hospital. The details of the in-patient and bill payment slip which is paid by the patient from the pocket are also provided by the hospital to the policy issuer. The patient can get the benefits from the policy which is subscribed after getting a discharge. The claimed bill will be added to the policyholder account which has been given at the time of subscribing the policy.

In this pattern [20], when the patient got admitted he/she should request to the policy issuer to provide the benefits of the policy. This request is done by the patient with the help of the request pattern and it will be simple since the patient has registered the policy id in the reception. When the patient requests the policy issuer, they will verify that the hospital where the policyholder is admitted is a network or non-network hospital with the help of verification pattern. The bills are provided by the reception from the date of admission until the date of discharge to the patient or policy issuer. It is done with the help of the bill/payment pattern. If it is a network hospital the details are given to the policy issuer directly by hospital and patient can enjoy the benefits of the policy. If it is non-network hospital the patient pays the bill from his/her pocket and the necessary details will be passed to the policy issuer to make the reimbursement to the patient with the help of the reimbursement pattern.

5 Verification of Pattern-Oriented Healthcare System

Coloured Petri Nets (CPNs) [21] mainly focus on consistency, transmission, and synchronization; it is a terminology for the de-signing and verification of systems. Colored Petri Nets (CPNs) are the combination of the discrete-event modelling language and functional programming language Standard ML. Petri nets provide the fundamental for designing consistency, transmission, synchronization and implement the base of the graphical notation. Standard ML implements the fundamental for the interpretation of data types, defining data manipulation, and for designing compact frameworks in prospective models.

CPN language [22] forms the modelling and execution of events, it consists of time perception and time taken in interpreting the events and CPN language is available to set up models as a prescribed module. CPN Tools is a technical stability tool for designing and evaluating CPN designs. CPN tools help to analyze the performance and simulation of system models, authentication is done by a technique of model analyzing, and state-space process, and it manipulates simulation analysis based on performance.

CPN models are prescribed in the logic to facilitate the syntax plus semantics, CPN modelling language has a mathematical description. They can validate system functions, confirm that required functions are rewarded, or definable undesired functions are assured to be there not present. The set of state-space methods supports the verification of system properties. The fundamental design principle of state spaces is to calculate every reachable state along with state modification of the CPN model, which represents it as a directed graph everywhere; nodes signify states and arcs signify taking place events. State spaces can be formed completely mechanically. The state space is possible to respond with a huge set of confirmation questions regarding the performance of the system methods such as lack of deadlocks, (1) the chance of constantly being capable to attain a specified state, (2) assured delivery of a set state. CP-nets were able to apply towards timed CP-nets. For this reason, timed CP-nets are used for the validation of efficient exact-ness of the system.

5.1 *Verification Design*

The verification design is done by creating a scenario for the diagnosis pattern, decision support pattern and health insurance pattern. These patterns are verified with the help of CPN tool. The diagnosis pattern is done by giving the place and transitions, in this design the place in diagnosis pattern are patient, reception, policy issuer and doctor, the transitions are register, notification, authentication, check-up.

The patient details are given as input and the patient report is obtained as output by check-up transition. In the decision support pattern, the report which is generated by the check-up is given as input to decide to the status of the patient. The places in this pattern are patient, decision, status and in/out-patient. Likewise, the health insurance pattern is also drawn as a scenario which has the in-patient, policy issuer, reception and network/non-network hospitals are the places, and the transactions are request, verification, and reimbursement. In this pattern, the patient request for the policy benefits the policy issuer and gets the payment through the reimbursement. Thus, the system is designed with three scenarios and verifies the pattern explicitly to get the quality of care and to claim the health insurance easily in pervasive healthcare.

5.2 *Verification Result*

The verification results are given separately for each scenario which has been done using the CPN tool. There are three scenarios; they are diagnosis pattern, decision support pattern and health insurance pattern. The diagnosis pattern is done by giving the input as patient details such as patient name, patient id, policy id, age, and registers to the reception. The reception gets the input and authenticates the policy id with the policy issuer and notifies the doctor about the patient. The doctor check-up the patient with the check-up pattern by getting the symptoms and problems from the patient and gives the patient report, to confirm the care given is accurate. The decision support pattern gets the input from the patient as the patient report to get the decision with the help of laboratory test and pharmacy. The laboratory test provides status by checking the patient as normal or abnormal. If the patient is normal, he takes the medication and leaves as an outpatient. Otherwise, the patient is admitted as an inpatient in the hospital. The insurance pattern gets the request from the patient as policy id, patient id and hospital name. The policy issuer verifies the hospital and gives a network or non-network hospital. If its network the payment is done to the recipient directly by policy issuer, otherwise it is done through reimbursement.

5.3 Result of Diagnosis Pattern

```
CPN Tools state space report for:
/cygdrive/C/Users/Sriram/Desktop/HEALTH CARE PATTERN/DIAGNOSIS
PATTERN.cpn
Report generated: Fri Jan 24 14:45:26 2020
Statistics
------------------------------------------------
State Space     Nodes:  15251   Arcs:   57082
Secs: 300       Status: Partial
Scc Graph       Nodes:  11818   Arcs:   44671
Secs:   91
Boundedness Properties
------------------------------------------------
Best Integer Bounds              Upper      Lower
DIAGNOSIS'Doctor 1               4          0
DIAGNOSIS'Patient 1              179        112
DIAGNOSIS'Reception 1            7          0
New_Page'Doctor 1                2          0
New_Page'Patient 1               60         0
New_Page'Reception 1             5          0
Best Upper Multi-set Bounds
DIAGNOSIS'Doctor 1
     4`(1,"pinky","Daily Nagar",4321,"typhoid")++
     3`(1,"ram","TM Nagar",3453,"I")++
     3`(1,"sri","FM Nagar",236,"cough")++
     3`(1,"sriram","RM Nagar",123,"Fever")
DIAGNOSIS'Patient 1
     71`(1,"pinky","Daily Nagar",4321,"typhoid")++
     36`(1,"ram","TM Nagar",3453,"I")++
     54`(1,"sri","FM Nagar",236,"cough")++
     18`(1,"sriram","RM Nagar",123,"Fever")
DIAGNOSIS'Reception 1
     7`(1,"pinky","Daily Nagar",4321,"typhoid")++
     6`(1,"ram","TM Nagar",3453,"I")++
     6`(1,"sri","FM Nagar",236,"cough")++
     6`(1,"sriram","RM Nagar",123,"Fever")
New_Page'Doctor 1
     2`(1,"pinky","Daily Nagar",4321,"typhoid")++
     2`(1,"ram","TM Nagar",3453,"I")++
     2`(1,"sri","FM Nagar",236,"cough")++
     2`(1,"sriram","RM Nagar",123,"Fever")
New_Page'Patient 1
     24`(1,"pinky","Daily Nagar",4321,"typhoid")++
     12`(1,"ram","TM Nagar",3453,"I")++
     18`(1,"sri","FM Nagar",236,"cough")++
     6`(1,"sriram","RM Nagar",123,"Fever")
New_Page'Reception 1
     4`(1,"pinky","Daily Nagar",4321,"typhoid")++
     4`(1,"ram","TM Nagar",3453,"I")++
     4`(1,"sri","FM Nagar",236,"cough")++
     3`(1,"sriram","RM Nagar",123,"Fever")
Best Lower Multi-set Bounds
DIAGNOSIS'Doctor 1          empty
DIAGNOSIS'Patient 1
     40`(1,"pinky","Daily Nagar",4321,"typhoid")++
     18`(1,"ram","TM Nagar",3453,"I")++
     30`(1,"sri","FM Nagar",236,"cough")++
     6`(1,"sriram","RM Nagar",123,"Fever")
DIAGNOSIS'Reception 1       empty
```

```
New_Page'Doctor 1          empty
New_Page'Patient 1         empty
New_Page'Reception 1       empty
Home Properties
-----------------------------------------
Home Markings              None
Liveness Properties
-----------------------------------------
Dead Markings              10488 [9999,9998,9997,9996,9995,...]
Dead Transition Instances None
Live Transition Instances None
Fairness Properties
-----------------------------------------
Impartial Transition Instances     None
Fair Transition Instances     None
Just Transition Instances     None
Transition Instances with No Fairness
     DIAGNOSIS'DIAGNOSIS 1
     DIAGNOSIS'ENQUIRY 1
     DIAGNOSIS'REGISTER 1
DIAGNOSIS'getpatient_details 1
New_Page'DIAGNOSIS 1
New_Page'ENQUIRY 1
New_Page'REGISTER 1
New_Page'getpatient_details 1
```

5.4 Result of Decision Support Pattern

In the decision support pattern, the diagnosis report of the patient provided by the doctor is given to take the decision or to confirm the problem of the patient. The decision is taken by running some necessary tests from the laboratory or prescribing the medication to the patient with the help of a decision pattern. The status of the patient is noted by taking tests from the laboratory and medications from the pharmacy. If the patient is suffering a moderate level of the problem, then he/she will be the outpatient. Otherwise, the patient will be admitted and be an in-patient in the hospital, due to a serious problem. The admission pattern provides with the conditions and it helps to decide to make the patient in/out with the help of decision patterns.

```
CPN Tools state space report for:
/cygdrive/C/Users/Sriram/Desktop/HEALTH CARE PATTERN/DECISION SUPPORT
PATTERN.cpn
Report generated: Fri Jan 24 14:46:37 2020
Statistics
------------------------------------------------
State Space     Nodes:  16408     Arcs:    64968
Secs:   300     Status: Partial
Scc Graph       Nodes:  16408     Arcs:    64968
Secs:    7
Boundedness Properties
-----------------------------------------------
Best Integer Bounds                       Upper      Lower
DECISION_SUPPORT_PATTERN'Decision 1       11         0
DECISION_SUPPORT_PATTERN'InPatient 1      3          1
DECISION_SUPPORT_PATTERN'OutPatient 1     3          0
DECISION_SUPPORT_PATTERN'Patient 1        106        8
DECISION_SUPPORT_PATTERN'Status 1         5          0
Best Upper Multi-set Bounds
DECISION_SUPPORT_PATTERN'Decision 1
     7`(1,"pinky",4321,"typhoid","lab")++
     7`(1,"ram",3453,"I","lab")++
     9`(1,"sri",236,"cough","pharm")++
     6`(1,"sriram",123,"Fever","pharm")
DECISION_SUPPORT_PATTERN'InPatient 1      3`"lab"
DECISION_SUPPORT_PATTERN'OutPatient 1     3`"pharm"
DECISION_SUPPORT_PATTERN'Patient 1
     43`(1,"pinky",4321,"typhoid","lab")++
     22`(1,"ram",3453,"I","lab")++
     31`(1,"sri",236,"cough","pharm")++
     10`(1,"sriram",123,"Fever","pharm")
DECISION_SUPPORT_PATTERN'Status 1
     4`(1,"pinky","typhoid","lab")++
     4`(1,"ram","I","lab")++
     5`(1,"sri","cough","pharm")++
     3`(1,"sriram","Fever","pharm")
Best Lower Multi-set Bounds
DECISION_SUPPORT_PATTERN'Decision 1       empty
DECISION_SUPPORT_PATTERN'InPatient 1      1`"lab"
DECISION_SUPPORT_PATTERN'OutPatient 1     empty
DECISION_SUPPORT_PATTERN'Patient 1        empty
DECISION_SUPPORT_PATTERN'Status 1         empty
Home Properties
--------------------------------------
Home Markings    None
Liveness Properties
--------------------------------------
Dead Markings                 8631 [9999,9998,9997,9996,9995,...]
Dead Transition Instances     None
Live Transition Instances     None
Fairness Properties
----------------------------------
No infinite occurrence sequences.
```

5.5 Result of Health Insurance Pattern

In the health insurance pattern, when the patient got admitted he/she should request to the policy issuer to provide the benefits of the policy. This request is done by the patient with the help of the request pattern and it will be simple since the patient has registered the policy id in the reception. When the patient requested the policy issuer, they verify the hospital where the policyholder is admitted and checks whether the hospital is network or non-network hospital with the help of verification pattern. The bills are provided by the reception from the date of admission until the date of discharge is provided to the patient or policy issuer. They are done with the help of the bill/payment pattern. If its network hospital the details are given to the policy issuer directly by hospital and patient can enjoy the benefits of the policy. If its non-network hospital the patient pays the bill from his/her pocket and the necessary details will be passed to the policy issuer to make the reimbursement to the patient with the help of the reimbursement pattern.

```
CPN Tools state space report for:
/cygdrive/C/Users/Sriram/Desktop/HEALTH CARE PATTERN/HEALTH INSURANCE
PATTERN.cpn
Report generated: Fri Jan 24 15:11:39 2020
Statistics
---------------------------------------
State Space     Nodes:  9885     Arcs:   37630
Secs:   300     Status: Partial
Scc Graph       Nodes:  9885     Arcs:   37630
Secs:   2
Boundedness Properties
---------------------------------------
Best Integer Bounds                                        Upper      Lower
HEALTH_INSURANCE'InPatient 1                               726        648
HEALTH_INSURANCE'Network 1                                 19         15
HEALTH_INSURANCE'Non_Network 1                             45         39
HEALTH_INSURANCE'POLICY_ISSUER 1                           16         1
HEALTH_INSURANCE'Patient_directly_pays_payment 1           0          0
HEALTH_INSURANCE'Reception 1                               0          0
HEALTH_INSURANCE'network_amount 1                          0          0
HEALTH_INSURANCE'non 1                                     0          0
Best Upper Multi-set Bounds
HEALTH_INSURANCE'InPatient 1
     74`(1,123,"A")++
     221`(1,236,"C")++
     145`(1,3453,"B")++
     286`(1,4321,"D")
HEALTH_INSURANCE'Network 1
     8`"A"++ 14`"B"
HEALTH_INSURANCE'Non_Network 1
     18`"C"++30`"D"
HEALTH_INSURANCE'POLICY_ISSUER 1
     7`(1,123,"A")++
     9`(1,236,"C")++
     9`(1,3453,"B")++
     12`(1,4321,"D")
HEALTH_INSURANCE'Patient_directly_pays_payment 1   empty
HEALTH_INSURANCE'Reception 1                       empty
HEALTH_INSURANCE'network_amount 1                  empty
HEALTH_INSURANCE'non 1                             empty
Best Lower Multi-set Bounds
HEALTH_INSURANCE'InPatient 1
     60`(1,123,"A")++
     193`(1,236,"C")++
     124`(1,3453,"B")++
     250`(1,4321,"D")
HEALTH_INSURANCE'Network 1
     5`"A"++ 10`"B"
HEALTH_INSURANCE'Non_Network 1
     14`"C"++ 25`"D"
HEALTH_INSURANCE'POLICY_ISSUER 1                      Empty
HEALTH_INSURANCE'Patient_directly_pays_payment 1      empty
HEALTH_INSURANCE'Reception 1                          empty
HEALTH_INSURANCE'network_amount 1                     empty
HEALTH_INSURANCE'non 1                                empty
Home Properties
---------------------------------------
Home Markings      None
```

```
Liveness Properties
---------------------------------------
Dead Markings           5333 [9885,9884,9883,9882,9881,...]
Dead Transition Instances
     HEALTH_INSURANCE'BILLING 1
     HEALTH_INSURANCE'PAYMENT 1
     HEALTH_INSURANCE'REIMBRUSMENT 1
HEALTH_INSURANCE'store_details 1
Live Transition Instances     None
Fairness Properties
------------------------------------
No infinite occurrence sequences.
```

6 Conclusion

By using Pervasive Computing technologies, Healthcare is accessible all over the places. The health care structure is formed by patterns because context-aware computing is the section of pervasive computing. As of the literature survey, it is found that there is no explicit design pattern to support the development of healthcare systems. The diagnosis pattern, decision support pattern and health insurance pattern are proposed and used in this framework. The framework provides the quality and accuracy of caring, along with the efficient way to claim the health insurance policy by the patient who holds insurance policy. Any healthcare applications can be developed using the proposed pattern.

References

1. Satyanarayanan, M.: Pervasive computing: vision and challenges. IEEE Pers. Commun. **8**(4), 10–17 (2001)
2. Zhang, D., Huang, H., Lai, C.-F., Liang, X., Zou, Q., Guo, M.: Survey on context-awareness in ubiquitous media. Multim. Tools Appl. **67**(1), 179–211 (2013)
3. Lo, C.-C., Chen, C.-H., Cheng, D.-Y., Kung, H.-Y.: Ubiquitous healthcare service system with context-awareness capability: design and implementation. Expert Syst. Appl. **38**(4), 4416–4436 (2011)
4. Wang, H.: Trust management of communication architectures of internet of things. J. Trends Comput. Sci. Smart Technol. (TCSST) **1**(02), 121–130 (2019)
5. Pandian, A.P.: Enhanced edge model for big data in the internet of things based applications. J. Trends Comput. Sci. Smart Technol. (TCSST) **1**(01):63–73 (2019)
6. Korhonen, I., Bardram, J.E.: Guest editorial introduction to the special section on pervasive healthcare. IEEE Trans. Inf. Technol. Biomed. **8**(3), 229–234 (2004)
7. Strobbe, M., Van Laere, O., Dhoedt, B., De Turck, F., Demeester, P.: Hybrid reasoning technique for improving context-aware applications. Knowl. Inf. Syst. **31**(3), 581–616 (2012)
8. Yuan, B., Herbert, J.: Non-intrusive movement detection in cara pervasive healthcare application. In: The 2011 International Conference on Wireless Networks, WORLDCOMP, vol. 11, pp. 360–366 (2011)
9. Yuan, B., Herbert, J.: Web-based real-time remote monitoring for pervasive healthcare. In: 2011 IEEE International Conference on Pervasive Computing and Communications Workshops (PERCOM Workshops), pp. 625–629. IEEE (2011)
10. Strobbe, M., Van Laere, O., Ongenae, F., Dauwe, S., Dhoedt, B., De Turck, F., Demeester, P., Luyten, K.: Novel applications integrate location and context information. IEEE Pervasive Comput. **11**(2), 64–73 (2011)
11. Ongenae, F., Claeys, M., Dupont, T., Kerckhove, W., Verhoeve, P., Dhaene, T., De Turck F.: A probabilistic ontology-based platform for self-learning context-aware healthcare applications. Expert Syst. Appl. **40**(18), 7629–7646 (2013)
12. Yuan, B., Herbert, J.: Context-aware hybrid reasoning framework for pervasive healthcare. Pers. Ubiquit. Comput. **18**(4), 865–881 (2014)
13. Yuan, B., Herbert, J.: Fuzzy cara-a fuzzy-based context reasoning system for pervasive healthcare. Procedia Comput. Sci. **10**, 357–365 (2012)
14. Shamsa, H., Zamanifar, K.: MVCC: an architectural pattern for developing context-aware frame-works. Procedia Comput. Sci. **34**, 344–351 (2014)
15. Sanni, L.: Distribution pattern of healthcare facilities in Osun State, Nigeria. Ethiop. J. Environ. Stud. Manag. **3**(2) (2010)

16. Yuan, B., Herbert, J.: Transparency issues in a hybrid reasoning architecture for assistive healthcare. AASRI Procedia **4**, 268–274 (2013)
17. Jhanjee, A., Kumar, P., Srivastava, S., Bhatia, M.S.: A descriptive study of referral pattern in department of psychiatry of a tertiary care hospital of North India. Medicine (Baltimore) **674**, 46 (2011)
18. Desalegn, A.A.: Assessment of drug use pattern using WHO prescribing indicators at Hawassa University teaching and referral hospital, south Ethiopia: a cross-sectional study. BMC Health Serv. Res. **13**(1), 170 (2013)
19. Aziz, S., Ejaz, A., Alam, S.E.: Mortality pattern in a trust hospital: a hospital based study in Karachi. J. Pak. Med. Assoc. **63**, 1031–1035 (2013)
20. Chiang, J.-H., Yang, P.-C., Tu, H.: Pattern analysis in daily physical activity data for personal health management. Pervasive Mob. Comput. **13**, 13–25 (2014)
21. http://cpntools.org
22. Christensen, S., Jørgensen, J.B., Kristensen, L.M.: Design/CPN—A computer tool for coloured Petri nets. In: International Workshop on Tools and Algorithms for the Construction and Analysis of Systems, pp. 209–223. Springer, Berlin, Heidelberg (1997)

Analysis of Groundnut Based Bio Modified Liquid Insulation for High Voltage Transformer

B. Pooraja, M. Willjuice Iruthayarajan, and M. Bakrutheen

Abstract The vegetable oil-based natural ester is gaining more attention as alternate liquid insulation. But natural esters have high viscosity value which affects the heat dissipating capacity of the transformer. There are some processes used to reduce viscosity such as thermal cracking, blending, etc. This research work aims to investigate the groundnut oil (crude and refined oil) for developing low viscous bio modified liquid insulation with the transesterification process. The important properties of the insulating oil samples like viscosity, breakdown voltage, flash point, pour point and density are measured as per standard for validating the effectiveness as liquid insulation. From the results, it is found that after the transesterification process, the bio modified liquid insulation has low viscous nature and also possesses the characteristics to be alternate liquid insulation. This bio modified oil may be further studied for direct utilization as liquid insulation in the transformer.

Keywords Power transformer · Insulating oil · Natural ester · Viscosity · Transesterification · Bio liquid insulation

1 Introduction

The electrical energy consumption is used all over the world for different purposes. In the electrical power system network, the transformer plays a major part in transmitting that energy to the consumers from the power generation station in the power system network. The decisive part of requirement on present technological development in the electrical power network is the service life management of high voltage power transformer. The life of the transformer is majorly decided by the insulation mediums of a transformer. The insulating oil provides one of the vital roles in the proper working of a power transformer as coolant and insulation and it also determines the

B. Pooraja (✉) · M. W. Iruthayarajan · M. Bakrutheen
Department of Electrical and Electronics Engineering, National Engineering College, K.R.Nagar, Kovilpatti, Tuticorin, Tamil Nadu 628503, India
e-mail: poorajaeee12@gmail.com

P. Karuppusamy et al. (eds.), *Sustainable Communication Networks and Application*, Lecture Notes on Data Engineering and Communications Technologies 55,
https://doi.org/10.1007/978-981-15-8677-4_34

lifetime of power transformers. Thus, the dependability of a power transformer is primarily determined by the insulation used in it [1].

The petroleum-based mineral oil is used in service insulating oil in the power transformer for many years [2]. Splendid dielectric and cooling characteristics offered by traditional mineral oil are the factors of its dominance in the global market of consumption of oil [3].

Non-biodegradable nature and shrinkage of available resources of mineral oil have increased the necessity for finding the alternate liquid among the research community. This has given rise to a new class of alternate biodegradable fluids as liquid insulation in the form of vegetable oil derived natural esters which has been focused as a suitable substitution to traditional mineral oil. Natural esters are naturally available vegetable oils which are straight derived from plant sources which can be utilized as dielectric fluids in the power transformer [4–7].

All-natural ester-based oil products demonstrated superior electrical properties for usage as oil insulation [8–10]. Furthermore, the flash point temperature is about more than 300 °C, which is much higher than that of traditional mineral oil (about 150 °C). Most of the natural esters have high viscosity due to its composition which is mainly of the triglyceride composition of fatty acids. Hence, it is considered higher viscous nature of natural esters which will not lead to a lesser service life of through dielectric material while replacing with mineral oil [11]. For the vegetable oil derived natural ester filled transformers, if the viscosity of natural esters reduced, working life will be longer because of a good heat dissipating capacity [11–13].

The low viscosity of insulating fluid could be more effectively assist the flow of oil to cool the windings and material surfaces inside the transformers in service [13]. Hence, the research work on reducing a viscous nature of natural ester is important in liquid insulation field. There are many ways of reducing the viscosity of oil such as thermal cracking, blending, dilution, etc. Based on the literature, the transesterification process is one of the feasible methods to reduce the viscosity of natural esters [14–16]. Trans-esterification is the chemical reaction of triglycerides with alcohol and catalyst to form esters and glycerol. In the chemical term, it is an action of one alcohol in the sample displacing another from an ester with oil, referred to as alcoholics [17]. Modified esters have demonstrated the good deal potential as a less viscosity value for natural esters which could stretch the life of power transformers [14–17].

In this work, to develop low viscous bio liquid insulation, crude and refined forms of groundnut are treated with transesterification process. The effectiveness of the process in developing low viscous liquid insulation is determined by the standard procedure of measuring the important properties of oil insulation such as viscosity, breakdown voltage, flash point, pour point and density with the specified standards for the transesterification process.

2 Sample Descriptions

Two different forms of groundnut oil (crude and refined forms) are selected for this investigation. The groundnut oils are bought from domestic/local oil refinery. For the analyze purposes, the following oil samples (as given in Table 1) are used in this investigation.

For transesterification process to produce low viscous natural esters, the main raw materials would be the chemicals of methyl ester, alcohol and accelerates. The entirety of the transesterification process arrangement is the flow methodology chart of a synthesis procedure is shown in Fig. 1.

The process begins with the mixing of 100 ml of methanol and 4.5 g potassium hydroxide and it is mixed for 30 min in the magnetic stirrer (Fig. 2) at the uniform mixing rate of 700 rpm with an oil temperature of 65 °C.

The above-mixed chemical solution is added to the 500 ml oil sample, the solution of the mixture is treated for mixing process for 3 h with the help of magnetic stirrer with the rate of 700 rpm and it is maintained at 65 °C temperature. The oil mixture

Table 1 Samples taken for investigations

Samples	Sample name
Sample 1	Crude groundnut oil (CGO)
Sample 2	Refined groundnut oil (RGO)
Sample 3	Transesterified crude groundnut oil (TCGO)
Sample 4	Transesterified refined groundnut oil (TRGO)

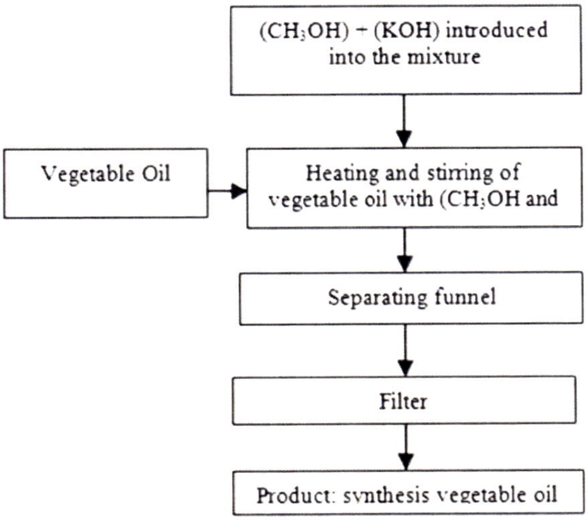

Fig. 1 Flowchart of synthesis vegetable oil procedure

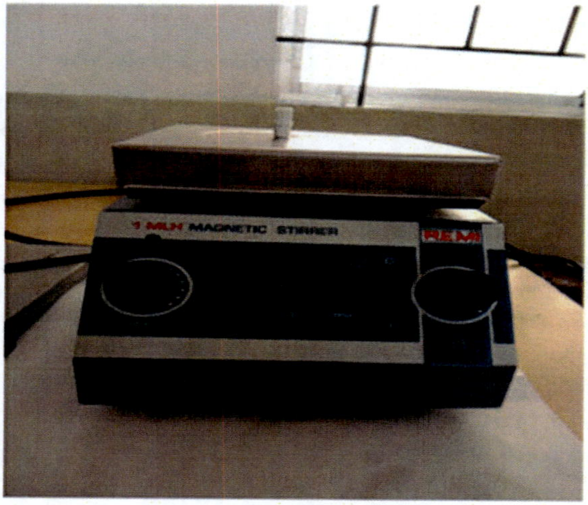

Fig. 2 Magnetic stirrer setup

sample is then stored in to separating funnels and is constantly tilted till the separation cloudy oil mixture of solution into two layers.

The funnel is maintained in the same position for 48 h after which the glycerol is drained while the methyl ester is separated. The methyl ester is the transesterified oil which is termed as bio modified liquid insulation as per ASTM D6871 standard for natural ester as dielectric [18–20].

3 Experimental Section

The important properties of oil insulation in the process of transesterification have measured to ensure the reliable nature of analyzed oil samples as liquid insulation.

3.1 Viscosity

The viscosity is calculated as per standard procedure based on ASTM D445 standard using redwood viscometer which is shown in Fig. 3.

Normally low viscous nature of the oil enhances flow rate and heat transferring ability. The viscometer consists of a test oil cup with a hole for measuring the time required for collecting the 50 ml of the sample. From the time noted, the viscosity of the sample is calculated.

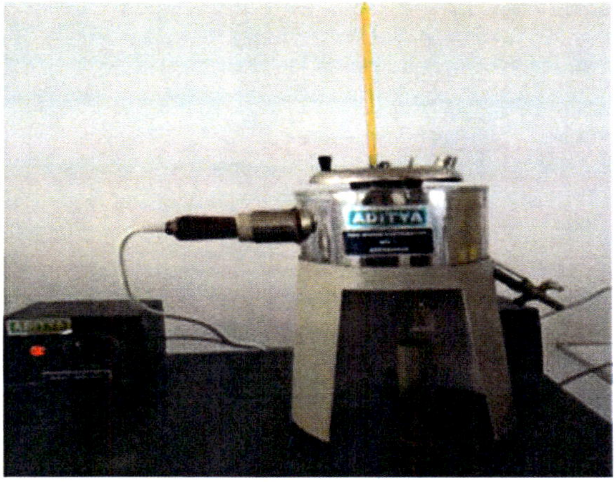

Fig. 3 Redwood viscometer

3.2 *Breakdown Voltage*

The breakdown voltage (BDV) is measured according to the IEC60156 and. The test kit consists of electrodes of identical hemispherical shapes with standard interspacing of 2.5 mm. The test oil is filled in the cup above the electrode level. Five breakdown voltages in successive steps are taken with the relaxation time of one minute between each measurement. The average of five BDV values is considered as the BDV of the samples. Breakdown voltage kit used for the measurement is shown in Fig. 4.

Fig. 4 Breakdown voltage kit

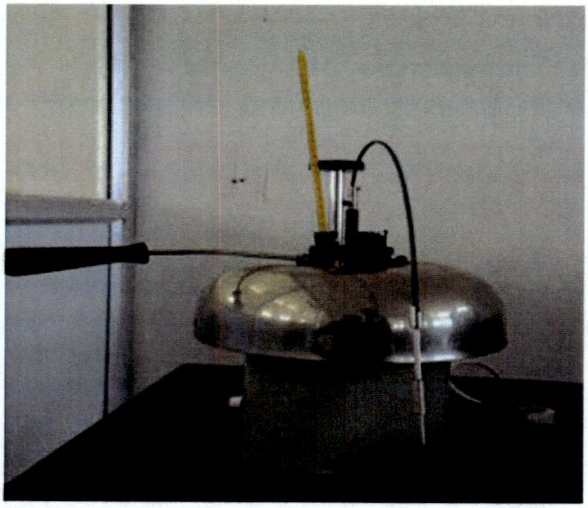

Fig. 5 Pensky-Martin closed cup method

3.3 Flash Point

The flash point is measured using standard pensky martin flash point closed cup tester according to ASTM-D93. The apparatus of measurement of the flash point is shown in Fig. 5. The flash point condition is indicated by a small flame over the oil surface while initiating a test flame in the opening on the surface.

3.4 Pour Point

It is the lowest temperature of the oil when it just stops the flow. As per ASTM D97 standard, pour point of an oil is mainly useful at the cold climatic places. The pour point measurement setup is shown in Fig. 6.

3.5 Density

A density meter, also known as a densometer as shown in Fig. 7, is equipment that measures the density as per the ASTM D1217 standard. Density may be useful for the indication and analysis of oil's reliability for use in power transformer.

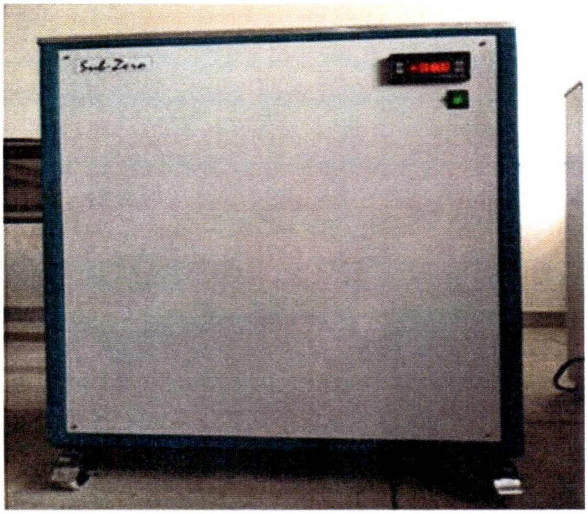

Fig. 6 Pour point kit

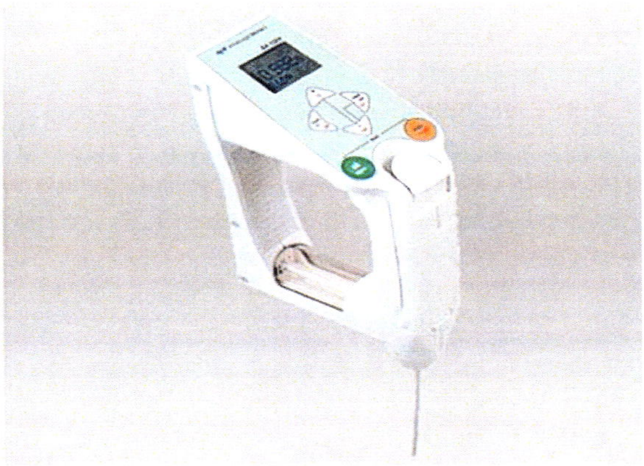

Fig. 7 Densometer

4 Result and Discussion

In this part of the paper, the properties of the groundnut oil (crude and refined oil) and transesterified groundnut oil (crude and refined oil) are presented and compared to find the insulating and cooling capability of transesterified oil for the application in transformers. Performance parameters of samples are tabulated in Tables 2 and 3.

- The viscosity value of crude groundnut oil and Transesterified crude groundnut oil varies from 103.83 to 18.65 cSt. The viscosity value of refined groundnut oil and transesterified groundnut refined oil varies from 93.14 to 43.78 cSt. From results,

Table 2 Properties of crude groundnut oil and transesterified crude groundnut oil

Performance parameters	Sample 1	Sample 3
Viscosity (cSt)	103.83	18.65
Breakdown voltage (kV)	35	31
Flash point (°C)	340	320
Pour point (°C)	-8	-16
Density (kg/cm^3)	0.906	0.871

Table 3 Properties of refined groundnut oil and transesterified refined groundnut oil

Performance parameters	Sample 2	Sample 4
Viscosity (cSt)	93.14	43.78
Breakdown voltage (kV)	33	30
Flash point (°C)	320	305
Pour point (°C)	−4	−7
Density (kg/cm^3)	0.905	0.875

it is observed that The transesterified oil has lower viscosity when compared with base groundnut oil. This shows how far effectiveness achieved in developing low viscous bio liquid insulation with the transesterification process.

- The BDV of all the samples has the values above the standard value of 30 kV which shows the better dielectric nature of oil samples. But the transesterified oil sample has low breakdown voltage than the base oil due to the modification in the composition of oil in the transesterification process.
- The flash point of all the samples is above 300 °C which is better than the standard specified values.
- The pour point value of crude groundnut oil and transesterified crude groundnut oil varies from −8 to −16 °C. The pour point value of refined groundnut oil and transesterified refined groundnut oil varies from −4 to −7 °C. The transesterified groundnut oil has a lower pour point when compared with base groundnut oil.
- The density value of crude groundnut oil and transesterified crude groundnut oil varies from 0.906 to 0.871 kg/cm^3. The density value of refined groundnut oil and Transesterified refined groundnut oil varies from 0.905 to 0.875 kg/cm^3. The transesterified groundnut oil has a lower density when compared with base groundnut oil.
- The changes in the properties are observed mainly because of the chemical reaction occurred during the transesterification process which modified the composition of oil samples and removal of fatty acids content of glycerol.
- Overall the transesterification process has yielded fruitful results in developing the liquid insulation which has low viscous nature, good dielectric characteristics, better flash point, pour point and density.

5 Conclusion

In the present scenario, the entire world is hugely witnessing the development of moving from the petroleum-based products that are depicting in resource and environment un-friendly into vegetable oil derived products that are naturally available and environmentally friendly. In this work, an experimental attempt has been carried out to develop and quality characterization of low viscous bio insulating fluid. Transesterification process is used to develop liquid insulation with the groundnut oil-based ester fluid. The transesterification process yields low viscous bio modified liquid insulation. From the result, it is concluded that the groundnut oil and transesterified ground oil under both crude oil and refined form have enough potential as liquid insulation. Bio modified vegetable oil extends the hope for the development of low viscous liquid insulation. However, a lot of investigations are needed to be conducted before the practical implementation related applications and commercialization of the low viscous insulating fluid in the high voltage field.

References

1. IEEE Standard C57.91: IEEE, Guide for loading Mineral-oil-immersed Transfonners. In: Annex I:Transfonner Insulation Life, (1995).
2. Sokolov, V., Bassetto, A., Oommen, T.V., Haupert, T., Hauson, D.: Transformer fluid a powerful tool for the life management of an ageing transformer population. In: Asia Pacific Technical Conference Proceeding (2003)
3. Lelekakis, N., Wijaya, J., Martin, D.: Effect of acid accumulation in power transformer oil on the aging rate of paper insulation. IEEE Electr. Insul. Mag. **30**, 19–26 (2014)
4. Asano, R., Page, S.A.: Reducing environmental impact and improving safety performance of power transformer with natural aster insulating fluid. IEEE Trans. Dielect. Electr. Insul. **50**, 134–141 (2014)
5. Senthil kumar, S., Willjuice Iruthayarajan, M., Bakrutheen, M.: Analysis of vegetable liquid insulating medium for application in high voltage transformers. In: IEEE International Conference on Science, Engineering and Management Research, ICSEMR, pp. 1–5 (2014).
6. Karthik, M., WilljuiceIruthayarajan, M., Bakrutheen, M.: Suitability analysis of natural esters based liquid insulating medium for high voltage transformer. Inter. J. Appl. Engg. Res. **10**, 15331–15335 (2015)
7. Reffas, A., Idir, O., Ziani, A., Moulai, H., Nacer, A., Khelfance, I., Ouagueni, M., Beroual, A.: Influence of thermal ageing and electrical discharges on uninhibited olive oil properties. IET Sci. Meas. Tech. **10**, 711–718 (2016)
8. McShane, C.P.: Vegetable-oil-based dielectric coolants. IEEE Indu. Appli. Mag. **8**, 34–41 (2002)
9. Oommen, T.V.: Vegetable oils for liquid-filled transformers. IEEE Electr. Insul. Mag. **18**, 6–11 (2002)
10. Tenbohlen, S., Koch, M.: Aging performance and moisture solubility of vegetable oils for power transformers. IEEE Trans. Pow. Deli. **25**, 825–830 (2010)
11. Liang, S., Liao, R., Yang, L., Li, J., Sun, C., Sun, H.: Thermal aging characteristics of natural ester impregnated Kraft paper and thermally upgraded paper insulation. İn: International Conference on High Voltage Engineering and Application (2008).

12. Amanullah, M., Islam, S.M., Chami, S., Ienco, G.: Analyses of physical characteristics of vegetable oils as an alternative source to mineral oil-based dielectric fluid. In: IEEE International Conference on Dielectric Liquids, (2005).
13. Amanullah, M., Islam, S.M., Chami, S., Ienco, G.: Low viscosity vegetable oil-based dielectric fluids. In: Google Patents (2007).
14. Leung, D.Y., Wu, X., Leung, M.: A review on biodiesel production using catalyzed transesterification. Appl. Ener. **87**, 1083–1095 (2010)
15. Kanoh, T., Iwabuchi, H., Hoshida, Y., Yamada, J., Hikosaka, T., Yamazaki, A., Hatta, Y., Koide, H.: Analyses of electro-chemical characteristics of palm fatty acid esters as insulating oil. İn: IEEE International Conference on Dielectric Liquids (2008).
16. Rottig, A., Wenning, L., Bröker, D., Steinbüchel, A.: Fatty acid alkyl esters: perspectives for production of alternative biofuels. App. Microbio. Biotech. **85**, 1713–1733 (2010)
17. Tekin, A., Hammond, E.G.: Factors affecting the electrical resistivity of soybean oil methyl ester. J. Amer. Oil Chem. Soc. **77**, 281–283 (2000)
18. Chandramouli, A., Sivachidambaranathan, V.: Extract maximum power from PV system employing MPPT with FLC controller. power **1**, 4 (2019)
19. Ramadhas, A.S., Jayaraj, S., Muraleedharan, C.: Biodiesel production from high FFA rubber seed oil. Fuel **84**, 335–340 (2005)
20. Oommen, T.V., C. Claiborne, C., Walsh, E., Baker, J.:A new vegetable oil based transformer fluid: development and verification. In: Annual Report Conference on Electrical Insulation and Dielectric Phenomena (2000)

Identification of Key Parameters Contributing to Technical Debt in Software Using Rank-Based Optimization

Harmandeep Kaur and Munish Saini

Abstract In classical terms, the amount borrowed from the issuer causes interest. It is not usual that, interest rises above the principal amount. However, in the case of technical debt, the accumulated debt in the form of interest can rise to the sum, which is more than the efforts required to repay it than the initial technical debt. In this paper, the estimation of the breakpoint to find the reasons for the abrupt growth in the technical debt is tried. For the analysis of the breakpoint, the highest contributor factors towards the technical debt are identified. Further, the extent to which these attributes contribute to technical debt is also explored. The rank-based optimization mechanism (RF-based approach) is proposed to predict the technical debt. The mechanism obtains the dominating factors that contribute towards the technical debt. The practitioners indulged in coding also get benefited in reducing the impact of the technical debt, if they understand and determine the insight of these contributing factors in advance.

Keywords Technical debt (TD) · RF · Open-source software (OSS) · Rank-based optimization · Technical debt management (TDM)

1 Introduction

In quality management, Harrington [1] specified the statement "*cost of poor quality*" referring to the fact that heavy penalty has to be paid for poor or defective material [1]. The low-quality product incurs cost not only of filling the gap between the actual and optimal quality products but also requires the cost of rectifying the issue [2]. In software engineering, the aforementioned cost is included and described as technical debt [3]. Technical debt is an associative term collaborating with a large number of deficiencies present within the software. Technical debt requires early detection

H. Kaur (✉) · M. Saini
Guru Nanak Dev University, Amritsar, India
e-mail: Kaurharmandeep0915@gmail.com

M. Saini
e-mail: munish.cet@gndu.ac.in

P. Karuppusamy et al. (eds.), *Sustainable Communication Networks and Application*, Lecture Notes on Data Engineering and Communications Technologies 55,
https://doi.org/10.1007/978-981-15-8677-4_35

along with dependency information. Research on technical debt and management techniques becomes intense during the phase of the last five years, suggesting the interest of the programmers and other communities to address the related issues to quality management [4]. From the perspective of associated cost-benefit analysis, an overview of technical debt management (TDM) can be found in recent reviews [2, 5–7]. Despite mass research on the subject, the TDM state of research exhibits major challenges.

In this study, the use of a unique heuristic (rank-based optimization mechanism) of tackling technical debt is proposed by identifying its root cause. Afterwards, also aim at evaluating the amount of technical debt caused by these factors. More specifically, look forward to answering the following questions:

Identify the highest contributing attributes to the technical debt.

Perform the classification of Technical debt to generalize the mechanism of finding the attribute contribution toward technical debt.

The rest of the paper is organized as follows: Sect. 2 gives the related work of the techniques used for technical debt. Section 3 presents the proposed system along with a mathematical foundation. Section 4 depicts the performance analysis and results of the study. Section 5 provides a conclusion and future scope.

2 Related Work

The quality of the software must be improved to increase the chance of acceptance. The research already done towards this aspect is described in this section.

Lenarduzzi et al. [8] described a technical debt dataset that consists of 33 Java projects source code [9, 10].

The code must be understandable and easy to read and it can be measured using complexity metrics [11]. In 1976, to measure the complexity of code cyclomatic complexity metric was developed by McCabe. The cyclomatic complexity is a control flow graph based program that calculates linearly independent paths from the source code. Halsted defined complexity measurements that calculate the complexity of programming language [5]. In this code, the smell is used that indicates that there must have a problem with the program. These are not considered as bugs rather it indicates the weakness of designing of code without affecting its functionality. It also identifies failures that may occur in the future during code execution. In this research taxonomy of bad smells is created to analyze code quality [12]. It classifies the code into various categories and performs an initial correlation between these categories [6, 13]. In this field, the researcher investigated the code quality using the system-level indicator, and also it checks maintainability measures of code. It inspects code source code using an automated approach and identifies the reason for failures. In this review, the feature location and static code analysis are studied [14]. The process of locating the functional requirements in the source code is known as feature location. It surveys various techniques used for handling code quality and reducing technical debt and describes the various technical debt measurement metrics. It manages debt using these metrics that are based on principal, interest, and interest probability. The debt management is done with the help of DISS and incurred debt at one part of the system is likely to affect other parts. In addition to refactoring the database also the adaption of the software system and the necessary migration of the data should be considered [15].

The comparison of the literature suggests that during the technical debt and code quality determination, all the attributes or features from the dataset must be utilized. This will be a time-consuming and complex task. To tackle the issue, correlation analysis becomes compulsory. In this paper, a dataset is prepared that contains around 6000 projects that are being analyzed. By analyzing these projects, 10 parameters are extracted that affect the performance of the projects. The correlation between all parameters has been evaluated to find which parameter affects more on code quality. From these parameters, two parameters are selected: security and reliability which constitute the technical debt and number of bugs. Based on security and reliability the network is trained and after training the testing is performed. The testing is done using KNN, RF, SVM, random forest, and decision tree algorithms.

3 Data Analysis and Methodology

This section provides details on the data collection, algorithms, and tools that have been used in this study. The stepwise description of the whole process is as follows (see Fig. 1).

3.1 Dataset

This is a primary component that is used to perform training to the network. All the attributes present within the dataset are used to determine the highest correlated attributes. The dataset used within the proposed system is created by executing multiple projects with Java, Visual Studio, C, and C++. All of these projects are analyzed using SonarQube. The dataset is stored in CSV format. The work of finding the highest-ranked attribute contributing the maximum to the technical debt is implemented within the R tool.

3.2 Finding Correlation

This step is critical as it determines the inclusion and exclusion criteria. The attributes having the highest correlation with the technical debt are retained and all other attributes are rejected from the feature vector.

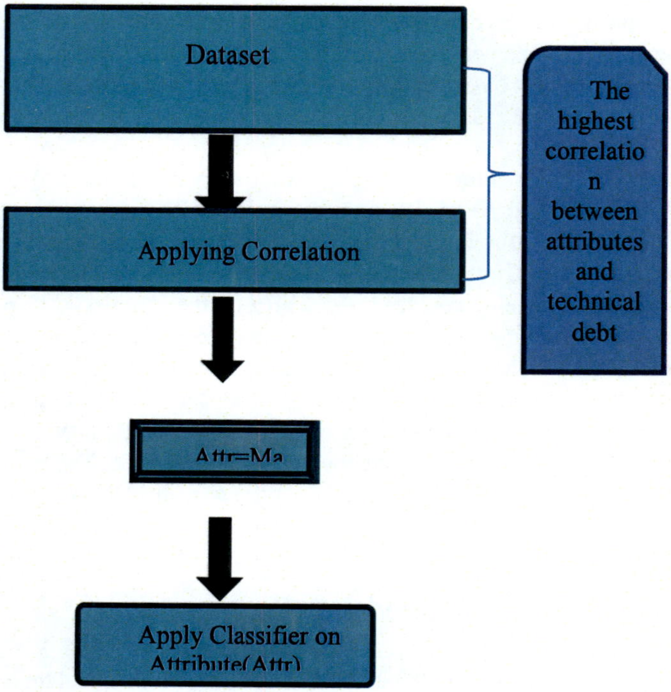

Fig. 1 Methodology for finding the highest-ranked technical debt attributes

3.3 *Checking Technical Debt*

Based on the highest correlation attributes, a comparison of test data attributes are made with feature vector to determine the technical debt.

Correlation

Correlation is taken between the attributes of the dataset and technical debt. The formula used for this purpose is given in Eq. 1

$$r_{xy} = \sum \frac{(x - x')(y - y')}{\sqrt{(x - x')^2 (y - y')^2}} \tag{1}$$

The bug attribute is found to have the highest correlation value with technical debt [16]. This attribute is assigned the highest rank. After applying correlation analysis, the dataset is organized to check the attributes and assigned ranks to individual attributes. Consider a training dataset of the form $(x1, y1)$ $(x2, y2)$ … (xn, yn). The linear hyperplane equation is given in Eq. 2

$$w \cdot x - b = 0 \tag{2}$$

'w' is the normal vector of the hyperplane. Parameter 'b' determines the offset for the hyperplane. This hyperplane is used within the support vector machine.

$$w \cdot x - b = 1 \tag{3}$$

Hyperplane Eq. 3 is followed in case strict boundaries are to be followed. The Euclidean distance is evaluated in the case of KNN. The Euclidean equation is given in Eq. 4

$$\text{Euc} = \sqrt{\sum (x - x_i)^2} \tag{4}$$

This distance is checked with every distinct point of the dataset. The threshold value is defined and if the distance is less than the threshold value than values are grouped within one group.

The random forest algorithm uses a bootstrap aggregation mechanism. The mechanism uses Eq. 5 for tree formation.

$$f = \frac{1}{B} \sum_{i=1}^{n} f(x_i) \tag{5}$$

where 'B' is the attributes within the dataset and 'n' is the size of the dataset. 'f' is the aggregate tree and 'x' is the cell value.

The activation function that determines the output of the network is given in Eq. 6. This activation function also gives the weight factor that must be adjusted to make convergence faster. Leaky RLU function is used as an activation function in a considered RF.

$$\text{Act}_L = \max(0.1 * x, x) \tag{6}$$

This will enable backpropagation even for the small negative values. Consistent prediction in the negative region is not supported through the RF.

3.4 Data Collection

This is the process of gathering information from relevant sources to answer the research questions. Two data collection mechanisms are in common: primary and secondary. The secondary mechanism is used to gather information from published contents, benchmark datasets, etc. the primary mechanism, however, includes strategies to gather real-time information. In the proposed work, data is gathered using a primary source [4, 6]. Attributes used within the dataset is given in Table 1.

Table 1 Attributes of the dataset

Attributes	Description
Project type	Indicates the category and name of the project
Bugs	Number of bugs present within the project
Coverage	Coverage associated with the project
Code smells	Logical problems within projects
Duplication	Duplication line of code
Complexity	Number of undue instructions
Reliability	Describe code reliability
Security	Critical security aspect associated with the project
Maintainability	Function point establishment is checked using this attribute
LOC	Number of a physical and logical line of codes

3.5 *Metrics*

The criteria used to measure and to perform evaluation is termed as the metric. The different metrics that have used within the proposed mechanisms are given in Table 2.

Rank-Based Optimization
This mechanism provides the highest ranked attributes that contribute maximum towards technical debt. By choosing the highest-ranked attribute, technical debt can be predicted accurately and at a rapid rate. The Rank-Based Optimization algorithm is formally defined as follows:

Table 2 Evaluation metrics

Metric	Description
Correlation	Evaluates the positive or negative aspect of attribute towards technical debt
Accuracy	Describes the validity of the mechanism used. Higher values prove worth of study

```
Algorithm: Rank Based Optimization
 Input:
Read the number of data (n) from the dataset
Read data X i and Y I from the dataset for i=1 to n
Output:
Calculate Correlation and assign Ranks to attribute
Create a decision tree of values based on the highest-ranked attribute to fed into layers
of RF
Initialize: yp
For i = 1 to n
        Set p = 1
                For j =1 to n
                        If I ≠ j then
                        Calculate p = p * (xp - X j)/(X i - X j)
                        End If
                Next j
        Calculate yp = yp + p * Y i
Next i
Display extracted project and add this project to yp as correlated value.
Compare the calculated values and predict technical debt which is an operation of the
output layer
```

4 Results and Discussion

To decide the components influencing technical debt, correlation analysis is directed. The correlation is applied to check for the credit having the most noteworthy commitment to technical debt. The clustering is done based on various characteristics of tuples within training data. Most elevated correlation group is chosen for deciding technical debt inside test data. The most elevated positive correlation is of Bugs. Classes of bugs considered and their examination with different kinds of bugs are given as under.

4.1 New Versus Old Bugs

Track what number of bugs are opened against what number of are shut to discover the rate at which your group clears bugs. This measurement is a pointer of your general tech obligation, just as whether your group is moving the correct way as far as general code quality.

4.2 Bug Burndown

Advancement and quality confirmation groups use bug burndown to comprehend open bugs and their anticipated terminations dependent on the normal bug conclusion rate. Groups that don't watch out for bug burndown can lose an idea about their general item quality and take on exorbitant measures of tech obligation in their push to fix bugs rapidly.

4.3 Level of High-Need Bugs

This direct figuring, dividing the quantity of current, high-need bugs by the complete number of bugs, is the level of bugs that your group has labeled as high-need (now and again high-seriousness) because of their effect on clients or the item in general. Followed after some time, this measurement enlightens some portion of the account of both item quality and tech obligation. An expanding pattern of all the more high-need bugs is regularly indicative of a group battling with item prerequisites, experiments, and test suites.

In addition to the bugs, Code smell is next to have the highest correlation with bugs. This means this attribute contributes to bugs and hence has to be considered for optimization.

4.4 Code Smells

Code smells mirrors the number of lines of code erased and included a similar line. It's a proportion of movement after some time that features hotspots in your code. With fresh out of the plastic new highlights, a great deal of movement in one territory isn't an issue, however after some time, code churn ought to lessen; on the off chance that it doesn't, you're doing an excessive amount of revamping and gathering a superfluous measure of tech obligation. High code agitates can likewise anticipate a drop in quality and speed.

Code coverage is the last attribute having a significant correlation with the technical smells.

4.5 Code Coverage

Additionally called test inclusion, Code coverage is the level of lines of code that are executed when running your test suite. In a general sense, Code coverage alludes to how successful your test procedure is at delivering a quality item. As a general

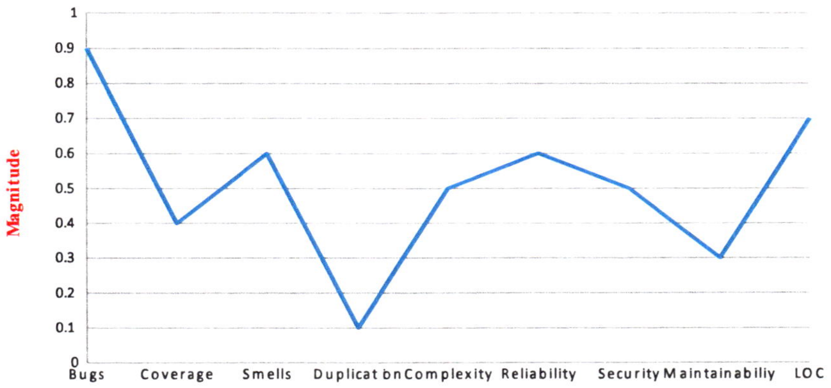

Fig. 2 Correlation between the attributes and technical debt

guideline, inclusion ought to be in the 80–90% territory. Code coverage doesn't gauge the inborn nature of your item; rather, it uncovers the procedure your group is attempted to accomplish a quality item. Code coverage features breakdowns in your test procedure, similar to when new code is included however not tried. On the off chance that your Code coverage rate drops after some time, commit more assets to the test-driven turn of events (TTD) and ensure untested regions are secured. A great test process takes off quality issues and pays off tech obligation, setting you to collaborate to be nimbler later on.

The result obtained corresponding to a contributing factor based on correlation is given in Fig. 2.

Figure 2 indicates the result of the correlation. The highest correlation attribute is a bug which is close to 1. The maximum value of the correlation analysis is 1. Value greater then zero indicates a positive correlation. This also suggests that attributes considered within the dataset possess a positive correlation with technical debt.

The most noteworthy position correlation property is bugged. Lessening bugs from the program pays off technical debt as it were. The connection between bugs and technical debt is straight. This is given in Fig. 3

Figure 3 gives the effort required to resolve the issue within considered projects. The value of bugs is directly proportional to the amount of effort required to resolve the issue. The average of technical debt by grouping projects of a similar sort is presented in Fig. 3.

Factor identification towards technical debt is performed using a different algorithm. The accuracy of these approaches is quite high. The accuracy result of different algorithms is given in Fig. 4. Technical debt categories must be classified using classification algorithms. Technical debt in label form is represented as high or low. In other words, two classes are defined for predicting technical debt high or low. This classification is used to determine the impact of the highest correlated attributes on technical debt. The linear relationship is obtained that indicates bugs and technical debt is directly proportional to each other.

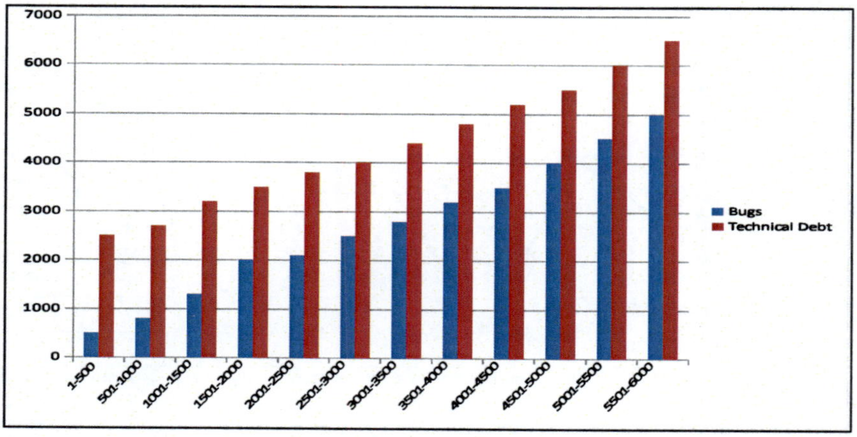

Fig. 3 Effect of bugs on technical debt

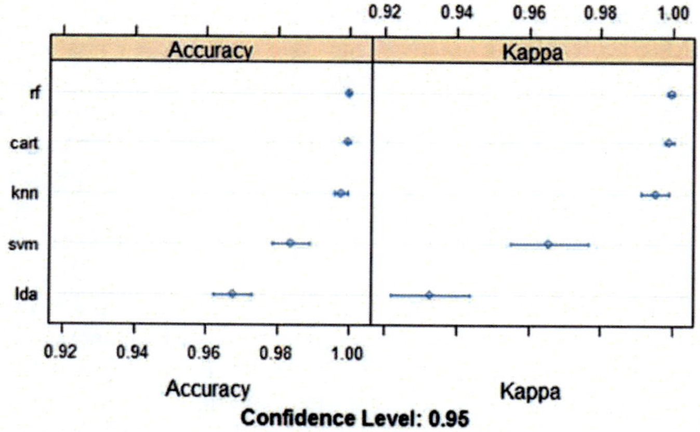

Fig. 4 Classification accuracy for different classifiers

The validity of the various algorithms in technical debt prediction using key attribute bugs is verified using Fig. 4. Test data includes 10–100 tuples. Several correct predictions associated with technical debt to the total predictions made by every algorithm gives the accuracy of the algorithm.

The outcome proposes that the random forest-based methodology is best in identifying factors adding to technical debt. The precision of 99% is accomplished when the bugs property is utilized for testing and preparing the activity. This implies improving the bugs' characteristics could straightforwardly affect technical debt.

4.6 Classification Accuracy Using Different Classifiers

The reproduction results related to Technical Debt acquired from SVM is given in Fig. 5.

The support vector machine is based on hyperplanes. SVM forms two hyperplanes. The first hyperplane is based on the least uncertainty and the second hyperplane is based on high uncertainty.

SVM (Support vector machine) used to detect the technical debt is partitioned into layers. These layers are known as hyperplanes. It is a supervised learning mechanism that is used in this work for detecting technical debt. The process of identification includes the selection of correct hyperplanes that classify the data better. First of all aggregation of highest ranked attributes is made. The aggregation leads to a central value. This value and corresponding labeling information are stored within the buffer.

Figure 6 indicates the hyperplanes identification by assigning weights that are calculated from aggregating the train data. The aggregated weights are termed as support vectors.

The hyperplanes are thus labeled with this aggregate value. This process is of training. The testing mechanism causes the identification of projects having the highest technical debt. The highest-ranked attribute value is matched against the aggregate value of train data. After this penetration is evaluated if test data highest ranked attribute penetrates the first hyperplane then the technical debt is detected otherwise no technical debt is detected.

The result of SVM is batter as compared to other approaches like PCA and KNN. The classification accuracy of SVM is better due to the existence of limited hyperplanes. The hyperplanes ensure that all the data must be classified in at least one of the hyperplanes. The degree of misclassification decreases due to this approach. The Technical Debt anticipated by the SVM from the framed dataset shows that in the greater part of the task Technical debts are high.

The research is suitable for the projects developed in visual studio, c++, c#.

5 Conclusion and Future Scope

Technical debt is a significant reason for programming failure. Increasingly, technical debt implies, more exertion is required to determine the issues presents inside the product framework. The target of this investigation is to decide the most noteworthy contributing elements to technical debt. Bugs ascribe, seen, as the most noteworthy contributing variable to the technical debt. Dataset arrangement utilizes SonarQube. The issue with SonarQube is non-consistency. Interesting modules are required for each language. The connection investigation is authorized inside classifiers. This relationship component helps in distinguishing significant ascribes adding to technical debt. The remainder of the characteristics inside the dataset is stamped optional. To arrange the outcomes, above all else, essential characteristics are coordinated, and

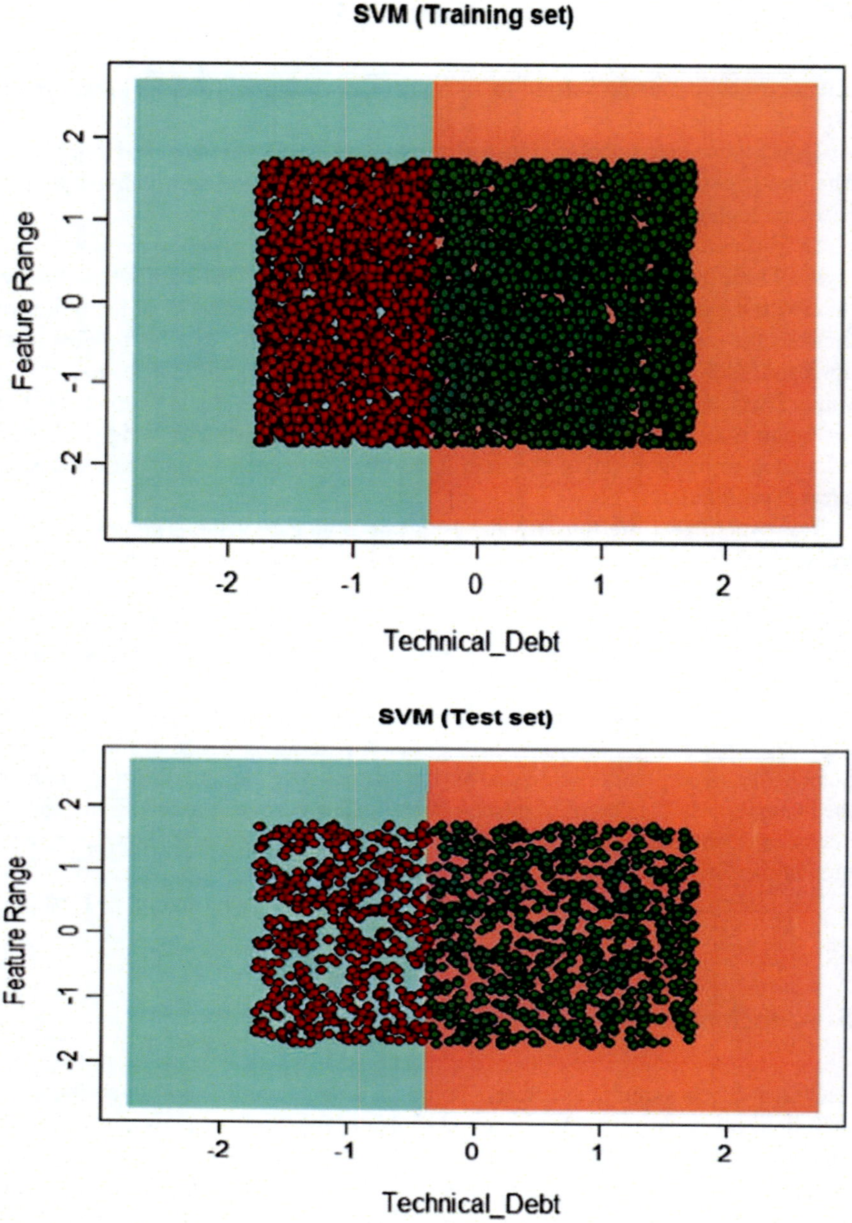

Fig. 5 SVM preparing and testing result plots

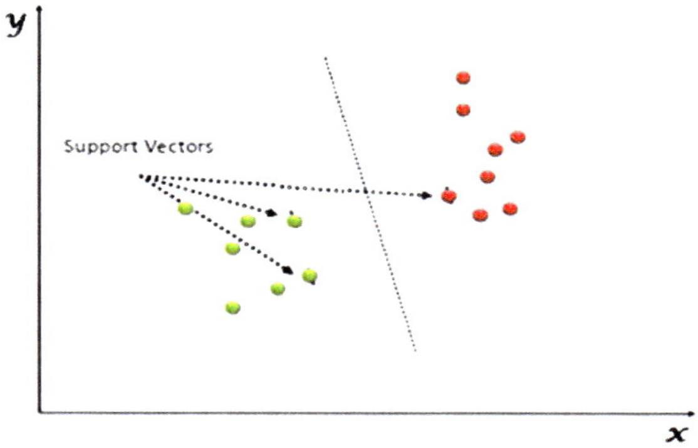

Fig. 6 SVM hyperplanes and support vectors

if there should be an occurrence of essential quality disregards the condition and neglects to group the test information then optional credits are utilized to anticipate technical debt. The grouping precision of RF classifiers arrives at 99%.

In the future, the quality recognizable proof stage Find Bugs can be utilized to anticipate technical debt and bugs from inside the product framework.

References

1. Seaman, C., Guo, Y.: Measuring, and monitoring technical debt. Adv. Comput. **82** (2011)
2. Oliveira, F., Goldman, A., Santos, V.: Managing technical debt in software projects using scrum: an action research. In: Proceedings—2015 Agile Conference, Agile 2015, pp. 50–59 (2015)
3. Nugroho, A., Visser, J., Kuipers, T.: An empirical model of technical debt and interest. In: Proceedings—International Conference on Software Engineering, pp. 1–8 (2011)
4. Griffith, I., Taffahi, H., Izurieta, C., Claudio, D.: A simulation study of practical methods for technical debt management in agile software development. In: Proceedings—Winter Simulation Conference, vol. 2015, January, pp. 1014–1025 (2015)
5. Codabux, Z., Williams, B.J., Bradshaw, G.L., Cantor, M.: An empirical assessment of technical debt practices in the industry. J. Softw. Evol. Process **29**(10) (2017)
6. Stopford, B., Wallace, K., Allspaw, J.: Technical debt: challenges and perspectives. IEEE Softw. **34**(4), 79–81 (2017)
7. Allman, E.: Managing technical debt. Commun. ACM **55**(5), 50–55 (2012)
8. Lenarduzzi, V., Saarimäki, N., Taibi, D.: The technical debt dataset. In: ACM International Conference on Proceeding Series, May, pp. 2–11 (2019)
9. Guo, Y., Seaman, C.: A portfolio approach to technical debt management. In: Proceedings—International Conference on Software Engineering, pp. 31–34 (2011)
10. Sökmen, N.: Model for software-intensive systems. **8**(9), 1628–1632 (2014)
11. Holvitie, J., et al.: Technical debt and agile software development practices and processes: an industry practitioner survey. Inf. Softw. Technol. **96**, 141–160 (2018)
12. Kuusinen, K., Gregory, P., Sharp, H., Barroca, L., Taylor, K., Wood, L.: Decomposition of US. **283**, 135–150 (2017)
13. Gat, I., Heintz, J.D.: From assessment to reduction: how cutter consortium helps rein in millions of dollars in technical debt. In: Proceedings—International Conference on Software Engineering, pp. 24–26 (2011)

14. Dos Santos, E.W., Nunes, I.: Investigating the effectiveness of peer code review in distributed software development. In: ACM International Conference on Proceeding Series, pp. 84–93 (2017)
15. Lerina, A., Nardi, L.: Investigating on the impact of software clones on technical debt, pp. 108–112 (2019)
16. Counsell, S., Hierons, R.M., Hamza, H., Black, S., Durrand, M.: Is a strategy for code smell assessment long overdue? In: Proceedings—International Conference on Software Engineering, pp. 32–38 (2010)

An e-Voting Model to Preserve Vote Integrity Employing SHA3 Algorithm

B. Patel and D. Bhatti

Abstract Democracy is a government "of the people, by the people, and for the people." But, in reality, the democratic processes are not trustworthy. One of the crucial democracy-oriented progression is an act of voting. When voting comes into consideration, it brings many feelings in reflection, like fraud, threatening, and cheating. But at some level, e-voting can introduce transparency in the process. Digitalization of any process always leads some security and privacy issues which further leads to democracy tempering. Voter authentication, privacy preservation, and integrity are the three essential principles of e-voting which needs to be indispensably implemented along with. Integrity plays a vital role to increase firm belief in such democratic processes. This research has been carried out to bring integrity in e-voting system to achieve security, trustworthiness, and coherence toward the system. To achieve the expressed objective, secure hash algorithm of family 3 has been implemented to generate message authentication code before final vote transmitted.

Keywords Integrity · E-voting · SHA3 512

1 Introduction

Election and act of voting are essential in favor of the modern community, for the well-being of democracy. Unlike Sensex, Nifty, and current affairs, a general election can have many consequences, as well as favorable advantages, due to awareness of citizens and their increased interest toward the nation and democracy. With the rapid advancement in emerging technologies, the traditional voting system has become antiquated and so, to improvise the election process, an e-voting system has been

B. Patel (✉)
Babu Madhav Institute of Information Technology, Uka Tarsadia University, Bardoli, India
e-mail: bhumika.patel@utu.ac.in

D. Bhatti
Shrimad Rajchandra Institute of Management and Computer Application, UTU, Bardoli, India
e-mail: dgbhatti@utu.ac.in

P. Karuppusamy et al. (eds.), *Sustainable Communication Networks and Application*, Lecture Notes on Data Engineering and Communications Technologies 55,
https://doi.org/10.1007/978-981-15-8677-4_36

implemented, wherein the phases of the whole election process are managed electronically through a ubiquitous network. On the other hand, this collaboration withstands a force to a substantial amount of security issues and infringement which leads to a haggled voting process with loss of privacy and integrity, resulting in the overall declining participation in the voting process and trust of the citizens. Thus, it becomes primal to introduce security measures to cope up with the pervasive situation at hand.

Moreover, due to the massive scale and distributed nature of any networks, guaranteed security, and privacy (e.g., security and safety-critical data, sensitive private information) are the major challenges [1]. In [2] it is mentioned that it is important to maintain integrity in WBSN, but it as well is equally essential, while communicating with a network of a ubiquitous nature especially in the domain of e-voting. If e-voting has no integrity checking, then it would result into an irreparable vulnerability which could cause consequences such as vote tampering and deletion of votes, thus reducing the total number of votes for a specific election candidate creating an overall impact on the result of the elections. Avoiding the implementation of integrity check, attacks like eavesdropping, spoofing, and code injection may also occur [3]. The integrity mechanisms which are based on MAC calculations, authenticate the source, verify integrity while in transit, and ensure freshness of the message. There are numerous integrity algorithms like MD5, SHA1, SHA2, and SHA3. But according to [4] until now, SHA3 has made most important efforts on providing information security, data integrity, and confidence of data origin. The SHA3 algorithm is efficient in providing information security in terms of integrity [5]. Furthermore, [1] says that the Keccak algorithm is faster than the Blake algorithm, the JH algorithm, and Skein algorithm to integrity verification of a large amount of data. It also provides better security in general when compared to the latter, as well as the SHA3 algorithm nicely works with hardware as well as software [6].

In general, information assurance is the main element which helps to achieve various objectives of information security. Information assurance helps to achieve the three main pillars of security namely confidentiality, integrity, and availability. However, to achieve any of these including integrity, it is extremely pivotal to make an infallible selection of security algorithm which would elevate the security of the system [7].

To resist against malicious tampering of the vote, in this work, an HMAC algorithm have been proposed to accomplish the aforementioned integrity requirement using SHA3 algorithm with slight modification in the algorithm.

2 Background Study

A considerable amount of work has been carried out in the area of e-voting from a different perspective. This section discusses the background study on e-voting. In paper CIEVS, the authors have used mailing and OTP concept in the voting process. When a user enters username and password, an e-mail notification is sent with voting for casting the vote. As the user clicks on the link, an OTP is sent to the registered mobile number, which upon verification allows the user to cast the vote [8]. In paper [9], the author has made a cloud-based system using multifactor authentication which takes username, password, and biological input from the user and upon verification allows the user to cast the vote. Authors of paper [10] have developed an SMS based voting system which only can be used by the mobile user, as well as the discussion is made on the hardware-based vulnerability and proposed embedded hardware to give a vote. In paper [11], authors have compared two privacy preservation algorithms—Pallier and Elgamal. They have also implemented a voting system which takes username and password, and upon verification, it allows the user to cast the vote which is then stored after being encrypted with Elgamal. In paper [12], authors have created a multimodal biometric-based e-voting system. In the system, the user is authenticated through fingerprint and facial recognition. After authentication, the user is allowed to cast the vote. Authors in [13] have created a voting system that uses customized voter IDs. The system will generate a voter ID for every user which is available for download from the mail. The users will be provided with an Aadhar-based voting machine. However, the user shall be allowed to cast the vote on the machine or online. To cast the vote, a user will be authenticated through his fingerprint and facial recognition, and after verification, he will be able to cast the vote. In these papers, authors have implemented their e-voting system, but they have not implemented any security approach that addresses integrity.

3 Proposed Model

A security framework for cloud-based e-voting has been previously proposed by us. The framework includes the component authentication, privacy preservation, and integrity. In the present study, the integrity component have been depicted along with innovative model, methodology, and implementation.

The proposed model is the component of the secured framework for cloud-based e-voting [14].

Anatomy of Integrity Component

The integrity component has been conceptualized on the SHA3 algorithms involving the theories of permutation boxes (P BOX) and substitution boxes (S BOX). Parallel to this also makes the use of salting thus proving that the proposed component follows the principles of diffusion and confusion by Claude Shannon. The integrity component being complex has been designed as a high-level view depicting the overall structure while the second design aka the low-level view presents the detailed anatomy of the encryption process.

The integrity component has two phases namely Phase I and Phase II; wherein the first phase focuses on preprocessing of data, whereas the second phase incorporates inner Keccak–Sponge construction (Fig. 1).

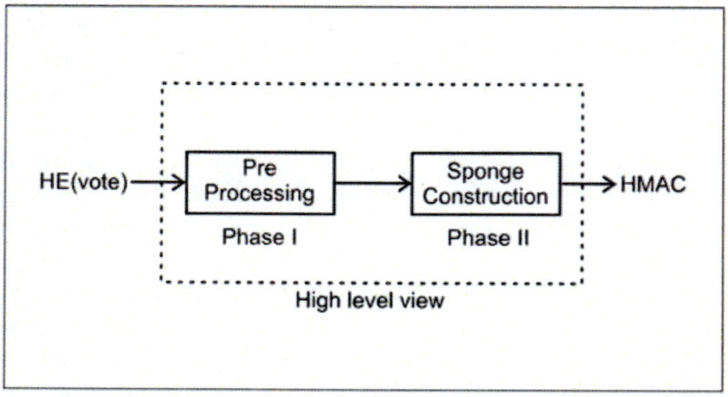

Fig. 1 A high-level view of integrity component

The high-level view of the integrity component has two phases—Phase I and Phase II. The below section explains in-depth the encryption process used in these two phases (Fig. 2).

A cloud-based EVM system using homomorphic encryption has been implemented which converts every vote into encrypted value using homomorphic encryption [15]. This encrypted vote will serve as an input in the preprocessing phase.

In the processing phase, a 34-bit encrypted vote shall be XORed with a 40-bit salt value. To do this, both the encrypted vote and salt will be first converted into binary. Here, the value taken for the sale shall be the Aadhar number of the user who has cast a vote. Padding is an important step in the preprocessing phase and must be properly dealt with to avoid attacks like the length extension [4] and so, after the values are XORed, the resultant value will be padded using the Keccak padding rule [16] where the resultant value will be padded with the pad value 0110 * 1. The final preprocessed data will be taken into phase II for sponge construction. The below section gives the mathematical notation of the preprocessing phase,

Mathematical Notation

Message (Encrypted vote)—*M*

Salt (Aadhar number)—*S*

Intermediate result—*R*

Predetermined string—*P*

Preprocessed data—*X*

Intermediate result $R =$ BIN (M) XOR BIN (S)

$$X = R||P10 * 1$$

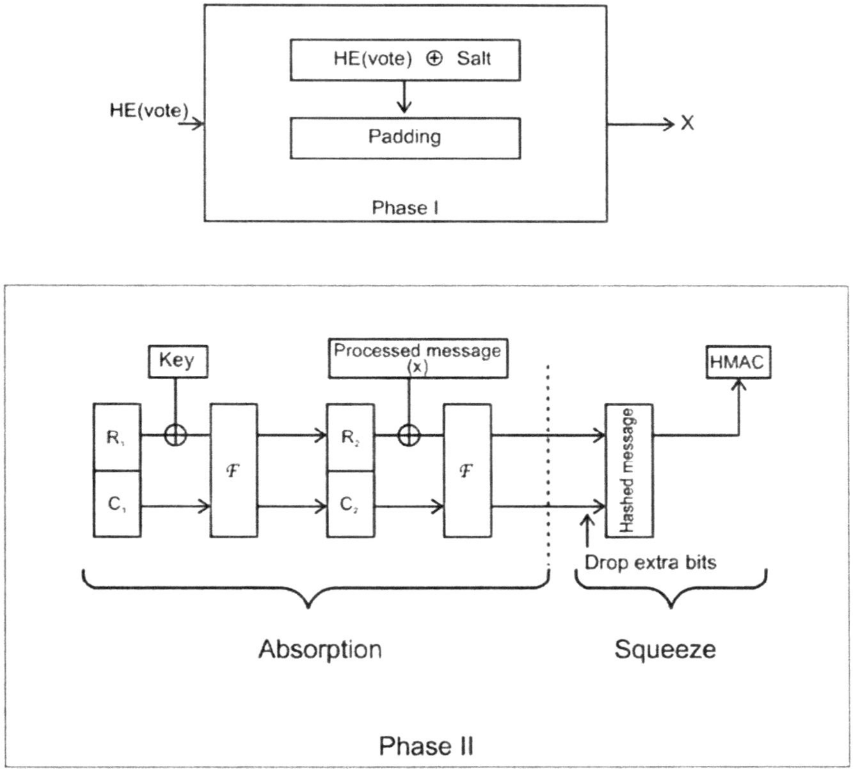

Fig. 2 Phase I preprocessing

Here *P* will change dynamically based on the mode in which SHA3 is being used. If the mode is SHA2 replacement, the minimum number of bits appended by the padding rules is four, and if the mode is SHA3, at least 6 bits will be added. After padding, message will be converted in the multiple of 576 bits.

In this phase, a three-dimensional state array of 5 × 5 × 64, i.e., 1600 bits, is taken as an initial vector. This is divided into two parts which are bit rate of 576 bits and capacity of 1024 bits. The first state is the root state in which the bit rate is XORed with a key of 576 bits. After XOR, the remaining bits of capacity are suffixed to the resultant value. In the next round, the output of the root state which is 576 bits will be XORed with the 576 padded bits of X (preprocessed data of phase 1). On this resultant value, processing of 24 rounds will be done which will incorporate five functions namely theta, pi, chi, rho, and iota as per the structure of the SHA3 algorithm.

After the processing of the rounds, a value of 576 bits will be obtained. From these 576 bits, 512 least significant bits of the result will be considered as the hash value, and the remaining bits will be discarded.

4 Results

The following shows the actual result obtained that is based on the vote data used in [15]. As discussed in the previous section of Phase I, the encrypted value of a user's casted vote is XORed with a salt value after converting both the values into binary. This salt value is the Aadhar number of the user's whose vote is being encrypted (Table 1).

The avalanche effect is evident if, when an input is changed slightly (for example, flipping a single bit), the output changes significantly. If the hash function does not exhibit the avalanche effect to a significant degree, then it has poor randomization, and thus, a cryptanalyst can make predictions about the input, being given only the output [17, 18].

Similarly, Table 2 shows the actual HMAC value after the phase II has been implemented. To check the efficiency of our approach, 1 bit in our XORed data is changed and cross-checked the corresponding HMAC of 1 bit change with the HMAC value of the original XORed data. Our analysis showed that there was a drastic difference between both the HMAC values. This, in turn, proves that the strong avalanche effect is successfully achieved.

Primal results also show that a 93% strong avalanche effect of our model have been achieved. This accuracy has been calculated based on the approach used in [17, 18].

According to NIST research [19], if SHA3 is implemented in 512-bit mode, then there are a few indirect advantages. Unlike other cryptographic hash functions, the Keccak sponge construction does not have the length extension attack [20], resistance against collision in 256 bits [21, 22], avoidance of preimage attack for 512 bits, and avoidance of second preimage attack for 512 bits.

5 Conclusion

This paper leverages the new MAC algorithm to preserve integrity in the e-voting system. SHA3 512 Keccak algorithm is somewhat modified to convert in HMAC. The latest invention of the SHA family is never been used in e-voting system to obtain integrity to the vote. Further, HMAC has been calculated as per our model and new HMAC has also been calculated after flipping the bit which proves the proposed model algorithm possesses a strong avalanche effect. Due to the sponge construction mechanism, proposed model achieves indirect benefits to the system like length extension attack, collision resistance, preimage attack, and second preimage attack.

Table 1 Result after applying phase I

Encrypted value	Binary	Aadhar number	Binary	XOR
5317389514	100111100111100001110110011001010 (33 bits)	568345362494	1000010001010100000001100000000000111110 (40) bits	1000010101101000111101101110110011110100 (40 bits)
13344258618	1100011011011000010011101000111010 (34 bits)	546236584094	111111100101110001111001101000010011110 (39) bits	111110000110101010111011110101010100100 (39 bits)
8480290938	111111001011101101111100001111010	354182364010	101001001110110111010011111101101101010	101001110001111100111110000001100010000
3871259671	11100110101111101011110000010111	304060892888	100011011001011011100010000011011011000	100011000101101110011111011101011001111
14899747940	1101111000000110000001110001100100	456897856523	110101001100001001111000000010000001011	110100100011001001001000001100001101111
4625828449	100010011101110001000101001100001	356245687845	101001011110001111001011100111000100101	101001111100010010111010100010001000100
5842821276	101011100010000100110000010011100	487965484578	111000110011101000000100110011000100010	111000011000001010000000000001101011110
3033281324	10110100110011000011001100101100	658989563254	1001100101101110110101101011110101110110	1001100111011010000110101000111001011010
15738470018	1110101010000101011111111010000010	535648499852	111110010110111001000110111100010001100	111111100011101001101101000011000001110
6755046122	110010010101000011101001011101010	568145362465	1000010001001000000110100011111000100001	1000010111011010101110111110110011001011

Table 2 Calculated HMAC and flipped data HMAC

Xored data	HMAC	HMAC after one bit flip
1000010101101000111101101110110011110100 (40 bits)	f324ae345b4e8aa453f91051602389c243f8fd320952670a75e06d54d3fd4431587d18ee0a6a774d8b9a2c6fec38ebe787465714b7356f4925e250b8724d0f00	1304934c281b9b55680989c69bc5a13c30c254a0923340a7288afcdfc7e27f01ebe129df25063d341ede5b7d0b9e0e29bff67e299e3768675ae1ae9e2c9d63fa
1111100001101010101110111101010101001000 (39 bits)	c73f26d91fd7a7106d27470ee6a6c1a8703d18e20ccc28f3dacc237c0522f8e36b9d522ef2b435efe1244923962b35ff3ead19be08a6eeef2f207c67f20ceadb	71b57201faf5a4e672fea97be2552f1e8203db4446d8827438b622edc47d9b7aab8073e46a086683311fc2a87e89296b2ec4b0fee558c526bb74f584e55aab39
1010011100011111001111100000011000100000	a02f876fa6ec6684bc784e211accad5793a9224caea864f9d08e760efb7f7b282c624a568be5d1b4208a54273689d1c6678e074f06135011c01ab888edd60e5d	136e2851ae22f48d02e9dd8bbe1a0ab11b3af371faf774a7170acf353ee1c535de45ae5634864e39b77c0c5fcab70a5b861b864ea8ee449516c54ba60666e541
1000110001011011100111110111010110011111	88dbac49cfa1ae31c7ac2bdf9d48564331e9f4af1b274dcfea1f4ca3c1b0c45b8b4b246bf41874438bb2a7af2bde70f5dd1c863a7921468509ebf42a6c929bef	a36b2f1f67e0d6851f55b03e4e63729fcede35a2a40fa4ed9121763230e8d98014f89b7f5bcce1e147f4c0474bee3fc728e5b20589f7d8491bad8344c5f303e1
1101001000110010010010000001100001101111	2f5ba9479626d84c1cac272de1ffbbb1867020d551eeae89db8e9dbf286b44da61c2564888a1a7ac1d1b788138f782156b3a57c143ecfe04b6be87b442dda012	e3e147c046654d95689a5a0ad973960fe89349eaaea3289fa7c383b69866710a75c926e254b54d97bbb948e3407ee78a56fa3c485bbc274b10fcbe8076b085e2
1010011111000100101110101000100010001000100	d8ee5c015345ee6c58de2e45edf7e20fc0cd4eeda620a1850f4938311135481ca25d0ab2dec8c99fad533be2bea491eb7b6755b940731384ecdd8f50f3f5ee21	8c985a8fa89f138e5f6e76ea8712147e5aa64140f64588941bf287e097ce24ba9af82071baf4ef685bd40c512558e8c7bfd0d89ae3d90c86e7887451ccb87d04
1110000110000010100000000000011010111110	73c495626daac955fc05a5a0cfc39cd50a84685bff17d4c4c8a2c1974867ef59c50370537323124552b9a9e1ba100627e21dbced8175428215a4ed1e28e3d8df	696600c4be2fd20c31f5e91f91f0dec5e27241da740e7595683685b91973ebdea365cfbac8c2e680b83e748f0e0d9772b4c57fa79246eae6d2f58b684ea91071
1001100111011010000110101000111001011010	3dca924b4b89ddc455b555c59882a64d1a93a2e105455b19b7a3f3cbf05fc1cf120c0e67fa039223e7b972320dc0fb96c90d5a4590900e84ec34aec5470c7b01	39c71631558e470b814429214f3c95b1b77040b6141275e44bef596eff5509a43f7fad38086ba4bc05d6492abc07bc28aa6a5384eececaca078fd34d8baa7baa
1111111000111010011011010000110000001110	0d4288ae2d55722a23486a1005cf35fae828ecaff9c0fc754641c620d5aa622a3794a2b37ad55e44914aaec83365fc6425f98211b36c6e3cfe3a51d32f5f5160	2a2b21951da56298bd7f1e0a274ddde240f977c3291fa9755cb173c8fbff49456031214085afc5f578e6af558eaea7287f1115d1c8fdcb57465d1208daa4aebb
1000010111011010101110111110110011001011	2b1a4dab378b9c3f68cfec2aeb8caa14b00d72faab10632a31eb8d3cb5c0e1e9a8faabb1a3bc5d477b40cbc58014f3e4fe7d594ff14e46046040a652f0eeb9de	4b6f5571ab96edf2d755b3d077dd93f9576dcc5dcdf5d6a13713cf8551800821348b8e6216413506d09e28210ad0a4684a4029ecd5ed3d2bc47b286484388541

References

1. Yang, Y., He, D., Kumar, N., Zeadally, S.: Compact hardware implementation of a SHA-3 core for wireless body sensor networks. IEEE Access **6**, 40128–40136 (2018). https://doi.org/10.1109/ACCESS.2018.2855408
2. Fan, L., Lei, X., Yang, N., Duong, T., Karagiannidis, G.: Secrecy cooperative networks with outdated relay selection over correlated fading channels. IEEE Trans. Veh. Technol. **66**(8), 7599–7603 (2017). https://doi.org/10.1109/TVT.2017.2669240
3. Wong, M., Haj-Yahya, J., Sau, S., Chattopadhyay, A.: A new high throughput and area efficient SHA-3 implementation. In: 2018 IEEE International Symposium on Circuits and Systems (ISCAS), 27–30 May 2018, Florence, Italy (2018)
4. Kuila, S., Chawdhury, D., Pal, M.: On the SHA-3 hash algorithms. J. Math. Inform. **3** (2015). ISSN: 2349-0632
5. Chandran, N., Manual, E.: Performance analysis of modified SHA-3. In: International Conference on Emerging Trends in Engineering, Science and Technology (ICETEST-2015), vol. 24, pp. 904–910. Elsevier (2016). https://doi.org/10.1016/j.protcy.2016.05.168
6. Wu, X., Li, S.: High throughput design and implementation of SHA-3 hash algorithm. In: 2017 International Conference on Electron Devices and Solid-State Circuits (EDSSC), IEEE, Hsinchu, pp. 1–2 (2017). https://doi.org/10.1109/EDSSC.2017.8126446

7. Chandrakar, O., Saini, J.: Empirical study to suggest optimal classification techniques for given dataset. In: 2015 IEEE International Conference on Computational Intelligence & Communication Technology. IEEE (2015)
8. Matharu, G., Mishra, A., Chhikara, P.: CIEVS: a cloud-based framework to modernize the Indian election voting system. In: 2014 IEEE International Conference on Computational Intelligence and Computing Research, Coimbatore, pp. 1–6 (2014). https://doi.org/10.1109/ICCIC.2014.7238454
9. Kishor, R.: Implementation of cloud for online election system. Int. J. Adv. Res. Comput. Sci. Manage. Stud. **3**(3) (2015)
10. Gawade, D., Shirolkar, A., Patil, S.: E-voting system using mobile SMS. IJRET: Int. J. Res. Eng. Technol. eISSN: 2319-1163|pISSN: 2321-7308
11. Azougaghe, A., Kartit, Z., Hedabou, M.: An efficient electronic voting system in a cloud computing environment. Int. Rev. Comput. Softw. (I.RE.CO.S) **10**(11) (2015)
12. Vidyasree, P., Raju, V., Madhavi, G.: Desisting the fraud in India's voting process through multi modalbiometrics. In: 2016 IEEE 6th International Conference on Advanced Computing (IACC), Bhimavaram, pp. 488–491 (2016). https://doi.org/10.1109/IACC.2016.97
13. Bhuvanapriya, R., Sivapriya, P., Kalaiselvi, V.K.G.: Smart voting. In: 2017 2nd International Conference on Computing and Communications Technologies (ICCCT), Chennai, pp. 143–147 (2017). https://doi.org/10.1109/ICCCT2.2017.7972261
14. Patel, B., Bhatti, D.: A proposed secured framework for cloud based E-voting. In: Proceeding of International Conference on New Frontiers of Engineering, Science, Management and Humanities (ICNFESMH-2018), pp. 364–369 (2018)
15. Patel, B., Bhatti, D.: Homomorphic encryption: privacy preserving amicable E-voting system. Int. J. Comput. Sci. Eng. **7**(12). E-ISSN: 2347-2693
16. Bertoni, G., Daemen, J., Peeters, M., Van Assche, G.: The KECCAK reference, Version 3.0, January 2011, https://keccak.noekeon.org/Keccak-reference-3.0.pdf
17. Roshdy, R., Fouad, M., Aboul-Dahab, M.: Design and implementation a new security hash algorithm based on MD5 and SHA-256. Int. J. Eng. Sci. Emerg. Technol. **6**(1) (2013). ISSN: 2231-6604
18. Raghuvanshi, K., Khurana, P., Bindal, P.: Study and comparative analysis of different hash algorithm. J. Eng. Comput. Appl. Sci. **3** (2014). ISSN: 2319-5606
19. National Institute of Standards and Technology, FIPS: SHA-3 Standard: Permutation-Based Hash and Extendable-Output Functions, Aug 2015. https://doi.org/10.6028/NIST.FIPS.202
20. Moreira, N., Astarloa, A., Kretzschmar, U.: SHA-3 based message authentication codes to secure IEEE 1588 synchronization systems. In: IECON 2013—39th Annual Conference of the IEEE Industrial Electronics Society, Vienna, 2013, pp. 2323–2328. https://doi.org/10.1109/IECON.2013.6699493
21. Paar, C., Pelzl, J.: SHA-3 and the hash function keccak an extension. In: Understanding Cryptography—A Textbook for Students and Practitioners. Springer, Berlin (2015)
22. Nugroho, K., Hangga, A., Sudana, I.: SHA-2 and SHA-3 based sequence randomization algorithm. In: 2016 2nd International Conference on Science and Technology-Computer (ICST), Yogyakarta, pp. 150–154 (2016). https://doi.org/10.1109/ICSTC.2016.7877365

Detection of Threshold Fall Angle of Elderly Patients for Protective Suit Purposes

Bibcy Thomas, A. Mahisha, X. Anitha Mary, Lina Rose, Christu Raja, and S. Thomas George

Abstract Falls are the second leading global cause for unintentional injury and deaths as per the world health organization [WHO]. Adults over 65 years undergo the highest number of fatal falls. This has led to the development many types of automatic fall detection systems. Here, a simple threshold angle detection method is proposed and is calculated by recording a person's falling movement. The falling movement is recorded with the help of a simple video recorder. By using the software, the time and distance of a person falling movement is calculated from its initial standing position to the ground. The result is obtained, and a 15° angle is considered as a threshold angle for elderly patients based on the detailed study of different BMI. Based on this approach, it will be feasible to establish the reliability of the development of the fall prevention system for the elderly and to be the basis for future growth.

Keywords Elderly · Fall · Threshold detection · Video recording · BMI

B. Thomas (✉) · A. Mahisha · L. Rose · S. T. George
Biomedical Engineering Department, Karunya Institute of Technology & Sciences, Coimbatore, Tamil Nadu, India
e-mail: bibcy@karunya.edu.in

A. Mahisha
e-mail: mahisha@karunya.edu.in

L. Rose
e-mail: linarose@karunya.edu.in

X. A. Mary
Robotics Engineering Department, Karunya Institute of Technology & Sciences, Coimbatore, Tamil Nadu, India
e-mail: anithajohnson2003@gmail.com

C. Raja
Electronics and Communications Department, Karunya Institute of Technology & Sciences, Coimbatore, Tamil Nadu, India
e-mail: christhuraja@karunya.edu

P. Karuppusamy et al. (eds.), *Sustainable Communication Networks and Application*, Lecture Notes on Data Engineering and Communications Technologies 55,
https://doi.org/10.1007/978-981-15-8677-4_37

1 Introduction

In some countries, the need for a convalescent home increases according to the rise of elderly patients. India is one such country. In a nursing home, some of the danger factors should be considered for elderly patients suffering from chronic diseases. In India, more deaths are caused by accidents than the illness (like high blood pressure, chronic diseases) for the elderly. However, 65% of these accidents could be avoided by enhancing the home atmosphere and providing fall prevention. Studies say that more than 32% of the elderly at the ages of over 70 experience fall accidents in a year. An accident causes emotional disturbance along with the physical injuries in the elderly. The reason that the fall accidents occur more in the elderly than other healthy populations are (1) The regular changes of aging, like poor eyesight or poor hearing, can make you more likely to fall. (2) Some medicines (for depression, high chronic diseases, diabetes, and heart patients) can upset your balance and make you fall. (3) Complex environments, unfavorable lighting situations, and slippery floors are also the cause.

Since some of the elderly are more prone to experience fall accidents than other age groups, there is need for special attention towards them [1]. Though nurses or other qualified individuals, nurses or, other qualified individuals are the best people to provide the care, the shortage of workforce is a common problem everywhere [2]. Figure 1 shows the elderly fall death rates in US which has increased to 30% from 2017 to 2016. If the rates continues to rise, by 2030 there will be 7 fall deaths every hour [3, 4].

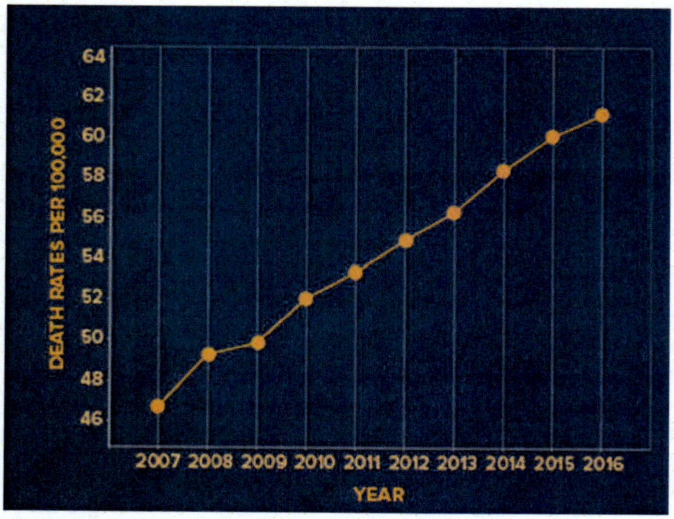

Fig. 1 Fall death rates in the elderly in the US (*Source* www.cdc.gov/HomeandRecreationalSafety)

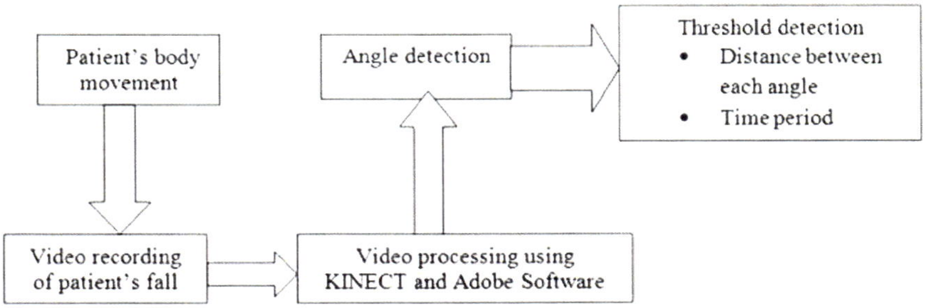

Fig. 2 Block diagram for fall angle detection

Due to more death rates of the elderly, researches focused on fall detection of the elderly. Some use special sensQ2ors (accelerometer or MPU6050), and some deploy active visual surveillance systems. Among them, an Omni-camera was used to detect the fall accident of the elderly in [5–7]. Here, we propose a new detection approach where we use a video camera to record the video and the simulation and editing software named as Adobe Illustrator [8] and Adobe Premiere Pro [9, 10] to edit the images and detect the fall angle to identify the threshold angle for fall protection systems [11–15].

2 Proposed to Enhance Fall Detection System

2.1 Block Diagram

The main objective is to detect the fall-angle threshold to prevent fall injuries for elderly patients, which could help the fall protection system to identify the falling threshold based on the BMI of the elderly people. The block diagram for fall angle detection is shown in Fig. 2. This system is designed to detect the accident fall of the elderly. The patient's falling movements were recorded using KINECT hardware and also with the help of a video recorder. The video recorded is then processed using Adobe premium pro and adobe illustrator to detect the threshold along with the time period.

The distance between each angle is measured with respect to the time period. Hence, when the patient moves beyond the given threshold value, then it is considered a fall.

2.2 Experiment

A person is made to fall in all different directions that could make an elderly patient fall (posterior fall, anterior fall, lateral falls). To affirm the chance of the proposed

Table 1 Tabular column of BMI calculation of different elderly people

Gender	Height		Weight		BMI	
	In cm	Tall/short	In kg	Thin/Obese		
Male	173	Tall	57	Normal	19 kg/cm	Normal
Female	159	Medium	64	obese	25.3 kg/cm	Obese
Female	150	Short	48	Normal	21.3 kg/cm	Normal
Male	178	Tall	88	at-risk	27.8 kg/cm	At-risk
Male	165	Medium	65	Normal	23.9 kg/cm	Normal
Male	160	Medium	96	obese class 2	37.5 kg/cm	Obese class 2
Male	176	Tall	80	obese	25.8 kg/cm	Obese
Male	184	Tall	59	mild thinner	17.4 kg/cm	Mild thinner
Female	160	Medium	70	obese	27.3 kg/cm	Obese
Female	155	Medium	60	normal	25 kg/cm	Normal
Female	178	Tall	55	mild thinner	17.4 kg/cm	Mild thinner

approach, 11 individuals were chosen with various statures and loads as our test targets, where 5 individuals are in 150–160 cm, 1 individual is in 161–170 cm, and 4 individuals are in 171–180 cm. Among them, one individual's weight is under 50 kg, 4 individuals are in 50–60 kg, 3 individuals are in 60–70 kg, 1 is 70–80 kg, and two are in excess of 81 kg. Among them, 3 individuals' BMI esteems are in 15–20, another three individuals' BMI esteems are in 21–25, and 5 individuals' BMI esteems are higher than 25, where it is discovered that every individual has an alternate BMI with the different physical structure of their body. They have categorized as obese, taller, shorter, healthy, and whether they are underweight, overweight, or normal based on the BMI calculations.

The below table visualizes the data of individuals that have been used for the project. Based on the formula given below, the BMI value for each individual has been calculated. The weight is calculated in Kilogram and the height is calculated in centimeters (Table 1).

$$\text{BMI} = \text{Weight}/(\text{Height} * \text{Height}) \tag{1}$$

2.3 Image Segmentation

When a person falls beyond a threshold angle, then it is considered a fall. Here, a person was made to fall in four different directions, which has been recorded using a video recorder. This recorded video is then converted into multiple different images of a person falling. Then each image of falling has been segmented into five different angles that are 0°, 10°, 15°, 30°, 35°. These angles help us to understand the primary

Fig. 3 Angle differentiation

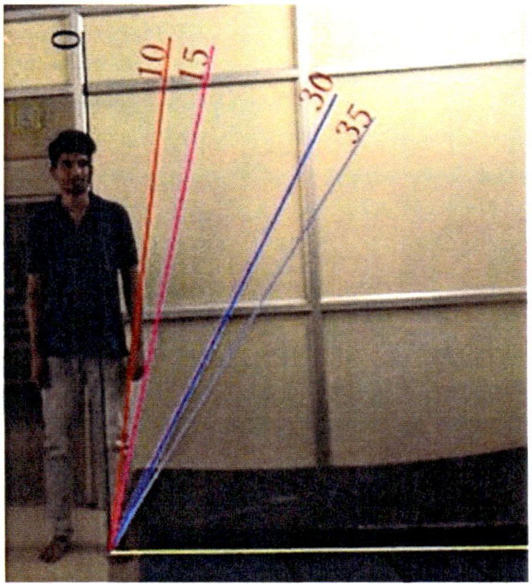

Table 2 The color used for different angle differentiation

COLOR	ANGLE
0°	Black
10°	Red
15°	Pink
30°	Blue
35°	Lavender

position when a person falls and make us known whether they could able to balance the fall quickly or not. Then with the help of adobe premiere pro and adobe illustrator, the frames and pixel of the images are calculated of a particular image.

The black line (0°) indicates the initial standing position of the person. The given box below shows the angle with its color representation.

Below given table indicates the color code for various angle that has been used in the images. These color variations indicate each line of various angle of 0°, 10°, 15°, 30°, 35° in order to differentiate different angles (Fig. 3; Table 2).

The images given below shows the various falling movement according to each segmented angle in four different directions to understand the falling position of the elderly person.

2.4 Threshold Angle Calculation

The time period is calculated based on the given images based on the pixel and frames using the software. Then these pixel and frames are then converted using the

Table 3 The table shows the calculated distance between two angles in order to calculate the threshold angle

Anterior Fall						
D(reference height) = 163, 0° pixel height = 590						
	d1		d2		Distance between two degrees	Time for each angle (sec/frames)
	Pixel value	Calculated value	Pixel value	Calculated value		
10°	99	27.32	580	160.08	162.3945	(11:24) 1:24
15°	149	41.12	570	157.32	162.6052	(11:32) 0:08
30°	300	82.8	518	142.97	165.2158	(11:46) 0:14
35°	349	96.32	497	137.17	167.6101	(11.49) 0:03

formula to obtain time period. The given table below shows the calculation of getting time period and speed from the given the values of pixel and frames that are derived from the video recording of a person falling from its initial standing position to the ground.

The same principle has been used in other directions of falling that are lateral (left and right) fall, posterior fall, etc. The time period is calculated based on the given images based on the pixel and frames using the software. Then these pixel and frames are then converted using the formula to obtain time period. The given table below shows the calculation of getting time period and distance from the given the values of pixel and frames that are derived from the video recording of a person falling from its initial standing position to the ground (Table 3).

Here, the distances between two angles are obtained by using Pythagoras theorem and thus distance is obtained. The time period is calculated by converting the pixel and frames into seconds. The time period is indicated in millisecond. These same conditions are used for all other types of fall that are anterior fall, lateral fall and posterior fall. Figure 4 explains the schematic diagram for calculating the distance between two angles. Here, D is the reference line (height of the falling patient), d_1 is a parallel line (towards falling angle) to the reference line and d_2 is a parallel line to the base line (the falling surface or ground). Based on these values the distances between the two angles are calculated with the help of PT. The calculation is shown below.

D (reference height of the patient) = 161.

0° (pixel height of patient in video) = 581.

Therefore,

$$x = \text{reference height/pixel height} \tag{2}$$

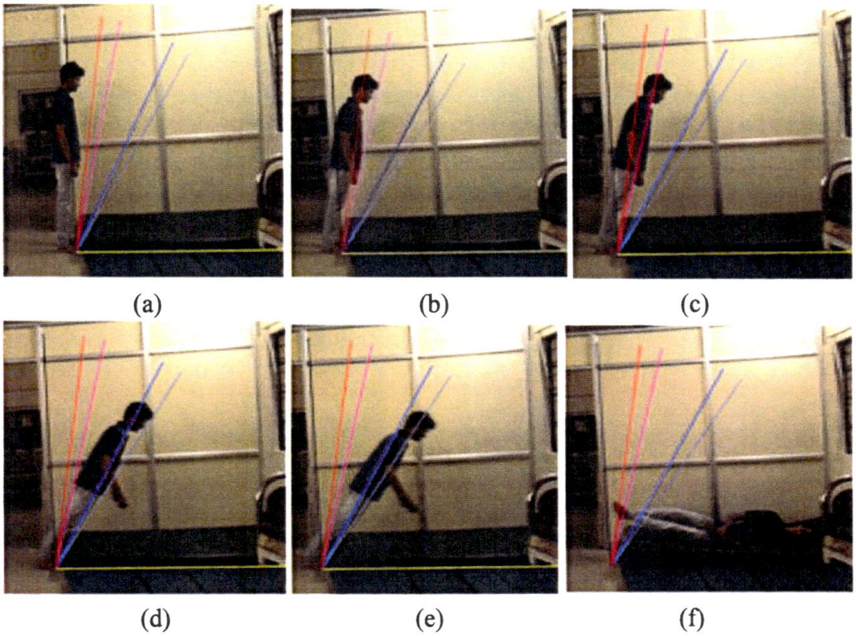

Fig. 4 The images show the different falling angles of a person while reaching the ground. **a** Person standing at initial standing position (0°). **b** Person falling slightly at an angle of 10°. **c** Person falling with an angle position 15°. **d** Person at an angle of 30°. **e** Person falling slightly with an angle position 35°. **f** Person falling and landing on the ground

Then, by substituting the values in Eq. (2) we get,

$$x = 161/581$$
$$x = 0.277$$

Now, by calculating the pixel value into the actual height of the person falling.

Then, multiply the x value with pixel value of each angle to obtain the calculated value of actual height of the person falling.

Pixel value (d_1) for 10° = 97

$$d_1 = 97 * 0.277$$
$$d_1 = 26.87$$

Similarly, we have to calculate the d_2 value.

Hence, by calculating the pixel value of d_2 value to obtain the distance.
Pixel value (d_2) for 10° = 572

$$d_2 = 572 * 0.277$$
$$d_2 = 158.4$$

Now, we will find the distance between two angles using the Pythagoras theorem.

$$c = \sqrt{a^2 + b^2} \quad (3)$$

Considering c as the distance between two angles and substituting the value of a and b as d_1 (calculated value) and d_2 (calculated value), respectively.

Therefore, by substituting the values in Eq. (3) then we get (Fig. 5; Tables 4, 5 and 6),

$$c = \sqrt{26.87^2 + 158.4^2} \quad (4)$$

$$c = 160.70$$

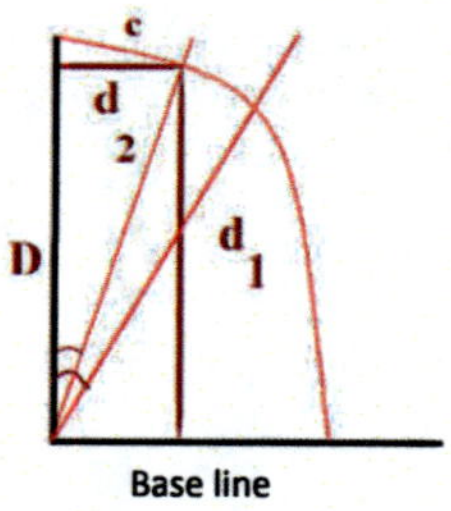

Fig. 5 Simple diagram for calculating the distance between two angles

Table 4 Calculation table of distance & time for Lateral right side fall

Lateral (Left) Fall						
D(reference height) = 161, 0° pixel height = 581						
	d1		d2		Distance between two degrees	Time for each angle (sec/frames)
	Pixel value	Calculated value	Pixel value	Calculated value		
10°	97	26.87	572	158.44	160.7023	(17:48) 0:28
15°	151	41.83	563	155.96	161.4722	(18:05) 0.57
30°	294	81.44	487	134.9	157.57691	(18:18) 0.13
35°	338	93.63	453	125.48	156.56247	(18:21) 0:03

Table 5 Calculation table of distance and time for Lateral right side fall

Lateral (Right) Fall						
D (reference height) = 163, 0° pixel height = 583						
	dl		d2		Distance between two degrees	Time for each angle (sec/frames)
	Pixel value	Calculated value	Pixel value	Calculated value		
10°	99	27.62	580	161.82	164.16022	(34:20) 1:06
15°	146	40.73	573	159.867	164.97391	(34:28) 0:08
30°	297	82.86	517	144.24	166.3459	(34:44) 0:16
35°	346	96.53	479	133.64	164.85658	(34:47) 0:03

Table 6 Calculation table of distance and time for Lateral right side fall

Posterior Fall						
D(reference height) = 163, 0° pixel height = 593						
	dl		d2		Distance between two degrees	Time for each angle (sec/frames)
	Pixel value	Calculated value	Pixel value	Calculated value		
10°	99	27.13	594	162.76	165.00562	(42:17) 0:87
15°	151	41.37	580	158.92	164.21645	(42:25) 0:08
30°	296	81.1	483	132.342	155.21474	(42:34) 0:19
35°	336	92.06	449	123.026	153.65689	(42:42) 0:05

We have also recorded the fall images of a person's movement falling from its initial stage until the person lands on the ground in KINECT format. KINECT helps us to study the posture, position, and physical condition of the body especially when the person falls. Below shows the image of a person falling using KINECT hardware (Fig. 6).

3 Result and Discussion

We have experimentally understood the fall angle detection from the height of a person and the image pixel value. Furthermore, the BMI rate of humans with different physiques is also calculated. Thus, the threshold value 15° angle has been obtained successfully by calculating the distance and time period between each angle and even by studying the fact of BMI of elderly people. This value is fixed as a constant set point during a patient's fall. The threshold angle can be considered as a set point for the inflation process for a self-inflating garment for elderly patients (Fig. 7).

Based on our research, we have seen that in many journals, they have considered a different recording method, such as an Omni-directional camera, a simple

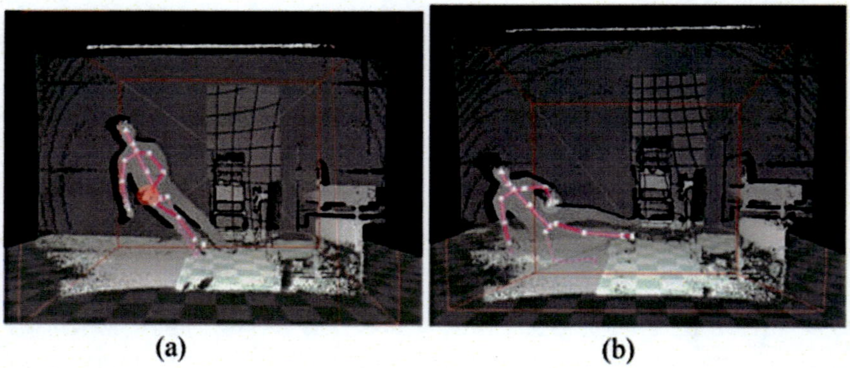

Fig. 6 The above images show the images of a person posture while falling to ground, a while falling, b while reaching ground

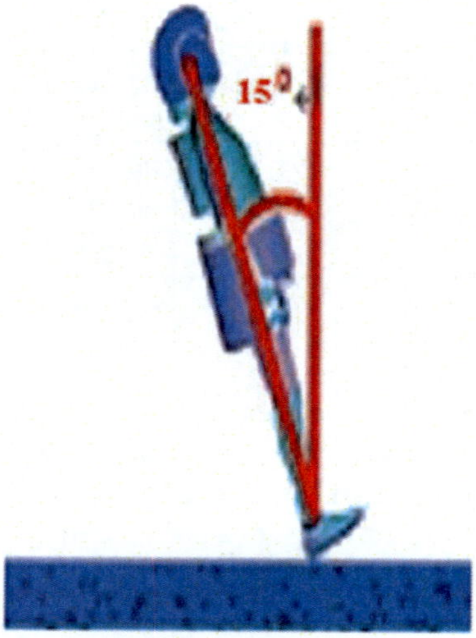

Fig. 7 15° angle is used as threshold value for fall detection

Omni camera, KINECT, etc. But we have used a basic video recorder to calculate the threshold value for fall detection. And also, there are many papers where they have considered a 20° angle as a threshold angle. But in our experiment, we have considered a 15° angle as the threshold value. This is because elderly patients, especially those who are suffering from PD or Alzheimer's disease or those with ear unbalance, vertigo, etc. cannot balance their movement, especially the fall since they are unstable in terms of fall.

4 Conclusion and Future Works

Fall detection is a significant focus area in elderly care and dramatically affects health, wellness, and disability. In this study, based on the BMI of different elderly people, an optimized fall detection threshold angle has been detected. The critical threshold angle has been determined by using the formulas and its time period and distance from its initial standing point to the falling surface. Based on the BMI value and other conditions such as physical structure and pressure required for the inflation process, the usage of cartridges has been considered. Moreover, in the future, a self-inflating garment that can be developed to protect not only the hip of an elderly patient but also the head and spine (back) of the body as spine and head are the essential parts of the body. Protection of these parts of the body can avoid the death of a person. For inflating, one can attempt to make a cylindrical form of the garment to cover the patient's body from shoulder to the knees and a helmet to protect the head with an updated false alarm system with an alert calling system.

References

1. Jo, B., Lee, J., Kim, J., Jung, S., Yang, D., Lee, J., Hong, J.: Design of wearable airbag with injury reducing system. In: ICT4AgeingWell, pp. 188–191 (2017)
2. Mcdowell, P., Dale, D.: Protective inflatable garment system for people with unstable balance. J. Emerging Trends Comput. Information Sci. **6**(1), 15–19 (2015)
3. Choi, W.J., Hoffer, J.A., Robinovitch, S.N.: Effect of hip protectors, falling angle and body mass index on pressure distribution over the hip during simulated falls. Clin. Biomech. **25**(1), 63–69 (2010)
4. Thuc, H.L.U., Tuan, P.V., Hwang, J.-N.: An effective video-based model for fall monitoring of the elderly. In: 2017 International Conference on System Science and Engineering (ICSSE), pp. 48–52. IEEE (2017)
5. Miaou, S.-G., Sung, P.-H., Huang, C.-Y.: A customized human fall detection system using omni-camera images and personal information. In: 1st Transdisciplinary Conference on Distributed Diagnosis and Home Healthcare, D2H2, pp. 39–42. IEEE (2006)
6. Huang, Y.-C., Miaou, S.-G., Liao, T.-Y.: A human fall detection system using an omni-directional camera in practical environments for health care applications. In: MVA, pp. 455–458 (2009)
7. Kepski, M., Kwolek, B.: Fall detection on embedded platform using kinect and wireless accelerometer. In: International Conference on Computers for Handicapped Persons, pp. 407–414. Springer, Berlin, Heidelberg (2012)
8. Faheem, S.S., Ahmed, Q.B., Nazir, A., Faris Iqbal, M.: Design and development of fall detector and prevention system for parkinson's patients. In: International Conference on Engineering and Emerging Technologies (2016)
9. Augustine, S.D., Augustine, R.S.: Inflatable Blast-Induced Brain Injury Prevention Device. U.S. Patent 10,001,346, issued June 19, 2018
10. Richards, S.S.: Wearable inflating system and apparatus for automatically arresting falls through gaps in structures, and method of use. U.S. Patent Application 14/941,224, filed March 17, 2016
11. Keyes, M.J.: Hip inflatable protection device. U.S. Patent 5,500,952, issued March 26, 1996
12. Tamura, T., Yoshimura, T., Sekine, M., Uchida, M., Tanaka, O.: A wearable airbag to prevent fall injuries. IEEE Trans. Inf Technol. Biomed. **13**(6), 910–914 (2009)

13. Al-Dahan, Z.T., Bachache, N.K., Bachache, L.N.: Design and implementation of fall detection system using MPU6050 Arduino. In: International Conference on Smart Homes and Health Telematics, pp. 180–187. Springer, Cham (2016)
14. Manoharan, S.: An improved safety algorithm for artificial intelligence enabled processors in self driving cars. J. Artif. Intell. **1**(2), 95–104 (2019)
15. Chandy, A.: A review on IoT based medical imaging technology for healthcare applications. J. Innov. Image Process. (JIIP) **1**(1), 51–60 (2019)

Toxic Comment Classification Using Hybrid Deep Learning Model

Rohit Beniwal and Archna Maurya

Abstract With the increasing availability of affordable data services and social media presence, our life is not untouched with 'cyber,' i.e., electronic technology. With it, various challenges and issues are faced, and the most sensitive among them is Cyberbullying. Cyberbullying is the form of 'abusive,' 'offensive,' 'inappropriate,' and 'toxic' comments that are present on the platforms. With the fear of online abuse and bullying, many people give-up on perceiving different opinions and stop expressing themselves. Nowadays, various online platforms like Quora, Wikipedia, Twitter, and Facebook have become part and parcel of everybody's life. These stages battle to viably encourage discussions, driving numerous networks to restrict or shutdown client remarks. Unfortunately, online comments with toxicity cause online badgering, bullying, and personal attacks. Therefore, toxic comment classification problem has attracted the attention of many organizations from the past few years. Hence, in this paper, we present a hybrid Deep Learning model that will detect such toxic comments and classify them according to the type of toxicity. As an outcome, we achieved the best results with an accuracy of 98.39% and an f1 score of 79.91%.

Keywords Bidirectional-gated recurrent unit · Convolution · Deep neural network · Multi-label classification · Toxic comments

1 Introduction

In today's digital era, social media provides a common platform that let users express their opinion in the form of online comments. People consider it their freedom of expression; however, many users use this fundamental right in a negative way, such

---

R. Beniwal · A. Maurya (✉)
Department of Computer Science and Engineering, Delhi Technological University, New Delhi 110042, India
e-mail: ms.archnamaurya96@gmail.com

R. Beniwal
e-mail: rohitbeniwal@yahoo.co.in

P. Karuppusamy et al. (eds.), *Sustainable Communication Networks and Application*, Lecture Notes on Data Engineering and Communications Technologies 55,
https://doi.org/10.1007/978-981-15-8677-4_38

as disrespecting other users, threatening other users, spreading fake news, cyberbullying, personal comments, toxic comments, etc. on online discussion platforms. Those comments, which are disrespectful and rude, and that force users to leave the conversation or online discussion are called "Toxic Comments." Nowadays, users face issues like abuse, harassment, cyberbullying online threats, and hate speeches, which can be classified as toxic comments. Therefore, such comments need to be recognized as quickly as possible and should be removed from the internet, but it is not that simple. It is a tedious task to filter and ban such comments. According to the Pew survey (2014) [1] about online harassment, some key findings are that every four in ten, i.e., 40% of internet users are victims of online harassment, purposeful embarrassment, and stalking. Both men and women experience a different kind of online harassment where women face it more frequently. Men experience fewer instances of verbal abuse, embarrassment, and threats, which are "less severe." In contrast, women experience badgering, such as being followed, inappropriate behavior, and threats on a more severe level.

Various online platforms are taking different initiatives to make their platform free from problems such as toxic comments, online harassment, and provide a safe online environment for their users. A few of the platforms even turn off comments for such posts based on crowdsourcing votes (upvotes/downvotes). Manually identifying such comments and flagging them is a time-consuming and challenging exercise. However, such an exercise is inefficient and not scalable. Comment classification is a classic example of Natural Language Processing (NLP) and a fundamental part of numerous applications such as web search, text mining, and sentiment analysis, etc. Hence, a wide scope of machine learning strategies has also been applied for comment classification.

The Deep Learning model identifies whether or not a comment is toxic. In the case of toxic, it further categorizes the comment in six different labels, namely toxic, severe toxic, obscene, threat, insult, and identity hate. All the listed labels are not mutually exclusive. Comment classification problems are generally also known as multi-class classification or multi-label classification [2]. Multi-class classification means data or comment belongs to only one out of the six labels. In contrast, Multi-label classification comment belongs to more than one label simultaneously. For example, a comment can be both insulting and threatening simultaneously.

In the recent past, Google and Jigsaw started a venture called "Perspective", which uses AI to distinguish toxic comments naturally [3]. The perspective API [4] score represents the impact of the comment in the discussion so that platforms can use this score to provide real-time feedback to users. In most of the cases, this model is not reliable, inclined to blunders, and the degree of toxicity is not determined.

Therefore, in this paper, we are using a hybrid Deep Learning model for improving the performance of toxic comment classification. To be particular, we analyze the dataset to understand how to process the data. Preprocessing and word embedding layer form a matrix, then we feed the matrix to Convolutional Neural Network (CNN) and Bidirectional Gated Recurrent Unit (Bi-GRU) layer respectively. Noise and important features are filtered out through CNN. After this, dense and dropout layers further perform the classification. Hence, we provide a multi-label classification model that is capable of recognizing different types of toxicity, such as severe toxic, threats, obscenity, insults, and identity-based hate. Moreover, we are providing probability estimates for each sub-type, which is conclusively strong enough to outperform 'Perspective' API's current models.

The rest of the paper is organized as follows: Sect. 2 deliberates the related work; Sect. 3 defines the proposed methodology followed by Sect. 4, which describes its implementation; Sect. 5 examines the result and analysis; finally, Sect. 6 concludes the research paper and provides direction for future work.

2 Related Work

Toxic comments have a profound impact on a user's health online as well as offline. There have been several research papers on detecting toxic comments in online discussions. Most of the work is based on machine learning, text classification, sentiment analysis, and Deep Learning neural network. Abusive comment classification work started with Yin et al. [5] paper in which they used Support Vector Machine (SVM) and Term Frequency-Inverse Document Frequency (TF-IDF) features and compared the performance with a simple TF-IDF model on a chat-style database. Nguyen [6] proposed a model for sentiment label distribution using a hybrid model of bidirectional Long Short Term Memory cell (LSTM) model with word-level embedding and Convolutional Neural Network (CNN) model with character embedding technique on Stanford Twitter sentiment corpus. This hybrid model achieved an accuracy of 86.63%. Chu and Jue [7] compared Deep Learning models such as Recurrent Neural Network (RNN) models as LSTM with word embedding, CNN model with character embedding, and CNN model with word embedding. CNN, with the character embedding model, performed best between them with an accuracy of 94%. This paper also specified that character level embedding has improved performance than word-level embedding for CNN. In the real-life for practical applications such as automatic comment moderation, CNN with word embedding was suggested.

Georgeakopoulos [3] proposed a Deep Learning approach using CNN for toxicity classification in the text classification and compare the performance with SVM, K-nearest neighbors (KNN), Naïve Bayes (NB) and Linear discriminated analysis (LDA). NB and KNN had the lowest precision and recall scores. It means that they classify some non-toxic comments to toxic comments and vice versa. CNN had the best precision and recall score. CNN also attained the best performance with an accuracy score of 92.7%. Khieu and Narwal [8] used different Deep Learning models for toxic comment classification. They used SVM, LSTM, CNN, multilayer perceptron in combination with word and character-level embedding models for toxicity detection. They evaluated their model on the Kaggle toxic comment classification challenge dataset. LSTM model achieved the best performance with an accuracy of 92.7% and an f1 score of 70.6%.

As far as the above models or approaches are concerned, we provide a model in this paper, combining both CNN and Bi-LSTM Deep Learning models with word embedding that increases the accuracy of toxic comments classification along with the F1 score.

3 Methodology

Our methodology for detecting toxic comments and classifying them according to the type of toxicity is divided into the following ten phases, namely dataset used, data preprocessing, embedding layer, convolution layer, max-pooling layer, Bi-GRU layer, global max-pooling layer, dropout layer, and two dense layers. Figure 1 represents the proposed methodology.

3.1 Dataset Used

In this research paper, we will be using the dataset available from the Kaggle Competition [9] known as the "Toxic Comment Classification Challenge". This dataset is a collection of comments from "Wikipedia's talk page edit". The dataset contains 159,571 comments that have been rated by humans for six sorts of toxicity labels such as toxic, severe toxic, obscene, threat, insult, and identity hate.

3.2 Data Preprocessing

Dataset is a collection of real-world data; however, such data is generally inconsistent and incomplete that requires data preprocessing. Data preprocessing helps us to clean, format, and organize the raw data. To achieve the same, firstly, we will remove Stopwords from the dataset. "Stop words are common English words such as, the, am, there; which do not influence the semantic of the review and removing them can reduce noise" [10]. Secondly, Tokenization will be performed. "Tokenization is the process of splitting the input into meaningful pieces" [11]. These pieces are called tokens of words. At last, the padding sequence will be used to make each comment of the dataset into the same length.

3.3 Embedding Layer

In the third phase, the Embedding layer will be used for mapping the words of comment on a vector of real numbers. For each unique word, the corresponding vector will be assigned in the space. There are various methods for creating word embeddings such as Glove, Word2vec, and FastText.

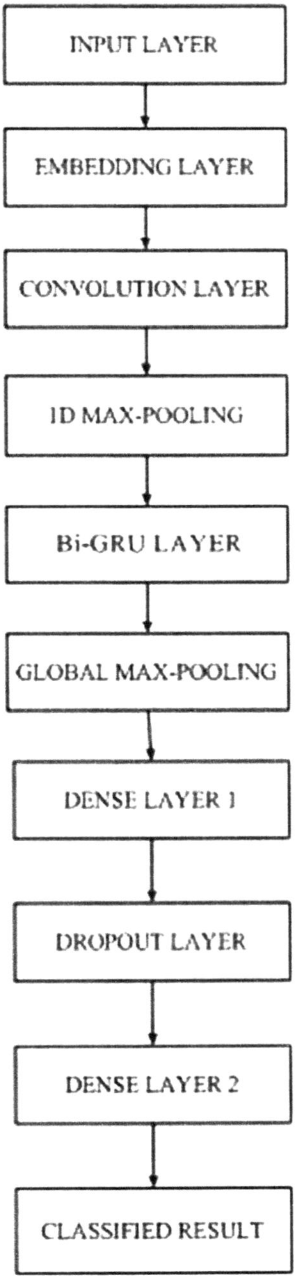

Fig. 1 Proposed hybrid deep learning model for toxic comment classification

3.4 Convolution Layer

In the fourth phase, the embedding layer output will feed into the 1D convolution layer. The Convolution layer is the center structure of the Convolutional Neural Network (CNN). The primary target of convolution will be to extract features from the input and pass its outcomes to the next layer.

3.5 Max-Pooling Layer

In the fifth phase, the yield of convolution will be transferred to the 1D Max pooling layers. The pooling layer will be used for reducing the dimension of processed data and only keeps important information. This layer will reduce the computation cost of the network. The pooling layer will diminish the features that decline the likelihood of overfitting.

3.6 Bidirectional-GRU Layer

The yield of Max pooling will be transferred into the bidirectional GRU layer. Bi-GRU is a kind of bidirectional recurrent neural network. It is almost similar to the bi-LSTM model. "GRU is faster than LSTM because it requires less calculation to refresh its concealed state" [12, 13]. GRU also overcomes the problem of vanishing gradient.

3.7 Global Max-Pooling Layer

In this phase, we will be using the 1D global max-pooling layer. The global constraint will yield the absolute most significant feature of the feature map rather than a feature window. The global max-pooling layer will reduce the dimension of input to one.

3.8 Dense Layer 1

In this phase, the dense layer will utilize the Relu (Rectified linear unit) activation function. A dense is only a normal layer of neurons in a neural system. This dense layer will receive the output from all the neurons of previous layers and help in refining the flow of gradient.

3.9 Dropout Layer

In this phase, we will use the dropout layer to dodge the issue of overfitting in the system. This layer will remove extra neurons from the neural network during the training phase and reduce the complexity of the model.

3.10 Dense Layer 2

Lastly, in the tenth phase, the last dense layer will utilize the sigmoid activation function for multi-label classification. The number of neurons in the last layer will be the number of classes in our dataset.

4 Implementation

The following sub-sections elaborate phase-wise implementation details of the proposed methodology. We used the Python programming language for the implementation of our model. TensorFlow and Keras libraries were used for building the neural network. To provide GPU support, we implemented our model on the Kaggle platform. After data preprocessing, we started building our model. We set up our input layer. As mentioned in the Keras documentation, we have to include the shape for the very first layer, and then Keras will automatically derive the shape for the rest of the layers.

4.1 Dataset Used

Exploratory Data Analysis (EDA) was performed on the dataset that gives us important information regarding the dataset and provides a way to handle the data for the model. Dataset was split into preparation and validation set into the 90:10 ratios. Table 1 defines the distribution of labels in the training set. Figure 2 shows the distribution of the number of comments vs. the length of comments that represent that the vast majority of comments were inside 500 characters length. Figure 3 shows sample instance of the dataset schema and database schema was represented as <id, comment, toxic, severe toxic, obscene, threat, insult, identity hate> that helps in understanding the dataset for further use in the model.

Table 1 Distribution of labels in the training set

Comment label	Number of comments
Toxic	15,294
Severe toxic	1595
Obscene	8449
Threat	478
Insult	7877
Identity hate	1405

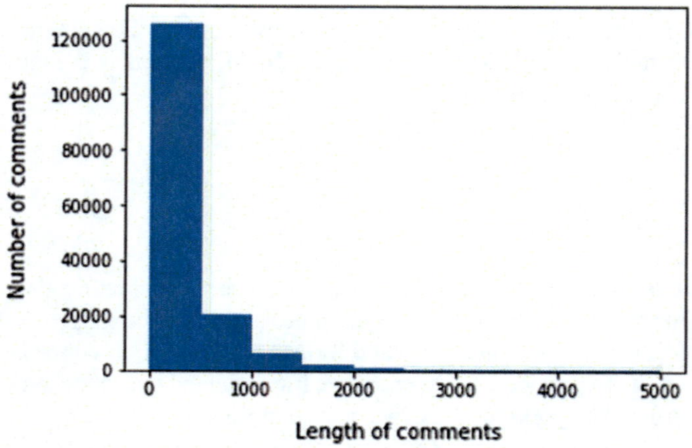

Fig. 2 Distribution of Comment length

	id	comment_text	toxic	severe_toxic	obscene	threat	insult	identity_hate
0	0000997932d777bf	Explanation\nWhy the edits made under my usern...	0	0	0	0	0	0
1	000103f0d9cfb60f	D'aww! He matches this background colour I'm s...	0	0	0	0	0	0
2	000113f07ec002fd	Hey man, I'm really not trying to edit war. It...	0	0	0	0	0	0
3	0001b41b1c6bb37e	"\nMore\nI can't make any real suggestions on ...	0	0	0	0	0	0
4	0001d958c54c6e35	You, sir, are my hero. Any chance you remember...	0	0	0	0	0	0

Fig. 3 Sample instance of the dataset

4.2 *Data Preprocessing*

Data preprocessing was used to increases the quality of the dataset and making it ready for model implementation. Firstly, we removed Stopwords using Python built-in dictionary of stopwords Nltk.corpus. Secondly, we performed Tokenization using keras.preprocessing.text() function. At last, we performed the padding sequence using keras.preprocessing.sequence() function. Padding sequence in Deep Learning was used to make each comment of the dataset into the same length. We assume the maximum length of a comment to be 500 and then add padding sequences at the end of shorter comments to make their length equal to 500. Beautiful soup Python library was utilized for hauling information out of HTML and XML records that exist in the dataset. This phase was mainly used for converting the dataset into the standard form for further implementation.

4.3 Embedding Layer

In our model, embedding will be made dependent on the tokenized text. TF-IDF tokenization will make a lattice of features which would be utilized to develop the embedding. Each comment of the dataset was changed over to a one-dimensional vector of numerical components paying little mind to the tokenization strategy. To handle the comment of variable length padding was used and the vector was filled with zeros at the end so that all the comments have equal length. We assume the size of the embedding word vector to be 240d.

4.4 Convolution Layer

In the 1D convolution layer, we used 100 filters with length 4. By 1D convolution, we understand that "the kernel used here for convolution was a one-dimensional vector" [13]. Adding CNN on the top of the GRU helps in the sense that CNN combines with the polling layer brings out the important temporal features devoid of any noise which bidirectional GRU can use more effectively. The yield of this layer is conveyed to the 1D Max pooling layers.

4.5 Max-Pooling Layer

Various types of pooling techniques could use, e.g., 1D max pooling, global, max, average, and sum that depends on the architecture of the model. We passed the output of convolution to the 1D max-pooling layer that applied the max pool operation on the window of every four characters. As the output, we get a matrix of size = number of sentences * 125 * 100, which was also called Extracted Features. These extracted features were then transferred into the bidirectional GRU layer.

4.6 Bidirectional-GRU Layer

Bidirectional GRU has only two gates, the reset and update gate. "The reset gate(r) was utilized to choose how much past state data was required to keep and to overlook" [12].

$$r^{(t)} = \sigma\left(W^{(r)}x^{(t)} + U^{(r)}h^{(t-1)}\right) \text{ (Reset gate)} \tag{1}$$

The update gate (z) worked similarly to the forget gate and input gate of the LSTM model. "The update signal $z^{(t)}$ is responsible for determining how much of the hidden state should be carried forward to the next state" [12].

$$z^{(t)} = \sigma\left(W^{(z)}x^{(t)} + U^{(z)}h^{(t-1)}\right) \text{ (Update gate)} \tag{2}$$

Bi-GRU had two units of the recurrent network, one unit to move the data in a forward way, and the second unit moved the data in a backward way with the help of reset and update gate.

4.7 Global Max-Pooling Layer

The next global max-pooling layer reduced the dimension of input to one. For example, if we had data [2,3,4,5,6,6,7] with pool length 3 yield was 4,5,6,6,7 respectively; however, if 1D global max-pooling was used, then yield equal to 7. We performed 1D global max-pooling using Keras.layers() function.

4.8 Dense Layer 1

This layer produced the yield of dimension 50. We performed this dense layer using Keras.layers() function and utilized the Relu (Rectified linear unit) activation function.

4.9 Dropout Layer

The yield of the dense layer passed to the Dropout layer, which impaired a few neurons in the following layer so that the entire system could conclude better. The dropout rate was set at 20% and performed using Keras.layers() function.

Table 2 Performance of hybrid deep learning model

Accuracy	Precision	Recall	F1 score
98.39	86.05	74.59	79.91

4.10 Dense Layer 2

The last dense layer utilized the sigmoid activation function for multi-label classification using Keras.layers() function that created six-dimensional vectors, defined as the labels of toxicity. The final output file contained the probability of labels occurring on the dataset.

We used binary cross-entropy loss function because this function is more effective on classification tasks compared to other loss functions. Adam optimizer is designed to improve the classic Stochastic Gradient Descent (SGD) optimizer. This model minimized the log-loss function using the SGD optimization algorithm. The model was trained using batch size 32 and run for 20 epochs. A 10% size validation dataset was likewise passed on along.

5 Result and Analysis

A confusion matrix is utilized to calculate the performance of the characterized model. We report the result of our model using standard Precision, Recall, Accuracy, and F1 measure [14]. "F1 score is defined as the harmonic mean of precision and recall score." Accuracy is defined as the ratio of exactly matched instances to total instances.

We partitioned the dataset into training and testing sets. The training set contains 143,613 comments and the testing set contains 15,958 comments. After training and testing, we got the best results as shown in Table 2.

6 Conclusion and Future Work

There have been ceaseless trials of experiments to detect the presence of toxic comments of various kinds on online platforms. This holds importance in the research field due to the tremendously growing online interactive communication among users. Toxic comment classification is used for detecting the toxicity in social media platforms. Our work is dedicated to finding the best possible solution for toxic comment classification. This paper has implemented a deep learning based model using convolution and bidirectional gated recurrent units that successfully performs the multi-label classification of different sorts of toxic comments. As an outcome, we achieved the best results with an accuracy of 98.39%, a precision score of 86.05%, and a recall score of 74.59%, and the computed F1 score of 79.91%.

"Common challenges for toxic comment classification among different datasets that contain out-of-jargon words with its long-range dependencies, and multi-word phrases" [15]. The limitation of this model is that it is trained on Google Jigsaw's toxic comment dataset and achieved a high accuracy on our test set. However, the proposed model may not be able to achieve the same level of performance on other datasets like twitter dataset. Another limitation is for handling the out of vocabulary words and our dataset is mostly a collection of English language comments, so our model works on the English language only. Google Jigsaw's toxic comment dataset is a collection of comments from "Wikipedia's talk page" and comments have been labeled by human raters [3, 7] because there is no standard definition of toxic labels, human raters rate the comments on their personal beliefs and therefore this dataset is skewed.

For future work, the proposed work will be extended to experiment with our model and the pre-trained word embedding techniques like Glove, Word2vec, and FastText trained on toxic comment dataset. Many enhancements may be possible to our model by adding consideration based instruments for better detection of toxic comments. Different users use different number of online platforms for discussions, so developing different models for each platform is not an efficient way to handle this problem; therefore, it is required to build a solitary framework that works over various platforms.

Reference

1. Duggan, M.: Online harassment (2017)
2. Risch, J., Krestel, R.: Toxic comment detection in online discussions. In: Deep Learning-Based Approaches for Sentiment Analysis, pp. 85–109. Springer, Singapore (2020)
3. Georgakopoulos, S.V., Tasoulis, S.K., Vrahatis, A.G., Plagianakos, V.P.: Convolutional neural networks for toxic comment classification. In: Proceedings of the 10th Hellenic Conference on Artificial Intelligence, pp. 1–6 (2018)
4. Perspective. https://perspectiveapi.com/
5. Yin, D., Xue, Z., Hong, L., Davison, B.D., Kontostathis, A., Edwards, L.: Detection of harassment on web 2.0. In: Proceedings of the Content Analysis in the WEB, vol. 2, pp.1–7 (2009)
6. Nguyen, H., Nguyen, M.L.: A deep neural architecture for sentence-level sentiment classification in twitter social networking. In: International Conference of the Pacific Association for Computational Linguistics, pp. 15–27. Springer, Singapore (2017)
7. Chu, T., Jue, K., Wang, M.: Comment abuse classification with deep learning. Von https://web.stanford.edu/class/cs224n/reports/2762092.pdf abgerufen (2016)
8. Khieu, K., Narwal, N.: Detecting and classifying toxic comments. Web: https://web.stanford.edu/class/archive/cs/cs224n/cs224n, 1184
9. Toxic Comment Classification ChallengelKaggle. https://www.kaggle.com/c/jigsaw-toxic-comment-classification-challenge/data
10. Maalej, W., Nabil, H.: Bug report, feature request, or simply praise? On automatically classifying app reviews. In: IEEE 23rd International Requirements Engineering Conference (RE), pp. 116–125 (2015)

11. Mullen, L.A., Benoit, K., Keyes, O., Selivanov, D., Arnold, J.: Fast, consistent tokenization of natural language text. J. Open Source Softw. **3**(23), 655 (2018)
12. Dey, R., Salemt, F.M.: Gate-variants of gated recurrent unit (GRU) neural networks. In: IEEE 60th International Midwest Symposium on Circuits and Systems (MWSCAS), pp. 1597–1600 (2017)
13. Saeed, H.H., Shahzad, K., Kamiran, F.: Overlapping toxic sentiment classification using deep neural architectures. In: IEEE International Conference on Data Mining Workshops (ICDMW), pp. 1361–1366 (2018)
14. Precision and recall. https://en.wikipedia.org/wiki/Precision_and_recall
15. Van Aken, B., Risch, J., Krestel, R., Löser, A.: Challenges for toxic comment classification: an in-depth error analysis. arXiv preprint arXiv:1809.07572 (2018)

Research and Development in the Networks of Cognitive Radio: A Survey

G. T. Bharathy, V. Rajendran, M. Meena, and T. Tamilselvi

Abstract The cognitive radio (CR) prototype that has been intended to scheme the future wireless communication structures is emerging progressively by utilizing its various features within the existing wireless system models. A considerable quantity of research attempts has been carried out for resolving CR disputes, and numerous technologies associated with CR in addition to vibrant accessibility of the spectrum have also been incorporated. Furthermore, software-defined radio [SDR] systems have progressed to a larger extent, where it can be utilized for implementing the CR networks. This paper is intended to provide wide-ranging investigation for deploying the increasing exploration in the field of CR systems by including all features like spectrum sensing, evaluations, numerical designing of spectrum utilization and concepts of physical layer including the modulation scheme, multiple access techniques, resource distribution, cognitive learning and strength and safety measures in CR networks. The evolving developments of CR research and disputes associated to the cost-effective CR systems are also summarized in this research study.

Keywords Cognitive radio networks · Dynamic spectrum access · Modulation schemes · Software-defined radio

G. T. Bharathy (✉) · V. Rajendran · M. Meena
Department of ECE, Vels Institute of Science, Technology & Advanced Studies (VISTAS), Chennai, India
e-mail: bharathy@jerusalemengg.ac.in

V. Rajendran
e-mail: director.ece@velsuniv.ac.in

M. Meena
e-mail: meena.se@velsuniv.ac.in

T. Tamilselvi
Department of ECE, Jerusalem College of Engineering, Chennai, India
e-mail: tamilselvi@jerusalemengg.ac.in

P. Karuppusamy et al. (eds.), *Sustainable Communication Networks and Application*, Lecture Notes on Data Engineering and Communications Technologies 55,
https://doi.org/10.1007/978-981-15-8677-4_39

1 Basics of Cognitive Networks

In free-space communications, the accessibility of the spectrum is usually explained with the help of distinct frequency, power required for transmission, method of usage, and the license interval. The above-mentioned features are also available in the existing command-and-control form meant for spectrum distribution and allotment controls the utility and outcome under the circumstances of low usage of the spectrum. The improvement in the radio spectrum utilization is required in the existing system to accommodate the novel services & applications which in turn demands the increase in the capacity and the speed in the transmission of data.

In order to achieve a better competence in the usage of the available spectrum, the aforesaid restrictions must be altered with the help of changing the existing scheme of licensing the spectrum by incorporating dynamic model in the management of the radio spectrum.

The basic concept is to incorporate more flexibility in the accessibility of the spectrum which allows CR users, i.e., the unlicensed users to have the right to utilize the radio spectrum with some amount of limitations. The heritage of the systems in wireless mode is intended to function in a devoted spectrum of frequency; it is not allowing the utilization of the flexible scheme for licensing. Consequently, the model of *Cognitive radio* (CR) evolved, to achieve the foremost objective to afford flexibility to wireless broadcast all the way incorporating the scheme of dynamic spectrum access (DSA) which enables utility of the spectrum with no loss in the benefits incorporated in the fixed spectrum distribution.

The cognitive radio is a "smarter radio" which senses the channels which consists of inputs available from a huge group of varied piece of equipments. Based on the method of sensing, the cognitive network employs complicated techniques which enable sharing of spectrum to accomplish competent wireless communication [1]. Figure 1 shows the diagrammatic representation of cognitive radio.

Spectrum access atmosphere in order to recognize and discover the spectrum atmosphere for managing and getting accustomed take the decisions on its accessibility [2]. The above mentioned network is to be operated in an environment in which several such equipments will try to access(either centrally or distributive) for making a judgment on the accessibility of the channel by means of suitable modulation method, power required for transmission, and other error control algorithms. Considerable disputes are still existing in the designing, analyzing, optimizing, developing, and deploying the Cognitive Radio Networks.

The CR represents a complicated network which imitates the brain of the human in a Dynamic

The foremost technological challenges comprises of: sensing of the spectrum; evaluations with numerical modeling of utility of the spectrum, types of modulation, schemes of multiple access techniques, spectrum allotment, power required for controlling the transmission for CRs; scattered knowledge, flexibility, collaboration for Cognitive Radios; schemes for toughness and safety in CR; broad-coverage region designing and investigation meant for networking worldwide; The disputes in

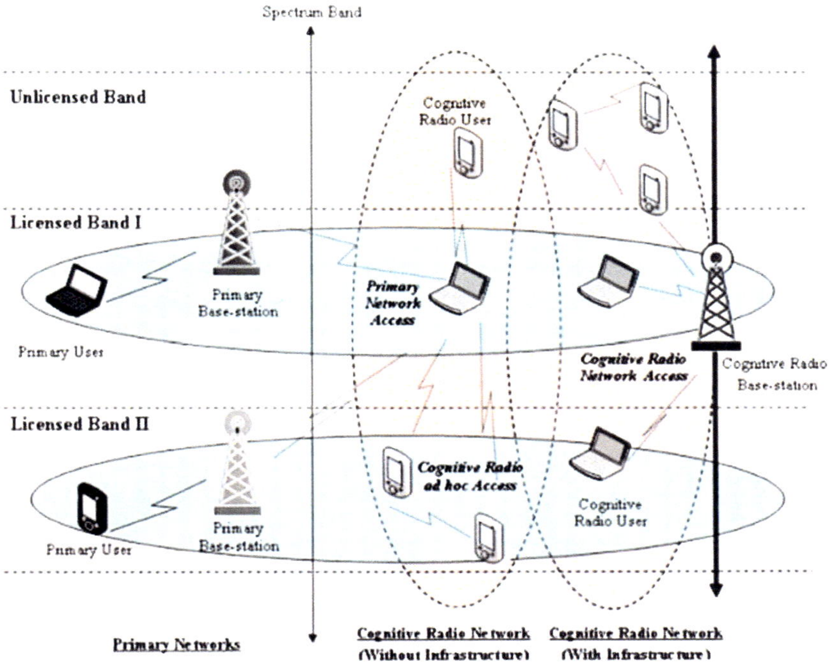

Fig. 1 Diagrammatic representation of cognitive radio [28]

the spectrum guiding principle comprise the following: expansion of realistic DSA guidelines which leads to resourceful spectrum usage, guard the privileges of primary users, and preserve the service of quality in addition to the improvement of documentation guidelines of Cognitive Radios [3]. The phrase cognitive radio be foremost utilized [4] in the year 1999 by J. Mitola III and G. Q. Maguire. In the year 2000, in his dissertation thesis [5], J. Mitola III mentioned that the cognitive networks is a expansion of the (SDR) theory in which cognitive networks are smart at the same time also conscious of its relevance required for customers.

The key cause leading to the spectrum insufficiency crisis is the conventional permanent spectrum allotment was initially recognized once Federal Communications Commission's (FCC's) Spectrum Guidelines Report [6] was made available in the year 2002. From that time onwards, FCC moved in the direction of improvements in techniques to supplementary complete utility of the spectrum, thoughts of developing Cognitive Radio techniques to assist the flexibility, competent, trustworthy employment in the field education and industry.

In the year 2004, Federal Communications Commission's provided a notice for planned guidelines, which pave way to the opportunity to allow the cognitive users to provisionally use the allotted spectrums under the circumstances that they doesn't provide dangerous intrusion to primary users. A vital landmark in the growth of cognitive radio is the endorsement by FCC in the year 2008 [7] of the unlicensed utility of the white space available in the TV depending on the sensing of spectrum in addition to discussion by means of mandated database of FCC [7]. The above mentioned white space is eye-catching for the reason that it has better-quality radio broadcast uniqueness. In the year 2010 [8], FCC made a public notice of novel guidelines to the usage of this white space by cognitive devices, wherein compulsory necessity of sensing was removed and accordingly made easy to access the spectrum with geo location- based channel allocation [8]. This has aggravated the improvement of novel wireless standards.

2 Ideology of Cognitive Radio and Dynamic Spectrum Access

The phrase cognitive radio is elaborated as follows [2]: "Cognitive radio is an intelligent wireless communication system that is aware of its ambient environment. A cognitive radio transmitter will learn from the environment and adapt its internal states to statistical variations in the existing RF stimuli by adjusting the transmission parameters (e.g., frequency band, modulation mode, and transmission power) in real-time and on-line manner."

The aforesaid characterization effectively incorporates elementary perception of cognitive radio. It facilitates the users to launch interactions between nodes of the cognitive networks. The interaction factors could be accustomed in accordance with the alteration working circumstances, necessity. There are two foremost goals of the CR; one is to develop the utility of the spectrum and also to accomplish extremely trustworthy and exceedingly competent communications in the free space.

Fundamentally the cognitive network has dual foremost operations, referred as *horizontal spectrum sharing* and *vertical spectrum sharing*. In the first category, each and every users/nodes has identical regulatory condition, while in the second category each and every users/nodes doesn't have identical regulatory condition. The process of symmetric sharing is defined as the heterogeneous network environment which has the cognitive/adaptive facility. Alternatively, if the cognitive/adaptive facility is absent then it is called as asymmetric spectrum sharing [9].

Some of the key functions of a CR device are sensing of the spectrum (spectrum sensing), managing the spectrum (spectrum management), and mobilizing the spectrum (spectrum mobility) [10]. With the help of spectrum sensing, the knowledge of the target radio spectrum can be acquired which enables the CR user to utilize it. The knowledge acquired by sensing the spectrum is utilized to analyze the availabilities to decide the accessibility of the spectrum. Figure 2 shows the cognitive cycle.

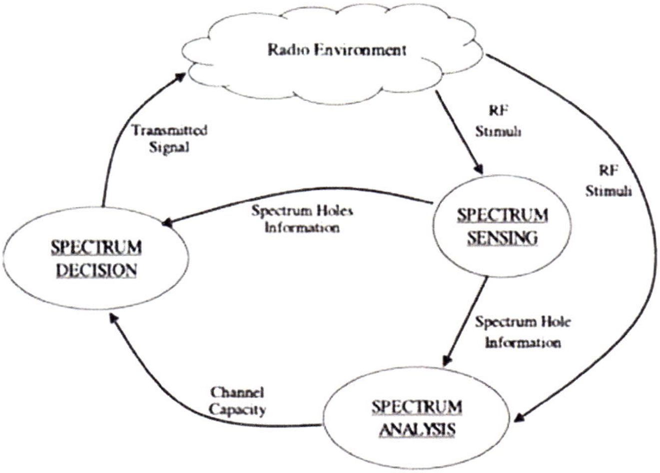

Fig. 2 Diagrammatic representation of cognitive cycle [50]

2.1 Spectrum Sensing

The main objective of the spectrum sensing method is to find out the condition of the spectrum and the action of the primary users by means of occasionally sensing the desired frequency. Meticulously, a CR transceiver identify a unoccupied spectrum called as spectrum hole in addition to it also finds out the technique to access it with no interference to the transmission by the licensed user.

The techniques utilized for sensing a spectrum are centralized or distributed. The method of centralized sensing of spectrum consists of a controller which senses the required frequency, knowledge acquired with help of this sensing is distributed to other nodes available system. The complexity will be reduced, since all the functions regarding the sensing are done by the controller. But, at the same time, it suffers from the diversity of location. The method of distributed sensing and sharing of spectrum enables the cognitive users to execute sensing of the spectrum autonomously.

2.2 Spectrum Analysis

The knowledge acquired with the help of spectrum sensing is utilized in scheduling and planning the accessibility of the spectrum by the unlicensed users. The analysis of the knowledge acquired by spectrum sensing is done to recognize the ambient RF environment and achieve information regarding the spectrum holes.

Machine learning algorithms of AI could be useful in understanding and information gathering. Consequently, a judgment on accessibility of the spectrum is done with the help of optimization of the performance of the system if the preferred goal if provided.

2.3 *Spectrum Access*

Based on the decision on the availability of the spectrum hole, the unlicensed user is allowed to access the spectrum. Access of spectrum is executed based on the cognitive medium access control (MAC) protocol which is determined to evade conflict/dangerous hindrance to the primary in addition to the cognitive users. One of the most favorable judgments on the accessibility of the spectrum is based on the ambient atmosphere and the supportive or aggressive activities of the cognitive users.

In a CRN, the cognitive users can use either a control of interference (*spectrum underlay*) methodology or a prevention of interference (*spectrum overlay*) methodology which makes the most of the spectrum holes. In the method of spectrum underlay, the cognitive users utilize the same spectrum used by the licensed users provided that the interventions provided to the main users do not go beyond the value of the threshold. Hence, this methodology necessitates a refined control schema for power of the cognitive transmitters. In the former method of spectrum overlay, the cognitive nodes requires the prior information regarding the spectrum holes to access the spectrum and to make sure that it do not offer any interference to the primary users.

2.4 *Spectrum Mobility*

Mobilizing the spectrum is the process associated with the alteration in the desired frequency of operation of CR nodes. The CR users will switch over to an idle band of spectrum, when a primary user accesses a radio channel which is currently used by the cognitive user. This process of altering the desired frequency of operation is defined as [9] spectrum hand-off [9]. When the above process occurs, the considerations/features of the protocol at various layers of the protocol stack will be altered accordingly so as to match the novel operating frequency.

The foremost principle of SDR comprises multiband process, and "plug-and-play" facility. Suppleness of software-defined radio is valuable in implementing, validating the technology of cognitive radio networks. The infrastructure of any network in CR must incorporate all the aforesaid principles. The network will assist communication among several cooperating CRs through suitable radio resource distribution (Table 1).

Table 1 Comparison of different spectrum sensing schemes

S. No.	Types of spectrum sensing	Algorithms	Key features
1.	Cooperative sensing	Centralized, distributed and external	Requires collaboration among the SU for finding the spectrum holes
2.	Interference-based modeling	Spatial modeling and SINR modeling	Algorithms are used to reduce the interference
3.	Single user sensing	Coherent, noncoherent	–
4.	Coherent sensing	Matched filter, feature detection waveform detection, correlation detection, cyclostationary detection	Needs prior knowledge of PU
5.	Non Coherent Sensing	Narrow band detection, wide band detection.	Do not require prior knowledge of PU
6.	Narrow Band Detection	Energy detection, Eigen value detection	Less efficiency
7.	Wide band detection	Sub-Nyquist, Nyquist	Efficiency is better
8.	Compressive Sensing	Distributed compressive sensing and jointly compressive sensing	Sparse representation is used

3 Cognitive Radio Network Blocks

The architecture of CRs occasionally shares the knowledge of their confined environment, communication necessities, and performances among themselves. The CRs utilizes both the knowledge obtained from their neighbor as well as the local knowledge to decide on the parameters of communication. Yet the network will be affected by the problem of hidden node huge control overhead issues [10]. The architecture of a cognitive radio system is illustrated in Fig. 3.

Fig. 3 Protocol architecture of CR system [10]

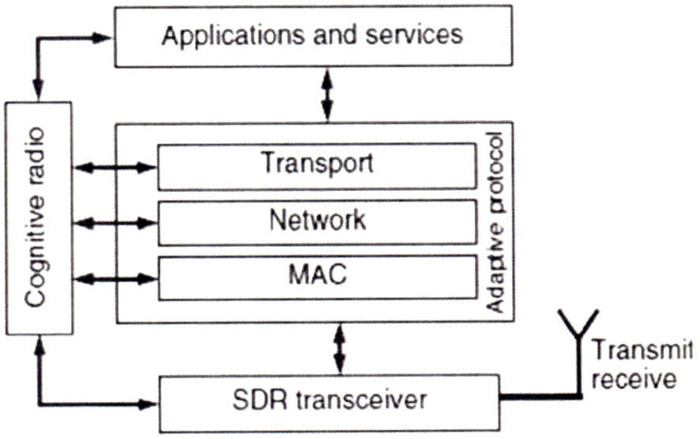

The protocols of communication available at various levels have to operate in a way that radio spectrum is utilized to its maximum; at the same time, all the policy restrictions also must be satisfied.

3.1 Physical Layer

The prime task of PHY layer is sensing of the spectrum incorporating the operations like identifying spectrum hole, interference, etc. Spectrum sensing obtains the spectrum utility features with respect to various aspects like time division, space division, frequency division, and code division. Numerous schemes like matched filter-based detection, energy-based detection, cyclostationary-based detection, etc., are available for identifying or detecting the presence of the primary users. The radio frequency front-end is realized in the PHY layer which in turn needs a large sampling rate, high-resolution, ADC with high vibrant ranges, several analog front-end circuitries, and high-end processors for computing and signal processing with less delay.

3.2 MAC Layer

The decision on transmission or consideration or exploiting the spectrum hole is done in the MAC layer. The MAC protocol operations comprise of the following: gathering the knowledge on the occupancy of the channel, synchronization of the time interval and the channel among the receiver & transmitter, initiate compromising strategy among the licensed and the unlicensed users for allotment of spectrum. The key issues in implementing the MAC protocols in cognitive radio networks comprise of finest sensing technique for the channel for accessing the multichannel main users, hidden node issues, synchronization adaptation of power. Features like counting of hop, quality of the link have to be incorporated in the routing statistics along with the parameters for management of spectrum such as spectrum holes and interference.

3.3 Network Layer

Building topology, addressing, and routing are the key functions done in the network layer. The building of topology comprises of detecting the spectrum and discovery of neighbor. Addressing could be fixed or dynamic. The routing decisions have to be done considering the topology, congestion in MAC, and reliability. The routing statistics have to incorporate features like counting of hop, quality of the link along with parameters for managing the spectrum such as interference and spectrum opportunities.

3.4 Transport Layer

The performance of MAC protocol along with the mobility of the spectrum depends on the congestion control and flow control performed in the transport layer. The performance throughput in the conventional TCP is found out by calculating RTT, probability of loss in packet probability, based on the spectrum opportunities, power required for transmission & reception along with the value of interference.

4 Spectrum Sharing in Cognitive Radio Networks

Spectrum metrics are vital data for cognitive users for accessing the spectrum of licensed users. Figure 4 shows the diagrammatic representation of primary and secondary users. To achieve this vital data, the following problems are analyzed in the literature. The performance metrics used are peak-to-average power ratio (PAPR), accuracy of detection probability (Table 2).

4.1 Spectrum Sensing

Sensing of the spectrum is a primary requirement for cognitive networks to identify the condition of the accessibility of the spectrum by the licensed users. Devoid of this vital data, the cognitive users will not be able to access the idle spectrum utility causing interference to the licensed users. Altrad et al. in [61] have explained about the Markov chain modelling considering both perfect and imperfect sensing scenarios.

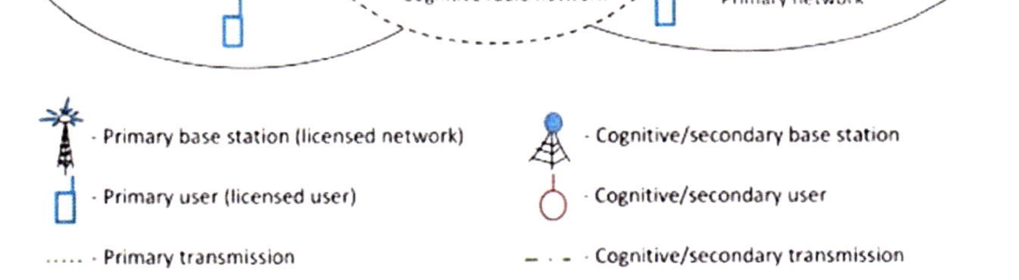

Fig. 4 Diagram representing primary and secondary users [10]

Table 2 Performance metric of CR [67]

S. No.	Parameter	Metric
1.	Situation awareness	PU probability of detection (PD) and probability of false alarm (PFA), PAPR
2.	Assessing cognitive functionality	Reasoning, decision making, planning, and learning
3.	Assessing Node Performance	Signal-to-(noise plus interference) ratio (SINR) or interference-to-noise ratio (INR), bit error rate (BER), bandwidth efficiency, and power efficiency

4.2 Interference Modeling

Cognitive users observe the spectrum interference for the following two reasons: First and foremost, for the cognitive users to ensure that their transmission does not provide any interference to the licensed users. Secondly, in the presence of interference, the cognitive users should be able to use the spectrum in a way that the broadcast necessities are fulfilled.

An outline of the regulatory requisites and the chief issues related to sensing of spectrum in cognitive radio networks is provided [11, 12]. A complete review of the various schemes for sensing the spectrum is done in [13, 14]. A complete outline of the various channel fading models is specified in [15, 16]. In [17], the authors have calculated the secondary channel capability for the average and peak power of interference power in several fading environments like Rayleigh fading and lognormal shadowing.

A random matrix theory [18] has been used to acquire the probability distributions of the test metrics. Mariani et al. calculated the effect of the unsure sequence [19] regarding the level of power of the noise in the energy-based detector category of sensing the spectrum. A discussion on a method of multitaper [21] has been provided by the authors, which is a nonparametric scheme for sensing the spectrum. In [20], a matched filter detection filter has been proposed by the authors to detect the spectrum in the presence of noise over fading channels. In [22], optimization of the duration of the spectrum sensing to increase the efficiency of the cognitive nodes along with the achievement of the desired detection likelihood has been done by the authors.

For enhanced output, cooperative-based sensing of spectrum could be utilized [23–30]. The process of cooperative-based sensing of spectrum operates as given below. Each and every cognitive user carry outs the sensing of its spectrum autonomously and decides the presence or absence of a licensed user. In a while, the cognitive users broadcast the judgments to a common node referred as the fusion node. In [23], numerous tough schemes on cooperative-based sensing of spectrum which depend on the idea of diversity in user and co-operation have been proposed by the authors.

In [26], the authors paid attention on the interpretations at the nodes are understood as to be interrelated, and consequently, a linear quadratic LQ fusion strategy has been developed. In [28], the authors offered best possible weighting which combines the

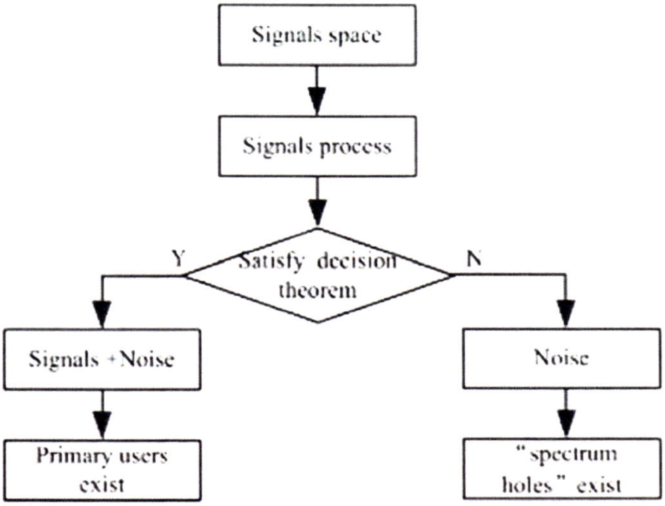

Flowchart 1 Spectrum sensing [68]

energies in a linear manner that is calculated at the cognitive nodes through which the detection probability is increased in addition to the limitation on the false alarm probability (Flowchart 1).

To decrease the overhead in communication Meng et al. has suggested [29] a matrix completion and joint sparsity algorithms for sensing the spectrum which depends on very few observations.

The authors in [30] have suggested a location-based cooperative spectrum sensing method to take a decision on the availability of primary users. The primary system of network is designed as a arbitrary geometric system, which is understood to have a prior knowledge on the locations of its users. The concept of truncated stable distributions is utilized for the designing [31]. The work in [32] explains the initial studies on occupancy of the spectrum and intrusion in cognitive radio systems. In [32], the authors explained the prospects, disputes, and interaction limits available in the cognitive radio systems. In [33], Datla et al. proposed a structure of spectrum calculation and preprocessing of calculated information. The structure is estimated by applying it in a real-time scenario. In [34], the authors have explained a numerical representation for occupation of the spectrum and the important features of the representation have been estimated using the measured information. In [35], the authors have offered in detail on an operation of measurement of the spectrum. Also, it has presented a stochastic representation which depends on modified beta distribution. The authors in [36] explained the probable shortcoming of Poisson representation for licensed user action and proposed a novel representation which is dependent on the metrics like temporal metrics, filter clustering. In [63], Manesh et al. have compiled a survey on different types of interference modeling like spatial modeling and SINR modeling and also evaluated the cognitive interference power. Figure 5 represents

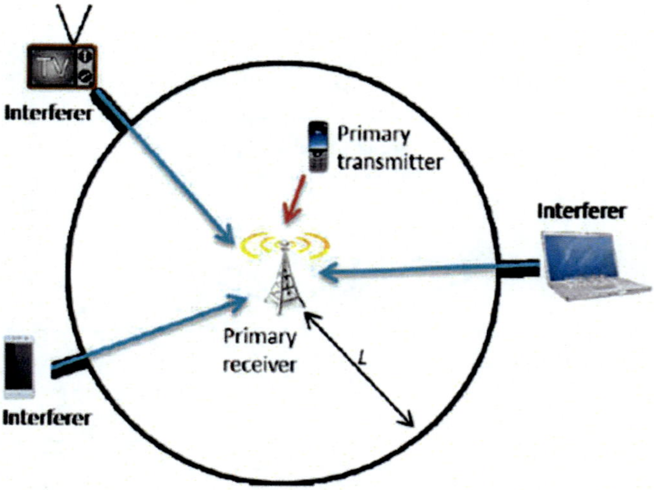

Fig. 5 Interference model [63]

the model of interference in CR. Maric S et al. in [64] have explained about algorithm based on the belief propagation to mitigate the PUEA.

5 Metrics for Performance Evaluation in CR

Zhao et al. IN [67] have given a detailed analysis of performance evaluation of cognitive radios.

6 Types of Modulation for Cognitive Radio Networks

In [37], Bogucka et al. presented an outstanding discussion on two types of multiple carrier modulation OFDM and non-OFDM to be used in wireless systems. In [38], Kollar and Horvath examined several modulation methods like DFT OFDM, OFDM with constant envelope and multicarrier with filter bank for cognitive networks along with the merits and demerits. In [39], a detailed analysis of non-OFDM and multi-carrier CDMA method was discussed in the perspective of a DSA cognitive network. In [40], the authors presented a detailed review of the up-to-date techniques particularly modeled to guard the usage of the spectrum by the primary users from likely interference provided by cognitive users with NC-OFDM technique.

In [41], a detailed analysis has been done taking into account the effects of sensing of spectrum and the traffic of licensed user for various channels under fading, the characteristics of adaptive modulation for CR with opportunistic access. The adaptive continuous-rate method and the adaptive discrete-rate method both have been taken into account. In [42], Budiarjo et al. proposed a modulation scheme which recognizes the existence among the cognitive radio dependent scheme by which the cognitive users are hidden for the main users.

In [43], various methods for identifying intrusive data, evaluating the distance linking victims and intrusive systems, have been explained. Zamanian et al. [44] modeled a classifier to be used for linear digital modulation whose operation depends on the k-means clustering procedure. In [45], Ye et al. proposed a original digital modulation categorization scheme for cognitive radio features. In [46], the author has explained a schema for modulation categorization whose operation depends on the SVM.

7 Developments and Security Considerations in Cognitive Radio Networks

Clancy and Goergen [47] presented a complete review and investigation on the issues of security in a cognitive radio networks. Chen et al. [48] methodically illustrated that an attack in which the attacker will imitate the licensed user causing serious interference issues which in turn reduces the utilization of the spectrum considerably. A novel transmitter verification method called the localization-based defense (LocDef) method has also been proposed by the authors to solve aforementioned attack issue. LocDef method can identify if the signal is from a licensed user or not with the help of an estimated location of the transmitter and from the features of the signal itself which is the received signal strength (RSS). In [49], the authors described attack called as data falsification as illustrated in Fig. 6, which reduces the efficiency of distributed sensing of spectrum in cognitive radio networks.

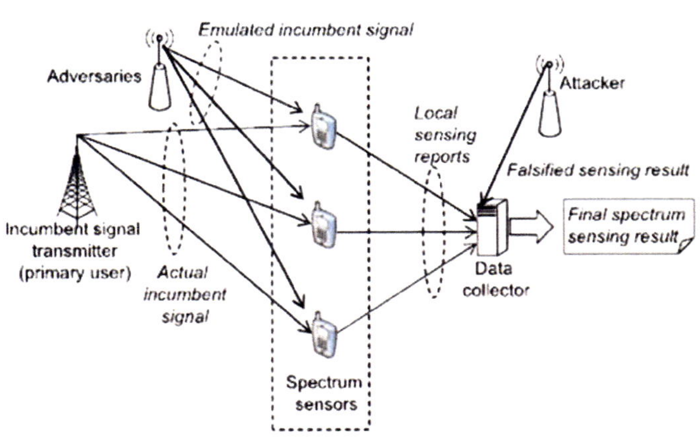

Fig. 6 Diagrammatic representation of emulation attack in spectrum [20]

In this method, an invader gains the spectrum access over other cognitive users by sending a signal which emulates the primary user signal. In [50], the authors have provided a detailed review of the cognitive network and various types of spectrum sensing schemes. Fatima Salahdine et al. in [62] have proposed a recovery algorithm for compressive spectrum sensing which increases the process of scanning the spectrum.

In [51] (different from [52]), the channel metrics were supposed to be well-known. In [51], the assault and the guard strategy are designed in the form of a zero-sum game. The defender performs the channel sensing process and the attacker jams the channel. Ajmery Sultana et al. in [65] has proposed and explained the energy harvesting in cognitive radio networks. Peter He et al. in [66] has proposed underlay and overlay method for managing the spectrum sharing.

8 Financially Viable CRN

Pricing is a significant concern in the CR system, which inspires the main and cognitive nodes to distribute the existing spectrum with a method referred as *spectrum trading* [53].

8.1 Spectrum Auction

Auction is a conventional however competent method to allocate goods and services to the consumers. The concept of auction could be obviously used in the cognitive radio networks since the cost of the spectrum could not be obtained accurately in prior hand [54]. In the concept of auction, the licensed nodes propose its requests, and the cognitive nodes propose its bids to vend and procure the spectrum. Several auction designs have been used in cognitive radio like single ended, double ended, and combinational auctions. An auction dependent procedure was proposed in [55], for helping the free space users in order to fight for the access of the channels in a very reasonable approach.

In [56], Kasbekar and Sarkar described a scheme based on the auction of spectrum which allows the selling of the channels to the licensed and cognitive users is done by the regulator. The major network in every channel is given superior precedence to send information in preference to the cognitive networks; at the same time, all cognitive networks are given similar priority.

In [57], Sodagari et al. explained the reliability of cognitive nodes to notify the estimation along with the appearance–disappearance time in the auction of the spectrum carried out via the licensed user.

In [58], the authors proposed a trading structure for the universal trading spectrum state of affairs as illustrated in Fig. 7. A hierarchical game design was proposed which obtains the stable way out for the licensed users for choosing the spectrum

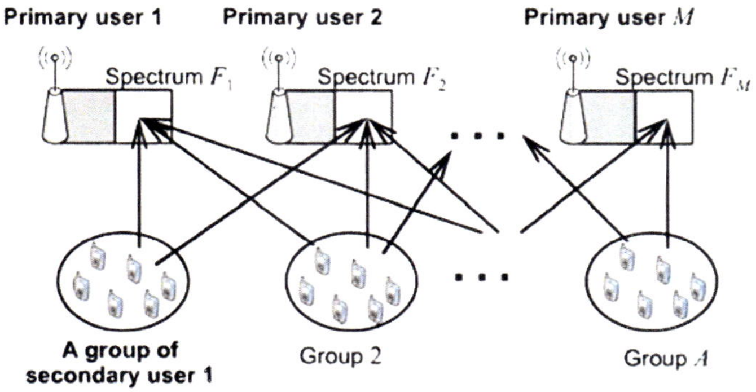

Fig. 7 Spectrum trading in cognitive radio [20]

cost to increase their gain for a given performance reduction of the primary users for distributing existing spectrum with the cognitive users. Ji and Liu [59] proposed a organized method which avoids the collision in cognitive network. In [59], the allotment of spectrum for multiple holders and cognitive nodes was designed as game to achieve the optimization in the efficiency. In [60], Yang et al. presented value-based control of spectrum access which allows the main users sell spectrum to the cognitive users.

9 Conclusion and Future Scope

Cognitive radio network affords a novel prototype for modeling the intellectual wireless systems to alleviate the spectrum insufficiency issue and offer a noteworthy profit in achieving the spectral effectiveness. A detailed analysis of the research actions in CR is reviewed in this paper. The key problems in the modeling of cognitive radio interaction networks have been analyzed and the associated literature review has also been presented. The chronological note of the CR has been provided to offer an inspiration to the vibrant and competent wireless systems. Several schemes for sharing of spectrum in cognitive network have been discussed. The safety and cost-effective problems have been considered and summarized.

A cost-effective and spectral efficient cognitive network can be designed and implemented by incorporating the best spectrum sensing schema, modulation method and multiple access techniques as a future work.

References

1. Guest Editorial: Special Issue on cognitive radio: a path in the evolution of public wireless networks. KICS/IEEE J. Commun. Netw. (JCN) (2008)
2. Haykin, S.: Cognitive radio: brain-empowered wireless communications. IEEE J. Selected Areas Commun. **23**(2), 201–220 (2005)
3. Steenkiste, P., Sicker, D., Minden, G., Raychaudhuri, D. (eds).: Future directions in cognitive radio network research. NSF Workshop Report, 9–10 March 2009 (Online) (Available at: http://www.cs.cmu.edu/~prs/NSF_CRN_Report_Final.pdf
4. Mitola III, J., Maguire Jr., G.Q.: Cognitive radio: making software radios more personal. IEEE Personal Commun. **6**(4), 13–18 (1999)
5. Mitola J, III.: Cognitive radio: an integrated agent architecture for software defined radio. Doctor of Technology Thesis, Royal Inst. Technol. (KTH), Stockholm, Sweden (2000)
6. Federal Communications Commission. Spectrum Policy Task Force report. ET Docket No.02-135 (2002)
7. Federal Communications Commission. In the matter of unlicensed operation in the TV broadcast bands. FCC 08-260, November 2008
8. Federal Communications Commission. Unlicensed operation in the TV broadcast bands and additional spectrum for unlicensed devices below 900 MHz and in the 3 GHz band. FCC 10-174, ET Docket No. 04-186 and 02-380, September 2010
9. Hossain, E., Niyato, D., Han, Z.: Dynamic Spectrum Access and Management in Cognitive Radio Networks. Cambridge University Press (2009)
10. Biglieri, E., Goldsmith, A.J., Greenstein, L.J., Mandayam, N., Poor, H.V.: Principles of Cognitive Radio. Cambridge University Press, December 2012
11. Ghasemi, A., Sousa, E.: Spectrum sensing in cognitive radio networks: requirements, challenges and design trade-offs. IEEE Commun. Mag. **46**(4), 32–39 (2008)
12. Ghasemi, A., Sousa, E.S.: Interference Aggregation in Spectrum-Sensing Cognitive Wireless Networks, IEEE
13. Meena, M., Monisha, M., Sharanya, C., Rajendran, V.: Modeling of cognitive radio system for the improvement of spectrum utilization. Int. J. Eng. Technol. **7**(2.33), 231–234 (2018)
14. Zeng, Y., Liang, Y.C., Hoang, A.T., Zhang, R.: A review on spectrum sensing for cognitive radio: challenges and solutions. EURASIP J. Adv. Signal Process. 2010; Article ID 381465
15. Sutton, P.D., Nolan, K.E., Doyle, L.E.: Cyclostationary signatures in practical cognitive radio applications. IEEE J. Selected Areas Commun. **26**(1), 13–24 (2008)
16. Molisch, A.F., Shafi, M., Greenstein, L.J.: Propagation issues for cognitive radio. Proc. IEEE Special Issue Cogn. Radio **97**, 787–804 (2009)
17. Ghasemi, A., Sousa, E.S.: Fundamental limits of spectrum-sharing in fading environments. IEEE Trans. Wireless Commun. **6**(2), 649–658 (2007)
18. Zeng, Y., Liang, Y.C.: Eigenvalue based spectrum sensing algorithms for cognitive radio. IEEE Trans. Commun. **57**(6), 1784–1793 (2009)
19. Mariani, A., Giorgetti, A., Chiani, M.: Effects of noise power estimation on energy detection for cognitive radio applications. IEEE Trans. Commun. **59**(12), 3410–3420 (2011)
20. Thangalakshmi, B., Bharathy, G.T.: Matched filter detection based spectrum sensing in cognitive radio networks. Int. J. Emerging Technol. Comput. Sci. Electronics (IJETCSE) **22**(2), 151–154, ISSN: 0976-1353, May 2016
21. Haykin, S., Thomson, D., Reed, J.: Spectrum sensing for cognitive radio. In: Proceedings of the IEEE, Special Issue on Cognitive Radio, vol. 97, issue 5, pp. 849–877, May 2009
22. Liang, Y.C., Zeng, Y., Peh, E.C.Y., Hoang, A.T.: Sensing throughput tradeoff for cognitive radio networks. IEEE Trans. Wireless Commun. **7**(4), 1326–1337 (2008)
23. Letaief, K.B., Zhang, W.: Cooperative communications for cognitive radio networks. Proc. IEEE **97**(5), 878–893 (2009)
24. Ganesan, G., Li, Y.: Cooperative spectrum sensing in cognitive radio, part I: two user networks. IEEE Trans. Wireless Commun. **6**(6), 2204–2213 (2007)

25. Ganesan, G., Li, Y.: Cooperative spectrum sensing in cognitive radio, part II: multiuser networks. IEEE Trans. Wireless Commun. **6**(6), 2214–2222 (2007)
26. Unnikrishnan, J., Veeravalli, V.V.: Cooperative sensing for primary detection in cognitive radio. IEEE J. Selected Topics Signal Process. **2**(1), 18–27 (2008)
27. Zhang, R., Liang, Y.C.: Exploiting multi-antennas for opportunistic spectrum sharing in cognitive radio networks. IEEE J. Selected Topics Signal Process. **2**(1), 88–102 (2008)
28. Quan, Z., Cui, S., Sayed, A.H.: Optimal linear cooperation for spectrum sensing in cognitive radio networks. IEEE J. Selected Topics Signal Process. **2**(1), 28–40 (2008)
29. Meng, J., Yin, W., Li, H., Hossain, E., Han, Z.: Collaborative spectrum sensing from sparse observations in cognitive radio networks. IEEE J. Selected Topics Commun. **29**(2), 327–337 (2011). (Special Issue on Advances in Cognitive Radio Networking and Communications)
30. Choi, K.W., Hossain, E., Kim, D.I.: Cooperative spectrum sensing under random geometric primary user network model. IEEE Trans. Wireless Commun. **10**(6), 1932–1944 (2011)
31. Lu, L., Zhou, X., Onunkwo, U., Li, G.Y.: Ten years of research in spectrum sensing and sharing in cognitive radio. EURASIP J. Wireless Commun. Netw. **28** (2012)
32. Roberson, D.A., Hood, C.S., LoCicero, J.L., MacDonald, J.T.: Spectral occupancy and interference studies in support of cognitive radio technology deployment. In: Proc. of First IEEE Workshop on Networking Technologies for Software Defined Radio Networks, pp. 26–35, September 2006
33. Datla, D., Wyglinski, A.M., Minden, G.J.: A spectrum surveying framework for dynamic spectrum access networks. IEEE Trans. Vehicular Technol. **58**(8), 4158–4158 (2009)
34. Ghosh, C., Pagadarai, S., Agrawal, D., Wyglinski, A.M.: A framework for statistical wireless spectrum occupancy modeling. IEEE Trans. Wireless Commun. **9**(1), 38–44 (2010)
35. Wellens, M., Mähönen, P.: Lessons learned from an extensive spectrum occupancy measurement campaign and a stochastic duty cycle model. Mobile Netw. Appl. **15**(3), 461–474 (2010)
36. Canberk, B., Akyildiz, I.F., Oktug, S.: Primary user activity modeling using first-difference filter clustering and correlation in cognitive radio networks. IEEE/ACM Trans. Netw. **19**(1), 170–183 (2011)
37. Bogucka, H., Wyglinski, A.M., Pagadarai, S., Kliks, A.: Spectrally agile multicarrier waveforms for opportunistic wireless access. IEEE Commun. Mag. **49**(6), 108–115 (2011)
38. Kollar, Z., Horvath, P.: Physical layer considerations for cognitive radio: Modulation techniques. In: Proceedings of 2011 IEEE 73rd Vehicular Technology Conference (VTC Spring), pp. 1-5, 15–18 August 2011
39. Rajbanshi, R., Chen, Q., Wyglinski, A.M., Minden, G.J., Evans, J.B.: Quantitative comparison of agile modulation techniques for cognitive radio transceivers. In: Proceedings of 4th IEEE Consumer Communications and Networking Conference (CCNC 2007), pp. 1144–1148, 11–13 January 2007
40. Kryszkiewicz, P., Bogucka, H., Wyglinski, A.M.: Protection of primary users in dynamically varying radio environment: practical solutions and challenges. EURASIP J. Wireless Commun. Netw. **23** (2012)
41. Chen, Y., Alouini, M.S., Tang, L.: Performance analysis of adaptive modulation for cognitive radios with opportunistic access. In: Proceedings of 2011
42. Ye, Z., Memik, G., Grosspietsch, J.: Digital modulation classification using temporal waveform features for cognitive radios. In: Proceedings of IEEE International Symposium on Personal, Indoor and Mobile Radio Communications (PIMRC 2007), pp. 1–5, 3–7 September 2007
43. Batra, A., Lingam, S., Balakrishnan, J.: Multi-band OFDM: a cognitive radio for UWB. In: Proceedings of IEEE International Symposium on Circuits and Systems (ISCAS 2006), 21–24 May 2006
44. Zamanian, M., Tadaion, A.A., Sadeghi, M.T.: Modulation classification of linearly modulated signals in a cognitive radio network using constellation shape. In: Proceedings of 2011 7th International Workshop on Systems, Signal Processing and their Applications (WOSSPA), pp. 13–16, 9–11 May 2011

45. Kostylev,V.I.: Energy detection of a signal with random amplitude. In: IEEE International Conference on Communications, ICC 2002, vol. 3, pp. 1606–1610 (2002)
46. Ramon, M.M., Atwood, T., Barbin, S., Christodoulou, C.G.: Signal classification with an SVM-FFT approach for feature extraction in cognitive radio. In: Microwave and Optoelectronics Conference (IMOC), SBMO/IEEE MTT-S International, pp. 286–289, 3–6 Nov. 2009
47. Clancy, T.C., Goergen, N.: Security in cognitive radio networks: threats and mitigation. In: Proc. of 3rd International Conference on Cognitive Radio Oriented Wireless Networks and Communications (CrownCom) 2008, pp. 1–8, May 2008
48. Chen, R., Park, J.-M., Reed, J.H.: Defense against primary user emulation attacks in cognitive radio networks. IEEE J. Selected Areas Commun. **26**(1), 25–37 (2008)
49. Chen, R., Park, J.-M., Hou, Y.T., Reed, J.H.: Toward secure distributed spectrum sensing in cognitive radio networks. IEEE Commun. Mag. **46**(4), 50–55 (2008)
50. Thangalakshmi, B., Bharathy, G.T.: Review of cognitive radio network. Int. J. MC Square Sci. Res. **7**(1), 10–17 (2015)
51. Li, H., Han, Z.: Dogfight in spectrum: combating primary user emulation attacks in cognitive radio systems, part I: known channel statistics. IEEE Trans. Wireless Commun. **9**(11), 3566–3577 (2010)
52. Li, H., Han, Z.: Dogfight in spectrum: combating primary user emulation attacks in cognitive radio systems, part II: unknown channel statistics. IEEE Trans. Wireless Commun. **10**(1), 274–283 (2011)
53. Niyato, D., Hossain, E.: Spectrum trading in cognitive radio networks: a market-equilibrium-based approach. IEEE Wireless Commun. Mag. **15**(6), 71–80 (2008)
54. Huang, J., Berry, R., Honig, M.: Auction-based spectrum sharing. Mobile Netw. Appl. **11**(3), 405–418 (2006)
55. Sun, J., Modiano, E., Zheng, L.: Wireless channel allocation using an auction algorithm. IEEE J. Selected Areas Commun. **24**(5), 1085–1096 (2006)
56. Chen, Y., Wu, Y., Wang, B., Liu, K.J.R.: Spectrum auction games for multimedia streaming over cognitive radio networks. IEEE Trans. Commun. **58**(8), 2381–2390 (2010)
57. Sodagari, S., Attar, A., Bilén, S.G.: On a truthful mechanism for expiring spectrum sharing in cognitive radio networks. IEEE J. Selected Areas Commun. **29**(4), 856–865 (2011)
58. Niyato, D., Hossain, E., Han, Z.: Dynamics of multiple- seller and multiple-buyer spectrum trading in cognitive radio networks: a game theoretic modeling approach. IEEE Trans. Mobile Comput. **8**(8), 1009–1022 (2009)
59. Ji, Z., Liu, K.J.R.: Multi-stage pricing game for collusion- resistant dynamic spectrum allocation. IEEE J. Selected Areas Commun. **26**(1), 182–191 (2008)
60. Yang, L., Kim, H., Zhang, J., Chiang, M., Tan, C.W.: Pricing-based decentralized spectrum access control in cognitive radio networks. IEEE/ACM Trans. Netw. **21**(2), 522–535 (2013)
61. Omar, A., Muhaidat, S., Al-Dweik, A. Shami, A., Yoo, P.: Opportunistic spectrum access in cognitive radio networks under imperfect spectrum sensing. IEEE Trans. Vehicular Technol. **63**, 920–925 (2014). https://doi.org/10.1109/tvt.2013.2281334
62. Salahdine, F., Kaabouch, N., El Ghazi, H.: A survey on compressive sensing techniques for cognitive radio networks. Phys. Commun. **20**, 61–73, September 2016
63. Manesh, M.R., Kaabouch, N.: Interference Modeling in Cognitive Radio Networks: A Survey (2017)
64. Maric, S., Reisenfeld, S.: Mitigation of primary user emulation attacks in cognitive radio networks using belief propagation. In: Weichold, M., Hamdi, M., Shakir, M., Abdallah, M., Karagiannidis, G., Ismail, M. (eds.) Cognitive radio oriented wireless networks. CrownCom 2015. Lecture Notes of the Institute for Computer Sciences, Social Informatics and Telecommunications Engineering, vol. 156. Springer, Cham ((2015))
65. Peter He, Lian Zhao, Xavier Fernando: 'Transmission Time Minimization of an Energy Harvesting Cognitive Radio System' IEEE Symposium on Signal and Information Processing for Smart Grid Infrastructures, IEEE GlobalSIP'16, Greater Washington DC, December 7–9, 2016

66. Ajmery Sultana, Lian Zhao and Xavier Fernand: 'Efficient Resource Allocation in Device-to-Device Communication Using Cognitive Radio Technology' IEEE Transactions on Vehicular Technology, Nov. 2017, Volume: 66, Issue11, Page(s): 10024–10034
67. Youping Zhao, Shiwen Mao, James O. Neel, Jeffrey H. Reed,: 'Performance Evaluation of Cognitive Radios: Metrics, Utility Functions, and Methodology' Proceedings of the IEEEl Vol. 97,No. 4, April 2009
68. Shi Y., Guo J., Jian Y.: (2012) Spectrum Sensing Algorithms in the Cognitive Radio Network. In: Huang DS., Gupta P., Zhang X., Premaratne P. (eds) Emerging Intelligent Computing Technology and Applications. ICIC 2012. Communications in Computer and Information Science, vol 304. Springer, Berlin, Heidelberg

Ms.G.T. Bharathy was born in the year 1979 in India. She completed B.E degree in Electronics and Communication Engineering from Easwari Engineering College, Chennai, Madras University, India, in the year 2000 and M.E degree in Communication Systems in Anna University in the year 2005 from Shri Venkateshwara College of Engineering, Chennai, India. She is currently a Research Scholar (part-time PhD) at Department of ECE, Vels Institute of Science, Technology & Advanced Studies (VISTAS), Chennai and also working as an in Jerusalem College of Engineering, Department of Electronics and Communication Engineering, Chennai, as an Associate Professor. She is a member of IEEE and also a life time member in ISTE. She has worked in the department of ECE at Anand Institute of Higher Technology and Prince Shri Venkateshwara Padmavathi College of Engineering, Chennai. She is interested in research areas such as RF and Microwave Circuits, Communication Systems and Wireless Communication & Networks. She has published five papers in Scopus-Indexed Journal, one paper in Springer Scopus-Indexed Journal and one paper in IEEE Xplore Digital Library and more than 20 papers in various other reputed National and International Journals.

Dr. V. Rajendran received his MTech in Physical Engineering from Indian Institute of Science, Bangalore, India, and received his PhD degree on Electrical and Electronics Engineering from Chiba University, Japan, in 1981 and 1993, respectively. He is currently, a professor and Director Research, Department of Electronics and Communication Engineering in Vels Institute of Science and Technology, Pallavaram, Chennai, India. He was awarded MONBUSHO Fellowship, Japanese Govt. Fellowship (1988–1989) through the Ministry of Human Resource and Development, Govt. of India. He was elected twice as Vice Chairman–Asia of Execution Board of Data Buoy Cooperation Panel (DBCP) of Inter-Governmental Oceanographic Commission (IOC)/World Meteorological Organization (WMO) of UNSCO, in October 2008 and September 2009, respectively. He was a Life fellow of Ultrasonic Society of India, India (USI) in January 2001. He was a Life fellow of Institution of Electronics and Telecommunication Engineering (IETE), India, in January 2012. His area of interest includes cognitive radio and software-defined radio communication, antennas and propagation and wireless communication, under water acoustic signal processing and under water wireless networks. He has published 52 papers in Web of science and Scopus-Indexed Journal.

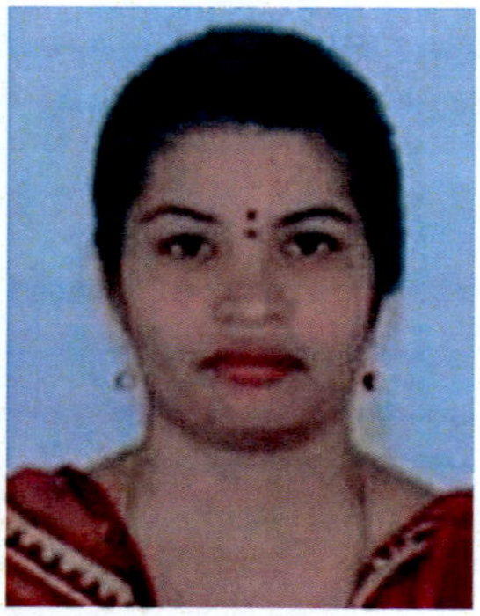

Dr. M. Meena received PhD degree in Electronics and Communication Engineering from Vels University. She completed her M.E. degree in Applied Electronics from Anna University. She is currently an Assistant Professor in Department of ECE at Vels University. Her research interests includes computer networks, wireless networking and software-defined radio. She has published 11 papers in Scopus-Indexed Journal.

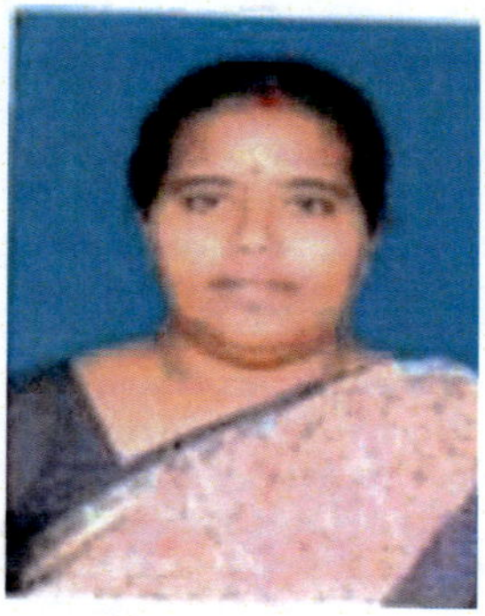

Ms.T. Tamilselvi was born in India the in the year 1978. She completed B.E degree Electronics and Communication Engineering from Adhiparasakthi Engineering College, Madras University, India, in the year 2000 and M.E. degree in Embedded System Technologies from College of Engineering, Guindy (CEG Main Campus), Anna University, India, in the year 2006. She is now working as Associate Professor in Jerusalem College of Engineering, Dept. of Electronics and Communication Engineering, Chennai. She is a life member in ISTE. Her research interest is VLSI and embedded design and wireless communication & networks. She has published five papers in Scopus-Indexed Journal, one paper in Springer Scopus-Indexed Journal and two papers in IEEE Xplore Digital Library and more than 20 papers in other National and International Journals.

Ensemble Method for Identification and Automatic Production of Related Words for Historical Linguistics

G. Sajini and Jagadish S. Kallimani

Abstract Language change throughout time and space is one of the major issues in linguistic history. The paper deals with new methods for the study of language evolutions to help researchers and experts. Firstly, a method is used to determine, if the words are cognate or not. A linguistic information algorithm is proposed to derive cognates from online dictionaries. Then a dataset is created of similar terms and machine learning techniques are used to focus on spelling to classify the cognates. The aligned subsequences are used to identify standards and guidelines for language change in newly created languages mainly to distinguish between non-cognate and cognates which are used for classification algorithms. Secondly, for identifying the sort of association between those words that humans expand the method to a simpler level. Discriminating cognates and debts gives an insight into a language's history and allows a clearer understanding of the linguistic relationship. The spelling characteristics have discriminative features and analyze the linguistic factors underlying this classification task. This is considered as the first such effort, to linguistic knowledge. Thirdly, a machine learning technique is developed for producing similar words automatically. One should concentrate on proto-word reconstruction to address issues related to it to generate the modern words which are not synonyms and another one is generating cognates. The task of reconstruction of proto words is to recreate words from its modern daughter languages in an ancient language. The method is based on the regularity of words and uses knowledge from many modern languages to build an ensemble method for proto-word reconstruction. This method is applied to multiple datasets to improve from the previous dataset accuracies.

Keywords Cognates · Proto-word · Linguistic borrowing · Diachronic linguistics · Etymons · Ensemble learning

G. Sajini · J. S. Kallimani (✉)
Department of Computer Science and Engineering, M S Ramaiah Institute of Technology, Bangalore, India and affiliated to Visvesvaraya Technological University, Belagavi, Karnataka, India
e-mail: jagadish.k@msrit.edu

G. Sajini
e-mail: sajini.narayana@gmail.com

P. Karuppusamy et al. (eds.), *Sustainable Communication Networks and Application*, Lecture Notes on Data Engineering and Communications Technologies 55,
https://doi.org/10.1007/978-981-15-8677-4_40

1 Introduction

The natural languages are like symbiotic systems because they constantly change and upgrade over the years. The two basic issues arise in linguistic history. Firstly, what is the relation between languages? Secondly, how are languages evolving over time and space? In both cases, the comparative linguistic instruments have historically been researched. The core concept behind the comparative approach is to compare several sister languages based on properties to determine the properties of their mutual ancestor [1]. The comparative reconstruction was a manual method that took a long time, involving comprehensive research. The first question involves designing strategies to identify cognates. The second issue includes investigating borrowing and studying the words movement when it comes to other languages. Cognates are nothing but words with the same meaning and shared origin in various languages. Cognates can be studied in many research fields, such as language learning, bilingual words comprehension, corpus linguistics, cross-lingual knowledge retrieval and computer translation, which are not only useful in historical but also in comparative linguistics.

The process through which words enter into another's vocabulary is called *linguistic borrowing*. There is no such thing as a *pure language*, which means without the need to borrow from any foreign languages. An empty word too is called borrowing. The term *loanword* is described as a lexical object that is borrowed from a different language, which was not originally in part with receiver language vocabulary but taken from a different language that became words of the vocabulary of the language that is borrowed [2]. The exceptional connections between languages and the proliferation of interactive technologies have resulted in language amplification by borrowing. The main question that arises is when and how the process of borrowing is carried out. This question requires their essence of experimental viewpoint. The effects of the borrowing process depend on several aspects such as duration and frequency of communication and the degree to which the communities in question are bilingual. The languages are primarily focused on needs and popularity in other languages.

Differentiating cognates from borrowings is critical in language classification problems. While the language classification that identifying cognates and word borrowings from different languages is undoubtedly regarded as important to the history of language borrowing, it is necessary that context of phylogenetical inference may increase the incorrect values to which the cognates are identical [3]. The fact that false cognates can draw wrong conclusions about the relationship between a different set of languages. Additional research issues are the production of cognates and production of modern languages (determining the pattern of a protocol in a contemporary daughter's language). One should highlight two directions of the research, which depend on these word-forms: diachronic linguistic analysis, which deals with the production of language over time and reviewing other languages by concentrating on learner during the acquisition of second languages [4]. Cognates can also contribute to poorly written languages with minimal resources to the lexicon generation process.

2 Related Works—Identification of Words

In historical linguistics, most research concentrate on recognizing pairs of cognates automatically. To classify cognates, there are three critical things commonly examined: semantic, phonetic and orthographic similarities. These were combined and used individually for finding cognate pairs. WordNet is used to determine semantic similarity and String similarity measures of cognate candidates, with phonetic or spelling word types as inputs these were different measures taken to examine and to compare. For cognates based on orthographic and phone forms, algorithms for string alignment should be used [5]. The changes undergone by words as they were entered from one language to another and the transition rules, they obey were effectively used to identify cognates in a variety of different ways.

Some of the metrics in this area are most widely used is edit size, XDice metric and longer common subsequence ratio. The SpSim is a more complex process to calculate the cognate pair similarities that tolerate learned transitions between words. One should use algorithms for basic sequence matching to obtain spelling alignments for candidates. A newly developed ALINE, which lines phonetics of words based on various phonetic characteristics and uses complex programming to measure similarity ratings [6]. Recent methods have been focused on neural networks and dictionary definitions to identify cognates reliably. The distinction between cognates and bonds is based on frequent sound changes, which produce frequent phonemic correspondences in cognates, in keeping with the regularity principle. Sound maps are in turn, to a certain extent, represented by alphabetical character maps.

3 Production of Related Words

The two main concerns and difficulties faces in linguistics of the diachronic are histories and comparative reconstruction. The historical derivation is that new forms of the words are derived from the former. The contrary method is the comparative reconstruction of words that are borrowed from a different set of languages that will be newly constructed. Language derivation has been a constant interest for researchers [7]. The first attempts to solve this issue were regular sound connections, with a proto-language, to construct modern forms of words or vice versa. For some early research, multiple alignments for historical comparison and proposal methods for cognitive alignment and recognition are being studied. Most of the previous approaches were based on phonetics. Mainly based on the theory that sound changes obey such regularities in a language's vocabulary in view of the phonological sense. The methods given included a list of known data, dictionary or published studies correspondences. More recently, the automation of the reconstruction process was developed.

When orthographic and phonetic forms of the words are applied, it was very effective to align the associated words with the extraction of spelling changes from one language to another. Spelling changes have also been used to produce cognates, which have not yet been studied as intensively but are very closely related to the task of identifying cognates [8]. Whereas cognates are identified as a word in the source language is used to generate cognates in a target language automatically.

An algorithm [9] developed to focus on editing distancing and recognizing orthographic signals. When the words enter a new language, a method is used to generate cognates that are mainly focused on statistical computer translations and learning orthographic patterns.

4 Building Dataset of Related Words

A cognate extraction algorithm from electronic dictionaries containing etymological information is created and get a set of corresponding words from dictionaries, from this an automatically develop machine learning methods to identify and produce related words [10].

4.1 Steps for Cognate Pair Identification

- *Find a selection of terms for defining the cognate pairs in a certain L1 language and apply the following strategy between L1 and the related language L2.*
- *Define the etymologies of the terms given then all words without etymology are translated into L2. The pairs of input terms and their representation are considered by cognate candidates.*
- *Extract the etymological details for the translated terms using electronic dictionaries.*
- *Compare their etymologies and etymons to classify cognates for each pair of candidates.*
- *Mark the words as cognates if they fit. Even if different forms of the same word are inflected, and presume that the Etymons match.* Figure 1*shows how the word-etymon pairs and cognate pairs are identified.*

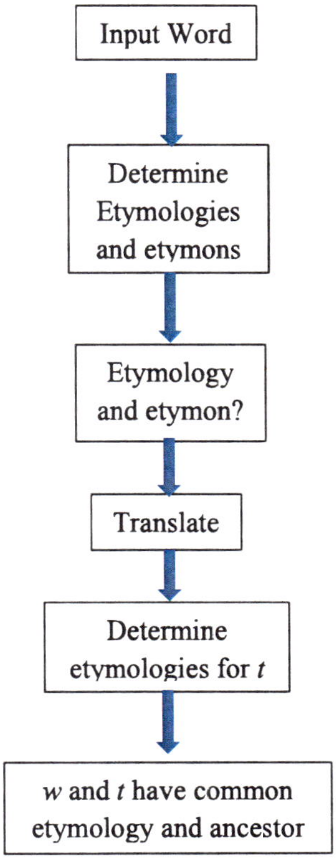

Fig. 1 Identification of Word Pairs and Cognate Pairs

4.2 Calculation of Accuracies for Different Linguistic Datasets

Randomly take 500 input words *w* from different other background languages such as German, French, Greek, Romanian and Latin, and along with the algorithm accuracies, one should manually calculate the cognates to test the above automatics method of extraction of etymology-related knowledge and for the detection of related words [11]. Obtain the accuracy rates of each language using different automatically result in generating machine learning algorithms and then compare the results with the etymologies obtained automatically and measure for each language the accuracy of the etymology extraction. The best accuracy in giving linguistic dataset will be chosen. Once the best linguistic dataset is obtained one should concentrate on training the model to improve the system performance by upgrading and modifying the dataset using online dictionaries and again manual results should be calculated and compared with the algorithmic results for achieving better accuracy [12].

5 Automatic Recognition of Related Words

The main objective is to define the association between the terms automatically and concentrate on cognates and borrowing by using an alignment approach and aligned subsequence [13]. The methodology used to identify cognates and discriminate between cognate and borrowing methods are as follows:

- *Align the related word pairs with a String alignment algorithm.*
- *Extraction of the related terms features.*
- *Machine learning distinguishing algorithms should be used to identify cognates and discriminate between cognate and borrowings.*

5.1 *String Alignment*

To fit pairs of terms, Needleman and Wunsch's algorithm is used. This algorithm attempts to decide the best alignment over the whole input sequence range. The algorithm guarantees optimum alignment and is effective [14]. The key principle is that the optimum path to this stage would be every intermediate sequence that leads to optimum paths. Thus, by increasing expansion of intermediate sub-paths, the optimum path can be calculated. One must regard terms as entry sequences for orthographic alignment and use a very simple substitution matrix that gives equivalent values, not taking diacritic factors into account.

5.2 *Feature Extraction*

The features are extracted around mismatches in alignments with the matched pairs of terms as an input. The insertion, deletion and substitution are three types of mismatches found while extracting the features [15].

Illustration with an example for String Alignment and Feature Extraction

Consider the word *exhaustiv* which is derived from the Roman language and word *esaustivo* which is derived from the Italian language, these two words are considered as cognate pairs. The String alignment is shown below:

e x h a u s t i v -
e s - a u s t i v o

- Mismatch 1: **x** & **s** occurred due to substitution.
- Mismatch 2: **h** & **-** occurred due to deletion.
- Mismatch 3: **-** & **o** occurred due to insertion.

5.3 *Learning Algorithms*

The spelling changes and the relationships between words are identified using Support Vector Machines and naive Bayes. With the Weka Laboratory, a collection of algorithms and tools to bring the program together. For SVM, the wrapper classes are used to supply to the LibSVM from Weeka and the RBF kernel is used, which can deal with the case where the relationship among the wrapper class attributes and kernel parameters are well maintained without any conflicts.

5.4 *Evaluation Measures*

The performance is assessed using the following evaluation measures:

- F-score
- Accuracy
- Recall
- Precision.

5.5 *Task Set-up*

- The data is divided into two subsets for each language pair, one for training and another one for testing with the ratio 3:1.
- The experiment is conducted for different values of n-gram of size n. The value ranges from one to three that belong to n.
- Three-fold cross-validation is carried out over the training datasets.
- The grid search is performed over the training datasets to optimize the evaluation measures.

6 Linguistic Factors

Below are the high predictive linguistic factors to improve system accuracy of the ML algorithms and to achieve better evaluation measures.

- **Part of speech**
  Part of speech plays a vital role in analyzing whether incorporating information about the word element leads to changes in results. Verbs, nouns, adverbs and adjectives come to language-specific ends; therefore, one should assume that this factor may be helpful during orthographic learnings. For the learning algorithm, one should use the POS feature as a second categorical feature.
- **Hyphenization**

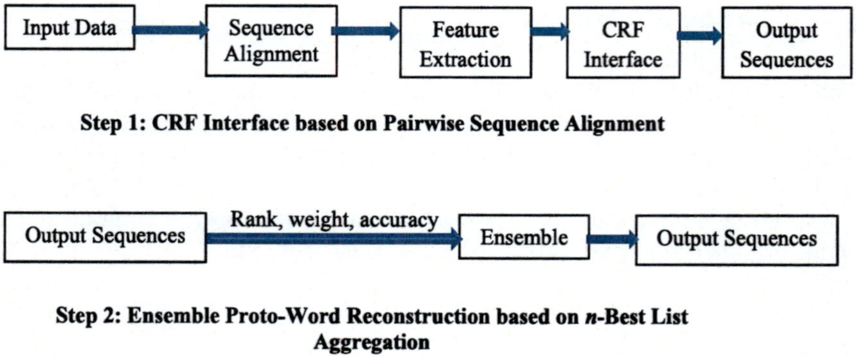

Fig. 2 Proto-words reconstruction methodology

A hyphenated type of words is used for the algorithm alignment to see how it increases cohesion and therefore feature extraction by establishing distinctions between the syllables and derive the term hyphenization from the Romanian RoSyllabiDict for knowledge strings also some Perl modules 19 accessible for certain languages. The hyphen constraints will be considered as the extra terms.

- **Consonants**
  System's effectiveness is analyzed by the matched consonant skeleton of the words (that is, the word form of which vowels are discarded) is qualified and checked to provide less detail that is useful for word recognition. As vowels are dropped, the efficiency of the system decreases. In data collection, one should also train and check the decision tree classifier and its output.
- **Stems**
  The first check is performed using stems rather than lemmas as data to understand that the associated appeals reflect the form of a relation between the two terms. One should use the Snowball Stemmer and note that when stems are used instead of lemmas, their efficiency is increased.
- **Diacritics**
  Diacritics elimination affects the system's efficiency and many words were transformed by an increase of linguistic diacritics when a new language was entered. Therefore, diacritics are expected to play a role in the classification task. One should note that the precision of the test set is lower in nearly any case when diacritics are excluded. The study of the classification of the characteristics derived from the misalignment in this direction offers much clearer evidence that is more than a quarter of the top 500 features contain diacritics for both languages.

7 Proto-words Reconstruction

A two-step approach is followed as shown in Fig. 2 to solve the problem of reconstruction of proto-words. Firstly, a system based on sequence labelling to recreate proto words. This approach is used separately in every modern language. Second, to pull together the optimum productions from different vocabulary, a variety of ensemble modes to incorporate knowledge from various structures and to calculate the system performance outcomes.

7.1 Fusion by Rank

The rank is calculated based on individual production. The output word is u and its rank weight w over the n lists is calculated as

$$w_r(u) = (1/k) * \sum_{i=1}^{k} w(u_i)$$

7.2 Fusion by Rank and Accuracy

Training accuracy is considered here for each language. The best-obtained accuracy is considered and multiplied with the weight obtained by training accuracy for each language I is calculated as

$$w_{ra}(u) = (1/k) * \sum_{i=1}^{k} w(u_i)\pi(i)$$

7.3 Fusion by Weight

The Confidence score is considered here based on individual productions. with the help of obtained confidence score, productions are reranked in sequence labelling system Li is calculated as

$$w_c(u) = (1/k) * \sum_{i=1}^{k} w(u_i)$$

7.4 Fusion by Weight and Accuracy

Training accuracy is considered here for each language. The best-obtained accuracy from all the languages is considered and multiplied with the weight obtained by training accuracy in sequence labelling system is calculated as

$$w_{ca}(u) = (1/k) * \sum_{i=1}^{k} w(u_i)\pi(i)$$

8 Modern Word Production

The hypothesis that the term v in language L1 derives from the text u (by means of a borrowing). One should use stems rather than lemmas to produce modern words. The form of inflected or derivative words reduces both lemmatization and stemming more drastically, to the common form of the base word, but stemming. In other words, the words are reduced to lemmatization. In addition, the stemming process eliminate suffixes and prefixes that are not lemmatized, this could make a difference in word production that leads to better results, instead of lemmas. The snowball stemmer is used and 9 of our 20 donor languages have stemmers.

One possible reason that stemming doesn't improve when new words enter the language and so root can't necessarily be more easily produced than the whole word (including the appeals) due to its morphological changes. Many morphophonological changes have been observed over the years and there are changes even in declinations and conjugations.

8.1 Cognates Production

This task is much like production. The interest in recognizing cognates using computation methods has been significant in the last couple of years but very few studies deal with automatic cognate pairs development. The main goal is to decide whether the system works for cognates differently from borrowing.

9 Automatic Related Word Production

The method of orthography-based production of related words is widely used and takes into account the form of association between words and discern specifically between proto words. Main of proto-word reconstruction without the use of any recipient's lexicon or data in the recipient's language, with less resources. A new approach to the production of related words is introduced by domain experts and a strategy based on random circumstances.

9.1 Conditional Random Fields

The system can learn orthographic patterns of spelling change between source and target language by harmonizing related words. The main approach is marking of sequences achieving transliterations that have proved helpful in our situation, the words are the numbers, and their characters are tokens. The aim is to get a sequence of characters composing its related word for each input word and use Custom Random Fields (CRFs) for this purpose. The CRF system uses the n-gram characters of the input words, extracted around the current token from a fixed window w.

9.2 Pairwise Sequence Alignment

The Needleman and Wunsch global alignment algorithm are mainly used. Consider two words *frumos* from the roman language and another word *formosus* from the Latin language. The Pair Sequence Alignment is as follows:

f - r u m o s - -

f o r - m o s u s

The associated label shall be the character that takes place in the same position in the target word for every character in the source word (after alignment). In the case of insertions, add the new character to the previous label because the inserted character could be associated as the label with no input character in the source language. Apply to each word and add two additional characters B and E, which represent the beginning and end of the word, for each input word. These specific characters are linked to the characters that are inserted in the target word at or at the end of the word. The sticker is removed to reduce the number of labels.

9.3 Evaluation Measures

The performance is assessed using the following evaluation measures:

Average edit distance

The editing gap between the words renders and the standard, to determine how close, the outputs are to the appropriate form of the associated language. The unstandardized and the normalized editing distance are recorded for normalization in [0,1] interval.

Coverage

Coverage is also referred to as accuracy of top n items in the language dictionary. The functional value of evaluating the top n findings to provide potential output terms to a smaller list that can be evaluated by linguists. $n = 1$ is known for its accuracy of the measurement.

Mean reciprocal rank

The mean mutual rank is a measure for systems that produce an ordered output list for each input instance. In the case of an input, word is calculated as:

$$\mathrm{MRR}(w_i) = \frac{1}{m}\sum_{i=1}^{m}\frac{1}{\mathrm{rank}_i}$$

where m indicates the total number of input instances, rank_i indicates the position of proto-word w_i in output lists.

10 Conclusion

A method for automatically sequence alignment identification of related words is identified. For the first time employed the tool to classify cognates for an automatically derived cognate dataset for four language pairs, for the detection of lexical shifts happening as words are joining new languages. Then the method of discrimination is based on their spelling between cognates and debt. The predictive analysis reveals that for cognates and borrowing the orthographic indications are different and that underlying language factors. One should look for more fine-grained vocabulary and more languages to broaden the studies and a framework which is language-relating, it might boost the efficiency of the program to integrate language skills. An automatic method to produce related words has been introduced and used a sequence labelling and sequence alignment approach which combined the results of the individual systems with assemblies. The benefit to the method is that less data is needed than previous approaches and that in historical linguists where resources are scarce, incomplete information is also accepted. First used the method to reconstruct Latin proto-words using multiple data sets in romantic languages to develop modern Romanian word forms as a receiver language. The languages, rather than language families in the light of their etymons (ancestors) and cognates, the recipient language could be a valid word and the created sequences also reflect the elderly shapes or the feminine shape of a word for the substantives.

References

1. Rama, T.: Siamese convolutional networks for cognate identification. In: Proceedings of COLING 2016, the 26th International Conference on Computational Linguistics: Technical Papers, pp. 1018–1027, Osaka (2016)
2. Pagel, M., Atkinson, Q.D., Calude, A.S., Meade, A.: Ultraconserved words point to deep language ancestry across Eurasia. Proc. Natl. Acad. Sci. **110**(21), 8471–8476 (2013)
3. Gomes, L., Lopes, J.G.P.: Measuring spelling similarity for cognate identification. In Proceedings of the 15th Portugese Conference on Progress in Artificial Intelligence, EPIA 2011, pp. 624–633, Lisbon (2011)
4. Gooskens, C., Heeringa, W., Beijering, K.: Phonetic and lexical predictors of intelligibility. Int. J. Humanities Arts Comput. **2**(1–2), 63–81 (2008)
5. Hall, D., Klein, D.: Finding cognate groups using phylogenies. In Proceedings of ACL 2010, pp. 1030–1039, Uppsala (2010)
6. List, J,M.: LexStat: automatic detection of cognates in multilingual wordlists. In Proceedings of the EACL 2012 Joint Workshop of LINGVIS and UNCLH, pp. 117–125, Avignon (2012)
7. Luong, M.-T., Brevdo, E., Zhao, R.: Neural machine translation (seq2seq) tutorial (2017). https://github.com/tensorflow/nmt
8. List, J.M., Greenhill, S.J., Gray, R.D.: The potential of automatic word comparison for historical linguistics. PLoS ONE **12**(1), 1–18 (2017)
9. Luong, T., Pham, H., Manning, C.D.: Effective approaches to attention-based neural machine translation. In Proceedings of EMNLP 2015, pp. 1412–1421, Lisbon (2015)
10. McMahon, A., Heggarty, P., McMahon, R., Slaska, N.: Swadesh sublists and the benefits of borrowing: An Andean case study. Trans. Philol. Soc. **103**(2), 147–170 (2005)
11. Tsvetkov, Y., Ammar, W., Dyer, C.: Constraint-based models of lexical borrowing. In: Proceedings of NAACLHLT 2015, pp. 598–608, Denver, CO (2015)
12. Simard, M., Foster, G.F., Isabelle, P.: Using cognates to align sentences in bilingual corpora. In: Proceedings of the Fourth International Conference on Theoretical and Methodological Issues in Machine Translation, pp. 67–81, Montreal (1992)
13. Shawe-Taylor, J., Cristianini, N.: Kernel Methods for Pattern Analysis. Cambridge University Press (2004)
14. Schuler, G.D.: Sequence alignment and database searching. In: Baxevanis, A.D., Ouellette, B.F.F. (eds.) Bioinformatics: A Practical Guide to the Analysis of Genes and Proteins, vol. 43, pp. 187–214. John Wiley & Sons, Inc. (2002)
15. St. Arnaud, A., Beck, D., Kondrak, G.: Identifying cognate sets across dictionaries of related languages. In: Proceedings of the 2017 Conference on Empirical Methods in Natural Language Processing, pp. 2519–2528, Copenhagen (2017)

A Robust Lightweight Algorithm for Securing Data in Internet of Things Networks

Abdulrazzaq H. A. Al-Ahdal, Galal A. AL-Rummana, and Nilesh K. Deshmukh

Abstract One of the modern technologies that link millions of different devices together, is the technology called Internet of things. The amount of data exchanged between these devices is very large. Therefore, it requires high protection for that data. Also, those devices are small size and limited resources. Therefore, conventional cryptography will take long process on the internet of things due to complex mathematical operations and an increase in the number of rounds, which leads to energy resource consumption for devices. In this paper, a lightweight cryptographic algorithm is suggested. The algorithm uses simple mathematical operations (XOR, XNOR, shifting, swapping). The algorithm works on a combination of Feistel and SP architectural methods to increase the complexity of the encryption. The algorithm has been compared to other algorithms in terms of structure, security, and flexibility.

Keywords IoT security · Khazad · Wireless sensor network WSN · Feistel · SP · Data encryption algorithm

1 Introduction

Internet of Things (IoT) technology has become an integral part of many activities in our daily life. Therefore, this modern technology has turned into a debate in the field of research and applications. Many different fields such as agriculture, industry,

A. H. A. Al-Ahdal (✉) · G. A. AL-Rummana · N. K. Deshmukh
School of Computational Sciences, S.R.T.M. University, Nanded, India
e-mail: alahdal201211@gmail.com

G. A. AL-Rummana
e-mail: Galal300z@gmail.com

N. K. Deshmukh
e-mail: nileshkd.srt@gmsail.com

A. H. A. Al-Ahdal
Faculty of Computer Science & Engineering, Hodeidah University, Al Hudaydah, Yemen

P. Karuppusamy et al. (eds.), *Sustainable Communication Networks and Application*, Lecture Notes on Data Engineering and Communications Technologies 55,
https://doi.org/10.1007/978-981-15-8677-4_41

medicine, and smart cities use this modern technology on a large scale in artificial intelligence, sensors and RFID, depending on Internet technologies [1].

These networks use very limited resources, low-energy, and very small devices that communicate with each other to transfer information between them. Therefore, the amount of information exchanged between devices is large and needs to be secured the exchange of information between devices is called conventional cryptography.

Conventional cryptography algorithms are not suitable for devices with limited resources on the Internet of things because they need long operation (process) and do not make a trade-off among memory, security, cost, power, and performance. Therefore, lightweight cryptography is the new direction for the Internet of Things because it concerns memory, security, cost, power, and performance. There are strict limitations and requirements in applications using IoT. On that basis, the requirements must be met when designing a lightweight cryptography algorithm [2].

In this paper, a lightweight algorithm is proposed for devices with resource-limited in the Internet of Things. It is described in Sect. 2. It is compared to various aspects with many algorithms in Sect. 3.

2 Proposed Algorithm

The proposed algorithm for resource-limited devices is designed to implement lightweight encryption in the Internet of Things. Some encryption blocks have the advantages of replacing, switching, and confusing to change the text. As an established fact, SP network architecture is used by AES [3], PRESENT [4] and Square [5] blocks. In addition, some of the other advantages of the encryption process are the decryption itself. While SF [6], Blowfish [7] and DES [8] blocks are used by Feistel network architecture. The proposed algorithm is a combination of Feistel and SP networks to obtain the security required to develop lightweight encryption algorithms that work on the Internet of Things.

In any cipher algorithm, cryptography consists of several rounds, each round consisting of mathematical operations to create confusion and diffusion. An increase in the number of rounds increases security but leads to the energy consumption of devices [9]. The optimum number of rounds is from 10 to 12 rounds, to be a robust algorithm. But the proposed algorithm is used to improve power on devices that operate on the Internet of Things. So it works in six rounds, each round contains logical operations working with 4 bits of data. This creates confusion for attackers and further complicates the encryption process, and this will be further discussed in Sect. 3. In general, the proposed algorithm has three main blocks:

- Key Expansion Block
- Encryption Block
- Decryption Block.

In the following subsections, these blocks will be further clarified in detail, and some of the essential notes followed in the interpretation are presented in Table 1.

Table 1 Notations

Notation	Function
$\oplus$	XOR
$\odot$	XNOR
$\|, \#$	Concatenation

2.1 Key Expansion Block

The key is the main component of the algorithm (encryption/decryption). The size of the encryption key is very important for security. That is how (key size) becomes a major obstacle for being known by the attacker. Accordingly, confusion and diffusion (key generation) reduces the possibility of weakening the key, increase the security, increase the complexity of the encryption, and lack of knowledge of the key by the attackers.

The proposed algorithm uses 80 bits cipher key (Kc) to encryption a 64 block of data. From the Kc, two keys will be produced. The first will be sent to open the encryption while the second will become an entry key for the keys expansion process as explained in sections a and b. Moreover, the key will generate six unique keys. In order to encrypt the block with six keys. These operations (confusion and diffusion) improve the strength of security and decrease the opportunity of attacking.

a. **Before Encryption Process**
   In this process, two keys will be created from the key entered by the user. The first key is the key that has been sent by the user for the process of opening the encryption as it will be explained in section b. The other key will be completely different from the key entered by the user and also will be the encrypted key for the data as in Fig. 1, and it will be discussed below:

- The 80-bit cipher key (Kc), It is divided into segment R0(40 bit) and segment L0(40 bit)
- For $i = 1$ to 40
  If Ri = 0 then.

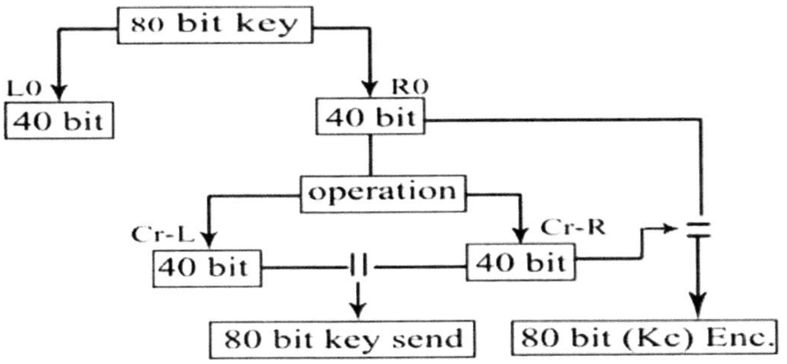

Fig. 1 Before encryption process

Cr-R[i] = 1, Cr–L[i] = 1.
Else.
Cr-R[i] = 0, Cr–L[i] = 1.

- To combine Cr-L, Cr-R to key send(Send Key 80 bit)
- To combine R0, Cr-R to Kc(encryption Key 80 bit).

b. Key Expansion Process

Key expansion of the component key is done in Section a. Therefore, it will be explained as in Fig. 2.

- The 80-bit cipher key (Kc), It is divided into 20 segments, each segment of 4-bits.
- The f -function used 4 segments, each segment 4 bit (16 bit) as illustrated in Fig. 2. Substitution can generate for cipher key (Kc) by f-function as shown in Eq. (1).

$$Kb_i f = \|_{j=1}^{5} K_{c4(j-1)+i} \tag{1}$$

where $i = 5$;

- $Ka_i F$ is output from Eq. (2)

$$Ka_i f = f(b_i f) \tag{2}$$

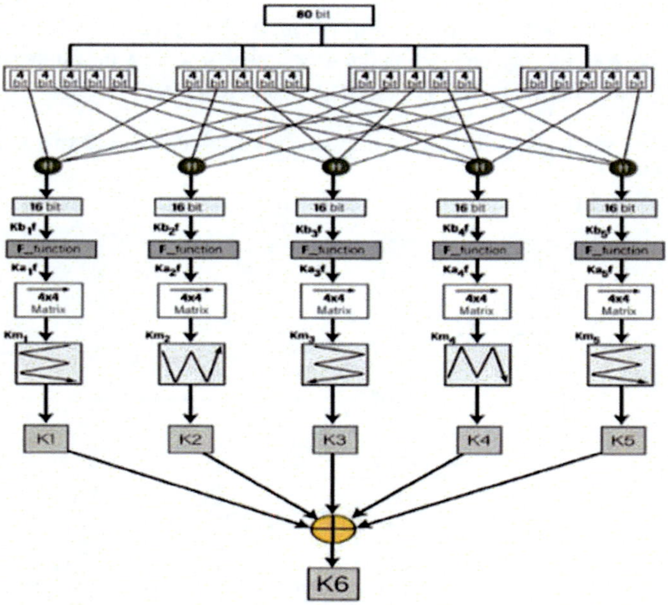

Fig. 2 Key expansion process

- f : f—function [10]. Is perform confusion and diffusion transformations linear and non-linear by comprised of P and Q tables are shown in the Tables 2 and 3 as illustrated in Fig. 3.
- Output, each f-function is generated one matrix 4*4 from (16 bit). Therefore it generates five arrays named (Km1, Km2, Km3, Km4, and Km5).
- The arrays generated K1, K2, k3, K4 and K5 are the round keys. The result of the rotation of the arrays is Km1, Km2, Km3, Km4, and Km5, respectively.
- To generate the sixth key, XOR works between the five keys generated by Eq. 3.

$$k6 = \bigoplus_{i=1}^{5} k_i \tag{3}$$

Table 2 P Table

Kci	0	1	2	3	4	5	6	7	8	9	A	B	C	D	E	F
P (Kci)	3	F	E	0	5	4	B	C	D	A	9	6	7	8	2	1

Table 3 Q Table

Kci	0	1	2	3	4	5	6	7	8	9	A	B	C	D	E	F
Q (Kci)	9	E	5	6	A	2	3	C	F	0	4	D	7	B	1	8

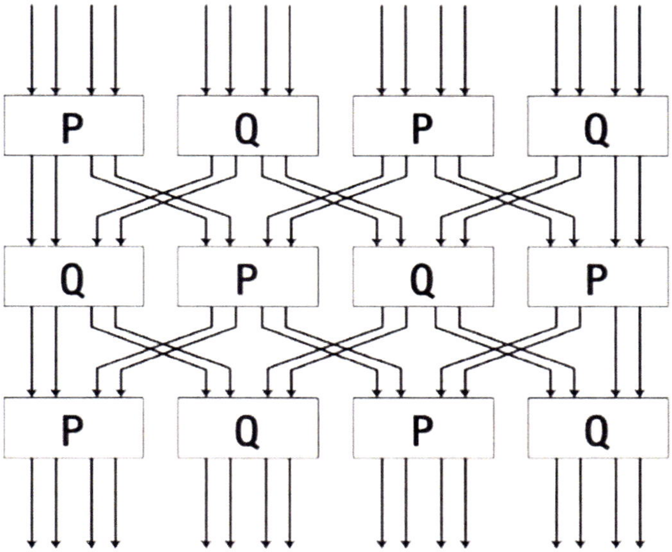

Fig. 3 F-Function of SIT algorithm

2.2 *Encryption Process*

The encryption process takes place after generating the sub-keys (K1, K2… K6) from the key expansion and also the plaintext to be encrypted as in Fig. 4. The encryption process contains simple logical operations of XNOR, XOR, shifting (left, right), replacing and switching. These operations increase complexity and create confusion for the attackers.

In each round, the blocks are divided into four segments, each segment is 16 bits (P_{0-15}, P_{16-32}, P_{33-47}, P_{48-63}), respectively. This is to produce segments (Ro11, Ro12, Ro13, Ro14).

Ro11 is the output of XNOR between P_{0-15} and K1, The product (Ro11) feeds F-Function to produce EfL1. Ro14 is the output of XNOR between P_{48-63} and K1, The product (Ro14) feeds F-Function to produce EfR1.

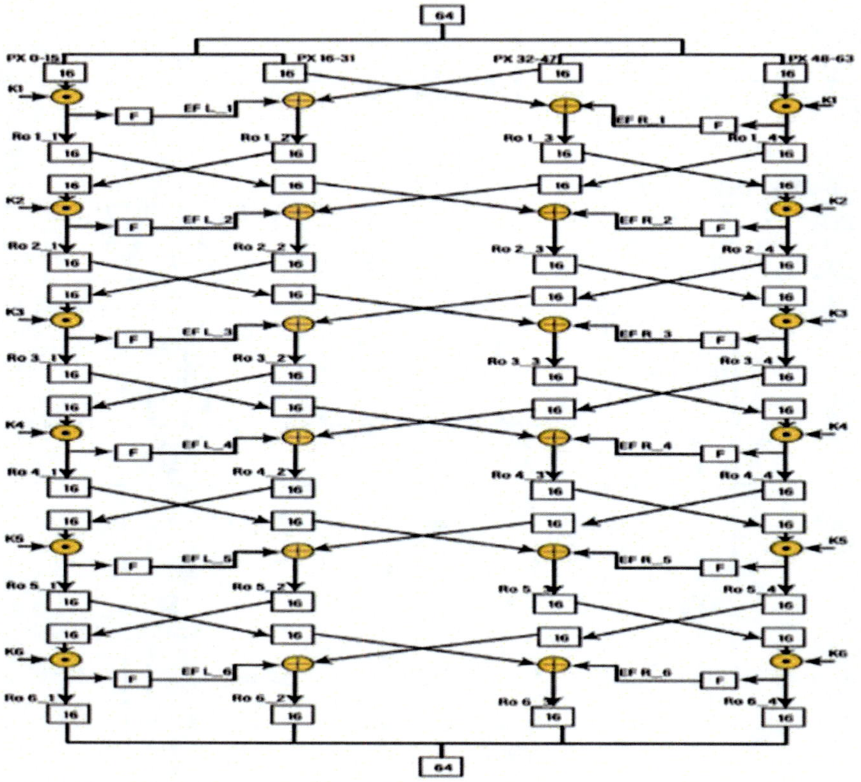

Fig. 4 Encryption process

Ro12 is the output of XOR between $P_{32\text{-}47}$ and EfL1. Ro13 is the output of XOR between $P_{16\text{-}31}$ and EfR that, the process of switching takes place during the encryption process between the two internal halves. Then, the switches are between the parts (Ro11, Ro12) and (Ro13, Ro14).

All the previous processes are to increase the complexity of the coding as shown in Fig. 4. The same steps for the rounds are repeated by Eq. (4).

$$Ro_{i,j} = \begin{cases} Px_{i,j} \odot k_i; & j = 1.5 \\ Px_{i,j+1} \oplus Ef_{li}; & j = 2 \\ Px_{i,j-1} \oplus Ef_{ri}; & j = 3 \end{cases} \tag{4}$$

After that, the encoded text is obtained by Eq. (5).

$$Ct = R_{51} +\!\!\!+ R_{52} +\!\!\!+ R_{53} +\!\!\!+ R_{54} +\!\!\!+ R_{55} \tag{5}$$

2.3 *When Decryption Process*

Before opening the encryption process, the encrypted key will be extracted from the key sent by the user. The process will be explained in Fig. 5.

- The 80-bit cipher key (send key), is divided into segment R0(40 bit) and segment L0(40 bit).
- K1 = R0 ⊕ L0
- To combine R0 and K1 (decryption Key 80 bit).

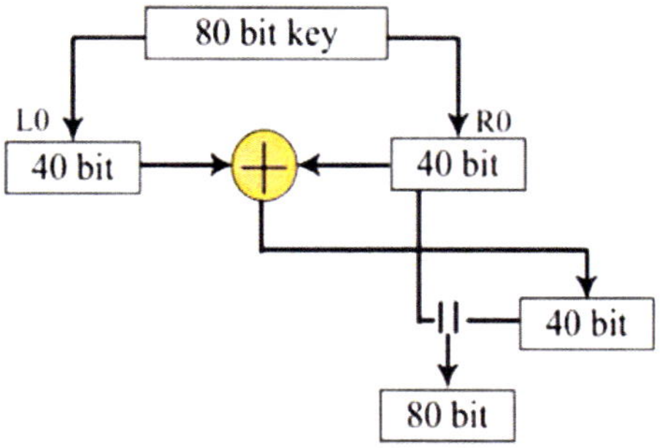

Fig. 5 When decryption process

3 Analytical Comparison of Symmetric Key Algorithm for IoT Proposed Algorithm

In Internet of Things technologies, there are many main symmetric Key algorithms, some of which offer sufficient security and some of them offer better efficiencies in this area. In this section, the proposed algorithm will be discussed, compared, and analyzed with the main symmetric Key algorithms.

Table 4, it shows the comparison process in the various parameters (structure, flexibility, security and limitations) between the proposed algorithm and many of the main symmetric Key algorithms.

3.1 Result and Analysis

The proposed algorithm for the Internet of Things contains a simple design to encrypt the data in six rounds; each round contains simple mathematical operations (free of mathematical complexity). Therefore, the algorithm reduces the process of processing and saving the energy of the devices. In contrast, the processing and power consumption of the algorithms increases AES, SKIPJACK, HIGHT, RC6 the number of turns 12, 32, 32 and 20, respectively, for block coding.

Moreover, the proposed algorithm is a lightweight algorithm for IoT not uses mathematically complex multiplication, small-sized S-boxes are used because most mathematical operations applied to 4-bit data and small-sized boxes. However, the proposed algorithm uses simple mathematical operations and small S-boxes are higher, the overall complexity is lower. Therefore AES and SKIPJACK use large size S-boxes, SEA use variable rounds.

The algorithm provides good flexibility compared to the rest of the algorithms to improve performance in the Internet of Things. Therefore, you use the 80-bit key for 64-bit encryption. The proposed algorithm's flexibility function aims to boost IoT's energy efficiency by reducing power usage, i.e. the energy needed to pad external bits to make the data block fit the size [2].

The proposed algorithm provides security because it merges the architectures SP and Feistel networks. As a result, it provides resistance to most attacks by attackers, as will be discussed below:

- **Linear and Differential Cryptanalysis**
  The proposed algorithm uses each bit in a similar way to maintain round transformation is uniform. Therefore, it provides resistance against this type.
- **Weak Keys**
  This condition occurs in [11] when the non-linear operations in the cipher blocks depend on the actual key value. However, the actual key in the proposed algorithm is not used and to increase the complexity, first uses XORed and then fed f-function. In the f–function all the non-linearity is fixed and there is no limitation on the selection of key.

Table 4 Some Symmetric Key Algorithms' Comparison in Terms of Flexibility, Architecture, Security & Limitation

	AES [16]	SKIPJACK [17]	HIGHT [18]	RC6 [19]	SEA [20]	Proposed algorithm
Overview	AES Rijndael developed by J. Daemen and V. Rijmen, was declared U.S.'s new encryption standard by the NIST. It uses variable key size making it extremely fast and compact cipher. Moreover, its symmetric and parallel structure provides great flexibility, with effective resistance against cryptanalytic attacks	The SKIPJACK algorithm was developed by NSA and is classified as SECRET. It was originally intended to be used with sensitive but unclassified information i.e. in the controversial clipper chip. Skipjack is a remarkably simple cipher, and consist of two different rounds A and B. The encryption with Skipjack consists of first applying 8 A-rounds, then 8 B-rounds	The HIGHT was developed by Korean researchers. The research was supported by MIC, Korea and was supervised by the IITA. HIGHT was especially designed for low resource computing device such as sensor nodes or RFID tag. HIGHT provides better security with simple operations to be energy efficient	RC6 a derivative of RC5, designed by R. Rivest, M. Robshaw, R. Sidney, and Y. Lisa Yin. It was design to congregate the requirements of the AES contest and was among the five finalists. It was also presented to the NESSIE and CRYPTREC projects. It is patented by RSA Security. RC6 offers good performance in terms of security and compatibility	SEA (Scalable Encryption Algorithm) was designed in 2006 by François Xavier and Mace. The design criterion of SEA was focused for low-cost embedded environments with limited resources (memory size, processor capacity). To meet the designing criteria, SEA algorithm makes use of basic bit operations such as XOR, bit/word rotations, modular addition, and s-box	The proposed algorithm is a symmetric key algorithm based on a mixture of the Festial and SP architecture that uses simple mathematical operations and fewer turns to provide better safety and lower energy consumption

(continued)

Table 4 (continued)

	AES [16]	SKIPJACK [17]	HIGHT [18]	RC6 [19]	SEA [20]	Proposed algorithm
Architecture						
• Structure	Substitution-Permutation	Unbalanced Feistal	Festial	Festial	Festial	Festial + SP
• Block size	128 bit	64 bits	64 bits	128 bits	48, 96, 144	64
• Key size	128, 192, 256 bits	80 bits	128 bits	128, 192, 256 bits	48, 96, 144	80
• No. of Round	10, 12, 14	32	32	20	Variable	6
• No. of operations/Rounds	5	4	3	4	4	4
• Mathematical Operations	XOR, Mixing, Substitution, Shifting, Multiplication, Addition. (16 bits)	Permutation, XOR, Shifting, Substitution. (16 bits)	Modular Addition, XOR, Modular subtraction, Shifting. (8 bits)	Variable Rotation, XOR, Modular Addition (2′s comp). (16 bits)	XOR, rotations, 2n mod addition, substitution (8 bits)	XOR, XNOR, Shifting, Substitution.(4 bits)
• S-P Structure	1 S-Box	1 S-Box	N/A	N/A	1 S-Box	4 S- Boxes
• S-Box Size	16 * 16 (16 bits)	16 * 16 (16 bits)	N/A	N/A	3 bits	4 × 4 (4 bits)
Flexibility	The layout can be extended to the 64-bit multiple, with the same amount of subkey as that of the main key	The framework does not allow modifications of any sort	The framework does not allow modifications of any sort	Can extend key length to 2048 bits	The size of the block and the key-length has to be in 6 bits and must not be independent of the processor bits	The framework does allow modifications of any sort

(continued)

Table 4 (continued)

	AES [16]	SKIPJACK [17]	HIGHT [18]	RC6 [19]	SEA [20]	Proposed algorithm
Security and Limitations	Increasing the size of the key generates security. Therefore, it provides resistance against some attacks (collision attacks and possible quantum algorithms)[13]	The diffusion process provides security. The use of a smaller number in the rounds represents optimal uses of energy[14]. However, impossible differential and differential attacks are possible	Security depends on the number of rounds (mathematical operations are performed in each round) in order for the block to be strong and resistant to some different attacks such as differential, linear, and saturation. However, to implement the iteration it uses the lookup table which increases the memory requirement. Moreover, a larger number of rotors and a larger key size require additional execution time [13]	Security lies in their totally random sequence of output bits of 15 rounds or less[15] operating on 128-bit input blocks[13]. It takes more than 17 rounds for a single set of low keys to reach maximum arbitrariness[15]. Since of any procedure, such as vector rotation, it is computationally complex, so multiplication takes a longer processing time. Furthermore, it has a large key size and number of round [13]	Key size and variable number of operations represent security. Therefore, they are strong against linear and differential analysis attacks. However, (number of rounds, key size and table of rules used) require longer to implement [13]	The level of confusion and the publishing of data and mathematical operations represent an increase in the complexity of the encryption, and this all generates security, so the main expansion algorithm consists of different mathematical operations, so it is complex
	The algorithm uses a large block, complex calculations in each round, large s-boxes(16 × 16). This needs additional memory	It took advantage of 16 × 16 F-BOX which should be stored in either RAM or memory of program, which leads to additional memory usage				

- **Related Keys**
  Using different keys (unknown or partly unknown) to launch an attack during the encryption process with a selected relation. The attack depends on the symmetry and diffusion relationship in the expansion block. The proposed algorithm for the key expansion process is designed to be resistant to this type of attack because it possesses a high diffusion and non-linearity.
- **Interpolation Attacks**
  Simple coding structures (S-box) had polynomial or rational expressions controllable. It is subject to interpolation attacks that cannot be performed on the proposed algorithm when expressing an S-box along with a diffusion layer.
- **Square Attack**
  Square attack was presented in [12]. The attacker could get the last byte in the last round of the key. In addition, repeats the attack eight times to obtain the rest of the key. Therefore, to get one a byte requires key guessing from the attacker 2^8 by 2^8 chosen plaintexts = 2^{16} S-box lookups.

4 Conclusion and Future Work

Many different devices and sensors in the Internet of things are interconnected to each other to exchange data. These devices are limited resources and require high data security. Therefore, the proposed algorithm is a lightweight which adopted on Internet of things networks with high flexibility and efficiency. Due to its light of computations, it providing low energy consumption of devices, providing high security, and low computation cost. The performance of the proposed algorithm was analyzed and compared to some symmetrical algorithms to various parameters.

The performance analysis resulting in the superiority of the proposed algorithm which can be represented in low complexity, high security, and saving energy. Briefly, balancing between security, performance, and complexity which leads to making it proper and suitable for devices and sensors in the IoT networks.

The future work is to implement the proposed algorithm in both hardware and software environments and comparing the performance in both environments, i.e., hardware and software.

References

1. Gubbi, J., Buyya, R., Marusic, S., Palaniswami, M.: Internet of things (iot): a vision, architectural elements, and future directions. FutureGeneration Comput. Syst. **29**(7), 1645–1660 (2013)
2. Al-ahdal, H.A.A., Deshmukh, N.K.: A systematic technical survey of lightweight cryptography on Iot environment. Int. J. Sci. Technol., March 2020
3. Standard, A.E.: Federal Information Processing Standards Publication 197, FIPS PUB, pp. 46–3 (2001)

4. Bogdanov, A., Knudsen, L.R., Leander, G., Paar, C., Poschmann, A., Robshaw, M.J., Seurin, Y., Vikkelsoe, C.: Present: an ultra-lightweight block cipher. In: International Workshop on Cryptographic Hardware and Embedded Systems. Springer, pp. 450–466 (2007)
5. Daemen, J., Knudsen, L., Rijmen, V.: The block cipher square. In: International Workshop on Fast Software Encryption. Springer, pp. 149–165 (1997)
6. Ebrahim, M., Chong, C.W.: Secure force: a low-complexity cryptographic algorithm for wireless sensor network (wsn). In: 2013 IEEE International Conference on Control System, Computing and Engineering (ICCSCE). IEEE, pp. 557–562 (2013)
7. Schneier, B.: Description of a new variable-length key, 64-bit block cipher (blowfish). In: International Workshop on Fast Software Encryption. Springer, pp. 191–204 (1993)
8. Coppersmith, D.: The data encryption standard (des) and its strength against attacks. IBM J. Res. Dev. **38**(3), 243–250 (1994)
9. Chandramouli, R., Bapatla, S., Subbalakshmi, K.P.: Battery power-aware encryption. ACM Trans. Information Syst. Security **9**(2), 162–180 (2006)
10. Usman, M., Ahmed, I., Aslam, M.I., Khan, S., Shah, U.A.: SIT: a lightweight encryption algorithm for secure internet of things. arXiv preprint arXiv:1704.08688. 27 Apr 2017
11. Daemen, J.: Cipher and hash function design strategies based on linear and differential cryptanalysis. Ph.D. dissertation, Doctoral Dissertation, March 1995, KU Leuven (1995)
12. Barreto, P., Rijmen, V.: The khazad legacy-level block cipher. Primitive submitted to NESSIE, vol. 97 (2000)
13. Koo, W.K., Lee, H., Kim, Y.H., Lee, D.H.: Implementation and analysis of new lightweight cryptographic algorithm suitable for WSN. In: International Conference on Information Security and Assurance, IEEE (2008)
14. Lu, M.C., Bayilmi, C., Özcerit, A.T., Çetin, Ö.: Performance evaluation of scalable encryption algorithm for WSN. Sci. Res. Essays **5**(9), 856–861 (2010)
15. Stankovic, J.A.: Research challenges for WSN. ACM SIGBED Rev. **1**(2), 9–12 (2004)
16. National Institute of Standards and Technology (NIST): Advanced encryption standard (AES). Federal Information Processing Standard (FIPS) 197, Nov. 2001
17. National Institute of Standards and Technology: SkipJack and KEA algorithm specifications (Version 2.0), May 1998
18. Hong, D., Sung, J., Hong, S., Lim, J., Lee, S.: HIGHT: a new block cipher suitable for low-resource device. Cryptographic Hardware Embedded Syst. **4249**, 46–59 (2006)
19. Pavan, R.L., Robshaw, M.J.B., Sidney, R., Yin, Y.L.: The RC6 Block Cipher, Ver 1.1, August 1998
20. Huang, S.I., Shieh, S.: SEA: secure encrypted data aggregation in mobile WSNs. In: International Conference on Computational Intelligence and Security, IEEE (2007)

UAV Communication Network: Power Optimization and End-To-End Delay

D. Vidyashree and M. K. Kavyashree

Abstract Unmanned aerial vehicle (UAV) has massive potential in the community of civilian approaches, where humanoid preferably be in the danger of extinction. A large swam of UAVs is utilized to provide a flexible and effective wireless access network with a large coverage area and better transmission rate for Internet of Things (IoT) services via relay nodes. The flying Ad-hoc networks comprise of a cluster of small swam UAVs associated in an ad-hoc manner, where these integrations line up into a squad to accomplish powerful goals. In event of disastrous circumstances, if a regular communication setup is not available, these flying networks are utilized to afford a speedy deployment, versatile, self-configuring, and moderately trivial operational expense or at the same time involving numerous UAVs in an ad-hoc network and power consumption are remaining as significant research challenge. In this paper, power optimization and the end-to-end delay are proposed by using an optimal sum of upper-level layer UAVs as the nodes move apart and analyses it. Furthermore, low latency routing algorithm (LLRA) is utilized here to optimize energy essential for the UAV nodes and MAC layer tune-up in transmission range and improves the faster data processing rate and the end to end delay. Finally, simulation results show the power-optimized in nodes and average link delay which improves the packet delivery ratio than the traditional routing algorithm.

Keywords Unmanned aerial vehicle (UAV) · Quality of services (QOS) · Flying ad-hoc networks (FANETS) · Power optimization · End to end delay

D. Vidyashree (✉) · M. K. Kavyashree
Department of Electronics and Communication Engineering, JSS Science and Technology University (Formerly SJCE), Mysuru 570006, India
e-mail: Shreevidya634@gmail.com

M. K. Kavyashree
e-mail: kavyashreemk@sjce.ac.in

P. Karuppusamy et al. (eds.), *Sustainable Communication Networks and Application*, Lecture Notes on Data Engineering and Communications Technologies 55,
https://doi.org/10.1007/978-981-15-8677-4_42

1 Introduction

The unmanned aerial vehicles (UAVs) is an emergent technology that can be employed for armed surveillance, forest fire detection, and other civilian applications in order to capture and deliver an accurate set of data or information. Inappropriate disaster caution, where the common community transmission network system would be immobilized. These UAVs comes into existence to serve by increasing the momentum of operation processes for saving and retrieval purpose. In conditions like a toxic smoke infiltration, wildfires, and wild bodily tracing the UAVs are often organized by software via a remote regulator, towards the supply of medicinal aid to zones which are unapproachable and are less complicated to cover a massive area without exposing the security and protection of the peoples entangle.

As there is an upward in the requirements for a definite Quality of Service (QoS), the traditional cellular network takes a sufficient amount of period and they lack in versatile in the exposure allowance and potentiality improvements. IoT is one more evolving technology and some innovative developing Information Communication Technology (ICT) such as the 5G devices/sensors cover a variety of application fields. These sensor devices capture the data monitoring at the lower level layers that is ground level. So, these UAVs take the ability to support the portable sensor to accomplish a virtuous exposure and divest data flow of traffic end-to-end delay and optimize the power consumed by UAV nodes in the upper layer that is sky level.

This paper proposes a UAV network with more enhanced efficient power optimization and considering end-to-end delay quality of service (QoS) parameter in the link of the network improves lifespan, power utilization, and lag in packet reachability. Shows in what way it is possible to improve the steadiness of route. It also gives the advantages and applications of UAVs in a small celled network.

1.1 Advantages and Applications of UAVS

Advantages of UAV

1. The UAV nodes save time in capturing the accurate set of information from the disaster area.
2. These types of network can reduce potential risks.
3. Improves the data density at each node while capturing data.
4. Fastest data acquisition while nodes are in movement.

Applications of UAV

1. Military Surveillance
2. Forest Fire Detection
3. Detecting deadly gas leakage
4. Environmental Monitoring
5. Agriculture Sites Monitoring.

2 Related Work

In the literature survey, there are many algorithms used across the world to improve the performance Quality of Services Parameters and energy optimization. In one paper Aforementioned Algorithm has been used for diverse learning contrivances in order to capture live audiovisual streaming data transfer which plays a major part in the Industrial-Internet of Things (IIOT) applications and this also raises the number of routes that is essential to exchange of data communication. They have used the 2-tired heterogeneous wireless network which includes more sensors with limited battery life. To estimate the parameter constraints, they have utilized the MATLAB tool. The proposed procedure improves the energy consumption and usual delay on the discovered routes on the way to their endpoint. Furthermore, gives a brief study for 5G/IoT integration technologies with sensors [1]. This survey paper presents an outline for the caching of UAV-enabled small-cell network systems. The disseminated caching policy for UAVs has also been presented. This paper mainly explains the 3 issues on how UAV offers data facilities to mobile users with backhaul relations. Then battery-operated restriction as UAV need not depend on the fixed power supply. This paper combines the UAV and caching to progress the performance parameter issues of small-cell network grids. A case study of Interference also offered [2]. This article portrays a unified agenda for UAV supported zones like an ecological disaster. It focusses on how trajectory communication scheduling on UAV is minimized for the mobile user surviving on the ground level. A multi-hop Device to Device and multi-hop transmitting scheme is utilized to enhance locations of UAVs. This paper clarifies the broadcast efficiency of UAV standard optimization Algorithms and the Shortest Path Routing (SPR) Algorithm used to establish and expand the unsurpassed route and to improve the transmit issues [3]. Low Latency Routing Algorithm (LLRA) is utilized to attain the optimum direction path with the least delay (lag) and enthusiastically allocate data traffic flows. This Layered UAV group of swam network topology architecture projected and the best quantity of UAVs are examined. The routine of this LLRA is proved to advance the packet transfer ratio [4]. The Fast Explore and Exploit Learning (FEEL) Algorithm is executed to VR immersion and this paper validates the networked VR recital effectiveness and in transported VR immersion fidelity, application interactivity/play-out potential, and broadcast power ingesting [5]. TPA-FCP Algorithm and Polynomial-time Randomized Approximation Scheme are utilized to resolve the overall time in optimal trajectory calculation but the UAVs are connected base station over a straight network connection and the projected system here gives 4 combined components depending upon the scope to next-generation shows the development in the UAV energy consumption proficiency to postulate accusing route optimization problem in a large-scale grid system [6]. Multi-Layered Planning is utilized to permit energy effective uplink communication for empowering public protection and security communications, which has the capacity to permit communication at ground level by means of wired or wireless communication. Polynomial-time Randomized Approximation Scheme (PRAS) presented to discover the least number of hovering locations [7]. Time Division

Multiple Access (TDMA) is utilized to permit analogous broadcasts and performances in the UAV-assisted system. This scheme gives an intention to minimalize the overall power consumption of UAV [8]. The iterative algorithm is utilized to resolve the subproblems and the overall communication power for devices is minimalized below the Signal-To-Interference-Plus-Noise-Ratio. The overall communication power of the IoT devices is condensed by 45% [9]. To advance the QOS parameters by considering the 3-tire hybrid network as UAV cooperates with satellite broadcasting and base station. Low-complexity greedy search algorithm and Lagrange dual breakdown and concave-convex procedure (CCP) technique are utilized and performance of reserve provision harvests an enhancement in the UAV network's overhead and throughput [10]. Fly-hover-and-communication Technique is introduced to determine the trade-off issue in the existing scheme and dissimilar guidelines for scenarios ideal altitude and beamwidth standards in diverse multiuser replicas and it affords beneficial perception for UAV communication network [11]. Non-Orthogonal Multiple Access (NOMA) and combination schema representation. Here it also gives the feasibility investigation and procedure for energy-efficient altitude optimization and energy power distribution, and this scheme accomplishes up to 18% power saving on signal broadcast with extreme power redeemable [12]. The two-stage algorithm and a three-stage alternative algorithm are used in paper [13] to crack the problem of badly-behaved partial and binary computation offloading methods and it also gives the structure for resource distribution outlines outperforms and the outlines converge is reckless and has a less computational complication.

3 Problem Statement

As there are numerous direction-finding protocols proposed for Mobile Ad-hoc networks each of them has its features, merits, and demerits for individual application circumstances. When the situations are too complex to accomplish and if the network topologies configuration changes quickly, the individual direction-finding protocol will not encounter the valuable demands of varied adjustable nodes nor can guarantee the network routinely. Precise insight of adjacent location and suitable change approaches of UAV nodes have developed the tendency to adopt different direction-finding protocols with collective supplies of Quality of Services (QoS) parameters.

3.1 Existing System

By considering all the referenced papers in the literature survey. The existing system is concluded as in the current century, the abilities and roles of Unmanned Aerial Vehicles (UAVs) have quickly changed, and the practice in armed and civilian zones is tremendously popular as a result of the development in the technology of machinelike robotic mechanical systems like workstations, sensors devices, transport network, and networking bits of knowledge. Although Mobile Ad-hoc Networks takes a multi-purpose application there's a requirement of a certain knowledge which may be overwhelmed from true where outdated MANET doesn't seem to be usable like disaster situations like sinking or armed battle zone. It's not possible to put in transportable nodes (which travel outward) in such a communication zone.

3.2 Proposed System

A covered UAV layered network upper-layer architecture is used and investigated in this proposed system and the optimal sum of UAVs in the superior layer is examined with adjacent-form exposure borders. Moreover, the total number of 30 UAV nodes is placed at a height of 50–90 m for network architecture construction. The low latency routing algorithm (LLRA) is utilized by using information about the location symmetries of nodes and network connectivity of UAVs in FANETs. Based on the minutest lag in delay and the projected connection steadiness of UAVs, the used LLRA can accomplish the optimum path with the less delay and vigorously allocate innumerable data traffic flows, which can proficiently exploit the packet transfer ratio and advance the steadiness of direct path and improves the average amount of energy-optimized.

4 Methodology

Nowadays, most of the Internet of Things (IoT) devices use the current technology the Unmanned Aerial Vehicles (UAVs) that are installed or deployed efficiently to deliver high Quality of Services. This paper concentrates on the Flying Ad-hoc Networks (FANETS) communication technology for multiple swam of UAVs to cover massive coverage area of IoT devices services. Therefore, the UAVs can guarantee communication performance during a disaster condition without the participation of personal. The UAV nodes placed at different geographical locations are used to capture the information or data from the ground level from any IoT devices/sensors as that is possible using a normal traditional network. As the low latency service requirement while sharing the information in those emergency situations bring UAV network topology as it changes every time and energy consumed by each node gets increased as topology changes now and then.

To address the end-to-end delay and energy or power optimization performance. Figure 1 shows the layered UAV nodes use the Low Latency Routing Algorithm (LLRA) to monitor the disaster situation by sharing among each UAV nodes and finally sending it to the Base Station (BS) using wireless communication. The IoT sensor devices or UAVs capture the information from the lower ground level during disaster situations. Those capture information from ground level is then transferred

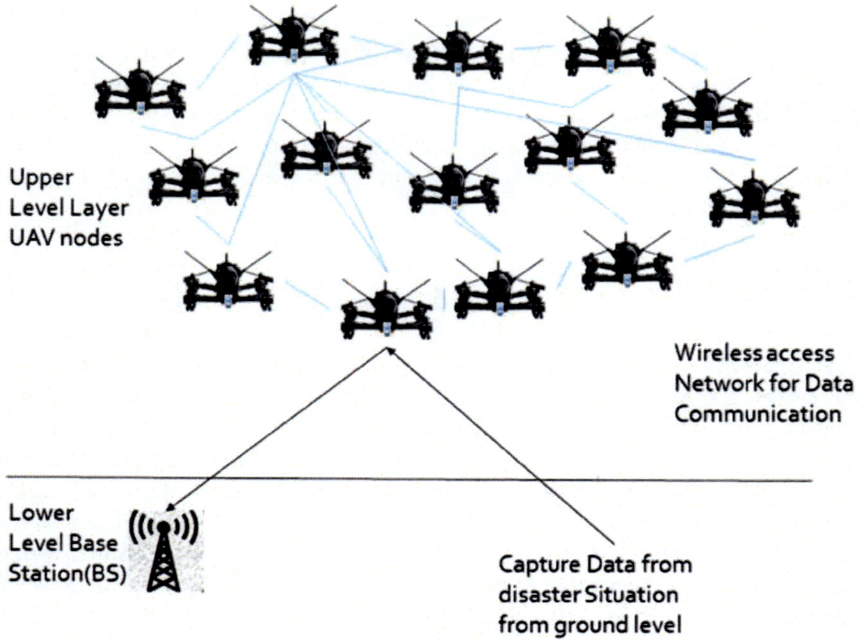

Fig. 1 Layered UAV nodes communicating with BS for data traversal

to the nearest routing UAV node for data communication and finally processed to the destination UAV node and then will be transferred to the Base Station (BS). Where the location in the disaster environment needs assistance (medical if needed).

The Low Latency Routing Algorithm (LLRA) is utilized to use the information about the location based on the symmetries of nodes and network connectivity of UAVs. This algorithm mainly determines the method to discover the route to send data packets from the initial to final destination UAV nodes in the upper layer network of FANETs. Based on the tuning up of the MAC layer is used in a transmission range to improve the fastest data processing rate and minimize the delay in UAVs nodes link stability. The projected system uses LLRA, which is used to accomplish the optimal route with the minimum delay and dynamically balance the data flows, which can efficiently minimize the average amount of energy utilized by UAV nodes and improve the stability of a route. This algorithm is used here to control the overcrowding of packets in the network and boosts the performance of end-to-end delay, as not all the UAV nodes of the network contribute to the route discovery for a specific initial source node to final destination node pair up. Delay insensitive standards such as 802.11b and 802.11g can work with numerous data rates for QoS-constrained communication to utilize the partial resources of UAV more competently.

Figure 2. is the flow chart that explains the flow of the working model from the initialization step to the output result of data routing to the destination using the

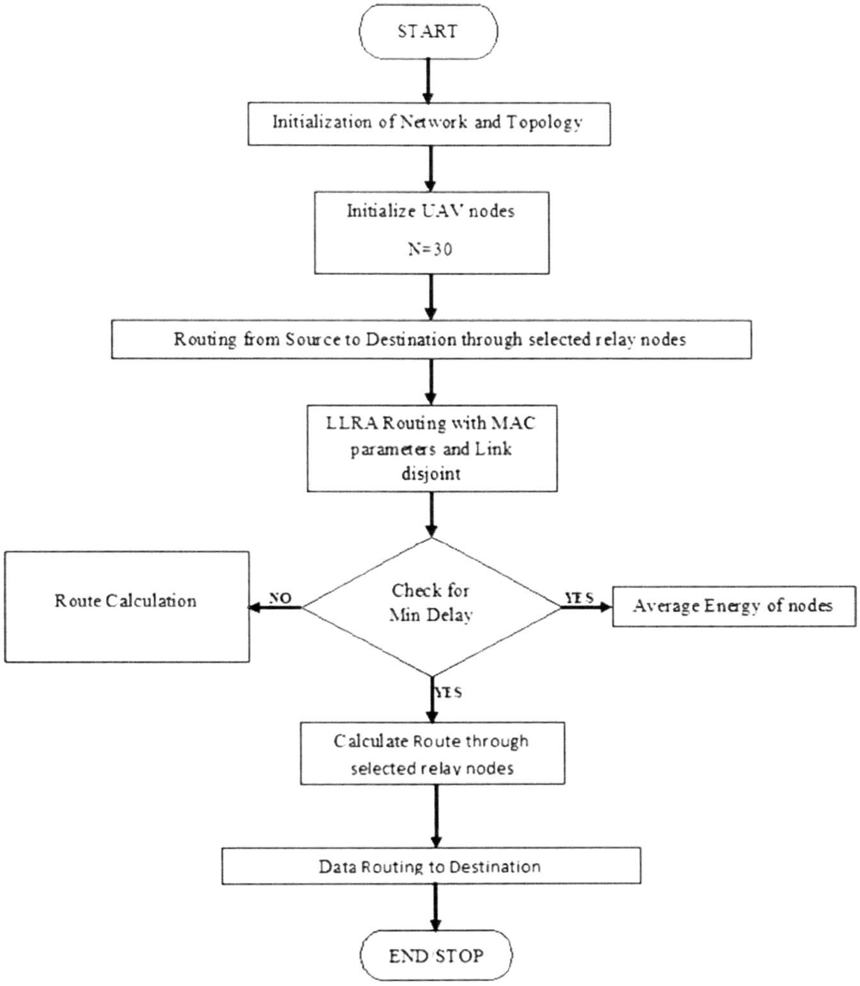

Fig. 2 Flow chart of proposed system

LLRA algorithm. Firstly, all the initialization simulation parameters will be initialized along with the no of UAV nodes used. Then the routing starts from the source node initialized. While routing using the LLRA routing algorithm with the MAC parameters checks for the shortest path to the destination node by the selected relay nodes. Checks for the minimum delay if yes then calculates the average energy by the nodes and the End-to-End Delay by calculating the route with minimum selected relay nodes through this data processes faster with efficient delivery of data to the destination node.

The Implementation of the proposed system in the NS2 Platform has simulation parameter settings that need to be done before setting up the system. Figure 3 and

```
###################################################################
#            Setting the Default Parameters                       #
###################################################################

set val(chan)           Channel/WirelessChannel
set val(prop)           Propagation/TwoRayGround
set val(netif)          Phy/WirelessPhy
set val(mac)            Mac/802_11
set val(ifq)            Queue/DropTail/PriQueue
set val(ll)             LL
set val(ant)            Antenna/OmniAntenna
set val(x)              800
set val(y)              800
set val(ifqlen)         80
set val(nn)             30
set val(stop)           100.0
set val(rp)             LLRA
set val(cp)             "30"
set val(sc)             "tcp"
set val(energymodel)    EnergyModel
set val(initialenergy)  100             ;# Initial energy in Joules
```

Fig. 3 Experimental settings of NS2 platform

Table 1 shows the simulation parameter values. Where the channel is set to the wireless channel, the TwoRayGround is a radio propagation model that depicts the path loss between the source and the destination node. Then the network interface is also wireless. The MAC layer used is 802.11 with the Drop-Tail interface queue. The overall area considered for the implementation of the proposed system is 800 × 800. The total number of UAV nodes is 30 and the impact height of UAV will vary from 50–90 m. 256 K bytes of packets will be forwarded with the queue length of 80 bytes using TCP protocol using a wireless network interface at a constant bit rate.

Table 1 Default simulation parameters

Simulation parameters	Values
NS Version	NS2
Network area	800 × 800
Channel	Wireless Channel
Interface Queue	Drop Tail
No. of nodes	30
MAC layer	802.11
Traffic	CBR
Transport protocol	TCP
Network interface	Wireless
Packet size	256 Kb
Queue length	80

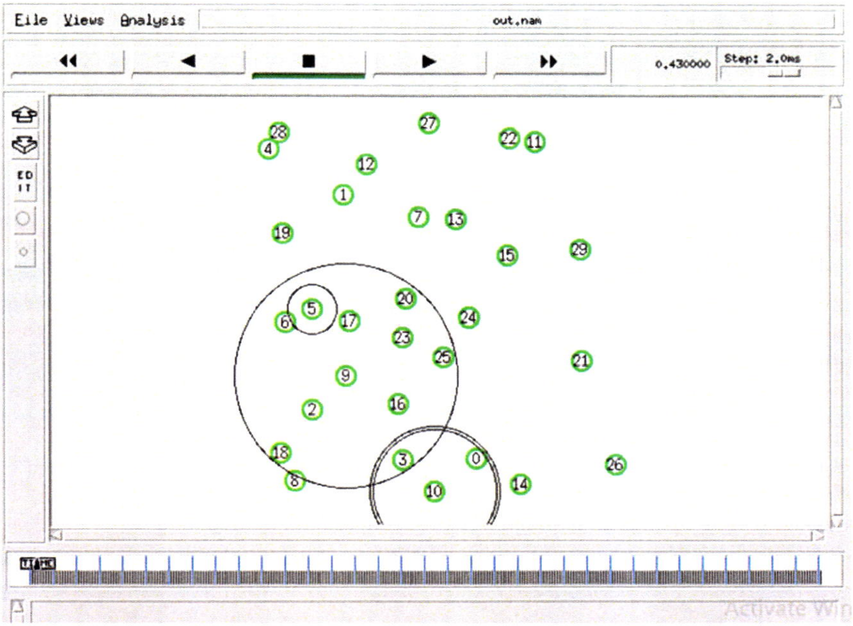

Fig. 4 It shows the. source UAV upper layer nodes transmitting data to destination UAV node

5 Results

5.1 Snapshots

The snapshots Fig. 4 shows the upper layer (sky level) UAV nodes transferring data from source to destination node and Figs. 5 and Fig. 6 shows the average amount of power-optimized and end-to-end delay, throughput and packet delivery ratio calculated using the Network Simulator (NS-2). Figures 7 and Fig. 8 are the X-graph which shows the graphical representation of the values achieved by the proposed system along with the existing system.

Table 2 gives the values that are predicted by using the proposed layered network architecture by using LLRA algorithms this table also shows the average amount of energy-optimized, end-to-end delay, overall throughput, and the packet delivery ratio (PDR) achieved.

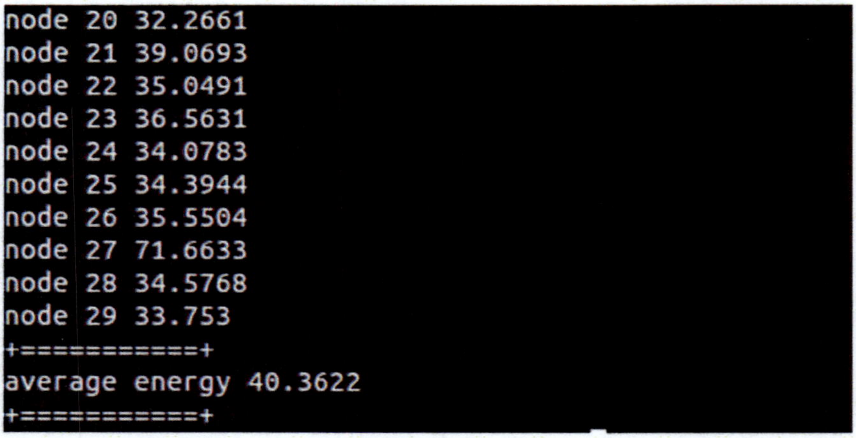

Fig. 5 Shows the average amount of energy or power optimized in the UAV nodes while transfer of data

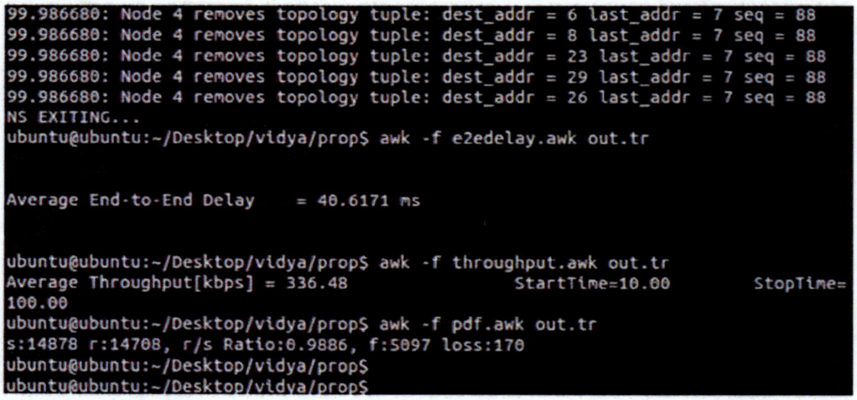

Fig. 6 It shows the average amount of end-to-end delay, throughput and PDR (Packet Delivery Ratio) accomplished at the multiple relay UAV nodes and traversal of data from one node to other

6 Conclusion

This paper proposed to avoid a dynamic technique to fetch an unfailing network performance with a FANET centralized procedure for network performance. To achieve the best possible routes in network environment the lively UAV nodes in the network architecture are utilized. In this projected method the MAC layer parameters are used tune-up in transmission range for faster data processing rate which also guarantees reliability on routing, and load distribution is thru to reduce the traffic overload among the neighboring UAV nodes and communication links by analyzing end-to-end delay. Finally, the average energy consumption is about 40.36 J and

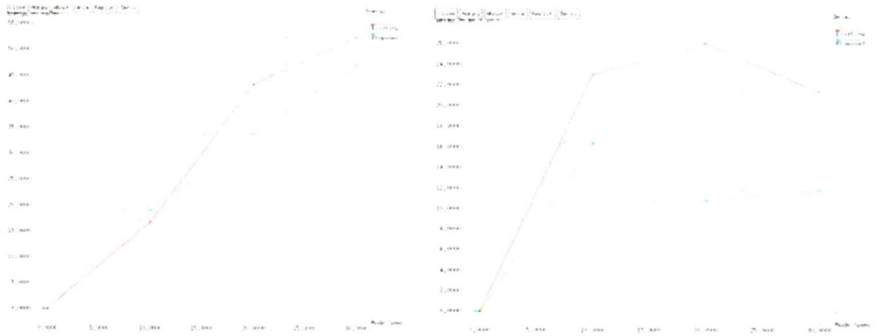

Fig. 7 The X-graph shows variance in average amount of energy or power optimized from existing to proposed system and the end-to-end delay achieved by the proposed system

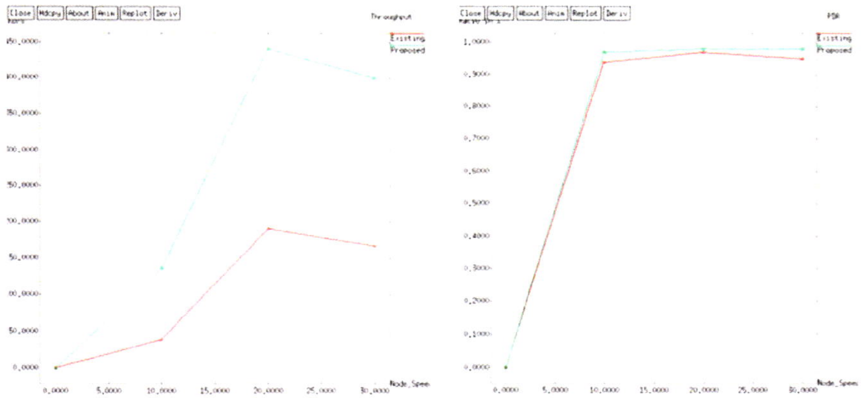

Fig. 8 The X-graph shows the average amount of throughput achieved from existing to proposed system and the PDR achieved by the proposed system

Table 2 Parameter values tabulated

Parameters	Values
Average Energy	40.3622 J
Average End-to-End Delay	40.6171 ms
Average Throughput	336.48 Kbps
PDR ratio	0.9886

end-to-end delay is reduced to 40.6 ms. In case of node failure self-recovering and self-adaptation are used to dominate the routes repeatedly and continue to broadcast. Energy-aware routing algorithm along with LLRA to improve the average amount of energy or power optimized.

Acknowledgements I tremendously thankful for Dr. Shankaraiah, Head of the Department of Electronics and Communication Engineering, JSS Science and Technology University (Formerly SJCE), Mysuru, for providing me timely suggestions, encouragement and support throughout the project. I'm indebted for Asst. Prof. Kavyashree M K, Department of Electronics and communication, JSS Science and Technology University (Formerly SJCE), Mysuru, for providing her valuable guidance, constant assistance and constructive suggestions for effectiveness of project.

References

1. Al-Turjman, F., Alturjman, S.: 5G/IoT-Enabled UAVs for Multimedia Delivery in Industry-Oriented Applications. Springer Science (2018)
2. Zhao, N., Richard Yu, F., Fan, L., Chen, Y., Tang, J., Nallanathan, A.: Caching UAV-Enabled Small-Cell Networks (2018)
3. Zhao, N., Lu, W., Sheng, M., Chen, Y., Tang, J., Richard Yu, F., Wong, V: UAV-assisted emergency networks in disasters. IEEE Wireless Commun. **26**(1) (2019)
4. Zhang, Q., Jiang, M., Feng, Z., Li, W., Zhang, W., Pan, M.: IoT enabled UAV: network architecture and routing algorithm. IEEE Internet Things J. **6**(2) (2019)
5. Chakareski, J.: UAV-IoT for next generation virtual reality. IEEE Trans. Image Process. **28**(12) (2019)
6. Wu, P., Xiao, F., Sha, C., Huang, H., Sun, L.: Trajectory optimization for UAVs' efficient charging in wireless rechargeable sensor networks. IEEE Trans. Vehicular Technol. **69**(4) (2020)
7. Shakoor, S., Kaleem, Z., Baig, M.I., Chughtai, O., Duong, T.Q., Nguyen, L.D.: Role of UAVs in public safety communications: energy efficiency perspective. In: IEEE Special Section on Mission Critical Sensors and Sensor Networks, vol. 7 (2019)
8. Du, Y., Yang, K., Wang, K., Zhang, G., Zhao, Y., Chen, D.: Joint Resources and Workflow Scheduling in UAV-Enabled Wirelessly-Powered MEC for IoT Systems, vol. 68, issue 10 (2019)
9. Mozaffari, M., Saad, W., Bennis, M., Debbah, M.: Mobile Unmanned Aerial Vehicles (UAVs) for energy-efficient Internet of Things communications. IEEE Trans. Wireless Commun. **16**(11) (2017)
10. Wang, J., Jiang, C., Wei, Z., Pan, C., Zhang, H., Ren, Y.: Joint UAV hovering altitude and power control for space-air-ground IoT networks. IEEE Internet Things J. **6**(2) (2019)
11. He, H., Zhang, S., Zeng, Y., Zhang, R.: Joint altitude and beamwidth optimization for UAV-enabled multiuser communications. IEEE Commun. **22**(2) (2018)
12. Sohail, M.F., Leow, C.Y., Won, S.H.: Energy-efficient non-orthogonal multiple access for UAV communication system. IEEE Trans. Vehicular Technol. **68**(11) (2019)
13. Zhou, F., Wu, Y., Hu, R.Q., Qian, Y.: Computation rate maximization in UAV-enabled wireless powered mobile-edge computing systems. IEEE J. Selected Areas Commun. **36**(9) (2018)

Improving the QoS of Multipath Routing in MANET by Considering Reliable Node and Stable Link

Mani Bushan Dsouza and D. H. Manjaiah

Abstract Routing overhead in mobile ad hoc network (MANET) can be reduced by maintaining multiple paths between the communicating nodes. This way, when one path fails, an alternate path can be readily used without resorting to a fresh route discovery. Ad hoc on-demand multipath distance vector (AOMDV) protocol is a link disjoint, loop-free multipath protocol, well suited for MANET. It uses hop count to choose the path. As the nodes in MANET are powered by battery, the residual energy of the node is a crucial factor in deciding the lifetime of the path. Because, nodes with the lower residual energy may drain off earlier, resulting in link failure. Moreover, the protocol selects an alternate path only on link failure. When a link breaks, packets may be dropped and additional control packets are required to re-establish the route. Instead of taking action after a link failure, a node can monitor the strength of the received signal and take necessary action, when the signal strength falls below a threshold level. The proposed routing protocol, namely reliable energy and link AOMDV (REL-AOMDV) takes it to consider both these factor during the route discovery and maintenance. Simulation of the proposed protocol in NS2 shows that, with the increase in node mobility, REL-AOMDV shows a lower routing overhead and a lower delay as compared to the original protocol.

Keywords Multipath routing · Reliability · QoS · Signal strength · Residual energy · AOMDV · Delay · Throughput

1 Introduction

Mobile ad hoc network (MANET) is a wireless network in which nodes cooperatively communicate by sending messages along with multiple hops without any centralized control. The random movement of the nodes causes a frequent change in network topology and the wireless medium imparts a constraint on available bandwidth [1]. Ad hoc on-demand multipath distance vector routing (AOMDV) [2] protocol is centered

M. B. Dsouza (✉) · D. H. Manjaiah
Department of Computer Science, Mangalore University, Mangaluru, Karnataka, India
e-mail: mani_bushan@hotmail.com

P. Karuppusamy et al. (eds.), *Sustainable Communication Networks and Application*,
Lecture Notes on Data Engineering and Communications Technologies 55,
https://doi.org/10.1007/978-981-15-8677-4_43

along an existing protocol, namely ad hoc on-demand multipath distance vector (AODV) [3]. AOMDV creates multiple paths between the communicating nodes which are loop-free and link disjoint. In this protocol, every node maintains a routing table containing a sequence number of the current request, destination address and a list of next hops along with their hop counts for forwarding packets toward the destination. An advertised hop count is also maintained by the node for the given destination, which provides maximum hop count for all the paths.

The AOMDV considers only hop count while choosing the path for communication. If the path so chosen contains nodes with lesser residual energy, then nodes with the battery will lose their power quickly and fail. This leads to link failure. When the link fails, there is a chance of packet loss and extra control packets are required for maintenance. Hence residual energy of the node is a crucial parameter while selecting the nodes within a path. AOMDV does not predict the stability of a link and starts the route repair process, only after the link failure. By monitoring the received signal strength, link stability can be assessed. When the nodes involved in a link move away from each other, the received signal strength decreases. On the contrary, when two nodes involved in a link move close to each other, the received signal strength increases. Thus, if a node realizes that the received signal strength of the next-hop node is continuously decreasing and is below a threshold value, it can switch to an alternate path. This way it can avoid packet losses and delay in the transmission of a packet. By considering both these factors, quality of the path can be increased. A modified protocol, namely Reliable Energy and Link AOMDV. (REL-AOMDV) considers the energy of the node during data transmission and uses the stability of the link to switch to an alternate path before the link actually breaks. This way, it provides a better quality of service.

2 Related Work

Many solutions have been proposed to improve the quality of service in MANET. Most of these are based on multipath protocols. Some of the relevant protocols are discussed here. Stability-based partially disjoint AOMDV (SPDA) [4] uses stability of the path while selecting the route. Stability of the node is measured by counting the number of RREQ transmitted by the node to those received by the destination node. The protocol either selects partially disjoint route or maximally disjoint route. Such a technique is found to increase the throughput and reduce the end-to-end delay. Maximally spatial disjoint multipath routing protocol (MSDM) [5] is an enhancement of AOMDV protocol that discovers spatially disconnected, maximally disjoint, multiple paths between the source and destination. This technique reduces the collisions due to the packet transmission over spatially disjoint paths. It was found that MSDM provides better routing overhead in comparison with AOMDV and reduces the delay and overhead. Prediction of Link Stability-AOMDV (PLS-AOMDV) [6] protocol periodically predicts the stability of the link by measuring the mobility of the node and the energy consumed by the node. It then chooses the path with the highest link stability. Simulation results show that the protocol is found to provide better packet delivery ratio. Channel-aware AOMDV (CA-AOMDV) [7] considers an average nonfading period of the channel during route discovery and uses pre-emptive handoff to maintain a consistent link. It showed improved performance over AOMDV under practical conditions. Prediction of link stability-AOMDV (PLS-AOMDV) [8] regularly checks the stability of the link by considering the energy consumed by the node and its mobility. It then chooses the path with the highest link stability for data transmission. Results indicate that selecting stable link intern provides better packet delivery rate and network lifetime.

Adaptive multi-metric (AM)-AOMDV [9] exchanges information between its one-hop neighbors to achieving route longevity. It shows better delivery ratio and reduced delay. Prior Path Failure Recovery AOMDV (PPFR-CAODV) [10] protocol monitors the behavior of channel as the data is being transmitted and assesses the stability of the link based on the signal stability. Channel is sensed by comparing the signal-to-noise ratio. When the noise strength exceeds signal strength, the path is rejected. The protocol is found to reduce the delay and provide an increase in throughput. Load balancing ad hoc on-demand multipath distance vector (LBAOMDV) [11] protocol controls energy and available bandwidth usage by utilizing the multiple paths during data transfer. This improves the quality of service and reduces node breakdowns. It also improves the reliability of the network. Maximal minimal nodal residual energy ad hoc on-demand multipath distance vector (MMRE-AOMDV) [12] protocol is a multipath protocol that aims to achieve energy efficiency and load balance. In this protocol, minimal residual energy of every node within the path is noted. Then, the paths are sorted in descending order and path with the highest residual energy is considered. Such a scheme enhances the network lifetime and achieves load balancing by choosing a path which is energy efficient.

A protocol that considers both lifetime of the node and number of hops during packet transmission was proposed by Othmen et al. [13]. Here, the paths are distributed among nodes having higher battery power. When an existing path fails, the protocol switches to a backup path. It was observed that the protocol can perform better than stable path routing (SPR) [13] and modified ad hoc on-demand distance vector (MAODV) [14] in terms of throughput, end-to-end delay, and loss rate. Khan et al. [15] proposed a protocol based on AOMDV. This protocol takes into consideration the energy of the nodes. Each state is determined by the value of hop value and energy metric. Reliability of the path is decided by the weakest node in the path and such weak nodes are avoided during transmission. Another protocol, namely optimized minimal maximal nodal residual energy AOMDV (OMMRE-AOMDV) [16] was given by Sivaraman and Karthikeyan. It considers minimal energy while deciding the best path and chooses the path with the highest residual energy to transfer the data. It was able to decrease the energy consumption, average end-to-end delay, routing overhead and normalized routing overhead.

3 Proposed Work

AOMDV is an on-demand routing protocol that establishes loop-free, link disjoint, and multiple paths between the communicating nodes [2]. It is based on the similar principles of AODV [17]. It contains a table called route list that holds the details of alternate paths, such as last and next hop, hop count for the path, the expiration time for a given path. The last hop count helps in maintaining link disjoint path. The sequence number of RREQ is used to achieve a loop-free path. Similar to AODV, the reverse path is setup whenever a valid RREQ is received at a node. Duplicate RREQ is used to determine alternate reverse paths and only those paths which are loop-free and link disjoint are recorded. For a given sequence number, node computes the advertised hop count using (1). Where i is the intermediate node and d is the destination node.

$$\text{advertised\_hop\_count}_i^d = \begin{cases} \max(\text{hop\_count}_i^d), & \text{where } i \neq d \\ 0, & \text{otherwise} \end{cases} \quad (1)$$

Whenever an intermediate node does not possess a valid path toward the destination for which it has not broadcasted RREQ with the same sequence number before, it broadcasts the RREQ; provided, such an RREQ result in either creation or modification of reverse path. However, if the intermediate node has a valid path toward the destination, then it will reply back with an RREP along the reverse path. Such RREPs will contain a forward path toward the destination that was not used in previous RREPs of this route discovery. For every loop-free and link disjoint RREQ that reaches the destination, an RREP is generated by the destination and send back toward the source. As the RREP travels across the intermediate nodes, a forward loop-free and link disjoint path is established toward the destination. If a node receives duplicate RREP, it is forwarded in a path that was not used previously for the similar RREP. Otherwise, the RREP is simply discarded. Data is transmitted along a path until it fails and when the path fails, an alternate path is used for data transmission. During the transmission, if the last know path toward the destination breaks, then the node generates a RERR packet and send it toward the source. HELLO messages are used to set the timeout for a given path and remove stale paths from the route list.

One of the reasons for link breaks in ad hoc networks is the random movement of the nodes. When the link fails, the packet sent along the link may be dropped and a new path may need to be discovered. This requires additional control packets and may cause congestion. This problem can be solved by predicting link failure and taking action before the link actually fails. One of the techniques for predicting link failure is, by measuring received signal strength. As the two nodes involved in a link move close to other nodes, the strength of the signal received increases. Similarly, when the nodes move away from each other, the strength of the signal reduced. Thus, by continuously measuring the strength of the signal, whether the nodes are moving closer or far away from each other, can be predicted.

For node i, situated at distance x from its neighbor, the signal strength [18] is given as (2),

$$SS_{ix} = \frac{G_r * G_t * S_t}{(4\pi * x/\lambda)} \tag{2}$$

Here, G_r is the receiving gain of the antenna and G_t is the transmitting gain of the antenna. The wavelength of electromagnetic wave used for transmission is λ and S_t is the maximum transmitting power of the transmitting antenna. If an antenna covers a circular radius R, the average distance between the nodes is given as $0.9054R$ [19]. Using this, the threshold value of signal strength for a given link i can be computed as (3).

$$TSS_i = \frac{G_r * G_t * S_t}{(4\pi * 0.905R/\lambda)^2} \tag{3}$$

In Eq. (3), antenna gain G_r, cover range R and used wavelength λ for a given node is known. Default values for G_r and G_t are taken as 1.0 and S_t is taken as 0.28183815, $\lambda = 3.0e8/914.0e6$. The transmitting antenna gain G_t and maximum transmitting power S_t are piggybacked along with the HELLO messages that are exchanged regularly with the neighboring nodes. Using all these values, a node can calculate the threshold, TSS_i for a given link. As the threshold is not dependent on the position of the node, its value is fixed for a given neighbor along with the link. By comparing the received signal strength in Eq. (2) with the threshold value in (3) a node can switch to an alternate path based on the following assumptions. Three consecutive received signal strengths are measured. For a given link i, that connects two nodes say i and j, suppose if the received signal strength is successively decreasing and $RSS_i > TSS_i$, then switch to an alternate path. If no alternate path exists, then when $RSS_i \leq 0.5 * TSS_i$, the link is assumed to be broken and RERR message is sent back to the previous node.

Energy consumed by the node [20] in-between time slots t_1 and t_2 is given as (4)

$$E_\delta = |E_{t_2} - E_{t_1}| \tag{4}$$

where E_{t_1} and E_{t_2} are the residual energy of the node at time t_1 and t_2, respectively. Rate of change in the energy [21] is given as (5)

$$R = \frac{E_\delta}{E_{t_1}} * 100 \tag{5}$$

Every node computes the rate of change in energy before sending a HELLO packet and piggybacks the rate of change in energy to all the adjacent nodes along with the HELLO message. Whenever a node wants to transmit the packet, it checks the value of the rate of change in energy for all the adjacent links and select, a link with least

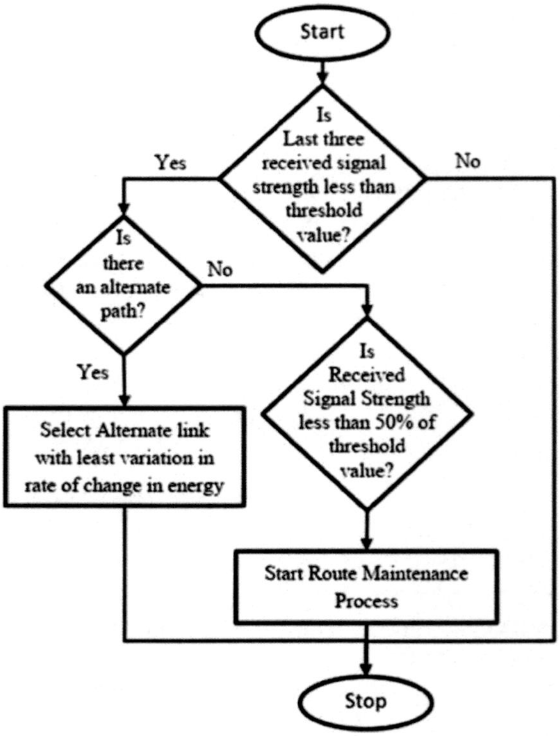

Fig. 1 Path selection process in REL-AOMDV

variation in the rate of change in energy. The path selection process is summarized in a flow diagram shown in Fig. 1.

4 Result Analysis

Nodes in MANET move randomly in all directions. As the speed of the node increases, there is a greater possibility of link breaks between the nodes. So, the simulation of AOMDV and REL-AOMDV was carried out with varying speeds using NS 2.34 simulator. The simulation was carried over a 1000 × 1000 Square meter flat area with 100 nodes. Details of configurations are discussed in Table 1.

Following metrics were used during result analysis.

- **Packet delivery ratio (PDF)**: It is the ratio between the numbers of packets delivered to the actual packets received. An effective protocol desires to achieve a higher value of PDF, as it indicates the rate of packet loss during transmission.
- **Packet loss ratio (PLR)**: It is the ratio of the difference in the generated and received packets divided by the total number of generated packets.

Table 1 Configuration details

Parameter	Value
Dimensions	1000 m * 1000 m
Number of nodes	100
Traffic type	CBR
Simulation time	100 s
Mobility model	Random waypoint model
Radio propagation model	Two ray ground
Antenna model	Omni
MAC type	802.11
Pause time	10 ms
Network load	4 packets/s
Packet size	512 byte
Initial energy	100 J
Transmission power	0.7 J/packet
Receiving power	0.3 J/packet
Speed	5, 10, 15, 20, 25 m/s
Routing protocols	AOMDV, REL-AOMDV

- **The end-to-end delay**: it is the total delay experienced by a data packet during its transmission. The delay may be due to queuing, propagation and retransmission of packet during its travel.
- **Normalized routing load**: It is ratio of total number data packets received at the destination to the total number of routing packet transmitted.
- **Throughput**: It measures the rate at which the packet is received at the destination. It is measured in bits/second.
- **Total energy**: It is the sum of the energy consumed by all nodes during the tenure of simulation.

As the speed increases packet delivery ratio decreases in both the cases, this is due to rapid link break downs that may occur due to the increase in speed. When the link fails, packets are dropped, thereby reducing the total number of packets being received at the destination. It is observed that the REL-AOMDV shows better PDR in comparison with AOMDV. This is because REL-AOMDV can predict the link breaks and switches to an alternate path before the link actually breaks. Also, only nodes with higher residual energy are chosen, this will further reduce the link breaks. This observation is evident in Fig. 2.

The packet loss ratio increases with an increase in speed. This is in connotation with the PDR. As the speed increases, there is a greater possibility of packet loss due to link failure.

Figure 3 shows that REL-AOMDV can provide lesser packet loss as compared to AOMDV. This is due to the fact that the REL-AOMDV can predict link failure and switch to the alternate path.

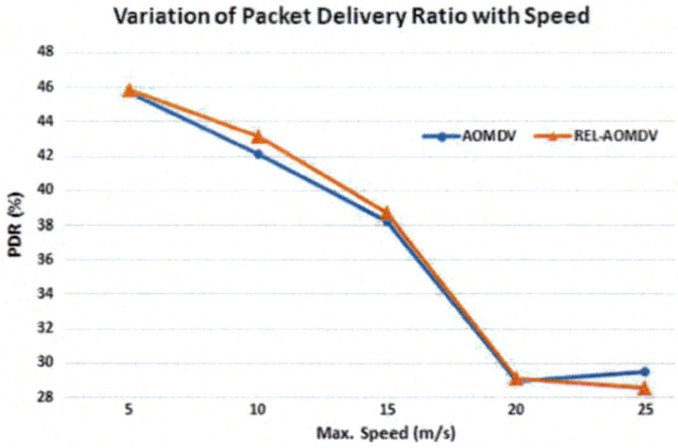

Fig. 2 Change in PDR % with Max. speed

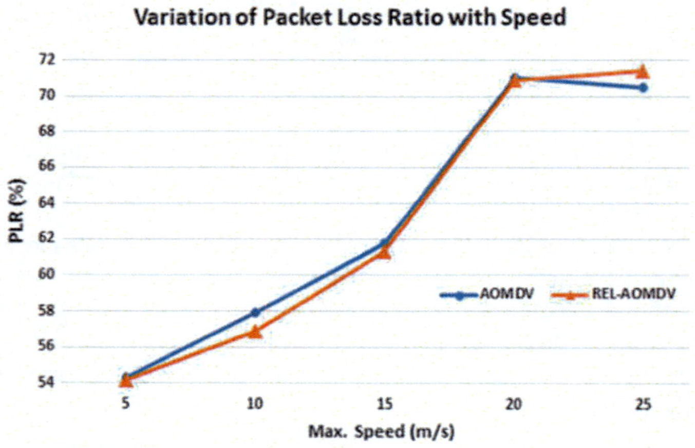

Fig. 3 Change in percentage of PLR with max. speed

End-to-end delay increases with the speed of the node. Even though REL-AOMDV is able to predict and use alternate path before link breaks down, it chooses the nodes with higher residual energy.

Such a selection may lead to higher congestions at selected nodes having higher residual energy. For the same reason, Fig. 4 indicate that REL-AOMDV is not been able to provide a considerable lower delay in comparison with AOMDV.

It is observed that the REL-AOMDV can reduce the control packets in comparison with AOMDV. This is as shown in Fig. 5. The reason for this is because the REL-AOMDV can predict and use the alternate link, before actual link failure. Also, it is able to prolong the path failure due to the use of node having higher residual energy.

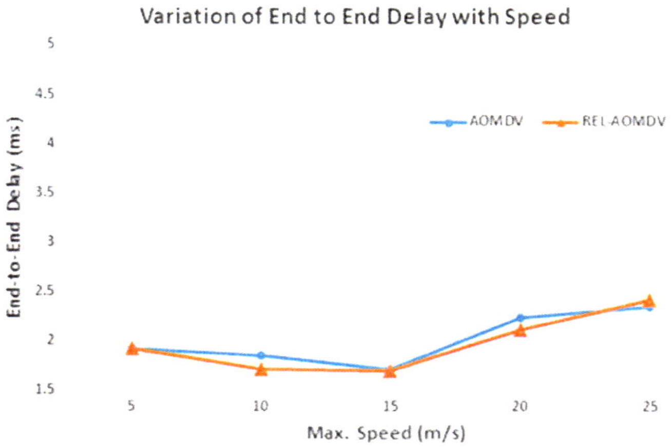

Fig. 4 Change in end-to-end delay with max. speed

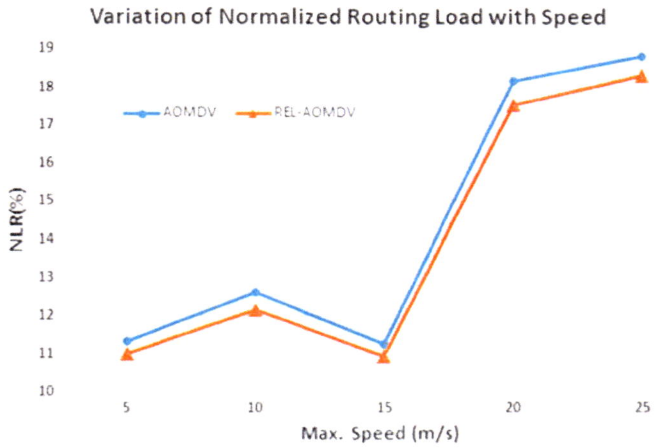

Fig. 5 Change in normalized routing load with max. speed

Both these factors result in a reduction in control packets, thus leading to a lower routing load.

Even though the delay is in comparison with the AOMDV, the REL-AOMDV is able to provide a better throughput with speed. This is evident from Fig. 6. Which shows higher throughput values for REL-AOMDV. The contribution factor for this is the lower packet losses which result in the reduced number of control packets and thus allowing more packets to reach to the destination.

The REL-AOMDV is able to prolong the path failure by selecting the nodes with higher residual energy. This is evident from Fig. 7 that shows lower total energy consumption in REL-AOMDV in comparison with AOMDV.

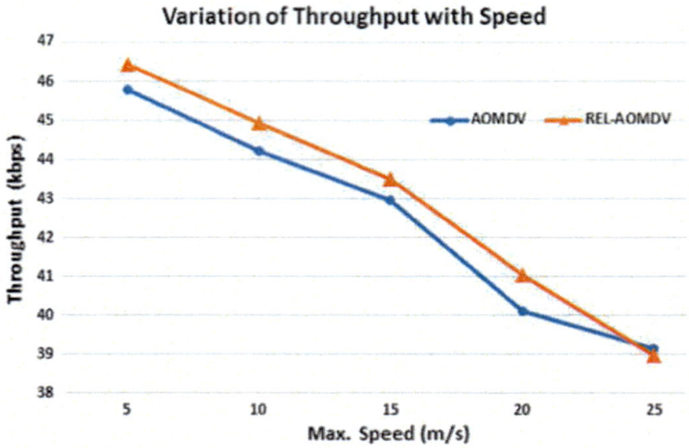

Fig. 6 Change in throughput with max. speed

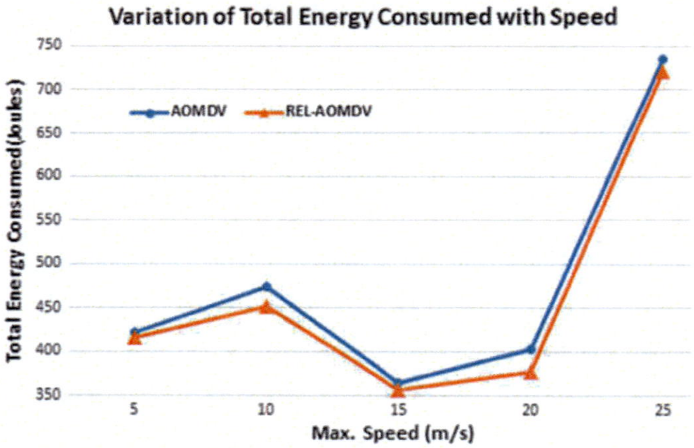

Fig. 7 Change in total energy consumed with max. speed

5 Conclusions

One of the critical issues in MANET is the issue of routing. By multiple paths between the communicating nodes, reliability of the route can be increased by switching over to an alternate path when the existing path fails. This prevents needless route discovery process and avoids extra control packets being sent out by the source. However, the effectiveness of the protocol can be enhanced further, by choosing the nodes with sufficiently higher residual energy. This will prolong the usability of the path as nodes with higher energy will stay alive for a longer time. By analyzing the received signal strength, link failure can be predicted. This way, nodes can switch over to an alternate path before, the actual link failure. This reduces the delay in packet delivery. Simulation results indicate that REL-AOMDV is able to reduce delay and deliver more packets as well as provide better throughput in comparison to AOMDV. Thus, it can be concluded that modification helps in providing a better quality of service to the communicating nodes.

References

1. Rathi, P.S., Mallikarjuna Rao, C.H.: Survey paper on routing in MANETs for optimal route selection based on routing protocol with particle swarm optimization and different ant colony optimization protocol (2019). https://doi.org/10.1007/978-981-13-9282-5_51
2. Marina, M.K., Das, S.R.: Ad hoc on-demand multipath distance vector routing. Wirel. Commun. Mob. Comput. **6**, 969–988 (2006). https://doi.org/10.1002/wcm.432
3. Perkins, C., Belding-Royer, E., Das, S.: Ad hoc on-demand distance vector (AODV) routing. RFC Editor (2003). https://doi.org/10.17487/rfc3561
4. Almobaideen, W., AlKhateeb, D.A.Z.: CSPDA: contention and stability aware partially disjoint AOMDV routing protocol (2015). https://doi.org/10.1109/aeect.2015.7360548
5. Almobaideen, W., Al-Soub, R., Sleit, A.: MSDM: maximally spatial disjoint multipath routing protocol for MANET. Commun. Netw. **05**, 316–322 (2013). https://doi.org/10.4236/cn.2013.54039
6. Singh, J., Sharma, K.: Energy efficient AODV routing protocol for mobile ad-hoc network. Int. J. Eng. Comput. Sci. (2015). https://doi.org/10.18535/ijecs/v4i9.77
7. Chen, X., Jones, H.M., Jayalath, D.: Channel-aware routing in MANETs with route handoff. IEEE Trans. Mob. Comput. **10**, 108–121 (2011). https://doi.org/10.1109/tmc.2010.144
8. Cai, J., Liu, K.: An improved AOMDV routing protocol based on prediction of link stability (2011). https://doi.org/10.1117/12.920958
9. Yu, M., Liu, W., Xing, T.: Link availability modeling for routing algorithms to reduce the link break time in MANETs (2011). https://doi.org/10.1109/icnsc.2011.5874913
10. Jagadeesan, D., Narayanan, S., Asha, G.: Efficient load sharing using multipath channel awareness routing in mobile ad hoc networks. Indian J. Sci. Technol. **8** (2015). https://doi.org/10.17485/ijst/2015/v8i15/67729
11. Alghamdi, S.A.: Load balancing ad hoc on-demand multipath distance vector (LBAOMDV) routing protocol. EURASIP J. Wirel. Commun. Netw. **2015** (2015). https://doi.org/10.1186/sl3638-015-0453-8
12. Liu, Y., Guo, L., Ma, H., Jiang, T.: Energy efficient on-demand multipath routing protocol for multi-hop ad hoc networks (2008). https://doi.org/10.1109/isssta.2008.112
13. Othmen, S., Belghith, A., Zarai, F., Obaidat, M.S., Kamoun, L.: Power and delay-aware multipath routing protocol for ad hoc networks (2014). https://doi.org/10.1109/cits.2014.6878956
14. Zhong, M., Fu, Y., Jia, X.: MAODV multicast routing protocol based on node mobility prediction (2011). https://doi.org/10.1109/icebeg.2011.5881805
15. Khan, K., Goodridge, W.: Energy aware ad hoc on-demand multipath distance vector routing. Int. J. Intell. Syst. Appl. **7**, 50–56 (2015)
16. Periyasamy, P., Karthikeyan, E.: Energy optimized ad hoc on-demand multipath routing protocol for mobile ad hoc networks. Int. J. Intell. Syst. Appl. **6**, 36–41 (2014). https://doi.org/10.5815/ijisa.2014.ll.05
17. Perkins, C.E., Royer, E.M.: Ad-hoc on-demand distance vector routing (1999). https://doi.org/10.1109/mcsa.1999.749281
18. Chatterjee, S., Das, S.: Ant colony optimization based enhanced dynamic source routing algorithm for mobile Ad-hoc network. Inf. Sei. (Ny). **295**, 67–90 (2015). https://doi.org/10.1016/j.ins.2014.09.039

19. Bettstetter, C., Hartenstein, H., Pérez-Costa, X.: Stochastic properties of the random waypoint mobility model. Wirel. Netw. **10**, 555–567 (2004). https://doi.org/10.1023/b:wine.0000036458.88990.e5
20. Feeney, L.M., Nilsson, M.: Investigating the energy consumption of a wireless network interface in an ad hoc networking environment. https://doi.org/10.1109/infcom.2001.916651
21. Aouiz, A.A., Boukli Hacene, S., Lorenz, P., Gilg, M.: Network life time maximization of the AOMDV protocol using nodes energy variation. Netw. Protoc. Algorithms **10**, 73 (2018). https://doi.org/10.5296/npa.vl0i2.13322

A Novel Approach to Detect, Characterize, and Analyze the Fake News on Social Media

G. Sajini and Jagadish S. Kallimani

Abstract Due to the upsurge in the usage of social media in the recent past, communication between people has undergone a great change. Users interact with each other by sharing a lot of information about the ongoing trends. But most of the information shared recently is misleading with the spread of false news which is known as fake news. Spreading a huge volume of such fake news can lead to many complications. The research field has been more concentrating on the inception, spread, and consequences, nowadays. Detecting the truthfulness of the news is of great concern. It can face many technical difficulties on many grounds. Usage of online tools has made the generation of content easy and is expanded fast, where this can lead to a huge amount of data for analysis. The online content is divergent, which deals with numerous fields, contributing to the job complication. Only the computers will not be able to evaluate the truthfulness and purpose, where it is completely dependent on the human and computer interaction. Sometimes, the information that might be termed as fake by a specialist might delude people. Such availability will always remain in restricted quantity but it could be a base for the mutual endeavor. Here, a broad summary of the fake news discoveries pertaining to the news will be given. The unfavorable effects, particularly the ongoing work on the methods to detect such news are demonstrated. The readily available datasets are studied to classify the fake news. An assuring solution is recommended to analyze the online fake news.

Keywords Social media · Online fake news · Fake news detection · Target-based · Social context based · Content-based news · Creator-based fake news

---

G. Sajini
Department of Computer Science and Engineering, M S Ramaiah Institute of Technology, Bangalore, India
e-mail: sajini.narayana@gmail.com

J. S. Kallimani (✉)
Visvesvaraya Technological University, Belagavi, Karnataka, India
e-mail: jagadish.k@msrit.edu

P. Karuppusamy et al. (eds.), *Sustainable Communication Networks and Application*, Lecture Notes on Data Engineering and Communications Technologies 55,
https://doi.org/10.1007/978-981-15-8677-4_44

1 Introduction

Due to the invention of Internet, there is a tremendous change in the manner in which human communication is carried out. Social media applications are helpful in reaching out to people across the globe. Because of less money, easy handling, sudden increase, it has become a big place for communication and data transfer. With respect to volume, velocity and veracity, the fake news are characterized. Usually, fake news are available in big volume, increased real-time velocity, and uncertain veracity. Almost 67% US grown-ups acquire news from social media and the count is increasing at a large rate.

But due to the significant growth of social media, the net remains as a proper place to distribute online fake news, like content which misleads, false ratings, fake commercials, gossips, false dialogues made by politicians, etc. The popularity of online fake news is not more from media but is more in social media. In the industrial and academic sectors, there is extreme ambiguity because of the facts which are partial. For shopping online, and other ventures, it is affecting heavily.

1.1 Importance of Detecting Fake News

The online fake news is irritating, misleading, intrusion by spreading everywhere. On communities, there is a huge effect. Features of fake news are shown below [1]:

The Volume of Fake News
Since to cross verify, there is no method, anyone can create fake news. Many sites on the Web are developed explicitly to generate fake news. These are usually looking like legal sites. Just to generate false claims, improper words, usually for selfish reasons like money and power. Huge amounts of data are all over despite notice of the users.

The Variety of Fake News
In human interaction, major modifications are seen impacting the thinking of home sapiens. Teenagers and the elderly are brainwashed easily because of exponential growth. May them can be caused due to the incidents happening around the globe if the false news is spread a lot. In our day-to-day lives, online fake news creates a big keen extreme deep acute detail.

The Velocity of Fake News
Fake news generators live for less time span. Many fake news concentrates on the ongoing affairs so that the spread is easy. There are complexities to check the number of people related to the current content, the tremendous repercussions. Putting an end to it is extremely grueling and punishing.

2 Fake News Characterization

The broad area is reached in this day and age because of online fake news and it is obnoxious. Highlighting server varieties of online fake news not leaving the originality, fake ratings, and commercials are given [2]. Few essential features of online fake news are (Table 1).

Fake news can be viewed as four main types of elements. The elements are discussed below:

Creator: Those who generate fake news are may be people or computers [3]. There might be spelling mistakes, purposefully, done with bad intentions by hackers.

Target victims: These are those who are targeted by the online fake news. These are people who make use of social media and the rest of the environments. They can be academicians, guardians, old, etc.

News content: It is the main part of the news having things like videos, audios, gifs, animations, and many things.

Social context: This shows the way in which the news is spread throughout. It contains users analyzing the users and pictures.

Many of the fake news available over Facebook by various users, can be mapped to the above discussed four types of elements [4].

Table 1 Various categories of fake news

News content	Non-physical content	Main purpose
		Sentiment
		News topics
	Physical content	Image/video
		Body text
		Headlines
Social context	Platform	Social media
		Main streaming
	Distribution	Broadcast pattern
		Community of users
Target victims	Platform	Online users
		Main streaming users
	Potential risk analysis	Role-based analysis
		Temporal-based analysis
Creator/spreader	Real-human	Fake news creator
		Benign author and publisher
	Non-human	Cyborg
		social bots

2.1 Fake News Generators and Publishers

The person who generated and the reason for their idea to create fake news needs to be known, understood, found ou, and must be explored. Generators can be either people or non-living beings.

Non-living being: Many robots exist to produce fake news. Using different algorithms for showing features like people and generating on their own not making use of people. Few legal data distributing bots play a significant part. Tweeting can be done by cyborg and take part in the activities. Same to same because of the fake news, there can be many issues that might arise like trustworthiness.

People: People are very important to start point of fake news. The non-living things are the transformers of fake news online. Automatically it is trained to generate and unfurl the news. If it is by a human or computer, irrespective of that, a group creates fake news and rolls out to reach and expand to the maximum number of people. Detecting whether the new is false or real is a really tedious mundane task. Even though the normal users are not generators, several legit users unknowingly and unintentionally become the carrier of such news. Their close ones think it is a real news and share it further which it spreads widely within a less timeperiod. Since there is anonymity, nobody takes the blame for what is shared, replied, and pasted.

2.2 News Content

Every news contains two contents, physical and non-physical. They are described below one by one.

Physical news content: This shows the headlines, vital part, and many components likes' pictures and multimedia. Strong channel to convey information is Facebook, Twitter, etc. Since it is a happening event, it happens at a high speed. URL, #, smileys, multimedia, pictures are part of this class which are the significant factors [5].

Non-physical news content: The first-class passes on the content and the second class are suggestions, viewpoints, judgments, advice, feelings, which the generators are in desire to formulate. There are various types, classifications, varieties, kinds of fake news like unreal ratings, unreal commercials, unreal content in the field of politics et cetera.

Flipkart, myntra, snap deal, and many more e-commerce sites are being assessed rated appraised criticized day-to-day each day daily. If there is partiality in such things, it can give a substandard inferior poor unpleasant terrible name to these sites. This will cause an effect in the way the users think, manipulates their selection choice options, and will ruin shatter wreck tear down a company's honor fame [6].

On the same lines to unreal ratings, unreal and false commercials are posted to deceive the consumers by giving ads which are deluding fooling misguiding content. They two are treacherous and demolish the trust of these sites.

3 Fake News Detection—Practical-Based Approaches

Fake news detection entails the mission to check how true the news is. Resources in detecting fake news are classified, and they are of two types: practicality method, current research approach. Here, major discussion is on the former online fake news detection.

3.1 Online Fact-Checking Resources

Media organizations commonly perform fact-checking resources. As ongoing contents are usually mixed of content, two class will sometimes tell the whole issue completely. Currently, a lot of performance factors are employed in determining the truthfulness of the content.

Verifying fact is a good way to identify fake news, informing people the trueness, wrong, a combination of the two. In the following, evaluation and comparison of a few famous verifying facts online tools are carried out.

Classifynews shows an environment for online to identify fake news with a mindset of building a method for determining trueness for a news item. Machine learning algorithms are used to identify this, basing solely on the textual content labelled news articles are collected from *Open Sources* and everyday training are implemented to these correct, incorrect page examples. To determine, two methods exist, namely—"Context-only", "Content-only" using two classifiers of machine learning [7].

FackCheck.org is created with the aim of reducing the amount to fool, ambiguous US politics, and this webpage evaluates the actual trueness of sentences made by big figures. These sentences originate in a multitude of information platforms like television ads, talks, and fresh information. Factmata.com refers to statistically fact verification and detecting.

Hoaxy.iuni.iu.edu are platforms to fetch, find, and analyze online fake news.

Hoax-Slayer.com shows the method for mail spamming and hackers.

Snopes.com is the 1st verification of fact page to verify US tradition.

4 Fake News Detection—Research-Based Approaches

With the growing use of online networking sites and the need to digitalize things, it is very important to have a correct transmission of information on the Internet. A large amount of information is shared, stored on the Internet with the growing rise of the use of the Internet for the smallest things. Our daily activities are done on the Internet like searching for restaurants to order food, or searching a customer care number for your queries, or making payment for a purchase. It is important to ensure that the correct information is available on the Internet and it does not use any wrong information, which could lead to exposure of our details on the wrong website. Here, the discussion is initiated to mitigate the leaking of personal details by detecting any false information on the website. The proposed work analyzes various existing ways for detecting the false information, limitation, and how it integrates with the advantages of all the available methods and overcome the limitations [8]. Deep learning can be used to process the text in order to detect the fake news by analyzing the semantics of the text. There are various categories of fake news detection and it is classified into:

4.1 Component Established Type

This section analyzes the details like who is author of the news, who is the reader, what type of content it is, what country or location the news targets, etc., and then analyses these attributes and find if the news is fake news. Normally, there will be a pattern followed by online fake news authors, which can be detected by doing semantic analysis on the content. The content will be such that will have enormous emotions so that it attracts a large audience in a short time. Also, normally, this kind of fake news will come from a similar author previously detected in fake news detection, this can be checked by checking the user id of the author, location of the author, etc. Also, this research work analyzes the type of audience targeted which can help in checking if it can be fake news.

Another method to detect false information is by having a group of experts to evaluate the information being circulated and validate the authenticity of the information. This requires manual intervention to address the limitations when automated tools are used to detect FN. Automated tools require correct data sets and can fail if the pattern is not recognized from the previous models. Human validation can address these limitations.

4.2 Data Mining Established Type

Data mining algorithms are widely used these days for this kind of analysis. Fake information can be detected either by a supervised learning model or unsupervised learning model. Former however requires large correctly labelled dataset which can be a problem in our case as the fake information data is huge on the Internet and it becomes difficult to filter these data set because of various different patterns it follows every time. Hence, latter can help in this case over former as it can still detect fake information without the need of having a large dataset. Latter uses semantic analysis to find patterns in the information to detect fake news [9].

4.3 *Implement Established Type*

This involves two types of systems: Online-detection and offline-detection system. Former targets real-time data base and can be beneficial over offline-detection system. The models used in the latter can have a disadvantage as it cannot accurately represent current online fake news pattern. Hence, the online system is used to predict effectively the real-time news and this model can be used to improve the offline methods as well.

Even with all these methods of detecting fake information, it can still observe a large amount of fake news is still published and the audience being unaware of the truthfulness. There is a need for a solid model which can help in detecting fake news effectively on the Internet. Research is still in progress in various institutes to achieve this goal but with the growing technology and the use of the Internet and new ways and patterns used by fake authors for distribution fake information, the development of effecting fake information detection system becomes difficult.

Various features which are involved in fake news representations are: creator or user-based features, news content-based features, and social context-based features. Each of these features again involve features such as user profiling, user credibility, behavior-based, linguistic and syntactic, style-based, visual-based, network-based, impact-based, and finally, temporal-based features [10].

5 Future Research and Open Challenges

Along with some promising research in this area, a few challenges and open issues for automatic online hoax news detection are tested. Thus, here, a method is proposed to construct an effective online fake news detection ecosystem.

5.1 *Unsupervised Learning for Fake News Analysis*

The dataset in real-world available for practical analysis does not come with quality labelling. This is one of the major hurdles in the detection of online fake news. Unsupervised learning refers to a type in the machine learning algorithm. Three types of unsupervised learning models for determining fake news exist. The types are as follows: Cluster analysis (CA), outlier analysis (OA), semantic similarity analysis (SSA), and unsupervised news embedding (UNE) [11].

Cluster analysis: In cluster analysis, the information is classified on the basis of increasing within class commonness, decreasing the between class commonness instead of examining labels. Generation of labels is done for data. In fake news detection, this method is used to analyze and recognize the independent class of content and those who write while discarding the dissimilar ones.

Outlier analysis: Working of OA is based on detection of the uncommon actions. Making use of statistic, distance, and weight approaches, outlier analysis algorithm can expose false or forged information and suspicious authors [12].

Semantic similarity analysis (SSA): Semantic similarity analysis is used for determining almost copied content. Online fake news creators reuse the existing news content because of missing information, comprehension, consciousness, and understanding plus awareness. Through semantic similarity analysis, the replica, half replica news information impressionable by crooked writers, an instance, and a false commentator who changes just some phrases for feedback to deceive users can be detected. SSA are essential methods for online fake news detection due to their textual nature.

Unsupervised news embedding: Embedding is an important step in natural language processing, which is a procedure of drawing out decentralized picturisation of information. The number pattern is utilized for input to analyze in the future. Implanting mechanization are categorized into different types depending on the ways in which they acquire the features of information with another viewpoint. Choice of an embedding method depends on the underlying nature of the news, for successful exposure of faulty information.

5.2 Evaluation of Fake News Detection System

Here, to detect fake news, an environment can be given and the performance measures to construct an efficient device to detect online fake news to study. Following are the elements to find out how OFN performs detecting:

Accurate detection (AD): The sole purpose in a device to detect can be AD of false information. Because of difficulties that the FN figuring process proposes, several effective methods need to be built and employed to increase their efficacy in the systems.

Interactive visualization: Visualization is an essential element of online fake news detecting and watching systems. They facilitate and eases human understanding by bringing different views and directions to represent data which are time delicate.

Early warning and post intervention (EWPI): As concluded by the above discussions, EWPI has a lot of scope for research when it comes to the finding of online fake news. They are significant features checking which the device for detecting fake news can be unveiled.

Verifying third party: Involvement of third-party verification eases the detecting device and can make it self-reliant. System to figure out online fake news is discussed here. An effective fake news detection ecosystem contains a combination of an *alert system, detection system*, and *intervention system*. It covers analysis, alerting, intervention, and detection of fake news, which form the pillars of an efficient system. A typical comprehensive fake news detection system may involve: intervention, fact checking, fake news detection, suspicious analysis, and potential fake news prediction.

6 Conclusion

It has been concluded that, the social nerwoking platforms are detected with more number of harmful and threatening fake news. This helps malevolent entities to maneuver individual's recourse and pronouncement on salient daily pursuit, to name a few, like education, healthcare, and so on. To overcome this complication, the need to have a fake news detection is extremely magnitude but performs the exact task across all the works of life. To summarize this paper, specific key points are dispensed as follows:

- In-depth apprehension of the distinct aspects of fake news, as news content, social context, and news creator to name a few. This can favor an implicating role in social imparting data analysis and anomalous information detection. Through clear characterization of online fake news.
- The researchers are propounded in order to notice the open issues, look into existing assertion frameworks and ameliorate them, and establish a monitoring and detection opportune for online fake news.

References

1. Zhang, X., Ghorbani, A.A.: An overview of online fake news: characterization, detection, and discussion. Inf. Process. Manage. **57**(2) (2020)
2. Banerjee, R., Feng, S., Kang, J.S., Choi, Y.: Keystroke patterns as prosody in digital writings: a case study with deceptive reviews and essays. In: Proceedings of the 2014 Conference on Empirical Methods in Natural Language Processing (EMNLP), pp. 1469–1473 (2014)
3. Bojanowski, P., Grave, E., Joulin, A., Mikolov, T.: Enriching word vectors with subword information. Trans. Assoc. Comput. Linguist. **5**, 135–146 (2017)
4. Bordes, A., Chopra, S., Weston, J.: Question answering with subgraph embeddings. E-print (2014)
5. Baldwin, B.T.: Automatic satire detection: are you having a laugh? Proceedings of the ACL-IJCNLP 2009 Conference Short Papers, Association for Computational Linguistics, pp. 161–164 (2009)
6. https://www.businessinsider.com/here-are-the%20-most-and-least-trusted-news-outlets-in-america-2014-10. Accessed 2018-04-19

7. https://github.com/BuzzFeedNews/2016-10-facebook-fact-check/blob/master/data/facebook-fact-check.csv. Accessed: 2018-04-19
8. Cao, N., Shi, C., Lin, S., Lu, J., Lin, Y.R., Lin, C.Y.: Targetvue: visual analysis of anomalous user behaviors in online communication systems. IEEE Trans. Vis. Comput. Graph. **22**(1), 280–289 (2016)
9. Castillo, C., Mendoza, M., Poblete, B.: Information credibility on twitter. In: Proceedings of the 20th International Conference on World Wide Web, pp. 675–684. ACM (2011)
10. Castillo, C., Mendoza, M., Poblete, B.: Predicting information credibility in time-sensitive social media. Internet Res. **23**(5), 560–588 (2013)
11. https://www.cbc.ca/news/canada/toronto/scarborough-hijab-attack-1.4487716. Accessed: 2018-03-21
12. Cha, M., Mislove, A., Gummadi, K.P.: A measurement-driven analysis of information propagation in the flickr social network. In: Proceedings of the 18th International Conference on World Wide Web, pp. 721–730. ACM (2009)

A Novel Design for Real-Time Intrusion Response in Latest Software-Defined Networks by Graphical Security Models

L. Sri Ramachandra and K. Hareesh

Abstract In the current era, many of the application domains are adopting software-defined network (SDN). SDN provides a special functionality to the network flow by controlling it dynamically with more robustness. Traditional networks are high in cost whereas SDN is economical. Since there are more chances of cyber-attacks, security solutions are proposed to reinforce and retreat the SDN. These security solutions have to be compared with each other and select the optimized solution to provide the best security for SDN. Graphical security models like attack trees and attack graphs can be used to measure and estimate the safety of SDN. Due to computational complexity, it is hard to provide security for SDN against cyber-attacks in real-time cases. Using precomputations, this paper aims to detect the disturbance which causes in SDN and finds the possibility of future attacks in the future for real-time cases. Various SDN components are taken into considerations for conducting an assessment on security which was not accessible in the existing network. Experimental analysis of this paper estimates all the possible attacking path in an ongoing attack which can be mitigated in real-time cases. It also exposes the security metrics which depends on the flow table that includes SDN components. Hence, it is possible to provide security for SDN against cyber-attacks in real-time cases.

Keywords Software-defined networks (SDN) · Moving target defense (MTD) systems · Attack graph · Hierarchical attack representation model (HARM) · Remote triggered black hole (RTBH)

L. S. Ramachandra (✉)
Department of Computer Science and Engineering, Government Engineering College, Ramanagara, Karnataka, India
e-mail: dsram49@gmail.com

Visvesvaraya Technological University, Belagavi, Karnataka, India

K. Hareesh
Department of Computer Science and Engineering, Government Engineering College, K R Pet, Karnataka, India
e-mail: hareeshk.gec@gmail.com

P. Karuppusamy et al. (eds.), *Sustainable Communication Networks and Application*, Lecture Notes on Data Engineering and Communications Technologies 55,
https://doi.org/10.1007/978-981-15-8677-4_45

1 Introduction

Logical network topology can be dynamically changed in real time by administrators of network if and only if SDN permits [1]. This can be achieved when controls are separated from data flows on top of the data plane and control plane. When network topology [2] is reconfigured dynamically by SDN, impact on performance by network disruptions will be negligible. This efficiently helps in optimization of load by administrators in real-time cases. New architecture like moving target defense systems (MTD) has to be designed and deployed to provide security for SDN. New networking components like forwarding devices and controllers are introduced by SDN which provides attackers with an attack vector for exploiting the SDN. Some of the security techniques such as DELTA and Athena are developed to protect SDN from cyber-attacks. DELTA [3] is a framework for evaluating the security, whereas Athena [4] is a framework developed for anomaly detection. Athena uses machine learning for predicting the patterns of various attacks. Graphical security models will be used for the evaluation of the SDN. Attack trees (AT) and attack graphs (AG) are used as security models since they provide security in-depth and compute the optimal countermeasures. New SDN components have to be considered for the assessment of security when this approach is applied to SDN. As there will be a delay in initial attack and response, this causes difficulty for intrusion detection system to find out the ongoing attacks in real time. This eases the attacker to reach the target so there should be an effective focus on a countermeasure to avoid the attacker reaching the target than giving more importance in finding out the attack detection. All the probable attacking paths have to be predicted to know the targets of the attacker. Computing all these possible attacking paths will lead to adaptability and scalability problems [5, 6]. To avoid all these problems, an efficient technique is required which can take care of all the attacking paths meanwhile considering new SDN components for the evaluation of the security. To overcome the above-mentioned problems, the hierarchical attack representation model (HARM), a graphical security model is incorporated with the precomputed approach along with the components of SDN. This evaluates the security of SDN in real-time cases. HARM helps in evaluating all the possible attacking paths before an attack is detected. This helps in formulating effective countermeasures and helps in estimating the possible attacking paths from the detection point. Evaluation is carried out by generating precomputed possible scenarios of attack with the help of a full attack graph (AG).To know how an attacker is trying to steal data from the outside, a scenario of attack was used. After identifying the paths of attack, it is found why the intrusion detection system took a long time to detect the attack than the time of the attack. To identify the relevant paths of attack, full attack graph which is precomputed is used. By evaluating this full AG, related countermeasures are deployed. Some of the experimental analyses are conducted to demonstrate that the proposed approach can efficiently detect the attacks without any delay and can diminish the ongoing current attack in real-time cases.

The contribution of the paper are summarized as follows:

- When new SDN components are taken into account with their respective attack vectors, security assessment is conducted for SDN.
- To develop a countermeasure for SDN and to provide a security assessment for real-time cases, precomputed attack scenarios are generated using full attack graph.
- To avoid the delays in finding an attack in an ongoing intrusion, response and prevention mechanisms are proposed in attack detection systems.
- Experimental analysis is conducted for the demonstration of the proposed approach for resolving the attacks in SDN with delayed detection.

The organization of the paper is as follows, related work is presented in Sect. 2, and a framework is presented in Sect. 3. Real-time intrusion response in SDN is presented in Sect. 4. Precomputation and attack prediction for security assessment are presented in Sect. 5. Section 6 presents results and analysis and finally Sect. 7 presents conclusion of this work.

2 Related Work

- SDN security: There are certain security issues to SDN which are discussed in [1, 7–10]. Threat vectors of SDN were presented by Kreutz et al. which were not there in the traditional network system. Potential solutions were given to resolve the threats which were identified but a method for evaluating those threads was not mentioned. The paper aims to provide a solution to these problems. For the better enhancement of OpenFlow protocol security in SDN, a framework called FRESCO is presented by Shin et al. [11] which allows SDN to detect and resolve the attacks. This work demonstrates that SDN switches which are the part of SDN components are more disposed to cyber-attacks. SE-Floodlight was presented by Porras et al. [12] which was the extension of OpenFlow Floodlight controller. This protects the SDN control plane with added security features.
- SDN intrusion detection: An attack which is carried out successfully is defined to be intrusion which includes malicious activities in the system. It will be perfect if all the intrusions are detected with the accuracy of 100% but in practice, and it is not feasible. This is because of the true positive and false positive alarms of the intrusion detection systems. A framework was proposed by Dhawan et al. [13] which helps to detect attacks in data plane forwarding and network topology. Based on capabilities on traffic flow, detecting distributed denial of service (DDoS), a lightweight method is proposed by Braga et al. [14]. OpenFlow protocol was used to advance the features of remote triggered black hole (RTBH). This can mitigate the distributed denial of service upon apply of SDN. This is demonstrated by Giotis et al. [15]. By using an efficient mechanism, anomaly detection and mitigation were performed in the architecture of SDN which is presented in the paper.

- SDN security modeling: A selection framework NICE was presented by Chung et al. [16]. This security model was developed as an intrusion detection system in the network and as a countermeasure framework. Attack graph was used for the evaluation of security which was a core work of the framework. Usage of attack graph is limited because of its scalable difficulties. In the same way, different graphical security models grieve from adaptability and scalability problems. The constraints in real time are the typical reason for the problems like scalability. To overcome all these issues, our approach suggests that attack scenarios have to be precomputed in advance and use them whenever it is necessary. Therefore, a full attack graph is used to create the possible precomputed attack paths. To address the adaptability and scalable problems, HARM is also used in our approach. When there is a detection of intrusion in a network system and effective countermeasure is formulated in real time, it is possible to evaluate security position of the SDN quickly with the precomputation of all the possible attacking paths.

3 A Framework for Real-Time Intrusion Response in SDN

A graphical security model is proposed to solve the problems of IDSes. The security model is precomputed. This model is used in the real-time interruption reaction in SDN. Some of the general steps are as follows:

- Information regarding vulnerabilities in security and dependencies in node connection of SDN has to be collected which are related with configuration of SDN.
- GSM for security valuations are generated by gathering all the collected inputs.
- Data on intrusion detection from SDN will be collected.
- Effective attack response is calculated with the selection of optimum countermeasure.

Figure 1 represents the relationship between the above steps mentioned.

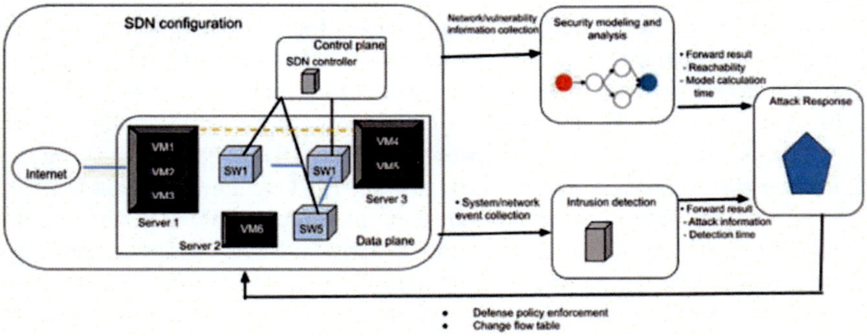

Fig. 1 Framework for real-time intrusion response in SDN

3.1 SDN Configuration

Necessary information regarding security which is associated with SDN has to be collected. Connectivity of each SDN component along with its respective dependencies and vulnerability of each component of SDN are the two main information. NESSUS and OpenVAS are the two vulnerability scanning tools used to collect the information of vulnerability associated with SDN components. With the help of settings in the SDN controller and flow table, dependencies of the components can be collected. Later the module, security modeling and analysis retrieve this information. IDSes are also present in SDN. If any of the intrusions are detected, those are directly sent to the intrusion detection module. Predicting the attack scenarios become more complex in nature as IDSes are not placed in every SDN components.

Figure 1 represents the framework for real-time intrusion response in SDN. In the SDN configuration module, SW acts as a connection between the data plane and the control plane. Six virtual machines are used which have connectivity to the Internet. All these virtual machines are placed in the server. SDN controller is placed inside the control plane. Security modeling and analysis module retrieve the information from the SDN configuration system to the network-related issues and vulnerabilities.

3.2 Security Modeling and Analysis

GSM can be generated with the inputs taken from the module of SDN configuration. For example, a model which is scalable and adaptable is HARM model which is used for demonstration purpose in this paper. Any of the GSM models can be used for the demonstration purpose but a HARM model has been chosen. Different security vulnerabilities are considered by GSM, and it computes diverse attack scenarios which have various dependencies. When the network size increases, the scalability problem also increases. Therefore, to achieve the response in the real-time attack, a technique of precomputation is required. Then, the module of attack response receives the information of precomputed assessment of security. This information will be used when there is a detection of intrusion.

3.3 Intrusion Detection

Intrusion logs are collected from the IDSes of SDN which will be then sent to the attack response module. Raw data regarding intrusion detection is processed in this module. Attack information such as attack type and metadata such as time of attack and location will be analyzed. This module will try its best to detect the attack fast and accurately. But in real time, its performance cannot be completely depend on.

3.4 Attack Response

This module is the heart of the architecture. Attack impact will be evaluated in this module by considering the time of intrusion detection and attack location. More than one computer will be affected if an attack takes place in a situation where a subnet is located. This module aims at reducing the attack impact by damage estimation, detecting the location of an attack and isolates the attack from getting progressed.

4 Real-Time Intrusion Response in SDN

4.1 SDN Configuration

By considering an example shown in Fig. 2, the usability of the proposed solution is demonstrated.

Nine nodes are included in the example of the toy where three switches and six virtual machines are considered. Assumed that the location of an attacker to be outside the SDN and Internet connection is provided only to the virtual machines that are located on the Web server. Since our proposed system takes care of both IDSes and security, the attackers who are inside the SDN can also be focused. Virtual machines aim at providing services to those who are located within as well as externally located. If a system does not encounter any problem, it works as follows. By considering the scenario from Fig. 2, when a user requests a data that is stored in the database (Virtual Machine 6), the virtual machine which is stored in the Web server (Virtual Machine 1) sends a request for the virtual machine in the application server (Virtual Machine

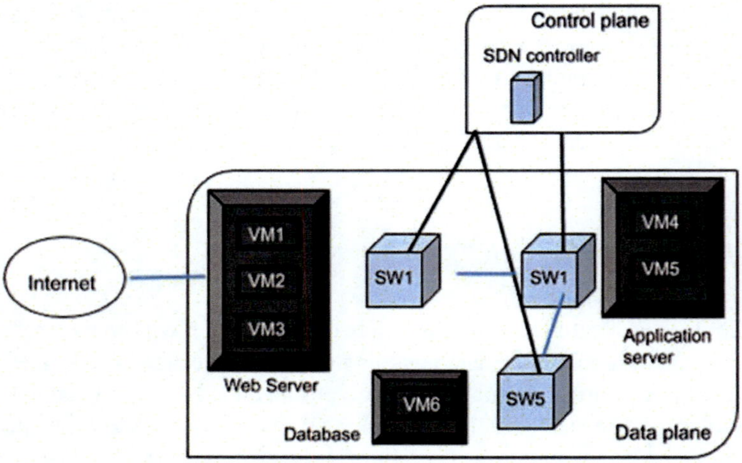

Fig. 2 SDN configuration example

4). From the application server, the requested data will be forwarded to the database for further processing. After processing the data, it is sent to the user. This procedure is followed only for the valid requests from the user. In this example, the data is processed from Virtual Machine 1 → Virtual Machine 4 → Virtual Machine 4.

Sometimes it leads to burst time when there are more requests from the user. Redundancies will be created to provide service in such case of emergency. In Fig. 2, redundancy is created between VM1 and SW2, another redundancy is between VM5 and SW1. If the attacker is successful in breaking the SDN, various attack paths will be generated using these redundancy connections.

4.2 *Security Modeling and Analysis*

4.2.1 Common Vulnerability Scoring System (CVSS)

The base score of CVSS is used for measuring the vulnerability severities. Physical metric, user interaction, and privileges required are added to the attack vector by considering the base vectors. The rank of integrity is changed from partial to low, availability rank is changed from complete to high whereas the rank of confidentiality remains same to be none. Probabilities of success of an attack and its impact have to be known for computing the security risk. The success of attack probability is represented using the exploitability metric which is associated with each vulnerability.

4.2.2 Attack Graph for SDN

Arbitrary code is executed on Virtual Machine 6. The definition of attack graph is as follows, an attack graph is a directed graph such that AG = (V, E), where V is the finite set of vulnerabilities present in the network system and E is the set of edges. Figure 3 represents the attack graph which is generated for mapping the various attack scenarios.

4.3 *Intrusion Detection*

Here, time is taken into consideration when an attack is detected. There will be chances of getting progress in the attack while the attack is getting detected. Therefore, considering the proper attack scenario plays a major role to mitigate the attack. The Bayesian theory has to be considered for the detection of an attack but here it is assumed that SDN mechanisms are correct for detection of an attack. Threshold random walk which is associated with the credit-based algorithm is similar to that of

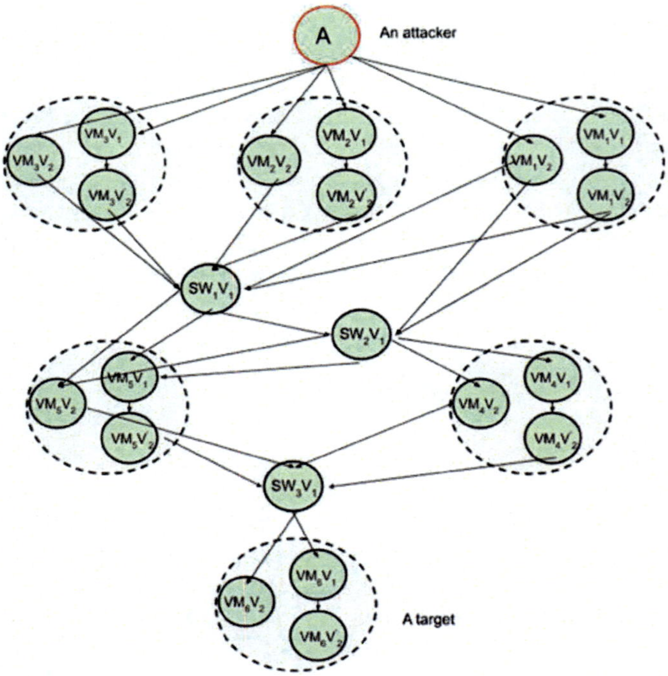

Fig. 3 An attack graph of SDN

applying Bayesian theory. If detection of an attack is delayed, then the attacker will be much progressed than the prediction. This scenario is represented in Fig. 4.

Figure 4 represents the location of the attacker at SW1 with the red dotted circle. Here, the attacker has successfully reached SW1 after passing through VM2. SDN administrator has been alerted only about the attack at VM2 and not SW1. To overcome this problem, the full attack graph is used which can be used to predict all the possible attack paths. This is represented in Fig. 4. As a countermeasure for the problem, flow table rule change is used. This limits the attack path from the number of hops from the node where the initial detection has taken place. For example, if the number of hops considered is 2, then the results are represented in Fig. 4. By following this, the attacker can be blocked by further accessing the paths. This proves that SDN functionalities can be maintained along with disconnecting the ongoing attacks.

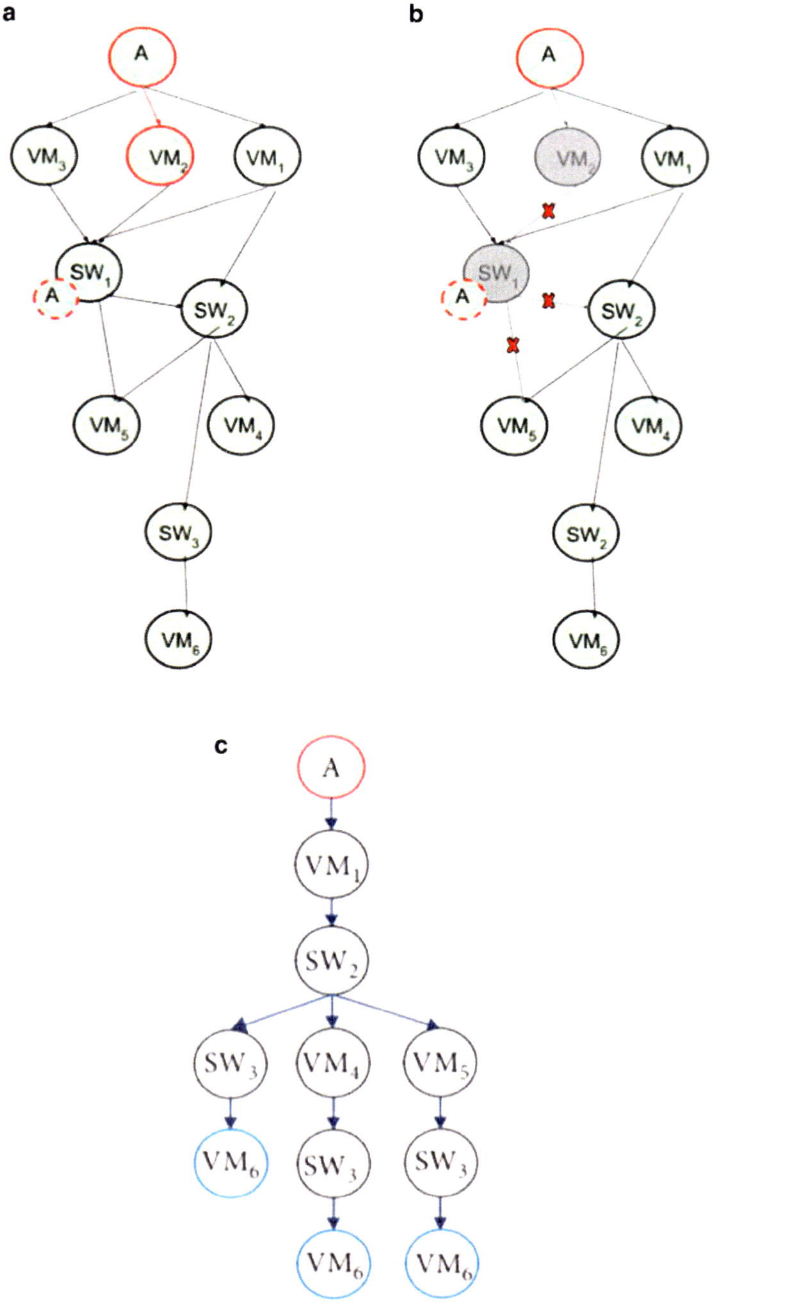

Fig. 4 **a** Representation of attack detection figures, **b** applying flow at Virtual Machine 2 table to block VM2, and **c** apply of full attack graph for countermeasure

5 Precomputation and Attack Prediction for Security Assessment

By considering the delays that occur in the detection of an attack, precomputation and prediction of attack are introduced. By precomputing the attack scenarios, the time taken to detect the attack can be reduced. Defense mechanism for SDN is enhanced to avoid the delay in finding the attack. Full graph and HARM are explained in this session.

- Full graph: Graphical security models cause a scalability problem while conducting the assessment of SDN security. Full attack graph is used to precompute the scenarios of all possible attack paths. This precomputed offline information can be later used in real-time cases whenever it is necessary.
- HARM: The above explained full attack graph is used in security assessment for particular attack scenarios of fast real-time attacks. In this case, the security problem arises since computing all the attack scenarios is not scalable. To provide more scalability, HARM is used through which SDN security is assessed. To reduce the scalability problem, network nodes of HARM are split into multiple layers.

6 Result and Analysis

By considering different security metrics, full attack graph is evaluated to check its effectiveness for precomputation. Any of the models can be used but the security metrics will not be changed. Since the computation results, full attack graph and HARM are same, and its result is not explicitly mentioned in this paper. Changes are observed in the security metrics by deploying the countermeasure and also without deploying it. Flow table rules are also changed to block the attack paths. Simulations are conducted to observe the difference between the performance of a full attack graph and HARM.

- Change in security metrics: Fig. 2 is used as the experimental scenario for the investigation in this session. Services are not available in the system until and unless the database receives the request. Therefore, the flow table rules of SW3 and Virtual Machine 6 are not changed by network administrator because of the system constraints. At least one path exists to connect the user requests to the network by ensuring the operability is made sure that. Minimal cost is considered for the improvement of security of SDN, by doing so, the performance of SDN will not be affected when the flows are modified.
- Numerical sensitivity analysis: Attacker can attack the node very fast when the response to the attack is very slow which results in the system loss. Comparison is made for the loss occurred due to slow response to an attack and to the cost required to respond to an attack in a fast way. The sensitive analysis method is applied since cost loss and cost spent to respond to an attack cannot be defined.

Response cost and attack cost are calculated based on attack time and the time taken for attack detection.
- Simulation: Evaluation time and generation time are simulated to examine the performance comparison of full attack graph precomputation to that of simple attack graph. Full attack graph precomputation is very important as it reduces the detection time in real ongoing attack and also reduces the evaluation time for security.

7 Conclusion

Network communications can be managed economically and more robustly by the SDN functionalities controls the network flow dynamically. New attack vectors and vulnerabilities were introduced which were not present earlier. Many security solutions were proposed for SDN to strengthen its power against the cyber-attacks. But still, the security problem of SDN and its functionalities are not solved which makes system administrator ensure a regular systematic check for the system security.

A framework for system modeling and Analysis was proposed in this paper which ensures the countermeasure for a real ongoing attack with the generation of precomputation for SDN. For the generation of possible attack scenarios to the current SDN, attack graph, full attack graph, and HARM use the precomputation method. These results are used as a combination with the precomputed attack scenarios of IDS and correlate the computation. When an attack detection mechanism fails, these models are later used to detect the potential attacks in real-time cases. To verify this, experiments are conducted and simulations were made on the SDN testbed which proved that our approach can be effectively used in the attack detection in a real-time ongoing attack and defend the attack.

Acknowledgements This research was supported by Visvesvaraya Technological University , Belagavi, Karnataka, India. We thank our principal and colleagues from Government Engineering College, Ramanagara who provided insight and expertise that greatly assisted the research and for assistance with particular technique for comments that greatly improved the manuscript.

References

1. Stabler, G., Rosen, A., Goasguen, S., Wang, K.-C.: Elastic ip and security groups implementation using openflow. In Proceedings of the 6th International Workshop on VirtualizationTechnologies in Distributed Computing Date, ser. VTDC'12, pp. 53–60, ACM, New York, NY, USA (2012)
2. Tariq, M., Koldehofe, B., Bhowmik, S., Rothermel, K.: PLEROMA: a SDN-based high performance publish/subscribe middleware. In Proceedings of the 15th InternationalMiddleware Conference (Middleware 2014), pp. 217–228, ACM, New York, NY, USA (2014)

3. Lee, S., Yoon, C., Lee, C., Shin, S., Yegneswaran, V., Porras, P.A.: Delta: a security assessment framework forsoftware-defined networks. In Proceedings of the 2017 Network and Distributed System Security Symposium, San Diego,CA, USA, March 2017
4. Lee, S., Kim, J., Shin, S., Porras, P., Yegneswaran, V.: Athena: a framework for scalable anomaly detection in software-defined networks. In Proceedings of the 2017 47th Annual IEEE/IFIP International Conference on Dependable Systems and Networks (DSN), pp. 249–260, Denver, CO, USA, June 2017
5. Hong, J., Kim, D.: Performance analysis of scalable attack representation models. In: Janczewski, L., Wolfe, H., Shenoi, S. (eds.) Security and Privacy Protection inInformation Processing Systems (SEC 2013), vol. 405, pp. 330–343. Springer, Berlin, Germany (2013)
6. Lippmann, R., Ingols, K.: An Annotated Review of Past Papers on Attack Graphs, Technical report ESC-TR-2005- 054. MIT Lincoln Laboratory, Lexington, MA, USA (2005)
7. Gude, N., Koponen, T., Pettit, J., et al.: Nox: towards an operating system for networks. ACM SIGCOMM Comput. Commun. Rev. **38**(3), 105–110 (2008)
8. Casado, M., Garfinkel, T., Akella, A., et al.: A protection architecture for enterprise networks. In: Proceedings of the 15th Conference on USENIX Security Symposium-Volume 15, ser.USENIX-SS'06, USENIX Association, Berkeley, CA, USA, July 2006
9. Matias, J., Garay, J., Mendiola, A., Toledo, N., Jacob, E.: Flownac: flow-based network access control. In Proceedingsof the 2014 'ird European Workshop on Software DefinedNetworks, ser. EWSDN'14, pp. 79–84, IEEE Computer Society, Washington, DC, USA, September 2014
10. Guang Yao, J.B., Xiao, P.: Source address validation solution with openflow/nox architecture. In: Proceedings ofthe 2011 19th IEEE International Conference on NetworkProtocols solution with openflow/nox architecture, pp. 7–12, Vancouver, Canada, October 2011
11. Shin, S., Porras, P.A., Yegneswaran, V., et al.: Modular composable security services for software-defined networks. In: Proceedings of the 20th Annual Network & Distributed System Security Symposium, e Internet Society, San Diego, CA, USA (2013)
12. Porras, P., Cheung, S., Fong, M., Skinner, K., Yegneswaran, V.: Securing the software-defined network control layer. In: Proceedings of the 2015 Network and DistributedSystem Security Symposium (NDSS), San Diego, CA, USA, February 2015
13. Dhawan, M., Poddar, R., Mahajan, K., Mann, V.: Sphinx: detecting security attacks in software-defined networks. In Proceedings of the 2015 Network and Distributed System SecuritySymposium, vol. 15, pp. 8–11, San Diego, CA, USA, February 2015
14. Braga, R., Braga, E.M.M., Passito, A.: Lightweight ddos flooding attack detection using nox/openflow. In: Proceedings of the 2010 IEEE 35th Conference on LocalComputer Networks, ser. LCN '10, pp. 408–415, IEEE Computer Society, Washington, DC, USA, October 2010
15. Giotis, K., Androulidakis, G., Maglaris, V.: Leveraging sdn for efficient anomaly detection and mitigation on legacy networks. In: Proceedings of the 2014 'ird EuropeanWorkshop on Software Defined Networks, ser. EWSDN '14, pp. 85–90, IEEE Computer Society, Washington, DC, USA, September 2014
16. Chung, C.J., Khatkar, P., Xing, T., Lee, J., Huang, D.: NICE: network intrusion detection and countermeasure selection in virtual network systems. IEEE Trans. Dependable Secure Comput. **10**(4), 198–211 (2013)

Implementation and Analysis of Dynamic Spectrum Sharing for Different Radio Access Technologies

Tejaswini G. Babajiyavar, R. Bhagya, and Amritash Kumar

Abstract With the evolution of telecommunication networks, there is a spectrum crunch leading operators to reform their 2G, 3G spectrum to support 4G and upcoming new technologies. However, refarming spectrum can be risky since this is a static or permanent reallocation of a spectrum between different radio access technologies (RATs). A technique called dynamic spectrum sharing (DSS) is proposed to resolve the refarming issue, where the overlapping spectrum is dynamically allocated between different RATs. This paper briefs the dynamic spectrum sharing implementation between 2G and 4G RATs in the laboratory environment. The DSS information is captured in the Abis interface using Wireshark. The BTS log tool captures the message exchange between peer RATs and gives information about PRB resource banking. Spectrum sharing between GSM and LTE is verified using spectrum analyzer.

Keywords LTE-GSM · DSS · PRB · SCF · HIT tool · Spectrum analyzer · BTSlog · Wireshark

1 Introduction

Telecommunication networks are in a state of constant change. The networks themselves change as the devices connected to them and the applications used change with the network and devices [1]. The change is enabled by the evolution of technology

T. G. Babajiyavar (✉) · R. Bhagya
Department of TCE, RV College of Engineering, Bengaluru, India
e-mail: tejalbabajiyavar@gmail.com

R. Bhagya
e-mail: bhagyar@rvce.edu.in

A. Kumar
Nokia Solutions and Network, Bengaluru, India
e-mail: amritash.kumar@nokia.com

P. Karuppusamy et al. (eds.), *Sustainable Communication Networks and Application*, Lecture Notes on Data Engineering and Communications Technologies 55,
https://doi.org/10.1007/978-981-15-8677-4_46

in devices, applications, and the network itself. Overall, can be seen that networks need to provide multifold increase capacity in mobile broadband capacity (MBB).

Refarming offers a unique, cost-efficient method to improve mobile broadband capacity and coverage without buying additional expensive spectrum. As GSM traffic is decreasing, the spectrum can be reallocated to 3G, 4G, and 5G RATs, ensuring improved mobile broadband performance. Based on long-term network traffic trends, efficient solutions can be deployed on legacy technologies such as GSM or UMTS to free up spectrum for new RATs such as LTE or 5G [2–6].

One of the key market demands today is to ensure that services are maintained during and after refarming to existing GSM customers, while at the same time building additional MBB capacity with the spectrum of legacy technologies such as GSM. However, without a continuous traffic analysis refarming spectrum can be risky, since this is a static or permanent reallocation of the spectrum between technologies. Such spectrum refarming decision can be flawed in some cases especially when there is a sudden surge of GSM traffic on some hotspot sites [1].

To solve these dilemmas the Spectrum Sharing is introduced as a solution, where separate RATs are provided with coordinate access to the same spectrum band [2, 7–9]. There are two types of sharing, namely dedicated spectrum sharing and dynamic spectrum sharing. In dedicated spectrum sharing, the different RATs have fixed spectrum whereas in dynamic spectrum sharing different RATs have both fixed and commonly shared spectrum [10]. To promote efficient use of radio spectrum, future wireless networks should be able to dynamically allocate resources to maintain the quality of service (QoS) and fulfill the rapid growth of spectrum demand.

2 Dynamic Spectrum Sharing

2.1 Overview

The dynamic spectrum sharing (DSS) allows the operator to allocate LTE and GSM carriers on an overlapping spectrum. Spectrum shall be dynamically shared between LTE and GSM by dynamically blanking or using the physical resource blocks (PRBs) that comprise the shared spectrum with GSM based on its traffic load. A GSM-LTE spectrum consists of the following parts: one dedicated only to GSM, used for the broadcast control channel (BCCH), one dedicated to LTE, and one shared, that is, dynamically assigned either to GSM or to LTE.

LTE allocates users either on LTE dedicated or shared carrier bandwidth. Shared GSM frequencies can overlap either with LTE bandwidth lower edge and/or upper edge. The LTE shared bandwidth can reach at most 40% of the maximum number of LTE PRBs, with a maximum of 20% shared bandwidth per each side. DSS can be enabled for LTE cells with 5, 10, 15, or 20 MHz nominal bandwidth. The LTE nominal bandwidth is divided into LTE transmission bandwidth and LTE guard band. The LTE

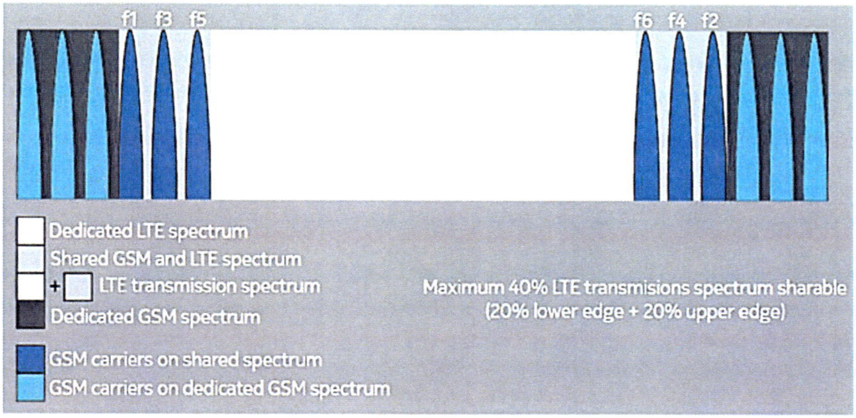

Fig. 1 Frequency spectrum usage by GSM and LTE

transmission bandwidth has a fixed PRB number, whereas each PRB is 180 kHz large and LTE guard band represents 2 × 5% of the LTE nominal bandwidth.

GSM allocates users on GSM dedicated carriers first. When there is higher traffic and there is no space left in the GSM dedicated area, GSM starts to allocate users in the shared part of the spectrum. GSM allocates users on shared TRX based on priority. The GSM calls are placed on carrier frequencies f1, f2, f3… which are farthest from the LTE center carrier frequency from the lower (f1, f3, f5) and upper (f2, f4, f6) edges of the LTE carrier bandwidth. Figure 1 shows the frequency spectrum usage by GSM and LTE.

2.2 *DSS GSM-LTE Solution*

The step-by-step dynamic shared spectrum (DSS) GSM-LTE solution is:

1. The background interference matrix (BIM) update process in the BSC monitors and collects statistical carrier-to-interference (C/I) data from all the cells, reported by the mobiles in both DFCA and non-DFCA sites. The DSS C/I threshold parameter is used for the BIM update process.
2. The BSC communicates both to its own BTSs and the nearby sites, the frequency-level shared spectrum usage to GSM baseband regarding the first call occupying and the last call releasing on any shared spectrum resources.
3. Shared spectrum usage frequencies which are received by GSM baseband from BSC needs to be further communicated to the LTE baseband and this could be done by means of an unreliable sys-com interface between basebands of GSM and LTE RATs. LTE scheduler uses this shared spectrum usage information and accordingly blanks PRBs of own site as depicted in Fig. 2.

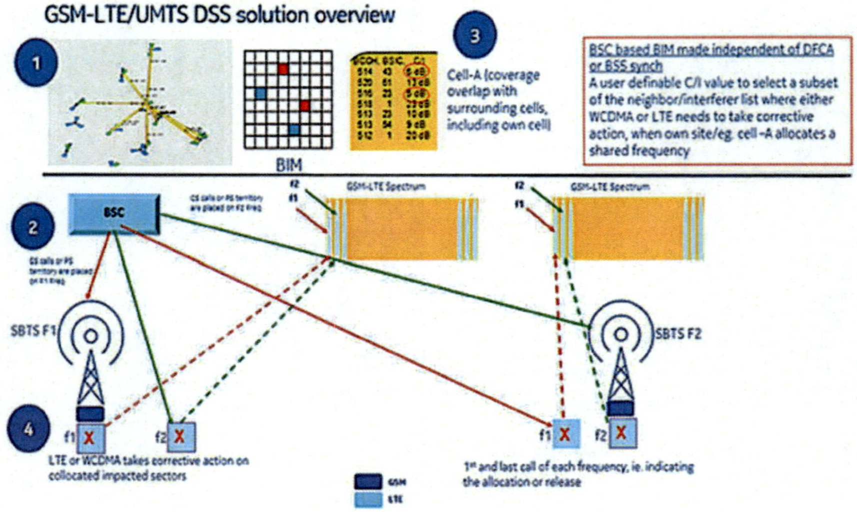

Fig. 2 DSS LTE + GSM solution

4. ARFCN-level shared spectrum usage is conveyed to LTE scheduler from GSM baseband based on updated received information from BSC, which can be used by LTE to accordingly schedule data on relevant PRBs/shared frequency blocks.

3 Implementation

This section presents a detailed explanation of the steps involved in the implementation of dynamic spectrum sharing feature in the laboratory environment. The minimum configuration (1 × BTS) system module called Airscale is connected to the radio module. Then, the LTE dedicated bandwidth is calculated when GSM sector frequencies overlap with LTE dedicated bandwidth. GSM and LTE cells are created by setting suitable managed objects in the Site Configuration File (SCF). The final step is to activate the DSS feature.

3.1 Frequency Calculation

The LTE paired cell and GSM paired sector must use at least one common antenna. The channel bandwidth for LTE cell is set to 10 MHz. The downlink and uplink carrier frequency is set to 950 MHz and 905 MHz respectively.

Equations 1 and 2 are the formulas to calculate LTE dedicated bandwidth in downlink when GSM sector frequencies overlap with LTE dedicated bandwidth.

$$\begin{aligned}\text{Downlink dedicated BW upper edge} &= \text{Central Frequency} + (\text{Nominal BW}/2)\\ &\quad - (\text{GB} - 0.01) - (\text{numBlankDLPrbsUp})\\ &\quad \times 0.18\end{aligned} \tag{1}$$

$$\begin{aligned}\text{Downlink dedicated BW lower edge} &= \text{Central Frequency} - (\text{Nominal BW}/2) + (\text{GB} - 0.01)\\ &\quad + (\text{numBlankDLPrbsDown})\\ &\quad \times 0.18\end{aligned} \tag{2}$$

where GB is guard band spectrum per edge, GB = 0.25 or 0.5 or 0.75 or 1 MHz per edge of 5, 10, 15, or 20 MHz carrier, respectively. Equation 3 gives formula for parameters numBlankDlPrbsDown and numBlankDlPrbsUp.

$$\begin{aligned}\text{numBlankDlPrbsDown} &= \text{numBlankDlPrbsUp}\\ &= (200\,\text{kHz} * \text{No. of GSM TRX})/180\,\text{kHz}\end{aligned} \tag{3}$$

The calculated LTE downlink (DL) lower and upper edge dedicated and shared frequencies are as shown in Table 1 and 2.

3.2 Site Creation

The GSM cell is attached to BSC through MML commands using HIT tool. The steps for attaching SBTS to BSC are as shown in Fig. 3a. The final site created for GSM cell (G-cell) at BSC is as shown in Fig. 3b.

3.3 GSM-LTE DSS Setup Configuration

The GSM-LTE DSS setup is bought up by creating an SCF. In SCF, GSM cell (i.e., LCELC) parameters and LTE cell (i.e., LCELL) parameters are set. Table 3 and 4 gives the parameters of LCELC and LCELL.

Channel configuration of GSM and LTE cell in SCF is defined. Each TRX in a GSM cell is configured with 1 transmitter and 2 receiver channels. LTE cell is configured with 2 transmitter and 2 receiver channels. The LCELL uses MIMO technique [11, 12].

3.4 Procedure for GSM-LTE DSS Feature Activation

The GSM cell at BSC level and TRX in the GSM cell are locked, and LTE cell from WebEM (GUI) is blocked. The DSS licence feature at BSC level enables and the

Table 1 LTE DL lower BW

GSM Lower Freq. (ARFN)	**GSM Lower Freq.** (MHz)
52	945.4
53	945.6
54	945.8
55	946
56	946.2
57	946.4
58	946.6
59	946.8
60	947
61	947.2
62	947.4
63	947.6
64	947.8
65	948
66	948.2
67	948.4
68	948.6
69	948.8
70	949
71	949.2

Shared (rows 52–60) LTE Dedicated (rows 61–71)

DFCA MODE is set to standby to calculate C/I ratio in the GSM cell. The DSS feature in the GSM cell is configured as shown in Fig. 4.

In WebEM, the automatic PUCCH object is created and automatic PUCCH allocation upper periodicity parameter is set to 20. Then, activation of automatic PUCCH allocation parameter is configured as true. DCI format 1A is configured for DCI format common channel transmission parameter in the LTE cell. The number of downlink PRBs blanked at the lower end of the spectrum (numBlankDlPrbs-Down) and number of downlink PRBs blanked at the upper end of the spectrum (numBlankDlPrbsUp) parameters are set to 10.

The DSS feature in the LTE cell is activated by setting activate dynamic shared spectrum (actDynamicSharedSpectrum) parameter to true. Then, the TRX in the GSM cell and GSM cell at BSC level are unlocked. In WebEM, the LTE cell is unblocked. Finally, the configuration is validated and activated. The detailed site view after activating the GSM-LTE DSS plan is as shown in Fig. 5.

Table 2 LTE DL upper BW

GSM Lower Freq. (ARFN)	**GSM Lower Freq.** (MHz)
78	950.6
79	950.8
80	951
81	951.2
82	951.4
83	951.6
84	951.8
85	952
86	952.2
87	952.4
88	952.6
89	952.8
90	953
91	953.2
92	953.4
93	953.6
94	953.8
95	954
96	954.2
97	954.4

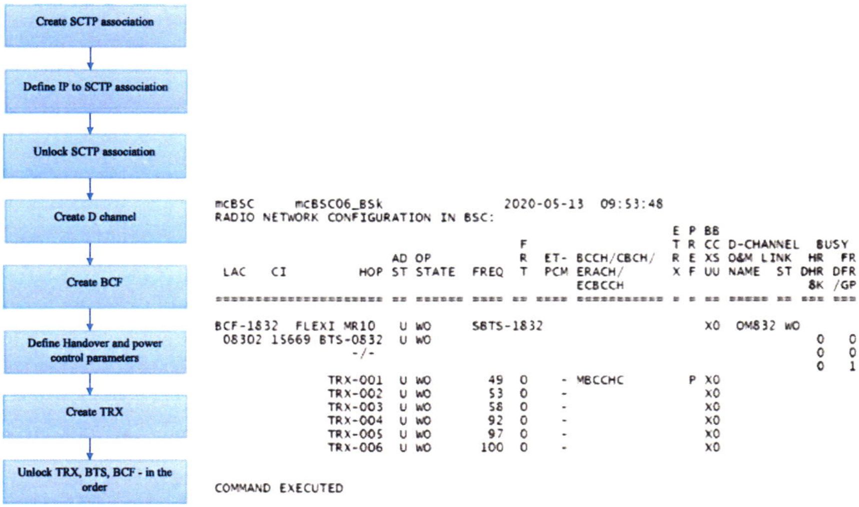

(a) (b)

Fig. 3 **a** Flowchart for attaching SBTS to BSC and **b** G-cell created using HIT tool

Table 3 LCELC parameters

Parameter	Value
Band number	8
Supported cell technology	FDD

Table 4 LCELL parameters

Parameter	Value
Band number	8
phyCellId	24
earfcnDL	3700
earfcnUL	21,700
dlChBw	10 MHz
ulChBw	10 MHz
Supported cell technology	FDD
Cell type	large
Expected cell size	15 km
maxBitrateDl	170,000
maxBitrateUl	60,000
dlMimoMode	Dynamic Open Loop MIMO

```
DSS PARAMETERS:
   DSS ENABLING ............................(DSSE).... 1 (DSS GSM+LTE)
   DSS C/I THRESHOLD........................(DSSCIT).. 35 dBm
   DSS MINIMUM FREQUENCY....................(DSSMIN).. 52
   DSS MAXIMUM FREQUENCY....................(DSSMAX).. 97
   DSS CENTER FREQUENCY.....................(DSSCF)... 75
   DSS NO OF LOWER SHARED FREQUENCIES........(DSSLSF).. 9
   DSS NO OF UPPER SHARED FREQUENCIES........(DSSUSF).. 9
```

Fig. 4 Configuration of the DSS feature in the GSM cell

4 Results and Discussion

This section presents the results of the dynamic spectrum sharing between GSM and LTE. Calls are placed on LTE and GSM mobiles and traces are captured in the Wireshark environment and BTSlog tool. Wireshark is a software for the analysis of open-source packets that allows examining packets moving through a network. BTSLog is a testing tool used for real-time monitoring and BTS system log recording. The GSM-LTE DSS is analyzed using a spectrum analyzer.

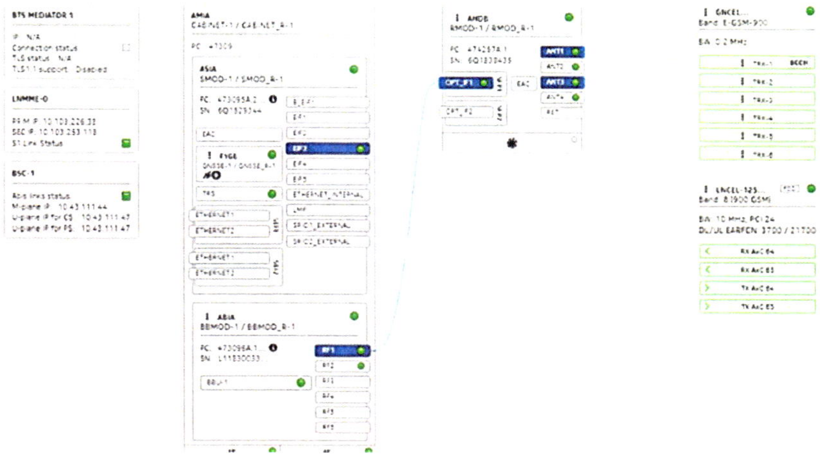

Fig. 5 Detailed site view after commissioning

4.1 Wireshark Log Analysis

BSC notifies the shared frequency usage on own site and nearby sites with new Abis L3 message (DSS GSM USAGE IND) containing the updated shared frequency usage bit map to GSM BTS, whenever CS calls and PS territory placed/released on shared TRXs.

DSS parameters information

BSC O&M sends "DSS Parameter" in BTS_CONF_DATA Abis O&M interface to SBTS when BCF restarts/unlocks, TRX and BTS unlock. The DSS sector information contains: BTS id, DSS value, DSS center frequency, and DSS shared frequencies and TRX information contains: TRX id and DSS TRX type and is shown in Fig. 6.

Notification of shared resource usage

"DSS_GSM_USAGE_IND" message is sent in the form of a bitmap of the shared ARFCNs. There can be a maximum of 40 ARFCNs which can be used by GSM in the shared spectrum. The ARFCN's 1–20 corresponds to lower shared frequencies [L1,L2,…,L20] and 21–40 corresponds to upper shared frequencies [U1,U2,…,U20] as shown in Fig. 7. The bitmap indicates usage of the ARFCN based on the value at bit position. "0"—indicates the ARFCN at the bit position is not being used by GSM. "1"—indicates the ARFCN at the bit position is being used by GSM.

No hopping mode

BSC allocates from shared TRXs in a non-hopping BTS only if it is not possible to allocate from dedicated TRXs in the BTS. The call is first time placed on shared TRX 2 as shown in Fig. 8a, in the order of priority associated with TRX. Hence, BSC

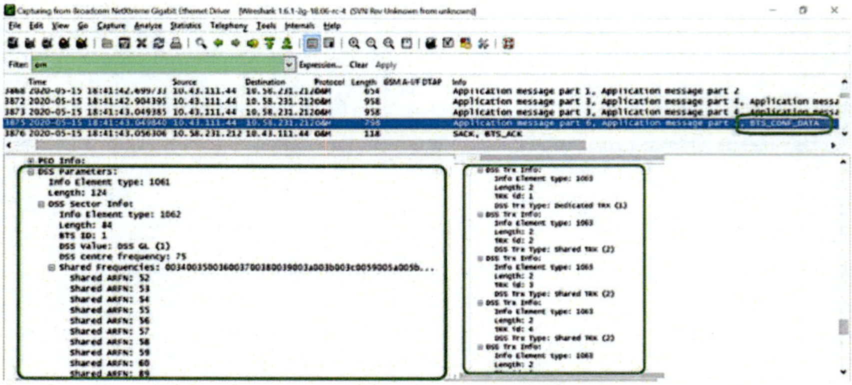

Fig. 6 DSS sector and TRX information

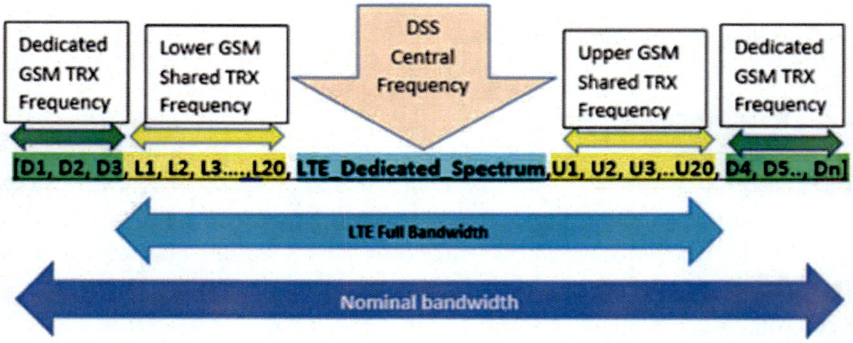

Fig. 7 Lower and upper shared frequencies

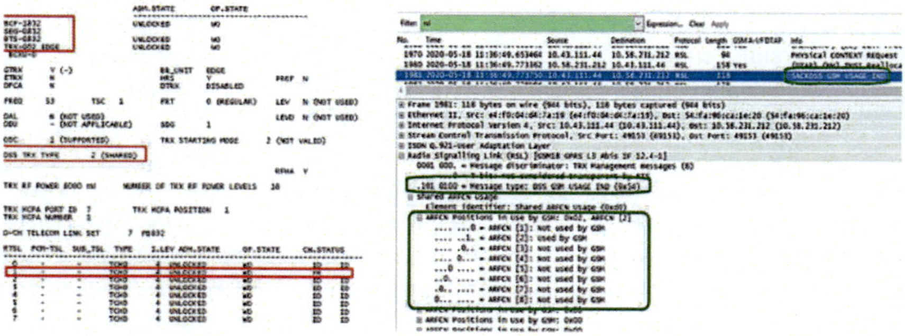

(a) (b)

Fig. 8 No hopping mode **a** Call on shared TRX-2, **b** corresponding Wireshark log sends DSS GSM USAGE INDICATION message with bitmap [0,1,0...,0,0,0,0,...0] as shown in Fig. 8b.

BB hopping mode

The shared frequencies are configured in the hopping group the BSC allocates resources on any TRX in a selected BB hopping BTS. Here, ARFCN 2, 7, 13, and 18 are in same hopping group; hence, BSC sends the Shared ARFCN Usage in DSS GSM USAGE IND with bitmap all set to "1" for the same hopping group frequencies when 1st call is placed on that sector as shown in Fig. 9. BSC sends the Shared ARFCN Usage in DSS GSM USAGE IND with bitmap all set to "0" for the same hopping group frequencies, when the last call released on that sector.

DFCA hopping mode

BSC selects shared DMALs in the associated priority order for the selected BTS for allocation when DFCA is configured in the BTS. BSC uses a lower priority shared DMAL for allocation in a BTS when it is not possible to allocate any higher priority ones in the BTS. The ARFCN is used when first call is placed in DFCA hopping enabled sector which is the same as BB hopping here.

RF hopping mode

BSC allocates resources on any hopping TRX in a selected RF hopping BTS when it is not possible to allocate resources on the BCCH TRX under the BTS and shared frequencies are configured in the attached MA list. Here, ARFCN 1, 3, 4, 5, 6, 14, 15, 16, and 17 are used in MAL list hence, and BSC sends the Shared ARFCN Usage

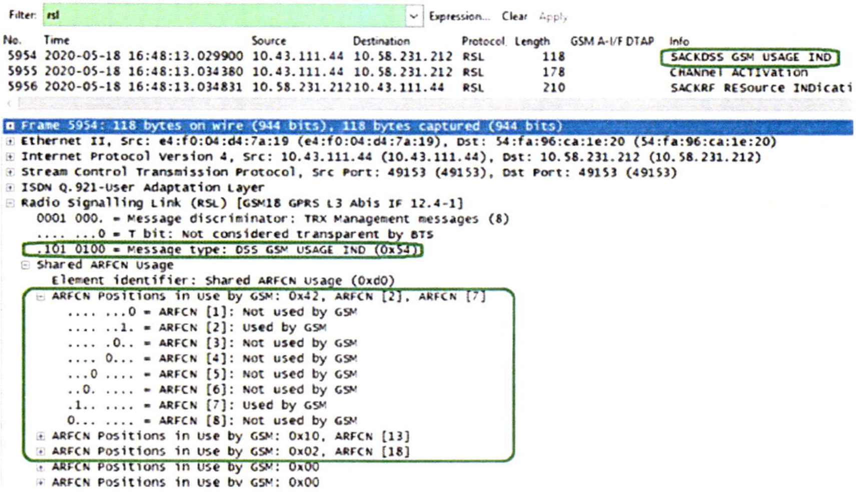

Fig. 9 Wireshark log for BB hopping mode

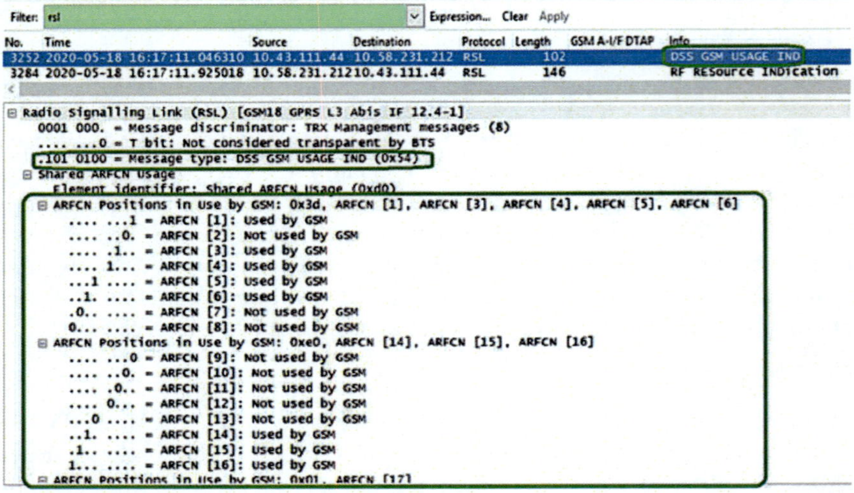

Fig. 10 Wireshark log for the first allocation with RF hopping mode

in DSS GSM USAGE IND with bitmap all set to "1" for the frequencies set in MAL list when first call is placed on that sector as shown in Fig. 10.

GPRS enabled

BSC uses the shared resources of a BTS for PS territory, in the order of priority for shared resource usage. BSC uses a lower priority shared resource for PS territory allocation only when all the possible higher priority shared resources are used for PS territory allocation. When hopping is enabled along with PS call BSC sends the Shared ARFCN Usage in DSS GSM USAGE IND with bitmap all set to "1" for the same hopping group frequencies for 1st PS call as shown in Fig. 11 and all set to "0" when last PS call released on that sector.

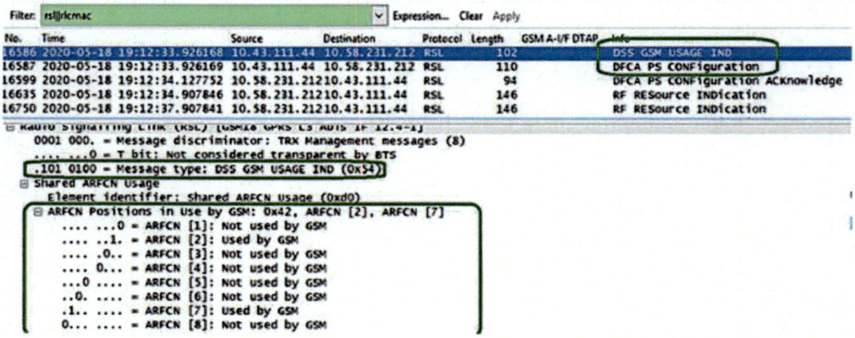

Fig. 11 Wireshark log for PS call with DFCA hopping

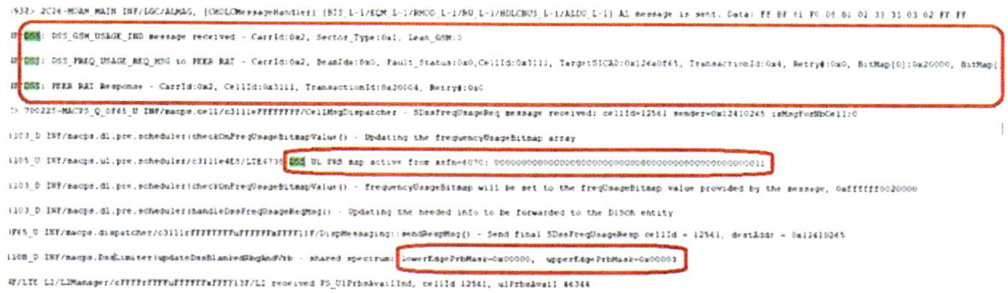

Fig. 12 DSS GSM usage request message to Peer RAT and PRB masking message

4.2 BTS Log Analysis

The Shared ARFCN Usage bitmap is sent from BTS to LTE baseband for PRB blanking process. BTS sends the DSS frequency usage request message to peer RAT and also gets a response from peer RAT. When GSM calls are placed on either lower edge or upper edge of the shared spectrum it sends PRB masking message. Figure 12 shows DSS GSM usage request message to peer RAT and PRB masking message. PRB masking removal is observed when the GSM call is released and is given back to LTE band.

4.3 Spectrum Analyzer Outputs

The total bandwidth assigned for LTE is 10 MHz. A theoretically upper and lower guard band is 0.5 MHz each, hence LTE must occupy 9 MHz of bandwidth. From spectrum analyzer output, it is observed that when LTE PS call of 100 MB is placed, it occupies approximately 8.94 MHz of bandwidth without any GSM call-in shared spectrum as in Fig. 13a.

When GSM call is placed in the lower edge of a shared spectrum, the LTE spectrum reduces to 8.4 MHz is observed. The lower edge PRBs are masked by LTE baseband when it is requested by GSM baseband and is given for GSM as shown in Fig. 13b. When GSM call is placed in both lower and upper edge, the LTE bandwidth adjusts itself by banking PRB for GSM usage as shown in Fig. 13c.

The GSM call in shared spectrum affects LTE throughput. When LTE PS call of 100 MB is placed, the average throughput is observed to be 12.5 Mbps. The LTE throughput drops to 11.31 Mbps when first GSM call is placed on the shared spectrum and further reduces to 10.77 Mbps when second GSM call is placed on the shared spectrum as shown in Fig. 14. The LTE throughput recovers when GSM calls are released on a shared spectrum.

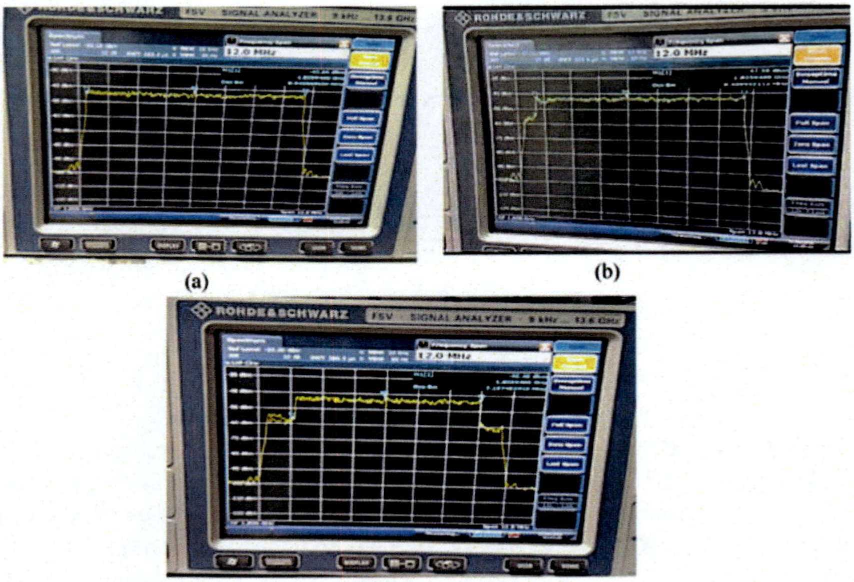

Fig. 13 LTE bandwidth **a** occupied when no GSM call is placed, **b** GSM call is placed on lower edge, **c** GSM call is placed both lower edge and the upper edge of shared spectrum

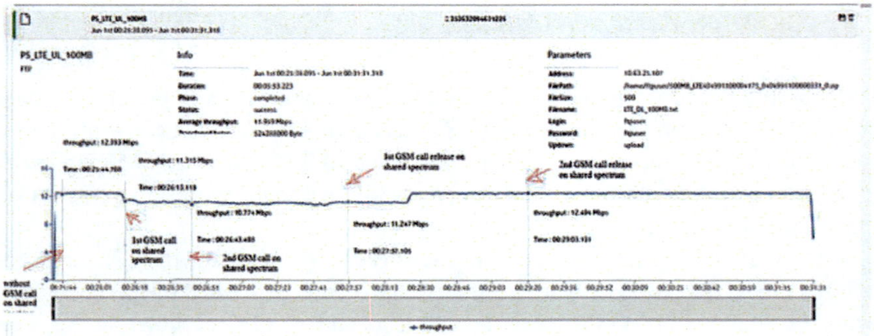

Fig. 14 LTE throughput when GSM calls are placed on shared spectrum

5 Conclusion

DSS allows the spectrum to be shared dynamically between 2G and 4G technology based on the needs of the end user to improve spectrum utilization. With DSS, operators do not need to split the available spectrum or have dedicated spectrum for either technology, but, instead, the spectrum is dynamically shared based on traffic demand. It is observed that spectrum efficiency becomes better by using DSS. LTE capacity is increased when it can run on high nominal bandwidth and dynamically use less only when GSM has a traffic peak. The DSS can be implemented for different radio access technologies (RATs) i.e., 3G, 4G, 5G, etc., in future.

References

1. Zhou, G., Deng, T., Yang, L.: A dynamic spectrum re-allocation scheme in GSM and LTE Co-existed networks. In: International Symposium on Wireless Personal Multimedia Communications (WPMC) (2015)
2. Li, Y., Zhang, Z., Li, F., Feng, Y., Chen, L.: An approach for intelligent spectrum allocating for multi-RATs networks. In: 2017 IEEE 9th International Conference on Communication Software and Networks (ICCSN) (2017)
3. Lin, X., Viswanathan, H.: Dynamic spectrum refarming of GSM spectrum for LTE small cells. In: IEEE Globecom Workshops (2013)
4. Guskov, P., Kozlovskiy, R., Maksymyuk, T., Klymash, M.: Methods and techniques of spectrum reforming for LTE network deployment. In: IEEE International Conference on Microwave & Communication Technology (CriMiCo'2013), pp. 474–475 (2013)
5. Linand, X., Viswanathan, H.: Dynamic spectrum refarming with overlay for legacy devices. IEEE Trans. Wireless Commun. **12**(10), 5282–5293 (2013)
6. Soroush, G., Malte, S., Markus, D., Egon, S.: An approach for automated spectrum refarming for multiple radio access technologies. In: 2011 Technical Symposium at ITU World (ITU WT) (2011)
7. Moawiah, A., Miguel, L.-B.: Dynamic contention window methods for improved coexistence between LTE and Wi-Fi in unlicensed bands. In: IEEE Wireless Communications and Networking Conference Workshop (WCNCW) (2019)
8. Jorswieck, E.A., Leonardo, B., Torsten, F., Eleftherios, K., Jian, L.: Spectrum sharing improves the network efficiency for cellular operators. IEEE Commun. Mag. **52**(3), 129–136 (2014)
9. Zhao, Q., Sadler, B.M.: A survey of dynamic spectrum access. IEEE Signal Process. Mag. **24**(3), 79–89 (2007)
10. Minho, J., Mykhailo, K., Taras, M., Ruslan, K.: Dynamic spectrum sharing algorithm for combined mobile networks. In: 20th International Conference on Microwaves, Radar and Wireless Communications (MIKON) (2014)
11. Ananda Murthi, L.S., Bhagya, R.: Design and performance analysis of MIMO-OFDM for WLAN standard. Int. J. Electronics Commun. Eng. (IJECE) **2**(3):77–84 (2013)
12. Sanjana, S., Lakshmeesha, D.D., Bhagya, R.: Development and enhancement of CFR estimate in WLAN and testing in Vector Signal Transceiver. In: 3rd IEEE International Conference on Recent Trends in Electronics, Information & Communication Technology (RTEICT), pp. 1851–1857 (2018)

TAMIZHİ: Historical Tamil Brahmi Handwritten Dataset

S. Dhivya and G. Usha Devi

Abstract This study focuses on creating offline benchmarks for handwritten characters in the local Tamil Brahmi language in India. This study has generated corpus of 1 lakh and 92000 character of brahmi script by getting offline handwritten scripts from 1600 writers. Hough transform and contour techniques are used to segment and extract scripts from the sheet. The offline data is also stored in csv format. The database is, however, intended to be useful for the identification of linguistic tasks. It will encourage the research work on Tamil Brahmi script handwriting recognition by providing free access to researchers. The designed dataset consists of 9 vowels, 18 consonants, and 209 classes so that the researchers can use machine learning algorithms to perform an extended research analysis. Handwritten data is collected and applied to create any linguistic frameworks for script analyzer.

Keywords Brahmi script · Recognition · Machine learning · Dataset

1 Introduction

The Tamil Brahmi ınscriptions are the most ancient of the extended written records of the Tamil language. The inscriptions furnish valuable information on many aspects of life in the ancient Tamil from a period somewhat anterior to the literary age of the Sangam. Now, a satisfactory decipherment of the cave inscriptions has also been achieved, and it is hoped that the scholars are interested in different aspects of ancient Tamil history and will make their further contributions based on a systematic analysis of the inscriptional material.

S. Dhivya (✉) · G. U. Devi
Department of Information Technology, School of Information Technology & Engineering, Vellore Institute of Technology, Vellore, India
e-mail: dhivyas@srmist.edu.in

G. U. Devi
e-mail: ushadevi.g@vit.ac.in

P. Karuppusamy et al. (eds.), *Sustainable Communication Networks and Application*, Lecture Notes on Data Engineering and Communications Technologies 55,
https://doi.org/10.1007/978-981-15-8677-4_47

Brahmi Script: There is no certain example of the occurrence of the Brahmi script before the third or perhaps the fourth century B.C. In the present state of study, it can be said that the Brahmi script was developed sometime around the middle of the First Millennium B.C. There has been a lot of controversy about the origin of the Brahmi script. However, the semitic origin proposed by Buhler and recently reiterated by Dani has never been seriously challenged. There are too many close resemblances between the North Semitic alphabet and the &Simi syllabary to bo dismissed as more coincidences. However, the aspirates, the medial vowel signs and the system of ligaturing letters are indigenous developments. To pursue the matter further would be beyond the scope of these papers which are concerned with the origin and the development of the Tamil-Bahl:al' script.

The existence of a dataset containing a substantial number of statistical samples is a critical part of any research into pattern/image recognition. The performance of character recognition research results is highly dependent on the reliability of the used database. The absence of standard databases for training and testing is a major obstacle to serious research work on handwritten character recognition of Indian scripts (Fig. 1).

This proposed research work has developed a database for Tamil Brahmi script, and that is named as Tamizhi. Only for Devanagari and Bangla scripts, Bhattacharya and Chaudhuri, 2005, extensive research on the creation of handwritten character database is found from the literature. Database, Sankaret al., 2011, consists of handwritten Bangladesh and Bangladesh script document file.

Developing a quality corpus for handwritten scripts is critical for research on developing a framework and applying to different algorithms, etc. The handwritten document is usually divided into various sections, namely the recognization of single characters, words, digits, text-line, special characters.

- **MNIST**-"Digits"
- **USPS**-"DIGITS"
- **CASIA-HWDB**-"Chinese"
- **Chars74K**-"ENGLISH AND KANNADA"
- **IRONOFF**-"French and english"
- **LAMIS-MSHD**-"French and Arabic"
- **ARDIS** - "SWEDISH"
- **PE92**-"Korean"
- **HIT-MW**-"Chinese"
- **PBOK** "Persian, Bangla, Oriya and Kannada Scripts"
- **ETL9**-"Japanese"
- **CEDAR**-"Arabic"
- **AHDB**--"Arabic"
- **Al-ISRA1** --"Arabic"
- **IFN/ENIT** --"Arabic"
- **CENPARMI**--"Arabic"
- **Khosravi and Kabir**-"Persian"
- **Mozaffari**-"Persian"
- **CMATERdb1**-"Bangla and English"
- **KHTD**-"Kannada"
- **OHR**-"Tamil and Kannada"

Fig. 1 List of few existing handwritten script dataset [1–8]

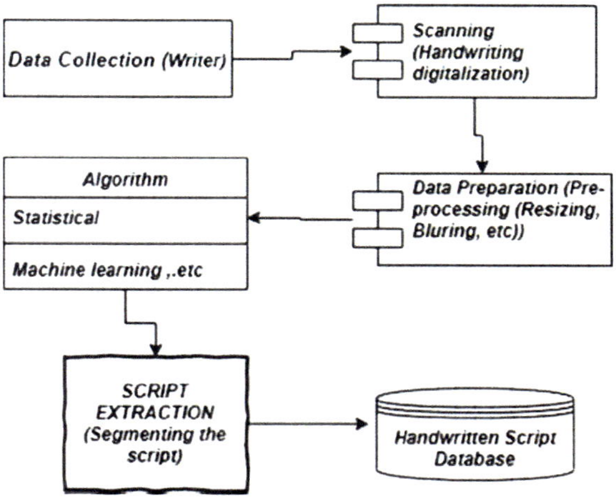

Fig. 2 Abstract methodology used for creating a database

2 Handwritten Script Recognition Methods

Major phase in the creation of a corpus is data collection that mostly constructs for specific purposes and analysis and therefore lacks standardization that contributes to the privacy of data. And the database is often used for no other purpose than the intended purpose. In order to solve this problem, the specification, the design of the corpus must be specified. Some of the elements that can be concentrated for the development of the corpus are the script list. The mode of collecting the data–device, regarding who will be writer their criteria, should be fixed like education, their native script which hand they use to write, etc. Figure 2 demonstrates the general sequence pattern to develop a corpus. Initial state is to collect the handwritten data. In next step, the raw data collected should be converted to digital data through scanning, data preprocessing is implemented to recreate the data such a way it is acceptable as input by the algorithm, final process to segment all the handwritten script and store it in a database [4–6, 9].

3 Data Collectıon Methodology

In this article, the methodology that has been followed in collecting handwritten data for Tamil Brahmi script is described. Figure 3.shows the specific flow of handwriting data collection. The primary step in the collection of data is to build the corpus of handwritten script [10]. The next step is to scan the raw data sheets and develop a new database as an input to the method. The subsequent step is to extract the scripts from the written document. The scripts are written in the grid pattern, and it is required to segment the sheets cell by cell for that Hough transform technique is used.

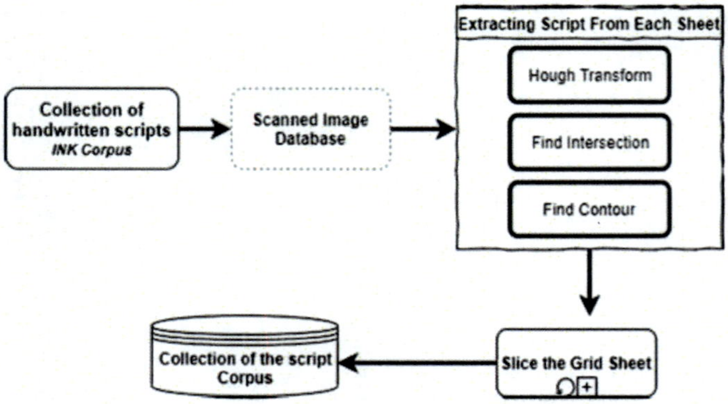

Fig. 3 Broadly used steps for handwritten data collection

A Hough transform is a technique used in handwritten data sheets to detect straight lines. Straightline Hough transformations requires a lot of raw materials because they have been built for straight lines of all forms. Hough transform is a method for extracting features to identify simple shapes such as circles, lines, etc. A line can be represented by two parameters, slope and intercept, while transform works effectively for finding lines in the image. The Hough transform take hold of binary edge as an input and tries to find edges positioned as straight lines. The edge point in the Hough space is converted into a line and the areas where most Hough space lines intersect are viewed as true lines. Contour is formed as a rectangle around esch script. The scripts are sliced and saved in the folder.

3.1 Layout of the Input Sheet

Instead of solving the complex issue for basic segmentation to extract the sample scripts, we customize an input page layout that allows segmentation. Traditionally, none of the Indic scripts are written in grid. Moreover, writing in grid pattern provides the recognition engine with useful segmentation indicator and thus simplifies writing segmentation and leads to improved accuracy. There are 9 vowels and 18 consonants in the Tamil Brahmi script. The input sheet of (11 $\times$ 19) is given to the writers to write the scripts. An informal writing by several Tamil writers showed that Tamil characters could be written in boxes with minimal training. Therefore, the primary section of the Tamil script data collection constitutes the collection of isolated Tamil Brahmi characters [11] (Fig. 4).

We decided to use all first- and third-year engineering students to produce samples for these experiments. According to the department, we randomly divided the users into two different groups, using one for training and the other for testing. We chose downsampling to pixels of 32 $\times$ 32. The scripts are classified into 209 class image is segmented to individual script using OpenCV.

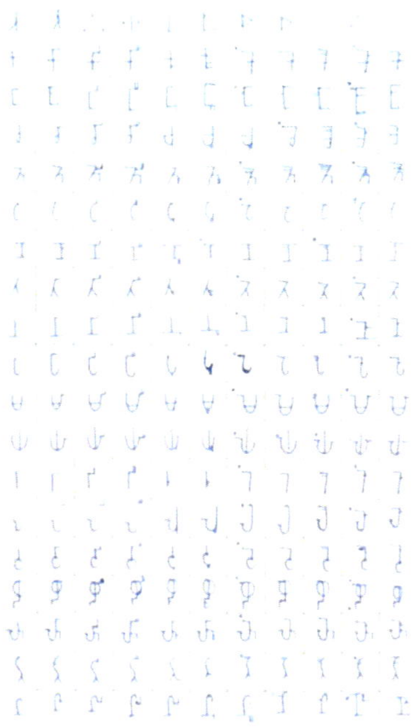

Fig. 4 Input page after scanning

3.2 *Grid Segmentation*

Isolated handwritten characters are extracted from the scanned data sheets by using the character segmentation algorithm. The segmented images of characters grouped into 209 classes to construct the image database of characters. This segment explains the process of extracting scripts from collected handwritten data sheets and the character image database structure generated from those segmented images.

3.3 *Hough Line Transform*

If the line can be represented in mathematical form, Hough transform is a best technique for detecting any line. Even if the line is broken or distorted, it can detect the line [12]. Representation of the line: $y = mx + c$ or $\rho = x\cos\theta + y\sin\theta$, perpendicular distance from origin to the line ρ. The angle of perpendicular line and the horizontal axis measured in the anti clockwise. This methodology is used in OpenCV [13].

The extraction process can be divided into the following steps to detect straight lines:

- Detecting the edge (x, y) [14]
- Accumulator array: Mapping of edge points
- The interpretation is carried out by thresholding
- Infinite lines are converted to finite lines [15].

Contours. Contours is defined as a graph that connects all the continuous points (along the boundary) with the same color or intensity. The contours are a useful tool to examine the shape and identify and recognize grid box spacing 30*50. For manipulation n- to create a new grid, b- to select a bulk rows and column data, to select the respective character, m- to move the grid. Finding contours in OpenCV is like finding a black background white object. Also, the object should be white and the background should be black [13].

Data set. The handwritten samples of documents are digitized and preserved as 8-bit gray images. The first step is to isolate every line of text in the sheet. For this function, a technique of Hough transform is used. The next step is to use the contours method as shown in Fig. 5 to slice each script within each grid. The segmented character images from a text image are sliced to fit into a minimum rectangle and stored in a database. The database is comprised of a total of 300,000 characters from 1600 writers.

4 Conclusion

This article has proposed the development of an offline Tamil Brahmi handwritten database. A well-structured and standard database is developed to perform a comprehensive study on Tamil Brahmi handwritten character. The key characteristics of this database are the samples that are collected to provide as much information as possible, and the structure is in such a way that it will be beneficial for the researches. Handwritten Tamil Brahmi database (Tamizhi), particularly after completion of data collection of about 3 lakh and 15 thousand handwritten scripts were achieved which was written by 1500 writers from students. This database addresses the research requirement to recognize ancient script or Brahmi script, identify and verify writers, analyze aspects, optimization, etc. for the development of any framework related to the text translation or transliterate.

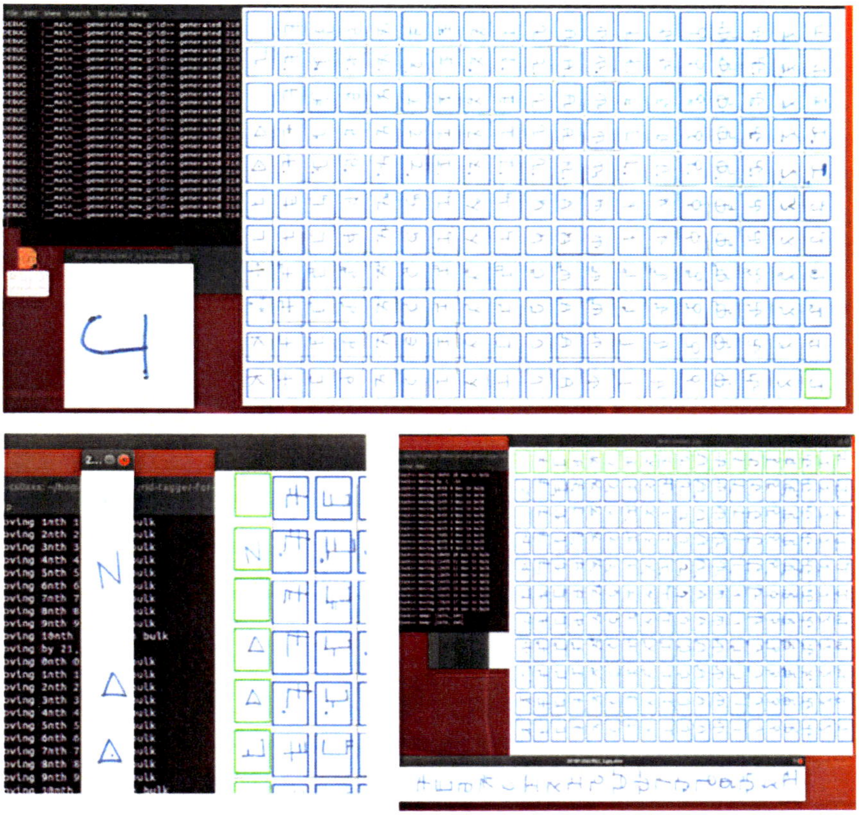

Fig. 5 Segmentation using Hough transform and contour using OpenCV

References

1. Das, N., Acharya, K., Sarkar, R., Basu, S., Kundu, M., Nasipuri, M.: A benchmark image database of isolated Bangla handwritten compound characters. Int. J. Document Anal. Recogn. (IJDAR) **17**(4), 413–431 (2014)
2. Hull, J.J.: A database for handwritten text recognition research. IEEE Trans. Pattern Anal. Mach. Intell. **16**(5), 550–554 (1994)
3. Jawahar, C.V., Anand Kumar, Phaneendra, A., Jinesh, K.J.: Building data sets for Indian language OCR research. In: Guide to OCR for Indic Scripts, pp. 3–25. Springer, London (2009)
4. Liu, C.-L., Yin, F., Wang, D.-H., Wang, Q.-F.: CASIA online and offline Chinese handwriting databases. In: 2011 International Conference on Document Analysis and Recognition, pp. 37–41, IEEE (2011)
5. Su, T., Zhang, T., Guan, D.: Corpus-based HIT-MW database for offline recognition of general-purpose Chinese handwritten text. IJDAR **10**(1), 27 (2007)
6. Dongre, V.J., Mankar, V.H.: Development of comprehensive devnagari numeral and character database for offline handwritten character recognition. Appl. Comput. Intell. Soft Comput. **2012**, 29 (2012)

7. Khan, H.A., Al Helal, A., Ahmed, K.I.: Handwritten bangla digit recognition using sparse representation classifier. In: 2014 International Conference on Informatics, Electronics & Vision (ICIEV), pp. 1–6. IEEE (2014)
8. Agrawal, M., Bhaskarabhatla, A.S., Madhvanath, S.: Data collection for handwriting corpus creation in Indic scripts. In: International Conference on Speech and Language Technology and Oriental COCOSDA (ICSLT-COCOSDA 2004), New Delhi, India, November 2004
9. Arvanitopoulos, N., Chevassus, G., Maggetti, D., Süsstrunk, S.: A handwritten french dataset for word spotting: CFRAMUZ. In: Proceedings of the 4th International Workshop on Historical Document Imaging and Processing, pp. 25–30. ACM (2017)
10. Chen, F., Chen, N., Mao, H., Hu, H.: Assessing four Neural Networks on Handwritten Digit Recognition Dataset (MNIST). arXiv preprint arXiv:1811.08278 (2018)
11. Mahmoud, S.A., Ahmad, I., Alshayeb, M., Al-Khatib, W.G., Parvez, M.T., Fink, G.A., Märgner, V., El Abed, H.: Khatt: Arabic offline handwritten text database. In: 2012 International Conference on Frontiers in Handwriting Recognition, pp. 449–454, IEEE (2012)
12. Al-Ohali, Y., Cheriet, M., Suen, C.: Databases for recognition of handwritten Arabic cheques. Pattern Recogn. **36**(1), 111–121 (2003)
13. Manjusha, K., Anand Kumar, M., Soman, K.P.: On developing handwritten character image database for Malayalam language script. Eng. Sci. Technol. Int. J. **22**(2), 637–645 (2019)
14. Daggumati, S., Revesz, P.Z.: Data mining ancient script image data using convolutional neural networks. In: Proceedings of the 22nd International Database Engineering & Applications Symposium, pp. 267–272. ACM (2018)
15. Bharath, A., Madhvanath, S.: Online handwriting recognition for Indic scripts. In: Guide to OCR for Indic Scripts, pp. 209–234. Springer, London (2009)
16. Shamim, S.M., Miah, M.B.A., Sarker, M.R.A., Al Jobair, A.: Handwritten digit recognition using machine learning algorithms. Glob. J. Comput. Sci. Technol. (2018)

Performance Analysis of Channel Estimation Techniques for High-Speed Railway Networks

A. J. Bhagyashree and R. Bhagya

Abstract Generalized spatial modulation orthogonal frequency division multiplexing (GSM-OFDM) for high-speed railways (HSR) wireless communication is proposed in this paper. Developing GSM-OFDM system for HSR has its challenges. First, since very few subcarriers are activated in GSM, there is difficulty in estimating the channel; secondly, the compressive sensing (CS) method and the matching method of signal detection tend to be ineffective as the unknown signals' nonzero elements count and the channel matrix's dimensions are large. To deal with these challenges, a channel estimation technique that uses block pilot pattern in the place of comb pilot pattern is proposed and a new interpolation method is employed considering the different symbols' time variation properties. Furthermore, for detection of the signal, decomposition and iteration process is used that reduces the channel matrix's dimension. Outcomes of simulation show that the suggested method of interpolation accomplishes better normalized mean square error (NMSE).

Keywords Signal detection · Channel estimation · High-speed railways (HSR) · Interpolation · Generalized spatial modulation (GSM)

1 Introduction

In modern times, wireless communication is considered as an important tool in the advancement of information technology. Due to the exponential growth of mobile devices and the increasing pace of implementation of high-speed railways (HSR) throughout the world, HSRs wireless communications systems have attracted much

A. J. Bhagyashree (✉) · R. Bhagya
Department of TCE, RV College of Engineering, Bengaluru, India
e-mail: bhagyashreeaj8@gmail.com

R. Bhagya
e-mail: bhagyar@rvce.edu.in

P. Karuppusamy et al. (eds.), *Sustainable Communication Networks and Application*, Lecture Notes on Data Engineering and Communications Technologies 55,
https://doi.org/10.1007/978-981-15-8677-4_48

research attention [1]. The main focus of research till now has been about the improvement of the HSR wireless communication systems' bandwidth efficiency. Multiple-input multiple-output orthogonal frequency division multiplexing (MIMO-OFDM) is the best-suited technology for high bandwidth efficiency [2]. Because of high IAI and ICI, wireless communication based on MIMO-OFDM for HSR cannot be achieved [3].

In spatial modulation OFDM (SM-OFDM) each transmitting antenna transmits data on only one subcarrier, unlike the MIMO-OFDM technique where all the transmitting antennas transmit data on all subcarrier. Accordingly in SM-OFDM, subcarrier which carries the information consists of two components: symbols which are modulated and the value of subcarrier's antenna index [4]. SM-OFDM has a major benefit of IAI-free communication as only one transmitting antenna is carrying one subcarrier. This effectively improves the receiver design by reducing the complexity of detection [5]. To obtain a further improvement in bandwidth efficiency, generalized SM-OFDM (GSM-OFDM) technique [6] is used in which each subcarrier requires many antennas to be activated, instead of a single antenna as in SM-OFDM.

A signal detection technique is proposed for GSM-OFDM system based on estimated channel coefficients. Further iteration and decomposition procedures are used to simplify the recovery and signal localization [7].

Notations: The notations $\approx$, Σ, diag(A), and $\log_a(b)$ represent approximately equal, summation, and diagonal of the matrix A and logarithm of 'b' to the base 'a', respectively. C_a^b represents the total number of combinations available to choose 'b' from a set of 'a' elements. $\lfloor a \rfloor$ stands for the highest integer smaller than a. $D(x, y)$ denotes xth row, yth column element of matrix D. $h(m)$ denotes the mth element of the vector h. $\|.\|_2$ represents a Euclidean norm. Bold upper letters denote matrix and vectors. $(\cdot)^H$, $(\cdot)^*$ and $(\cdot)^T$ denote Hermitian, conjugate, and transpose, respectively. $E_x\{\cdot\}$ denote the expected value with respect to the random variable x.

2 System Model

2.1 GSM-OFDM

Low complexity and high spectral efficiency make GSM prominent multiple antenna techniques. In GSM, the number of transmitting antennas required to send the information is reduced by sending the identical signal from a combination of two or more antennas. GSM system is having N_t transmitting antennas, out of which $N_K (N_K \ll N_t)$ antennas are operated in every time slot and residual antennas are kept inactive. Each subcarrier in N_K antenna carries $N_K \log_2 M$ number of bits.

Additionally, there are $C_{N_t}^{N_k}$ active antenna combinations which can carry $\left\lfloor \log_2 C_{N_t}^{N_k} \right\rfloor$ bits of information. The combined data rate in GSM-OFDM system having N subcarriers is given as $T = NN_b$ where $N_b = N_K \log_2 M + \left\lfloor \log_2 C_{N_t}^{N_k} \right\rfloor$ bps/Hz.

Considering D as incoming bitstream which is assumed as $D \in Z^{T\times 1}$ and remodeled to be a matrix $E \in Z^{N\times N_b}$ which is shown in Eq. (1). Every single row of remodeled matrix E represents the data transmitted by each subcarrier.

$$E = \begin{bmatrix} \underbrace{\begin{matrix} E_{1,1} & \cdots & E_{1,\log_2 M} \\ \vdots & \ddots & \vdots \\ E_{N,1} & \cdots & E_{N,\log_2 M} \end{matrix}}_{symbol1} & \cdots & \underbrace{\begin{matrix} E_{1,(N_K-1)\log_2 M+1} & \cdots & E_{1,N_K\log_2 M} \\ \vdots & \ddots & \vdots \\ E_{N,(N_K-1)\log_2 M+1} & \cdots & E_{N,N_K\log_2 M} \end{matrix}}_{symbolN_K} & \underbrace{\begin{matrix} E_{1,N_K\log_2 M+1} & \cdots & E_{1,N_b} \\ \vdots & \ddots & \vdots \\ E_{N,N_K\log_2 M+1} & \cdots & E_{N,N_b} \end{matrix}}_{index} \end{bmatrix} \begin{matrix} \text{sub}1 \\ \vdots \\ \text{sub}N \end{matrix} \tag{1}$$

The received signal in a GSM-OFDM system having N_t transmit and N_r receive antennas are represented as

$$y = Hx + e \tag{2}$$

where H is channel matrix, e is noise vector, and x is transmitted vector. The values of y, H, e and x are given below as

$$y = \left[y_1^T \ldots y_{Nr}^T\right]^T,$$

$$H = \begin{bmatrix} H_{11} & \cdots & H_{1Nt} \\ \vdots & \ddots & \vdots \\ H_{Nr1} & \cdots & H_{NrNt} \end{bmatrix},$$

$$x = \left[x_1^T \ldots x_{Nt}^T\right]^T,$$

$$e = \left[e_1^T \ldots e_{Nr}^T\right]^T$$

where $y_{n_r} \in C^{N\times 1}$ symbolizes n_rth antenna's received signal, e_{n_r} is the noise term and $x_{n_t} \in C^{N\times 1}$ represents n_tth antenna's transmitted signal, where $n_r \in [1, N_r]$, and $n_t \in [1, N_t]$. The channel matrix between n_tth transmit and the n_rth receive antenna in frequency domain is represented as $H_{n_r n_t} \in C^{N\times N}$, which is given by

$$H_{n_r n_t} = \begin{bmatrix} H_{n_r n_t}(1,1) & \cdots & H_{n_r n_t}(1,N) \\ \vdots & \ddots & \vdots \\ H_{n_r n_t}(N,1) & \cdots & H_{n_r n_t}(N,N) \end{bmatrix} \tag{3}$$

Based on Eq. (2), associated QAM symbols and indexes of nonzero elements of x can be detected jointly. After this, information bits are obtained by performing an inverse mapping process.

2.2 HSR Wireless Communication System

In a wireless communication system, the base station is a fixed device where a huge number of antennas are equipped than in the carriage, to increase the system capacity. Taking into account the high mobility situation and the setup amount, less antennas are employed in the train carriage in a high-speed train (HST) system, i.e., $N_r < N_t$, where N_t and N_r represent the number of antennas at the base station and carriage, respectively.

A standard wireless communication structure on high-speed railway comprises of three elements: the moving terminal points inside the carriage, the relay nodes on every compartment, and the base station [1]. The communication link contains two hops, one hop is between the moving terminals of the passengers and relay node and the other hop is from the relay node to the base station. Because of time-varying channel, the second hop experiences performance decline which is the main concern in this paper. In HST, the high speed induces inter-carrier interference, that leads to the channel matrix being full in the frequency domain, i.e., $H_{n_r n_t}$ in (3) is full, where $n_t \in [1, N_t]$, $n_r \in [1, N_r]$.

3 Channel Estimation Technique for GSM-OFDM System

3.1 Proposed System Model

The block diagram of the proposed GSM-OFDM system on high-speed railways is given in Fig. 1. The input data bits are grouped and mapped to QAM data symbols that are complex numbers representing the modulation constellation points. These complex source symbols are treated by the transmitter as though they are in the frequency domain and are the inputs to an N point IFFT block (hence the serial

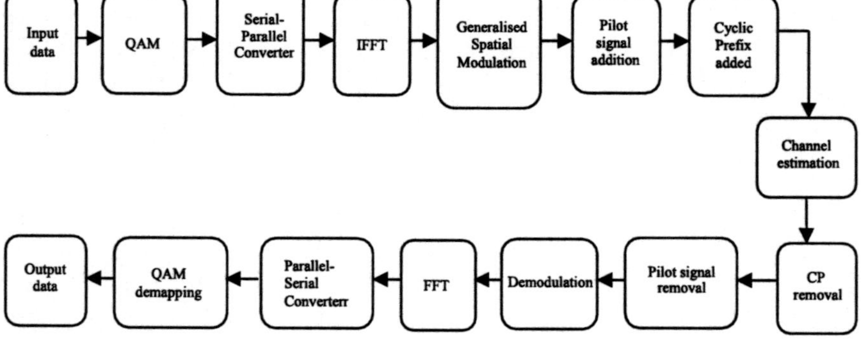

Fig. 1 Block diagram of GSM-OFDM in high-speed train system

to the parallel converter) that transforms the data into the time domain. The IFFT takes in N source symbols at a time and produces N orthogonal sinusoids, where N is the number of subcarriers in the system. The block of N output samples from the IFFT makes up a single OFDM symbol. These N output samples are further processed for GSM with the proposed pilot scheme for transmission. At the receiver side, GSM demodulation is done with the proposed signal detection scheme and the input data is recovered by performing the necessary inverse operations to the respective transmitter blocks.

3.2 Block Pilot-Based Channel Estimation

Block pilot-based channel estimation is accomplished by introducing pilot signals into all subcarriers in a specific period of time. In the HSR system, information is transmitted between two pilot symbols and one complete OFDM symbol is considered as a pilot symbol. Before transmitting information from the transmitter, block pilots are inserted inside the information to obtain the channel statistics. Based on the received pilot symbols, the receiver will get to know the characteristics of the channel.

The complete pilot arrangement is shown in Fig. 2 which contains N subcarriers. Subcarriers are orthogonal to each other and are divided into the number of groups denoted by G. Each block consists of four consecutive OFDM symbols out of which, the initial one is the pilot signal and others carry information. Then the number of subcarriers allocated in each group is represented by R. Subcarriers which are present in the mid position carry nonzero pilots and subcarriers present on both side (represented as Q) carry zero pilots. Subcarrier distribution is represented as $R = 2Q + 1$. The value of Q is obtained using the number of subcarriers, system bandwidth, and maximum Doppler shift.

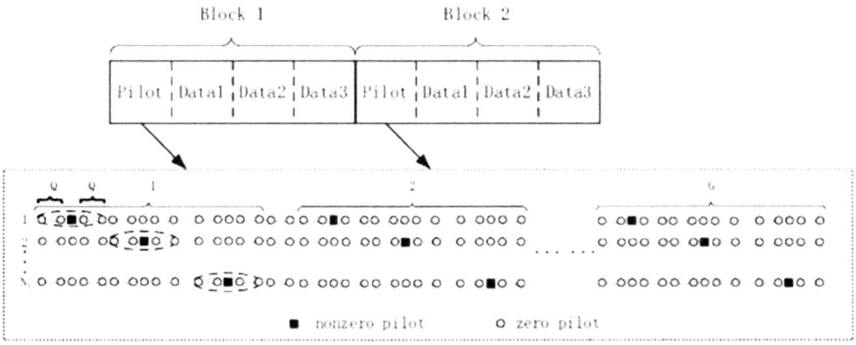

Fig. 2 The pilot pattern with $Q = 2$ [7]

Based on the pilot structure, complex exponential basis expansion model least square (CEBEMLS) [7] scheme is used to accomplish channel estimation.

3.3 Mixed Interpolation Method

For block pilot pattern in the OFDM system, interpolation is conducted to estimate the data symbol's channel coefficients. The interpolation techniques in common use are cubic spline interpolation, cubic interpolation, linear interpolation, etc. Nevertheless, the above techniques give poor performance in HST communication because the channel's properties vary quickly with time [9]. But mixed interpolation method is based on the block pilot, which gives better accuracy compared to the current techniques and forms the basis for signal detection in GSM-OFDM system on HSR.

For instance considering the linear interpolation, which gives an error and degraded accuracy of estimation under high SNR for not taking into account the channel variation effects accurately. A new interpolation technique referred to as enhanced interpolation [8] is presented to overcome this problem. This enhanced interpolation has three sequential steps: in the first step, each pilot symbol's estimated coefficients are linearly approximated. Secondly, in each block, the first and third data symbol's channel coefficients are predicted with the help of the neighboring pilot symbols' channel responses and in the last step, the second data symbol's channel coefficients are estimated by means of first and third data symbol's linear interpolation. But the precision of linear interpolation is better than that of the enhanced interpolation in regions having low SNR owing to the noise effect. Hence, at high values of SNR, the enhanced interpolation is used and at low values of SNR the linear interpolation is used, and this combination is named mixed interpolation technique [8]. The mixed interpolation method's details are given as follows.

In mixed interpolation method, SNR of the transmitted signal is considered as $\mathrm{SNR_t}$ and $\mathrm{SNR_{th}}$ represents the threshold SNR. At first, the value of $\mathrm{SNR_t}$ is compared with $\mathrm{SNR_{th}}$. If the condition $\mathrm{SNR_t} < \mathrm{SNR_{th}}$ satisfies, then channel estimation is performed using linear interpolation or else enhanced interpolation which is mentioned in algorithm 1 is implemented.

Enhanced interpolation algorithm uses to transmit, receive antenna pairs $h_{P_{l,b}}^{(n_r,n_t)}$ and estimated channel coefficients as input where $n_t \in [1, N_t]$, $n_r \in [1, N_r]$, $l \in [1, L]$ and $b \in [1, B]$. Here, the data symbols' count in each block is represented by D, the blocks' count is represented by B, and the multiple paths' count is denoted by L. The signal $h_{D_{l,d}}^{(n_r,n_t)}$ represents the output where $n_t \in [1, N_t]$,$n_r \in [1, N_r]$, $l \in [1, L]$ and $d \in [1, D]$. The estimation for the data symbols' channel coefficients is given by the enhanced interpolation.

Algorithm 1 is organized as two parts: the first part determines the linear approximation of the pilot symbols' channel coefficients and the method of estimating the channel is concentrated in the second part. Channel coefficient of $N/2^{th}$ and $3N/4^{th}$ point is calculated by taking the mean of the former and latter $N/2$ points, i.e.,

$$\overline{h}_{P_{l,b}}^{(n_r,n_t)}\left(\frac{N}{4}-1\right) \approx \frac{2}{N}\left(\sum_{n=0}^{N/2-1} h_{P_{l,b}}^{(n_r,n_t)}(n)\right) \tag{4}$$

$$\overline{h}_{P_{l,b}}^{(n_r,n_t)}\left(\frac{3N}{4}-1\right) \approx \frac{2}{N}\left(\sum_{n=N/2}^{N-1} h_{P_{l,b}}^{(n_r,n_t)}(n)\right) \tag{5}$$

The approximated $N/4$th and the $3N/4$th points define a straight line with slope given by

$$\beta_{l,b}^{n_r,n_t} = \frac{\overline{h}_{P_{l,b}}^{(n_r,n_t)}\left(\frac{N}{4}-1\right) - \overline{h}_{P_{l,b}}^{(n_r,n_t)}\left(\frac{3N}{4}-1\right)}{N/2} \tag{6}$$

Hence, an approximated straight line is given as

$$\overline{h}_{P_{l,b}}^{(n_r,n_t)}(n) = \left(n+1-\frac{N}{4}\right)\beta_{l,b}^{n_r,n_t} + \overline{h}_{P_{l,b}}^{(n_r,n_t)}\left(\frac{N}{4}-1\right) \tag{7}$$

Estimated channel coefficients are used for the channel estimation process. For instance, in block1 for obtaining data1, data2, and data3's channel estimations, block1 and block2 pilot symbols' estimated channel coefficients are used. A linear relation is expected between the sampling time and the channel coefficients in a data symbol. Therefore, the starting point of data1 in block1 is taken to be the endpoint of the approximate linear pilot symbol in block1, i.e.,

$$h_{D_{l,1}}^{(n_r,n_t)}(0) = \overline{h}_{P_{l,1}}^{(n_r,n_t)}(n-1) \tag{8}$$

Moreover, the slopes of data1 in block1 and the approximated linear pilot symbol in block1 are taken to be same. Hence, the relation of the channel coefficients of data1 in block1 with the respective sampling points is given as

$$h_{D_{l,1}}^{(n_r,n_t)}(n) = n\beta_{l,b}^{n_r,n_t} + \overline{h}_{P_{l,1}}^{(n_r,n_t)}(N-1) \tag{9}$$

The start point of the approximated linear pilot symbol in block2 is taken to be the endpoint of data3 in block1, that is

$$h_{D_{l,3}}^{(n_r,n_t)}(N-1) = \overline{h}_{P_{l,2}}^{(n_r,n_t)}(0) \tag{10}$$

The slopes of data3 in block1 and the approximate linear pilot symbol in block2 are taken to be same. The estimated channel coefficients of data3 in block1 can be obtained as

$$h_{D_{l,3}}^{(n_r,n_t)}(n) = -(N-n)\beta_{l,2}^{n_r,n_t} + \overline{h}_{P_{l,3}}^{(n_r,n_t)}(0) \tag{11}$$

Lastly, the linear interpolation of data1 and data3 in the block1 gives the second data symbol's estimation in block1, which is given as

$$h_{D_{l,2}}^{(n_r,n_t)}(n) = \left(h_{D_{l,1}}^{(n_r,n_t)}(n) + h_{D_{l,3}}^{(n_r,n_t)}(n)\right)/2 \tag{12}$$

Algorithm 1: The Enhanced Interpolation

Required: n_t, n_r, l, b, $h_{P_{l,b}}^{(n_r,n_t)}$

Ensure: $n_t \in [1, N_t], n_r \in [1, N_r], l \in [1, L], , b \in [1, D]$

Step I: Determining the pilot symbols' linear approximation

for $n_r = 1$ to N_r do
  for $n_t = 1$ to N_t do
    for $b = 1$ to B do
      for $l = 1$ to L do
        Getting two points, $N/4^{th}$ and $3N/4^{th}$ points of $h_{P_{l,b}}^{(n_r,n_t)}$ as (4) and (5).
        Using two points, calculating the line slope in, $\beta_{l,b}^{(n_r,n_t)}$ according to (6).
        Channel approximation $h_{P_{l,b}}^{(n_r,n_t)}$ is obtained as (7)
      ending for loop
    ending for loop
  ending for loop
ending for loop

Step II : Process of Channel estimation

for $n_r = 1$ to N_r do
  for $n_t = 1$ to N_t do
    for $l = 1$ to L do
    The channel estimation for 1st symbol inside the block is conducted as (9)
      Next the channel coefficients for the 3rd symbol inside the block is obtained by (11).
      Next symbol is obtained using (12).
    ending of for loop
  ending of for loop
ending of for loop

4 GSM-OFDM Signal Detection Scheme

As the wireless communication channel is unpredictable, it is necessary to implement signal detection scheme which works under imperfect CSI. Hence, for GSM-OFDM system on HSR, a signal detection scheme is proposed under imperfect CSI. This signal detection scheme endorses ML instead of the CS algorithm. ML algorithm is derived from the decomposed structure and it uses the iteration process. Using the ML algorithm [10], BER performance deterioration can be avoided easily.

Considering the huge value of channel error and decomposed structure, the ML detector is used to increase the accuracy of detection. Value in (2) is decomposed to (13), to get the value of y,

$$y = (D(H) + H - D(H))x + e, \tag{13}$$

in which

$$D(H) = \begin{pmatrix} diag(H_{11}) & \cdots & diag(H_{1N_t}) \\ \vdots & \ddots & \vdots \\ diag(H_{N_r1}) & \cdots & diag(H_{N_rN_t}) \end{pmatrix} \tag{14}$$

Block diagonal matrix $D(H)_{ad}$ can be derived by altering the rows and columns order of $D(H)$ which is shown as

$$D(H)_{ad} = \begin{bmatrix} \underbrace{\begin{matrix} H_{11}(1,1) & \cdots & H_{1N_t}(1,1) \\ \vdots & \ddots & \vdots \\ H_{N_r1}(1,1) & \cdots & H_{N_rN_t}(1,1) \end{matrix}}_{S_1} & & \\ & \ddots & \\ & & \underbrace{\begin{matrix} H_{11}(N,N) & \cdots & H_{1N_t}(N,N) \\ \vdots & \ddots & \vdots \\ H_{N_r1}(N,N) & \cdots & H_{N_rN_t}(N,N) \end{matrix}}_{S_N} \end{bmatrix} \tag{15}$$

All the nth diagonal elements of the matrix H_{n_r,n_t} are picked and grouped as matrix S_n having dimensions $N_r \times N_t$ in which $n_r \in [1, N_r]$ and $n_t \in [1, N_t]$. Then, the created matrix S_n is set as the nth block of $D(H)_{ad}$, where $n \in [1, N]$.

According to the subcarrier index structure, rearranging the model in (13) as

$$y_{ad} = D(H)_{ad}x_{ad} + \left[H_{ad} - D(H)_{ad}\right]x_{ad} + e \tag{16}$$

And value of H_{ad} is given as,

$$H_{ad} = \begin{bmatrix} H_{11}(1,1) & \cdots & H_{1N_t}(1,1) & & H_{11}(N,1) & \cdots & H_{1N_t}(N,1) \\ \vdots & \ddots & \vdots & \cdots & \vdots & \ddots & \vdots \\ H_{N_r1}(1,1) & \cdots & H_{N_rN_t}(1,1) & & H_{N_r1}(N,1) & \cdots & H_{N_rN_t}(N,1) \\ & \vdots & & \ddots & & \vdots & \\ H_{11}(N,1) & \cdots & H_{1N_t}(N,1) & & H_{11}(N,N) & \cdots & H_{1N_t}(N,N) \\ \vdots & \ddots & \vdots & \cdots & \vdots & \ddots & \vdots \\ H_{N_r1}(N,1) & \cdots & H_{N_rN_t}(N,1) & & H_{N_r1}(N,N) & \cdots & H_{N_rN_t}(N,N) \end{bmatrix} \tag{17}$$

The values of y_{ad} and X_{ad} is given in (18) and (19), respectively.

$$y_{ad} = \left[y_1(1), y_2(1), \ldots y_{N_r}(1), \ldots y_1(N), y_2(N), \ldots, y_{N_r}(N)\right]^T \quad (18)$$

$$X_{ad} = \left[x_1(1), x_2(1), \ldots y_{N_t}(1), \ldots x_1(N), x_2(N), \ldots, x_{N_r}(N)\right]^T \quad (19)$$

The value of $X_{ad}^{(i)}$ which is in the Eq. (16) is obtained by

$$X_{ad}^{(i+1)} = SEARCH_{alg}\left(y_{it}^{i}, D(H)_{ad}\right) \quad (20)$$

and

$$Y_{it}^{(i)} = y_{ad} - \left[H_{ad} - D(H)_{ad}\right]X_{ad}^{(i)} \quad (21)$$

where, $i \in [0, I-1]$ represents the index of the iteration. The iteration gets terminated when $\|y - Hx_2\|$ is identical to $\|y_{\text{ad}} - H_{\text{ad}}x_{\text{ad}}\|_2$. $x_{\text{ad}}^{(0)} = 0$.SEARCH$_{\text{alg}}$ mean the searching method. For SEARCH$_{alg}$ algorithm, the assumption made is $n \in [1, N]$,

$$X_{ad}^{(i)}[n] = \left[x_{ad}^{(i)}((n-1)N_t + 1), \cdots, x_{ad}^{(i)}((n-1)N_t + N_t)\right]^T \quad (22)$$

and

$$Y_{it}^{(i)}[n] = \left[y_{ad}^{(i)}((n-1)N_r + 1), \cdots, y_{it}^{(i)}((n-1)N_r + N_r)\right]^T \quad (23)$$

Thus,, $x_{ad}^{(i)}$ and $y_{it}^{(i)}$ can be viewed as the concatenate of $x_{ad}^{(i)}[n]$ and $y_{it}^{(i)}[n]$ respectively with $n \in [1, N]$, $n \in [1, N]$. Moreover, the value of $X_{ad}^{(i)}[n]$, $n \in [1, N]$ is obtained by solving the subsequent equation,

$$Y_{ad}^{(i)}[n] = S_n x_{ad}^{(i)}[n] \quad (24)$$

where, $X_{ad}^{(i)}[n] \in C^{N_t}$, denotes the nth subcarrier signal from the N_t transmit antennas. To ensure the correctness of the recovery, a search is conducted in the set of M-QAM symbols and in the position code book for the solution $X_{ad}^{(i)}[n]$ with a smaller $\left\|Y_{it}^{(i)}[n] - S_n\tilde{x}_{ad}^{(i)}[n]\right\|_2$. The value of $x_{ad}^{(i)}[n]$ is obtained by executing hard decision process. Hence, the value of $x_{ad}^{(i)}$ in (13) is computed according to (19) and (22). Algorithm 2 shows the complete process of signal detection in HSR system.

Algorithm 2: Signal Detection Procedure followed at receiver

Required: H, y
Ensure: x
Initialization part: where $x_{ad}^{(0)}$ and $i \leftarrow 0$ are initialized
while criterion is not halted, do
  for {the element in *[1, N]* is traversed by n}
  1st Step: Estimating $x_{ad}^{(i)}[n]$
      $\bar{x}_{ad}^{(i)}[n] \leftarrow 0$
  2nd Step: Searching for the symbol in the symbol set and code books
      for $(n_t^1, \ldots, n_t^{N_K})$ traverses the code books}
        for $(a^1, \ldots, a^{N_K})$traverses the M-QAM symbol set }
        $\tilde{x}_{ad}^{(i)}[n] \leftarrow 0$
        $\tilde{x}_{ad}^{(i)}[n](n_t^1, \ldots, n_t^{N_K}) \leftarrow (a^1, \ldots, a^{N_K})$
        if $\|y_{it}^i[n]\text{-}S_n\tilde{x}_{ad}^{(i)}[n]\|_2 < \|y_{it}^i[n]\text{-}S_n\bar{x}_{ad}^{(i)}[n]\|_2$
          $x_{ad}^{(i)}[n] \leftarrow \tilde{x}_{ad}^{(i)}[n]$;
        else
          $x_{ad}^{(i)}[n] \leftarrow \bar{x}_{ad}^{(i)}[n]$;
        ending of if statement
        ending of for loop
      ending of for loop
3rd Step: Taking hard decision of $x_{ad}^{(i)}[n]$and assigning decremented value to it
  ending of for loop
  Obtaining $x_{ad}^{(i)}$by (19) and (22)
  incrementing the value of i, i.e., $i \leftarrow i + 1$
end of while loop

5 Simulation Results

Parameters considered for simulations are given in Table 1.

Table 1 Parameters considered for simulation

Parameters	Values
Number of transmitting antennas N_t	12
Number of receiving antennas N_r	8
Number of activated antennas N_K	2
Number of subcarriers N	512
Central carrier frequency	2 GHz and 10 kHz
Number of groups G	10
Value of Q	2

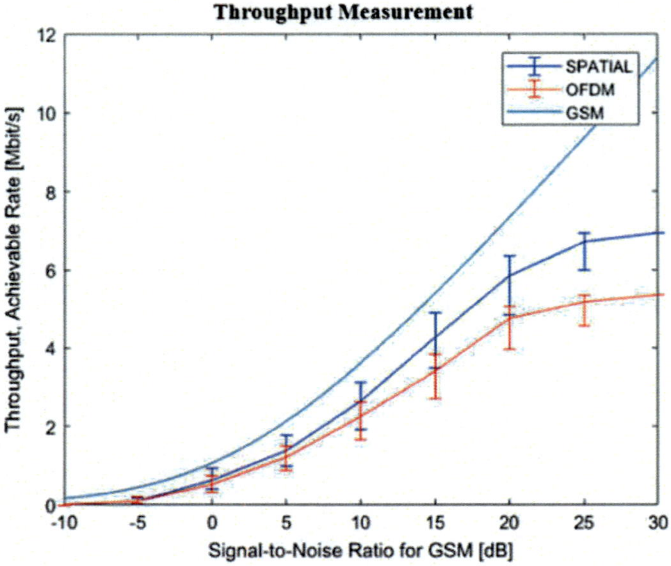

Fig. 3 Throughput of SM-OFDM, GSM-OFDM, and MIMO-OFDM

s.

Throughput measurement of three techniques SM-OFDM, GSM-OFDM, and MIMO-OFDM are compared in Fig. 3. As the SNR increases, the throughput of the system increases. At SNR of 15 dB, GSM-OFDM, SM-OFDM, and MIMO-OFDM achieve a throughput of 5.419Mbps, 4.67Mbps, and 3.79Mbps, respectively. Hence, when compared with the MIMO-OFDM and SM-OFDM techniques, GSM-OFDM provides better increment in the throughput value.

Figure 4 shows the pilot symbols channel estimation performance using CEBEMLS scheme, having a speed of the pilot tones as 500 km/h and 300 km/h. If an estimator is unbiased and achieves the Cramer–Rao Lower Bound (CRLB), then it is said to be efficient or least noisy. CRLB is bound to assess the efficiency of a given unbiased estimator. From the plot, it can be noticed that the pilot symbols' estimation characteristics are similar to that of the (CRLB). As speed increases, channel estimation accuracy decreases because of larger ICI caused by Doppler spread.

Comparison of the channel estimation for different interpolation techniques is shown in Fig. 5. If considered 25 dB as a threshold SNR, enhanced interpolation provides high performance gain above threshold SNR. Mixed interpolation which combines linear and enhanced interpolation provides better gain throughout the region. The NMSE performance is not satisfactory for conventional interpolation methods due to the channel's property of time variation. The proposed mixed interpolation method overpowers the NMSE level and provides high performance gain.

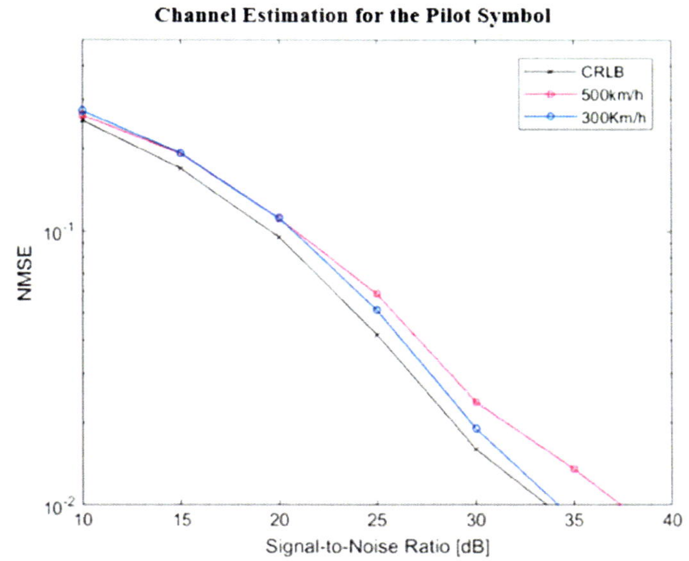

Fig. 4 Channel estimation of pilot symbols with different speeds

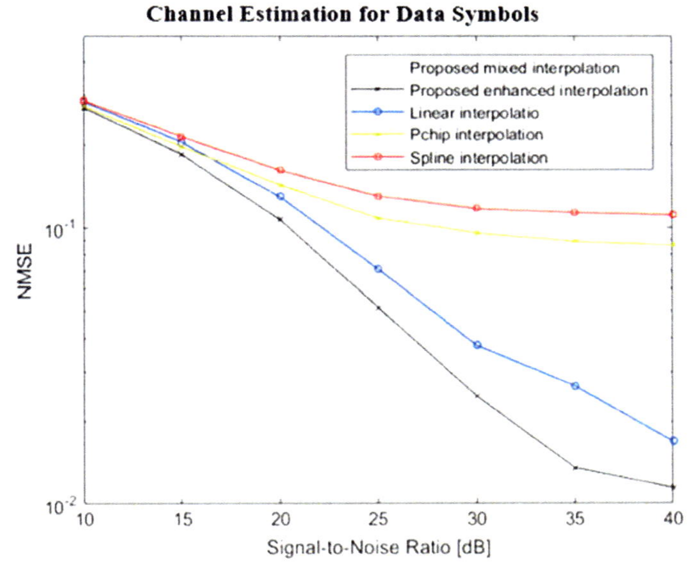

Fig. 5 Channel estimation of data symbols with different interpolation methods

6 Conclusion

The use of GSM-OFDM in HSR wireless communication is demonstrated to achieve a good balance between the spectral efficiency and receiver complexity. Block pilot-based channel estimation technique is a good fit for this case where each transmitting and receiving antenna pair is having few of the subcarriers that are actuated. Also, the channel estimation method built on block pilot pattern which considers the channel variation properties is offered to increase the accuracy in estimation. Further, a signal detection algorithm is developed with high accuracy. Outcomes of simulation show that the channel estimation scheme used performs better and provides good detection performance.

References

1. Ren, X., Chen, W., Tao, M.: Position based compressed channel estimation and pilot design for high-mobility OFDM systems. IEEE Trans. Vehicular Technol. **64**(5), 1918–1929 (2015)
2. Hassan, N., Fernando, X.: Massive MIMO wireless networks: an overview. MDPI J. Electronics **6**, 63 (2017) (Special Issue on Smart Antennas and MIMO Communications)
3. Bhagya, R., Ananth, A.G.: Performance studies of 2x2 MIMO system for different modulation and OFDM multiplexing techniques using ML detector. Int. J. Electronics Commun. Eng. (IJECE) **3** (2014)
4. Gong, B., Gui, L., Qin, Q., Ren, X.: Compressive sensing-based detector design for SM-OFDM massive MIMO high speed train systems. IEEE Trans. Broadcasting **63**(4), 714–726 (2017)
5. Ganesan, S., Mesleh, R., Ho, H., Ahn, C.W., Yun, S.: On the performance of spatial modulation OFDM. In: Fortieth Asilomar Conference on Signals, Systems and Computers, Pacific Grove, CA, pp. 1825–1829 (2006)
6. Di Renzo, M., Haas, H., Ghrayeb, A., Sugiura, S., Hanzo, L.: Spatial modulation for generalized MIMO: challenges, opportunities, and implementation. In: Proceedings of the IEEE, vol. 102, no. 1, pp. 56–103 (2014)
7. Ren, X., Tao, M., Chen, W.: Compressed channel estimation with position-based ICI elimination for high-mobility SIMO-OFDM systems. IEEE Trans. Vehicular Technol. **65**, 6204–6216 (2016)
8. Gong, B., Gui, L., Luo, S., Guan, Y.L., Liu, Z., Fan, P.: Block pilot based channel estimation and high-accuracy signal detection for GSM-OFDM systems on high-speed railways. IEEE Trans. Vehicular Technol. **67**, 11525–11536 (2018)
9. Tang, Z., Leus, G.: Identifying time-varying channels with aid of pilots for MIMO-OFDM. In: EURASIP J. Adv. Signal Process 1–19 (2011)
10. Xiao, L., Yang, P., Xiao, Y., Fan, S., Renzo, M.D., Xiang, W., Li, S.: Efficient compressive sensing detectors for generalized spatial modulation systems. IEEE Trans. Vehicular Technol. **66**(2), 1284–1298 (2017)
11. Li, T., Wang, X., Fan, P., Riihonen, T.: Position-aided large-scale MIMO channel estimation for high-speed railway communication systems. IEEE Trans. Vehicular Technol. **66**, 8964–8978 (2017)
12. Li, S., Xiong, J., Cheng, P., Gui, L., Xu, Y.: Low-complexity ICI cancellation based on BEM for OFDM systems over doubly selective channels. IEICE Trans. Commun. **E96.B**, 1588–1596 (2013)

Risk Assessment System for Prevention of Decubitus Ulcer

M. Nagarajapandian, M. Geetha, and P. Sharmista

Abstract The purpose of this paper is to develop a risk assessment system to prevent the occurrence of decubitus ulcers and to monitor the patient's conditions via the Braden scale assessment tool. This ulcer is a type of wound that happens to bedridden people admitted for different diseases for an extended period. To prevent and alleviate the occurrence of decubitus or pressure ulcers, a system has been proposed consisting of a sheet embedded with various sensors like pressure sensor, temperature sensor, tactile sensing array, and moisture sensor interfaced with PIC microcontroller programmed using MPLAB. This sheet continuously monitors the activities of the patient through the aid of the Braden scale. The result indicates the patient's pressure risk level calculated on basis of the Braden scale risk criteria and thereby updates it regularly to the hospital and the doctor's system through the display monitor and enables centralized control over the patient.

Keywords Braden scale · Decubitus ulcer · PIC microcontroller

1 Introduction

Health is one of the greatest resources for living a full life. It is necessary to maintain good health to prevent the human body from diseases and to bounce back from physical damages, illness, and other problems. It is essential to understand the necessity of future health care in the lives of human beings [1]. A decubitus ulcer (DU) is open damage or injury on the dermis and basal tendons, covering the bones of the humans, caused as a result of extended hospitalization or confinement to bed for

M. Nagarajapandian (✉) · M. Geetha · P. Sharmista
Sri Ramakrishna Engineering College, Coimbatore, Tamilnadu, India
e-mail: nagarajapandian.m@srec.ac.in

M. Geetha
e-mail: geethuprasanth6794@gmail.com

P. Sharmista
e-mail: sharmista2009@gmail.com

P. Karuppusamy et al. (eds.), *Sustainable Communication Networks and Application*, Lecture Notes on Data Engineering and Communications Technologies 55,
https://doi.org/10.1007/978-981-15-8677-4_49

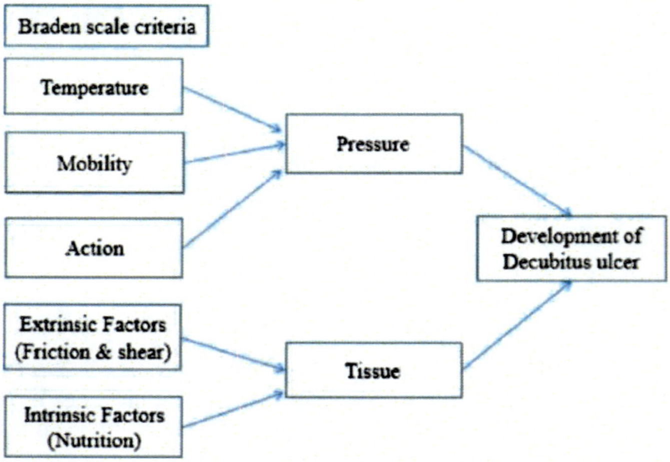

Fig. 1 Development of decubitus ulcer

other serious problems [2]. Decubitus ulcers are often recognized as ulcer pressure or bedsores or pressure sores. The decubitus ulcer often establishes on the feet, ankles, hips, backbone, and coccyx [3]. Pressure ulcers are caused mainly by extremely long skin [4]. The conceptual image on the formation of the ulcer including the criteria responsible for pressure on the tissues is shown in Fig. 1. This ulcer will happen to anyone but usually affects the people who are undergoing a long treatment process involving immobility of body parts and to people who are constantly cramped to beds or mattresses or constantly bounded to wheelchairs. It commonly occurs to elderly people, to people who have reduced or no mobility, to people who are treated for a longer period, to people who cannot move certain body parts, paralyzed patients, bedridden patients, and to people who have delicate skin. This ulcer develops first on the upper layers of skin and slowly extends toward the inward sides of the skin [5]. It limits the patient from altering the body positions. The healing process of these wounds consumes time [6]. This is treatable, but in severe cases like chronic deep ulcers are difficult to treat.

Decubitus ulcers cause severe burden to affected patients as well as to the caretakers and nurses who are tending to the problem. These ulcers increase the stay of affected patients and require high maintenance of patients along with increased hospital costs. This ulcer also tends to increase the complexity of healthcare costs and patients care [7]. Prevention of these ulcers through continuous monitoring of the patients and early detection of ulcers through regulated assessment of patient's mobility conditions will minimize the occurrence of these ulcers. The existing approach carried out to detect DUs completely relies on manual and timely check-up of patients by health care staff. It is interpreted from the available technologies that the caregiver shifts positions from left to right and right to left positions by making a change in the bed to treat the bedsores. The assessment is currently carried out with the aid of the Braden Scale through manual calculations and patients are observed continuously for skin breakdown or injuries.

2 Background

Decubitus ulcer (DU) is damage or injury on the dermis and basal tendons formed as a result of extended hospitalization or confinement to bed for other serious problems. Not all the pressure or decubitus ulcers are potentially curable and preventable some are difficult to tend because of the absence of circulation in the ulcer prone areas. The people who are at the risk of this ulcer are mostly people who have poor diet conditions followed by insufficient nutrition content and other risk factors include dehydration and diabetic patients because of the presence of less blood circulation in their bodies. The symptoms of these ulcers are a pain in a particular area, discoloration of the dermis, reddening of skin texture, and areas of open skin.

There are primarily four stages for the development of this ulcer. In the first stage, discoloration of the skin occurs. The second stage involves the breakage of the skin leading to open wounds and the formation of blisters. In the third stage, pus is formed in the deeper part of the skin. In the final stage, a brown or yellow scab is formed around the dermis.

The treating process of this ulcer involves the reduction of pressure in the affected area, frequent dressing of the wound, medications to relieve the pain, antibacterial drugs to treat the wounds, and in severe cases surgery is done to remove the dead tissues.

Risk factors have to be assessed once the doctor comes in contact with the immobile patient to avoid deterioration of health and to prevent the occurrence of this ulcer as it is potentially curable. As per the assessment of the risk factors of an individual, a risk profile is created and maintained to provide relief measures through body movements to the patient in time.

3 Existing Assessment Methodologies

Bedsore or pressure ulcers are localized skin and muscle injuries caused by insufficient blood flow in bedridden patients' weight-bearing areas [8]. High moisture and high temperature increase the rate of occurrence of bedsore. The extent of bedsore depends on the degree and duration of pressure exerted on the body of a patient.

The existing system has solved the issue by developing an anti-bedsore bed that is a medical device that eliminates or reduces the incidence of decubitus ulcers through softening the touch pressure imposed on a sufferer due to the impact of the mattress. The existing system [9] consists of a pair of air-filled tubes making up a bed whose pressure is changed regularly so the dimensions of the tubes are changed slightly. Arduino controller controls the air pressure given to the tubes. To evacuate the air within the pipe, a compressor is used by attaching into two separate solenoid valves. The time frame of pressure within the tubes is governed through the relay circuit that worked by the delay computed in the controller Arduino.

The paper [10] provides a wide-ranging analysis of the pressure ulcers by building an electronic system with suitable sensors. The device takes different snaps of an immobilized patient's area as well as performs an ideal assessment of potential and present bedsores alongside their symptoms using the patient's collected alcohol and thermal content data.

The plantar pressure monitoring system [8] solves the issue of ulcers by developing a feedback system for comparing normal pressure values to those of the threshold values of the patients affected by the plantar pressure monitoring system. In paper [11], the authors carry out a feasibility study on positions of patients for analyzation of ulcers and diseases associated with ulcers.

In the paper [12], an additional bed set to be mounted on a common hospital bed was built to avoid pressure ulcers with a device to adjust contact points between the patient's body and bed. The bed set is composed of four sections to support the head, hip, thighs, and heels of a patient. Increases in pressure on the patient's body due to the movement of the bed sets are assessed by pressure films, and tests are used in the research to test the bed's ability to avoid pressure ulcers by performance testing with a pressure ulcer risk curve. Test findings indicate that if the bed set works within a 15-min interval, the extra bed set will avoid pressure ulcers.

Existing wheelchairs are of major discomfort for the boney areas of the patients but the paper [13] addresses the problem by designing a wheelchair with air alteration to prevent the pressure ulcers. Pressure sensor bed made of force-resisting sensors is proposed as a solution to prevent and alleviate the pressure or bed sores [14]. This bed defines the pressure contour through the received values of the sensor.

There is currently no system in place to prevent decubitus ulcer from occurring. The existing system does not analyze the condition of the patient regularly and does not assess the risk level of the patients suffering from other ailments or disease and the existing ones tends to cure rather than prevention of pressure ulcer [15].

4 Proposed System

The proposed technique is the risk analysis approach for the prevention of decubitus ulcers based on the Braden scale. Figure 2 shows the proposed framework. This system consists of an anti-decubitus sheet or mat which is embedded with sensors like pressure sensors, temperature sensors, tactile sensing array, and moisture sensor. These sensors are interfaced with PIC microcontroller (PIC16F877A) and this entire system is given regulated power supply for the functioning of the system. The sensing sheet or mat continuously monitors the activities of the patient through the aid of the Braden scale for preventing and assessing the threat of decubitus ulcers.

According to the Braden risk level, the patient's position turning mannerism is automatically conducted out by activating the two pumps of the air or water supply

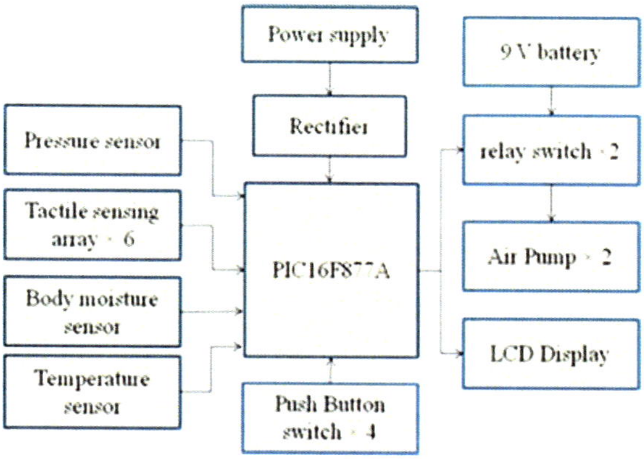

Fig. 2 Block diagram for decubitus ulcer prevention

for the inflation of the checkerboard mattresses and the sensors data required for pressure risk assessment is updated regularly to the doctors and the hospital's system through display monitor and thereby enables centralized control over the patient.

The LCD provided in the system shown in Fig. 2 acts as the display monitor to indicate the status of the patient to the caretaker and hospital staff or nurses who are in charge of the patient. According to the assessment of risk, the pumps are activated for aligning the checkerboard bed.

5 Technical Requirements

The technical requirements for the formation of the risk assessment system are shown in Table 1 and the specifications of the used hardware are also listed across the components.

Table 1 Technical requirements

S. no.	Components	Specification
1	Transformer	12 V/1A
2	Rectifier	–
3	Temperature sensor	10 kΩ
4	Moisture sensor	FC 37
5	Tactile switch sensor	6 × 6 × 2 mm
6	Thin-film pressure sensor	RP-L-170
7	PIC microcontroller	PIC16F877A
8	LCD display	16 × 2 display
9	Relay module	5 V DC
10	Pump	(3–5 V) mini-air pump

6 Software Requirement

MPLAB might be a software program for your embedded microcontroller design that runs on your development environment. Microchip incorporates a great suite of software integrated software and hardware development tools within one software package called integrated development environment (IDE) MPLAB. MPLAB IDE could be a free, integrated toolset for the event of embedded applications on microchip's PIC and dsPIC microcontrollers. It is called an integrated toolset for embedded application events on the PIC and dsPIC microcontroller of a microchip. MPLAB IDE runs on MS Windows as a 32-bit application, is direct to utilize, and contains a stock of free software components for fast application advancement and high-speed troubleshooting. MPLAB IDE also one, a unified graphical program for software and hardware development tools extra microchip and third parties.

Moving between devices could be a snap and upgrading to hardware troubleshoots and programming tools from the free software simulator is completed through a very flash because MPLAB IDE has an identical application for both tools.

MPLAB IDE is an optimized integrated development environment for the PIC MCU families and the dsPIC digital signal controllers based on Windows® operating system (OS).

The MPLAB IDE offers the capability of:

- Use the built-in editor to create and edit source
- Assemble and compile source code and connect it
- Debug the executable logic by looking program flow with the built-in simulator or with in-circuit emulators or debuggers in real time
- Carry out simulator or emulator timing measurements
- See watch windows for variables
- Firmware load into computer programming tools devices.

7 Braden Scale Risk Assessment

The scale of Braden scale evaluation score,

- Extremely high risk: score total 9 or low
- Serious risk: score total (10–12)
- Average risk: score total (13–14)
- Slight risk: score total (15–18)
- No danger: score total(19–23).

8 Hardware Setup

Risk assessment and monitoring system for the prevention of decubitus ulcer is the proposed system based on the Braden scale. The hardware setup of this system shown in Fig. 3 consists of an anti-decubitus sheet or mat which is embedded with sensors like pressure sensors, temperature sensors, tactile sensors, and moisture sensors. These sensors are interfaced with the PLC microcontroller, and this entire system is given regulated power supply for the functioning of the system. Sensing sheet or mat continuously monitors the activities of the patient through the aid of an assessment tool, the Braden scale for preventing an assessment of the risk of the ulcer. There is an LCD to display the current and average value of the Braden scale and sensor values of pressure, temperature, moisture, and tactile are displayed.

The output of LCD is sensors value such as temperature sensor, moisture sensor, pressure sensor and tactile sensor. The nutrition value is given in the keypad, and it shows the values of the current and average value of the Braden scale.

The LCD output is shown in Fig. 4. The output of various sensors is displayed such as T denotes temperature, A for action, M is for the motion, Mo is for the moisture, and P denotes the pressure value.

The nutrition value is shown in Fig. 5. The nutrition value can be increased or decreased depending on patient health. Assessing the nutritional status of a patient looks at its normal daily nutrition patterns. In this group, consuming the only portion of meals or possessing an unbalanced diet could suggest a maximum risk.

The current and average value of the Braden scale is shown in Fig. 6. Each category shall be defined on a scale of (1–4), precluding the category of friction and shear on a scale (1–3). This combines for a potential total of 23 points, with a greater score

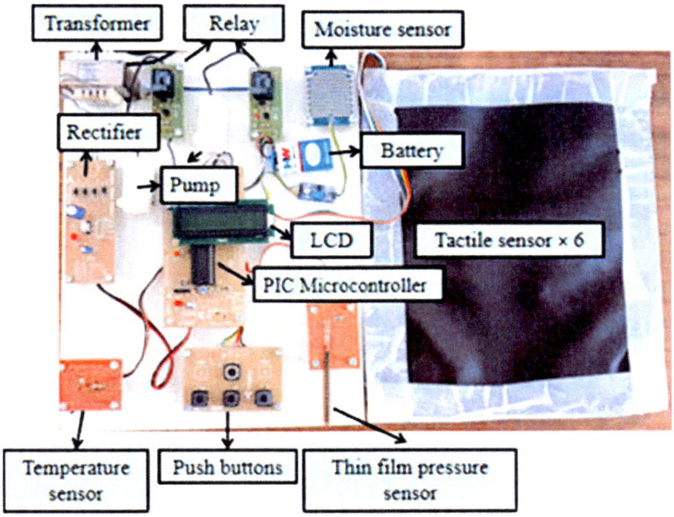

Fig. 3 Hardware setup

Fig. 4 Output of various sensors is displayed in LCD

Fig. 5 Nutrition value

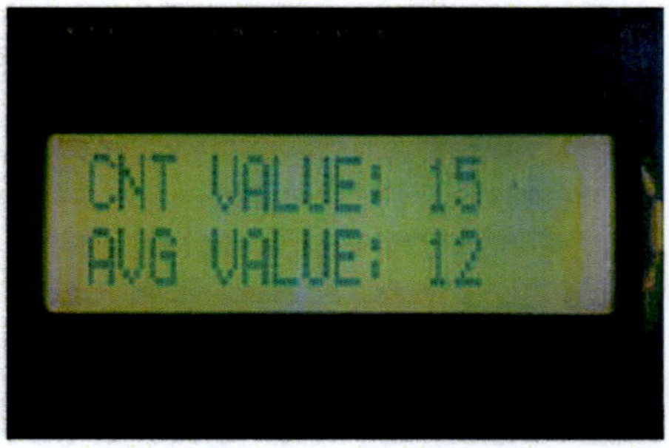

Fig. 6 Current and average value

signifying a reduced risk of developing a decubitus ulcer and contrariwise. A score of 23 defines nil risk for establishing a decubitus ulcer whereas the reduced minimum score of 6 points indicates a greater chance for establishing a decubitus ulcer.

9 Experimental Analysis

The experimental analysis on the prediction of decubitus ulcer was conducted out for different persons with the inference of the Braden scale. The analysis was carried out by examining the six mentioned criteria of the Braden scale.

All of these six criteria are rated out through a scale of (1–4). These values are acquired through the aid of various sensors of the risk assessment system.

Table 2 shows the classification of the assessed risk values. Here, Braden scale value 1 refers to completely limited functioning or activity, value to 2 corresponds

Table 2 Classification of assessed risk values

	Temperature		Action		Mobility		Moisture	
	Actual value	Braden scale value (1–4)	Actual value	Braden scale value(1–4)	Actual value	Braden scale value (1–4)	Actual value	Braden scale value(1–4)
Patient 1	35	4	4	3	14	4	175	3
Patient 2	34	3	4	3	8	3	325	2
Patient 3	36	4	2	2	7	3	142	3
Patient 4	34	3	1	1	5	2	98	4
Patient 5	35	4	2	2	12	4	345	2
Patient 6	34	4	2	2	7	3	437	1
Patient 7	32	3	5	3	6	3	82	4
Patient 8	36	4	4	4	15	4	92	4

to very limited functioning, value 3 defines slightly limited functioning, and value 4 refers to no impairment.

Table 3 shows the type of risk for the calculated total Braden value on the scale of (0–23).

Table 3 Type of risk

	Pressure		Nutrition	Average value	Current value	Type of risk
	Actual value	Braden scale value (1–4)	Constant value	Braden scale (0–23)	Braden scale (0–23)	
Patient 1	290	3	3	20	20	No risk
Patient 2	140	1	3	17	15	Slight risk
Patient 3	340	3	3	18	18	Slight risk
Patient 4	215	2	3	17	15	Slight risk
Patient 5	420	1	3	17	16	Slight risk
Patient 6	350	3	3	17	16	Slight risk
Patient 7	365	3	3	17	19	No risk
Patient 8	350	3	3	18	22	No risk

Figure 7 shows the analysis graph of decubitus ulcer conducted out for eight different patients. This graph examines the patient's temperature, mobility status, moisture content, actions, pressure, and nutrition intake. The graph readings drawn between the current and average value of decubitus ulcer prediction are shown in Fig. 8. This graph depicts the risk level of the patients for the ulcer with the Braden scale as an assessment tool. The current value lies between the range (0–23), as per the current value the risk level of the patient is categorized as no danger, slight risk, average risk, serious risk, or extremely high risk.

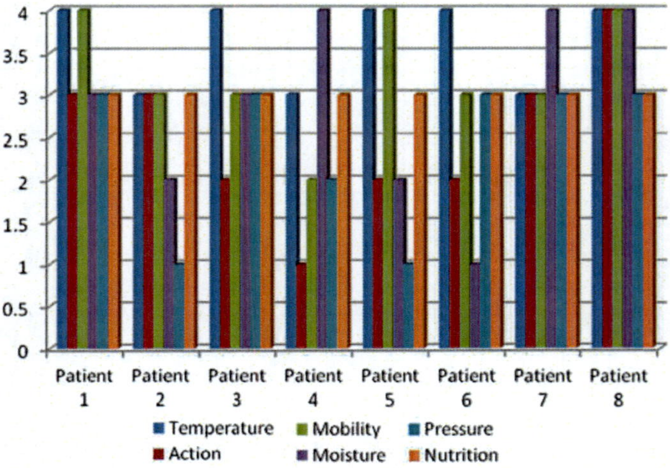

Fig. 7 Analysis graph of DU

Fig. 8 Graph between the current and average value of DU

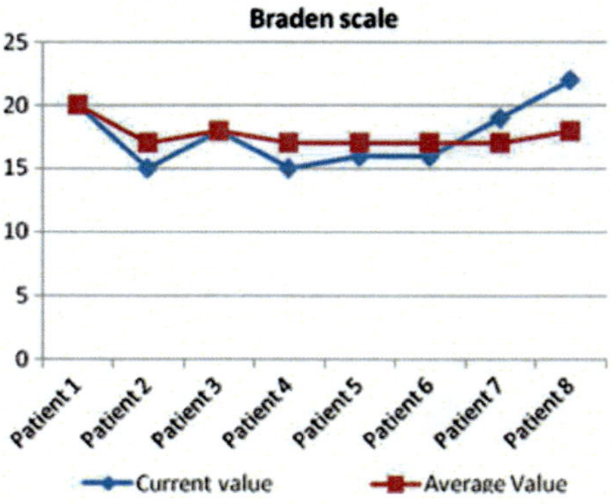

10 Conclusion

Risk of pressure ulcers is prevented during this project by automatically measuring the Braden scale assessment tool using appropriate sensors. The ultimate score of the Braden scale is 0 computed and assessed the chance. Then, the air mattresses are automatically changing its position and distribute its pressure and also the risk of pressure ulcers is rectified successfully. The outcomes or the deliverables of this paper are accurate pressure, temperature, motion, and tactile, and moisture detection is done, determining the Braden scale assessment of the sensors, displaying the average and current value of the Braden scale, activation of pumps to flow air in the air mattress by following the Braden scale assessment.

Due to caregiver obedience to turning guidelines, such a twisting operation is not routinely practised strictly in ICU and hospital wards. Difficulties in constantly monitoring the patient's status, shortage of a system which can cater turnover intimations or warning signals with inadequate staff numbers ratio of caregiver rise the occurrence of pressure injuries. The goal remains far away, but at least at an appropriate level, and this bed designed with a self-assessing feedback pressure dispersion and specially developed double-layered compact air cells meets the above appeals.

The future scope is to store the info of the Braden scale and sensor value within the cloud for future analysis and nutrition value to be calculated automatically.

Acknowledgements This research work was supported and carried out at Department of Electronics and Instrumentation Engineering and center for innovation Sri Ramakrishna Engineering College, Coimbatore, we would like to thank our management, director(academics), Principal, and Faculty friends for supporting us with the infrastructure to carry out the research work.

References

1. Smys, S., Chen, J.L.Z.: Special section on innovative engineering solutions for future health care informatics. J. Med. Imaging HealthInformatics **6**(7), 1570–1571 (2016)
2. Pressure Ulcers: Prevention, Evaluation, and Management. https://www.aafp.org/afp/2008/1115/p1186.html, 15 November 2008
3. What Should you Know About Decubitus Ulcer. https://www.healthline.com/health/pressure-ulcer, 2 July 2019
4. Bedsores or Pressure Sores: What you Need to Know. https://www.medicalnewstoday.com/articles/173972#prevention, 22 December 2017
5. Anders, J., Heinemann, A., Leffmann, C., Leutenegger, M., Pröfener, F., Renteln-Kruse, W.: Decubitus ulcers: pathophysiology and primary prevention. In: Dtsch Arztebl Int., pp. 371–382 (2011)
6. Shai, B., Maibach, H.: Wound Healing, and Ulcers of the Skin: Diagnosis and Therapy—The Practical Approach. Springer, Berlin (2005)
7. Palanisamy, R., Verville, J., Taskin, N.: Impact of trust and technology on interprofessional collaboration in healthcare settings. Empirical Anal. Int. J. e-Collaboration **13**(2), 10–44 (2017)
8. Gerlach, C., Krumm, D., Illing, M., Lange, J., Kanoun, O., Odenwald, S., Hubler, A.: Printed MWCNT-PDMS-composite pressure sensor system for plantar pressure monitoring in ulcer prevention. IEEE Sensors J. **15**(7), 3647–3656 (2015)

9. Hepsiba, D.,Vijay Anand, L.D.: Design of automated bed for prevention of pressure ulcer for patient with moving disability. Int J Innov Technol Exploring Eng **8**(11), 2252–2254 (2019)
10. Díaz, C., Zapirain, B., Castillo, C., Sierra-Sosa, D., Elmaghraby, A.: Simulation and development of a system for the analysis of pressure ulcers. In: IEEE International Symposium on Signal Processing and Information Technology, pp.453–458 (2017)
11. Barsocchi, P.: Position recognition to support bedsores prevention. IEEE J. Biomed. Health Informatics **17**, 53–59 (2013)
12. Yousefi, R., Ostadabbas, S., Faezipour, M., Farshbaf, M., Nourani, M., Tamil, L., Pompeo, M.: Bed posture classification for pressure ulcer prevention. In: Annual International Conference of the IEEE Engineering in Medicine and Biology Society, pp.7175–7178 (2011)
13. Yang, Y., Wang, J., Gao, Z., Zhou, Y.: Design and preliminary evaluation of an air-alternating wheelchair seating system for pressure ulcer prevention. In: International Conference on Bioinformatics and Biomedical Technology, pp. 239–243 (2010)
14. Chi-Chun, H., Yu-Wei, H., Yu-Hsien, C., Chia-Hao, K.: Bayesian classification for bed posture detection based on kurtosis and skewness estimation. In: International Conference on e-health Networking, Applications, and Services, pp.165–168 (2008)
15. Nagarajapandian, M., Ramprasanth, U., Selvakumar, G., Tamilselvan, S.: Automatic irrigation system on sensing soil moisture content. Int. J. Innov. Res. Electrical Electronics Instrumentation Control Eng. **3**(1), 96–98 (2015)

Faulty Node Detection Using Vertex Magic Total Labelling in Distributed System

Antony Puthussery and G. Muneeswari

Abstract Distributed system consists of huge number of nodes that are connected to a network, which is mainly intended and predominantly used for information sharing. Large users are prone to share data through the network and the stability and reliability of the nodes are remaining as the major concern in this system. Therefore, the inconsistent message transmission causes the nodes in the network to act differently, which would not be acceptable. A rapid method of malfunctioning nodes detection can improve the QoS of distributed computing environment. In this paper, a novel algorithm is proposed based on the calculation of vertex magic total labelling (VMTL) value for each and every node in the network. Upon receiving the message from the sender node, the receiver node will quickly detect the faulty node by comparing the VMTL pivot value (Pv). Experimental results show that the proposed approach leads to high true fault rate (TFR) detection accuracy compared to the false fault rate (FFR) detection. Finally, all the information related to the faulty nodes will be sent to the server node for further investigation and action.

Keywords Distributed system · Faulty node · Vertex magic total labeling · Adjacency matrix · Undirected graph · Client–server communication

1 Introduction

In the current modern world, distributed systems have been widely used in many of the client–server architecture, whose model is depicted in Fig. 1. It basically consists of huge number of nodes connected with a network that are mainly intended and predominantly used for information sharing. Large users are prone to share

A. Puthussery
Department of Computational Sciences, School of Science, CHRIST (Deemed to be University), Delhi, India
e-mail: frantony@christuniversity.in

G. Muneeswari (✉)
Department of Computer Science and Engineering, School of Engineering and Technology, CHRIST (Deemed to be University), Bangalore, India
e-mail: muneeswari.g@christuniversity.in

P. Karuppusamy et al. (eds.), *Sustainable Communication Networks and Application*, Lecture Notes on Data Engineering and Communications Technologies 55,
https://doi.org/10.1007/978-981-15-8677-4_50

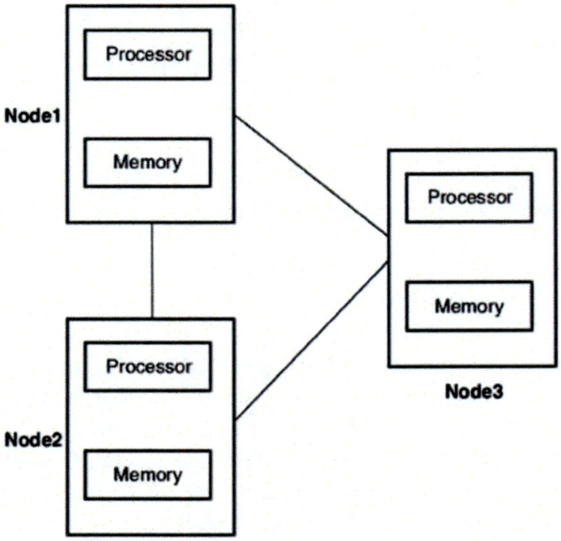

Fig. 1 Distributed system architecture

data through the network and the stability and reliability of the nodes are the major concern in this system. In a distributed system, every node must have an access to all the other nodes by using a dedicated communication line. Send and receive primitives are used for the actual communication between the sender node and the receiver node. The amount of information sharing in the large-scale network has been tremendously increasing everyday due to the huge number of users. In that case, there is a possibility that a faulty node in the system may send inconsistent data to the data requested node. Even though many algorithms are developed for detecting the faulty nodes in the network, there is a requirement for rapid identification of the faulty nodes. So, in this paper, a novel idea of calculating a vertex magic total labelling (VMTL) pivot value plays a key role in detecting the inappropriate faulty nodes in the system.

A vertex magic total labelling (VMTL) [1, 2] is a kind of labelling method used in graph theory for an undirected graph. Here, a magic constant value is calculated by adding the labels (weights) of all the edges incident on a particular vertex along with adding the label (weight) of vertex itself. Finally, this magic constant value is same for all the vertices in the graph. This concept is incorporated in the distributed system for faulty node identification.

The main organization of the paper is described as follows. Section 2 explains the review of the existing work, and Sect. 3 throws limelight on the working methodology incorporated. Experimental results are elaborated in Sect. 4, and conclusion and future enhancement has been given in Sect. 5.

2 Literature Survey

A new method of matrix [3] is given for identifying failure nodes in the network. In this approach, matrix row round trip and matrix column round trip delays are compared to detect the faulty nodes. This paper [4] proposed graph-based colouring method for server client communication request scheduling for distributed system. Based on the idea [5] of spatial and temporal correlation of the data, the permanent faults in the nodes can be identified in the proposed method. In this research [6], the parameters used for failure node detection are identified as malfunctioning of hardware components, change in the energy levels, physical systems, some unauthorized attacks, etc. Clustering-based approach [7] is proposed in a distributed system. Every large network is divided into small network of clusters and the cluster head is intended to identify the faulty nodes. Apart from the clustering method, a novel routing protocol is also implemented. In this research [8] of WSN faulty node detection, a fuzzy model is incorporated to group the nodes into four different categories as normal, traffic, end and dead node. Based on the status of the nodes in WSN, the protocol determines the active and malfunctioning nodes. A kind of data structure tree-based algorithm [9] is proposed to notify the faulty nodes in the network. If a node becomes faulty, then it will be deleted from the tree structure and all its children are pointing to a new parent. A new hard and soft faulty node [10] detection mechanism is implemented mainly in the large-scale WSN. If a particular node does not receive any message from the neighbour hop then it will be categorized as hard faulty node and so on. Akbari et al. work [11] proposed not only detecting the faulty nodes but also suggests a clustering based mechanism to recover from the faults which is energy efficient.

The research [12] proposed is basically a kind of security mechanism incorporated for the distributed system which could be taken as a sample for identifying faulty node in the network. In the work of Duche and Sarwade [13], the confused nodes are taken out from the existing WSN based on the RTT delays and acknowledgements from the receiver nodes during the communication. Once the nodes are detected, then the receiver will no more communicate with the faulty node. Naïve Bayes algorithm [14] is a probabilistic method depicted to find out the nodes that are faulty in the distributed WSN. Both training and testing phases are used to make the model more automatic detection. A new novel method of learning automata is deployed in the small-scale WSN to learn about the dead or active nodes [15]. This method consists of more complicated parser to learn about the network. Each node is intended to do the self-diagnosis mechanism [16] which could be done periodically to know about the malfunctioning. In this idea, all nodes will be performing a predefined test case and the result should be transmitted to the next hops in the network. In this method [17], classifying different activities in the network would be taken as a base for identifying faulty node and active node in the network. A survey [1] on graph labelling is proposed which deals with the graceful and harmonious labellings and its variants. Magic-type and antimagic-type labellings are discussed in a more elaborative way.

3 Working Methodology

In the proposed idea, a fault node identification algorithm based on VMTL value can be implemented in all the sites of the distributed system. Distributed system is a collection of nodes connected by a network. Every node in the system would be represented as a vertex in the weighted undirected graph $G = (V, E)$. The edge represents the communication link between the nodes in the system. Every link (edge) and the nodes (vertices) are assigned with a labelling (weight). However, all the weights in the graph are assigned to satisfy the vertex magic total labelling (VMTL) property.

In the example undirected graph shown in Fig. 2, if we calculate the VMTL pivot value (magic constant) it results with a value Pv = 20. VMTL pivot value is calculated for every vertex in the graph which represents the node in the distributed system. If the property of VMTL is satisfied then all the vertices in the graph will have the same constant value. This constant Pv value plays a vital role in determining the faulty nodes in the distributed system. The proposed method comprises of five algorithms. Algorithms 1 and 2 are used to construct the VMTL_Adjacency matrix and VMTL_Adjacency list, respectively. Algorithm 3 is used to compute the VMTL Pv value whereas Algorithm 4 incorporates the communication between the sender and the receiver node in the distributed environment. Finally, Algorithm 5 finds out the count of active nodes and faulty nodes in the network.

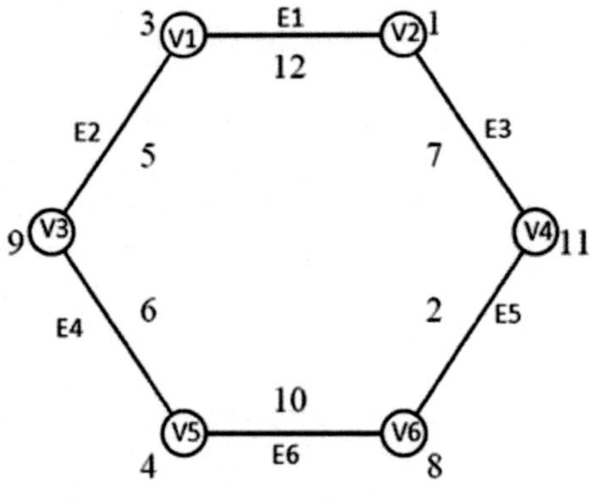

Fig. 2 Undirected graph with VMTL property

Algorithm 1: VMTL_Adjacency Matrix Construction

Input: Undirected Graph G= (V, E) with VMTL Property

V={V1,V2,V3,.....,Vn} set of vertices of a graph G= (V,E)
E={E1,E2,E3,.....,En}set of edges of a graph G= (V,E)

```
Procedure VMTL_Adjacency_Matrix(*G)
//VMTL Adjacency Matrix Construction
Begin
   For each vertex Vi from i=1 to N
   For each Edge Ej from j=1 to N
   // N is Number of Nodes in the Graph G
   Set the value of A[i,j] to Weight of Ej
End
```

Output:
A[i,j] is a two dimensional array of size n x n where i represents the vertex id and j represents the edge id incident on that vertex.

The above Algorithm 1 takes the undirected graph as an input and generates a two-dimensional array as an output which consists of the entry of labels of the edges incident on a particular vertex.

This entry has to be repeated for all the nodes in the network which is represented as a graph in Fig. 3. An example construction of the VMTL_Adjacency matrix is shown for the graph depicted in Fig. 2. If there is no edge from a defined vertex then the entry in the matrix is marked as 0. Otherwise the edge weight will be entered.

	1	2	3	4	5	6
1	0	12	5	0	0	0
2	12	0	0	7	0	0
3	5	0	0	0	6	0
4	0	7	0	0	0	2
5	0	0	6	0	0	10
6	0	0	0	2	10	0

Fig. 3 VMTL_Adjacency matrix representation of a graph

Algorithm 2: VMTL_Adjacency List Construction

```
Input: Undirected Graph G= (V, E) with VMTL Property

V= {V1, V2, V3,Vn} set of vertices of a graph G= (V,E)
E={E1,E2,E3,.....,En}set of edges of a graph G= (V,E)

Procedure VMTL_Adjacency_List(*G)
//VMTL_Adjacency List Construction
Begin
   For each vertex Vi from i=1 to N
   For each Edge Ej from j=1 to N
Call Function Insert (struct vertex** Node_ptr, int vertex_data)
// N is Number of Nodes in the Graph G
// Inserting a Node (Vertex) at the beginning of the linked list

Function Insert (struct vertex** Node_ptr, int vertex_data)
Begin
   // New Node (Vertex) Allocation
     struct vertex* Adj_node = (struct Node*) malloc(sizeof(struct Node));
   // Data assignment to the Node (Vertex)
     Adj_node->data  = vertex_data;
    // New Node (Vertex) allocation as a head of the linked list
      Adj_node->next = (*Node_ptr);
     // Linked list header will point to the created node
     (*Node_ptr)   = Adj_node;
End

Output:
For each vertex Vi, A different singly linked list will be created in the graph G =
(V, E).
```

In Algorithm 2, similar to Algorithm 1, an undirected graph is given as an input and generates a singly linked list as an output. This would be designated as VMTL_Adjacency list shown in Fig. 4. All vertices will be considered as a node in the singly linked list and insertion is of a new node is done at the beginning of the list. If there is an edge Ej from one vertex Vi to another vertex Vj then that vertex will be added as a new node in the list. The last pointer will be marked as a NULL pointer as there are no more nodes connected to the current vertex in Fig. 4. An example construction of the VMTL_Adjacency_List is shown for the graph depicted in Fig. 2.

The main novel idea of this paper lies in the calculation of the VMTL_Pv value that is the magic constant. In Algorithm 3, we give the input as the VMTL_Adjacency matrix and a vector V consists of all the labels (weights) assigned for the vertices.

Initially for each row in the matrix $A[i, j]$, the sum is calculated for all the columns and after the loop we calculate the VMTL_Pv value as the sum of entries in a row and vertex value Vi stored in the single dimensional array (Vector).

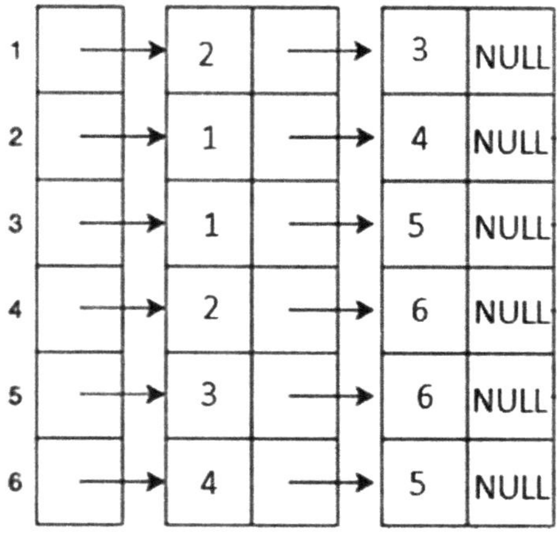

Fig. 4 VMTL_Adjacency list representation of a graph

Algorithm 3: Calculation of VMTL Pv Value

```
Input: VMTL_Adjacency Matrix:Two Dimensional Matrix:A[i,j]
Vertex Label Values: Single Dimensional Array:V[i]
Initialize the value of i,j,sum to 0

Procedule Pv-Calc(*A, *V)
Begin
//Finding Sum for each row in the VMTL_Adjacency Matrix
   For each Vertex Vi in the Adjacency Matrix A[i,j]
   For each row i from 0 to N
      Begin
          For each column j from 0 to N
             Begin
             sum=sum+A[i,j];
             End
       Pvi=V[i]+sum;
       Sum=0;
      End
End
```

Each sender node can look up the VMTL_adjacency list to find out the neighbouring nodes to which it can communicate by sending a message. In Algorithm 4, basically the faulty node status condition (NSC) is checked and marked as an "Active Node" or "Faulty Node" in the local cache and later forwarded to the server.

A sender node finds out the next hop from the VMTL_Adjacency list stored in the VMTL_IB (information block) and sends a message to the neighbouring node.

Algorithm 4: Sender Node and Receiver Node Communication

```
Set Sender Node to SN(Vi)
Set Receiver Node to RN(Vj)
Procedure Communication (SN(Vi), RN(Vi))
  Begin
  //At particular time instant T
  //SN(Vi) invokes the function call

  Function Send ({M(Vi), Pvi})
  //M(Vi) is the message coming from Vertex Vi
  //Pv is a VMTL Value calculated across different nodes
   // RN(Vi) invokes the function call

  Function Receive ({M(Vi), Pvi})
  //Node RN(Vj) access its internal local cache to check for the value of Pvj

      If  Pvi == Pvj then
      // NSC is Node Status Condition
        Set the value of NSC[Vi]=1
        Else
        Set the value of NSC[Vi]=0;
          If NSC[Vi] = 1 then
          Set "Active" State for the vertex Vi
          Else if NSC[Vi] = 0 then
          Set "Faulty" State for the vertex Vi
  End
```

Each message consists of the actual data along with the VMTL_Pv value calculated across the nodes in the network. Once the receiver node gets the message checks the Pv value received with its own stored Pv value If both are equal then the sender node will be identified as an active node otherwise it will be detected as a "Faulty Node". Further, NSC will be either set to 1 or 0 depends on the condition.

Algorithm 5: Calculation of number of Active Nodes and Faulty Nodes

```
Procedure Node_Count(*NSC)
// NSC is the Node Status Condition

Begin
  Set the value of FNC to 0
  //FNC is the value of Faulty Node Count
  Set the value of ANC to 0
  //ANC is the value of Active Node Count

  For each Node Vertex Vi in the network
    Begin
      If (NSC[Vi]==0) then
     //Increment the value of FNC by 1
      FNC=FNC+1;
      Else
     //Increment the value of ANC by 1
     ANC=ANC+1;
    End
End
```

As per the detection mechanism incorporated in Algorithm 4, finally a number of active and faulty nodes are detected in the distributed environment as shown in Algorithm 5. The parameter FNC and ANC will give the results of the count of detected active and fault nodes in the network.

4 Results and Discussion

Fault node identification using VMTL property of a graph is tested against MATLAB software with high end configuration processor. As a sample test case, a total of 250 nodes are deployed in the simulation model. As and when the test cases are incorporated, we started recording the range of node failure probability across various values of the number of nodes. The actual value of TFR and FFR is calculated using a simple formula given as below:

$$\mathrm{TFR} = \mathrm{FNC}/\mathrm{TN}\alpha \tag{1}$$

$$\mathrm{FFR} = \mathrm{ANC}/[\mathrm{TN}(1-\alpha)] \tag{2}$$

FNC indicates the fault node count that has been correctly identified as a "Faulty Node" in the network and ANC indicates the active node count that may be wrongly identified as a "Faulty Node" in the network. Sample simulation results are taken for the number of nodes $N = 10$, 20, 30 and 40 and the range of true fault rate (TFR) and false fault rate (FFR) is plotted against α value (ranges between (1 and 60%), which is the node failure probability in the network.

Figures 5 and 6 show the value of TFR and FFR with the network node failure rate α for the test cases ranges from the number of nodes $N = 10$ to 40. As a result, the major inference from the test cases would be the TFR increases with increase in number of nodes and FFR decreases with increase in number of nodes.

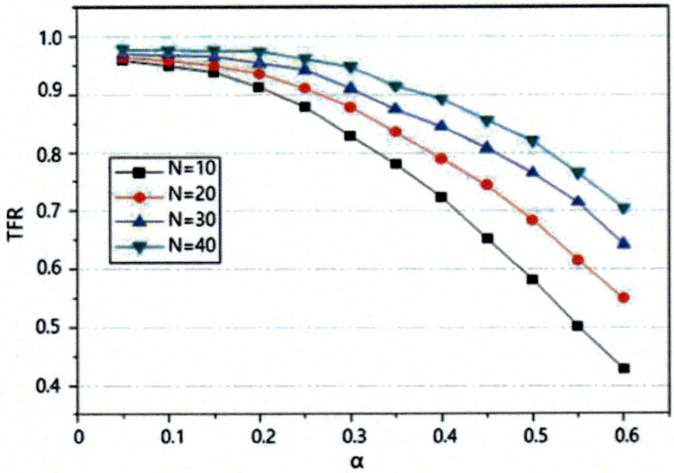

Fig. 5 TFR ratio for $N = 10, 20, 30, 40$

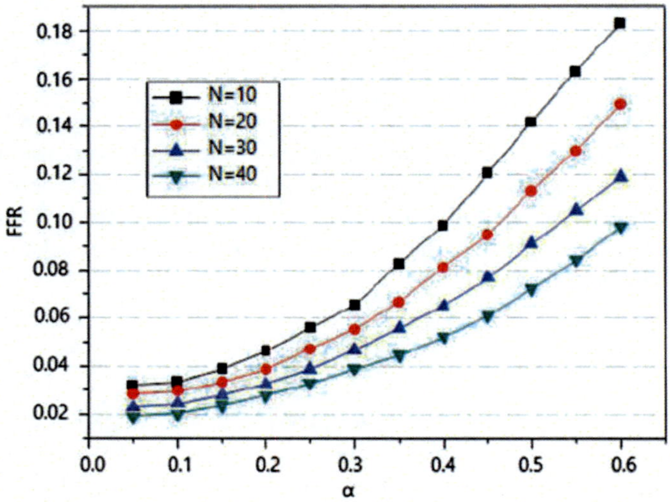

Fig. 6 FFR ratio for $N = 10, 20, 30, 40$

5 Conclusion

This paper proposes a novel simple mechanism based on vertex magic total labelling (VMTL) for detecting faulty nodes in the distributed system environment. We increased the number of nodes in the network to analyse the impact of true fault rate (TFR) and false fault rate (FFR). As a result, the percentage of TFR is about to be 95% with a minimum 5% FFR during the simulation analysis. All the information related to the number of active nodes and the number of faulty nodes are forwarded to the server node for taking a decision on whether to remove the node from the network or to correct the fault. A rapid method of detection of malfunctioning nodes can improve the QoS of distributed computing environment. This will be the future enhancement of proposed methodology.

References

1. Gallian, J.A.: A dynamic survey of graph labeling. Electron. J. Comb. 15 (2008)
2. MacDougall, J.A., Miller, M., Wallis, W.D.: Vertex-magic total labelings of wheels and related graphs. Utilitas Math. **62**, 175–183 (2002)
3. Palanikumar, R., Ramasamy, K.: Effective failure nodes detection using matrix calculus algorithm in wireless sensor networks. Cluster Comput. **22**, 12127–12136 (2019)
4. Muneeswari, G., Puthussery, A.: Process scheduling in heterogeneous multicore system using agent based graph coloring algorithm. Int. J. Recent Technol. Eng. **8**(3), 330–345 (2019)
5. Sharma, K.P., Sharma, T.P.: rDFD: reactive distributed fault detection in wireless sensor networks. Wireless Netw. **23**(4), 1145–1160 (2017)
6. Duche, R.N., Nisha, P.: Sensor node failure detection based on round trip delay and paths in WSNs. IEEE Sens. J. **14**(2), 455–463 (2014)
7. Azharuddin, M., Kuila, P., Jana, P.K.: Energy efficient fault tolerant clustering and routing algorithms for wireless sensor networks. Comput. Electr. Eng. **41**, 177–190 (2015)
8. Chanak, P., Banerjee, I.: Fuzzy rule-based faulty node classification and management scheme for large scale wireless sensor networks. Expert Syst. Appl. **45**, 307–321 (2016)
9. Suganthi, K., Vinayagasundaram, B., Aarthi, J.: Randomized fault-tolerant virtual backbone tree to improve the lifetime of wireless sensor networks. Comput. Electr. Eng. **48**, 286–297 (2015)
10. Panda, M., Khilar, P.M.: Distributed self fault diagnosis algorithm for large scale wireless sensor networks using modified three sigma edit test. Ad Hoc Netw. **25**, 170–184 (2015)
11. Akbari, A., Dana, A., Khademzadeh, A., Beikmahdavi, N.: Fault detection and recovery in wireless sensor network using clustering. IJWMN **3**(1), 130 (2011)
12. Muneeswari, G., Daniel, D., Natarajan, K.: Activity classifier: a novel approach using Naïve Bayes classification. In: International Conference on Inventive Computation Technologies (2019)
13. Duche, R.N., Sarwade, N.P.: Sensor node failure or malfunctioning detection in wireless sensor network. ACEEE Int. J. Commun. **3**(1), 57–61 (2012)
14. Lau, B.C., Ma, E.W., Chow, T.W.: Probabilistic fault detector for wireless sensor network. Expert Syst. Appl. **41**(8), 3703–3711 (2014)
15. Banerjee, I., Chanak, P., Rahaman, H., Samanta, T.: Effective fault detection and routing scheme for wireless sensor networks. Comput. Electr. Eng. **40**(2), 291–306 (2014)
16. You, Z., Zhao, X., Wan, H., Hung, W.N., Wang, Y., Gu, M.: A novel fault diagnosis mechanism for wireless sensor networks. Math. Comput. Model. **54**(1), 330–343 (2011)

17. Muneeswari, G., Puthussery, A.: Multilevel security and dual OTP system for online transaction against attacks. In: IEEE 3rd International Conference on IOT in Social, Mobile, Analytics and Cloud (2019)

Taxonomy of Diabetic Retinopathy Patients Using Biogeography-Based Optimization on Support Vector Machine Based on Digital Retinal Images

N. Vinoth, M. Vijayakarthick, S. Ramesh, and E. Sivaraman

Abstract Diabetic retinopathy is a sort of eye infection portrayed by retinal harm welcomed on by diabetes and is the significant reason for visual deficiency in the individuals of age group between 20 and 64 years. Picture processing procedures are utilized to identify and arrange retinopathy pictures viably. This paper has proposed a PC approach for the discovery of diabetic retinopathy stage by utilizing shading fundus pictures. The highlights are extricated from the fundus pictures with the help of picture processing methods, object order, and by considering the support vector machine (SVM), which consequently identify the veins as ordinary or abnormal. The incorporation of SVM strategy has contrasted the Naïve Bayes (NB) calculation and biogeography-based optimization (BBO). The BBO calculation has been proposed with the code age techniques in the current investigation to improve the binarization process. The created code lattice is considered as a best solution for a given issue since it is structured by considering the highlights of the code network, for example, by and large classifier exactness, least hamming separation and edge of grouping, and the issue highlights, for example, traits, tests, and classes. The produced numerous parallel classes were grouped using standard SVM. It is seen that BBO-based SVM beats the SVM and NB strategy for grouping.

Keywords Biogeography-based optimization · Classification · Diabetic retinopathy · Support vector machine · Naïve Bayes

---

N. Vinoth (✉) · M. Vijayakarthick
Department of Instrumentation Engineering, Anna University, MIT Campus, Chennai, India
e-mail: vinothbalaji@rediffmail.com

M. Vijayakarthick
e-mail: vijayakarthick@yahoo.co.in

S. Ramesh
Department of Electronics and Instrumentation Engineering, Annamalai University, Chidambaram, India
e-mail: sekaramesh06@gmail.com

E. Sivaraman
Government College of Engineering, Tirunelveli, India
e-mail: sivaraman2k7@gmail.com

P. Karuppusamy et al. (eds.), *Sustainable Communication Networks and Application*, Lecture Notes on Data Engineering and Communications Technologies 55,
https://doi.org/10.1007/978-981-15-8677-4_51

1 Introduction

BBO is a recently evolved heuristic calculation, with the promising and reliable presentation [1] by turning into a solid contender to other populace-based calculations like GA, PSO, and ACO. Biogeography is the investigation of the geological appropriation of natural organisms. The science of biogeography is considered as the reason for the improvement of this new field. BBO shares certain highlights for all intents and purpose with other science-based optimization techniques, for example, GA [2] and PSO [3]. Like GAs and PSO, BBO has a method of sharing data between arrangements. GA arrangements kick the bucket toward the finish of every age, while PSO and BBO arrangements endure always, in spite of the fact that their qualities change as the optimization process advances [4]. PSO arrangements are bound to bunch together in comparable gatherings, while GA and BBO arrangements do not really have any worked in propensity to frame groups. Also, BBO is a populace-based optimization calculation and it does not include multiplication or the age of kids. BBO arrangements straightforwardly share their traits with different arrangements by its suitability index variable (SIV). BBO is appropriate to a significant number of similar kinds of issues that GA and PSO are utilized for, to be specific, high-measurement issues with numerous nearby optima. Nonetheless, BBO likewise has a few highlights that are remarkable among science-based optimization strategies. For these novel highlights, adaptable ideas and the preferences, BBO is favored in the proposed work. The optimization is simply based on the habitat suitability index (HSI) [5] of the species in the island to assess the nature of every species in the territory. At the point when the HSI of the species is high then the arrangement yielded is the most precise one. Any optimization issue can be structured utilizing BBO by picking suitable incentive to HSI, SIV, movement rate, displacement rate, and the use of BBO administrators, for example, choice, relocation, and change. For every territory, map the HSI to the quantity of species S, the movement rate, and the resettlement rate. BBO administrators are applied to the species more than once to decide the best arrangement over successive ages. The most extreme relocation rate is E, which happens when the natural surroundings contains the biggest number of species that it can support. The equilibrium number of species is S0, so all in all the migration and displacement rates are equal. Be that as it may, there might be intermittent trips from S0 because of worldly impacts. Positive trips could be because of an abrupt spray of movement or an unexpected eruption of speciation. Negative outings from S0 could be because of infection, the presentation of some characteristic calamity. It can require some investment in nature for species checks to arrive at equilibrium after a significant irritation. Figure 1 shows the migration and displacement curves. These curves are shown by straight lines, yet when all is said in done they may be increasingly muddled curves.

The algorithm terminates with the predefined number of generations. The overall operations of BBO for efficient code matrix generation are shown in the form of a flowchart in Fig. 2.

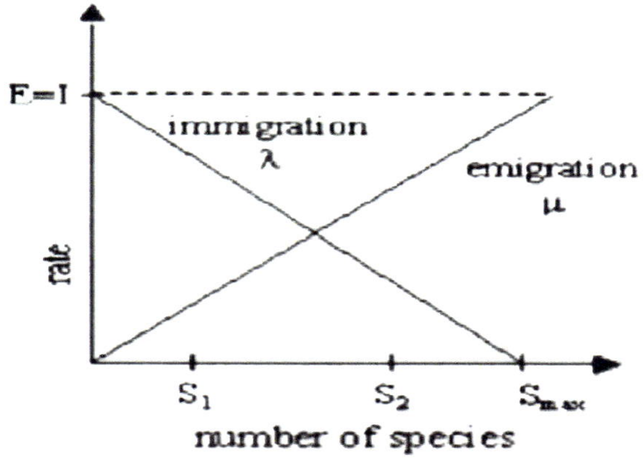

Fig. 1 Species of single habitat

In the proposed work, BBO algorithm has been utilized to order a fundus picture. BBO calculation is a fundamentally used to locate the ideal arrangement of an issue. However, fundus picture characterization is a grouping issue that requires each class to be extricated as a bunch. The first BBO calculation does not have the inbuilt property of bunching. In any case, here to extricate highlights from the picture, we have attempted to make the groups of various classes present in the picture and proposed a changed biogeography-based calculation arrange a fundus pictures.

2 Naïve Bayes Classifier

The Naive Bayes classifier [6] utilizes Bayes' theorem of likelihood to characterize pictures. When the highlights of the preparation set are taken care of into the classifier, the probabilities of individual highlights being available, given the result (e.g., the class—'Ordinary' or 'Glaucoma') just as the probabilities of every one of the two classes are determined. In light of these probabilities, the pictures of the testing set are arranged utilizing the recipe given underneath: Eq. (1)

$$P(\text{outcome}|\text{evidence}) : \frac{P(\text{Likelihood of Evidence}) \times \text{Prior probability of outcome}}{P(\text{Evidence})} \quad (1)$$

The Naive Bayes classifier (NBC) is generally mentioned as the maximum a posteriori (MAP) decision rule. It is jotted that the assumption with respect to each feature which is statistically independent at times causes problems in few practical instances and does not fit in certain occurrences [7]. However, the training strategies based on MAP decision rule with Naïve Bayes assumptions provide an optimal

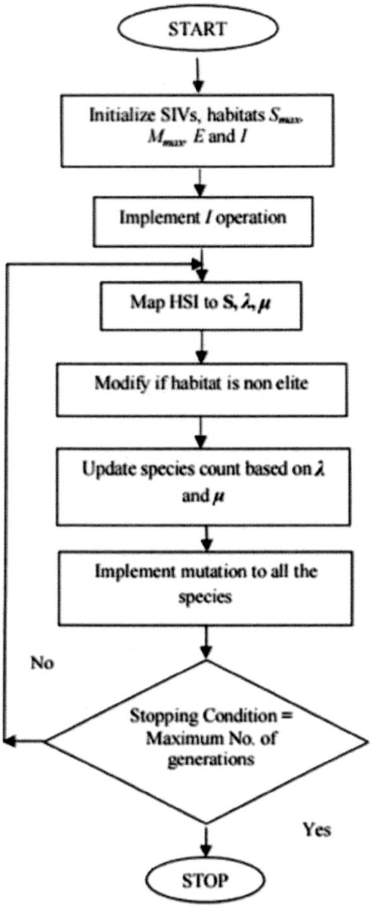

Fig. 2 Flowchart of BBO algorithm

classifier for various experimental studies and applications with unsuitable suppositions. NB algorithm is very simple and feasible training process and this process is possible with small amount of data to approximate the parameters. It does not use any iterative procedure for training process and the classification outcomes are found to be aggressively accurate.

Conventional classifiers compute the local classification decision probability, and a global classification decision is achieved by utilizing this data. Here, the proposed NBC enumerates the classification probability by using Eq. (1) when the model embraces a NBC as its local classifier [8]. The proposed NBC reckons the probability that a feature vector component is classified as its class with subsequent probability density function when the feature value is a continuous value: Eq. (2)

$$P(x_i|c_k) = \frac{1}{\sqrt{2\pi}\sigma c_k} e^{-(x_i-\mu_{c_k})^2/2\sigma_{c_k}^2} \quad (2)$$

where the likelihood thickness work is set up during nearby classifier preparing period with mean and standard deviation of each class information for each element vector segment. Likelihood can likewise be delineated from Eq. (2) with each measurement singularly in light of the fact that NBC obtains the Naive Bayes groups as its restricted classifiers [9]. While performing preparing system, the likelihood thickness capacity of each component vector for each measurement on every neighborhood classifier is created utilizing Eq. (2) [10]. After the assurance of likelihood thickness work, the preparation process for worldwide classifier is prematurely ended. The dynamic activity for a given information is that the element vectors are overcomes the undifferentiated from neighborhood classifiers and the class for the given information is determined utilizing the equation: Eq. (3)

$$\text{Class}(x) = \arg\max_k \frac{1}{N}\sum_{j=1}^{N} P(c_{ij}) \prod_{i=1}^{n} P(x_i|C_{jk}) \quad (3)$$

3 Results and Discussions

The classification of diabetic retinopathy (DR) was carried out by NBC, SVM, and BBO-based SVM as shown in Fig. 3. The performance of the methods is evaluated using sensitivity, specificity, and accuracy.

The classification of diabetic retinopathy is categorized into normal NPDR (mild) and PDR (severe). Figure 4 shows the normal eye without DR, optic disk, iris, and hemorrhages are segmented from the optic image as shown in Figs. 5, 6, and 7, respectively. From Fig. 7, it is observed that there are no hemorrhages present in the normal optic image.

The features are extracted from the hemorrhages image (image of interest), for DR, PDR, and NPDR images. They have been formed as data from the training phase of the algorithms. The proposed BBO-based SVM on 100 normal fundus images and 75 abnormal fundus images obtained from the benchmark DRIVE repository.

There are no hemorrhages present in the normal optic image. Similarly, Fig. 8 shows the affected DR image and Figs. 9, 10, and 11 shows the segmented images of optic disk, iris, and hemorrhages respectively. The features are extracted from the hemorrhages image (image of interest), for DR, PDR, and NPDR images. They have been formed as data from the training phase of the algorithms. The proposed BBO-based SVM on 100 normal fundus images and 75 abnormal fundus images obtained DRIVE. Three different classifiers SVM, Naïve Bayes classifier, and BBO-based SVM are evaluated to classify the diabetic retinopathy [11]. In Bayes classifier, typical Gaussian dispersion is utilized to fit or model the component base

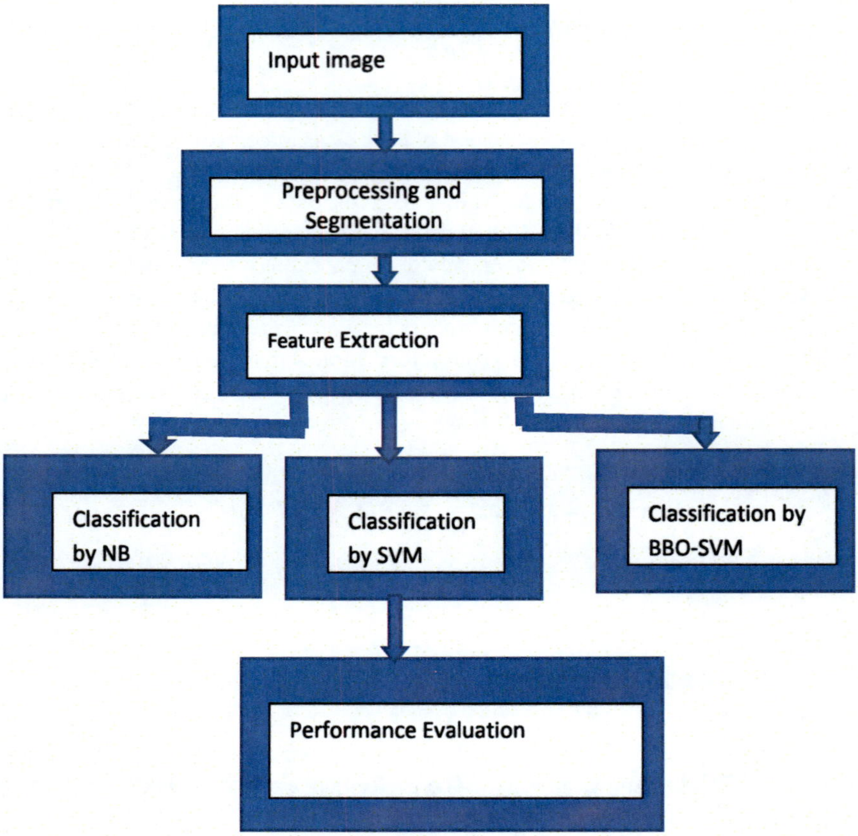

Fig. 3 Flowchart for classification

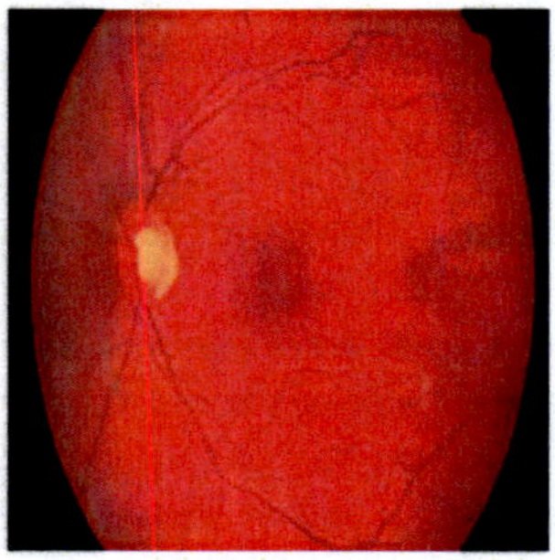

Fig. 4 Normal eye (without DR)

Fig. 5 Optic disk

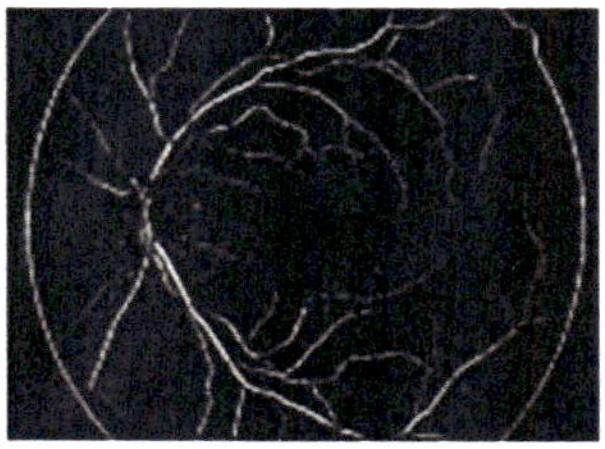

Fig. 6 Iris of eye

Fig. 7 No hemorrhages

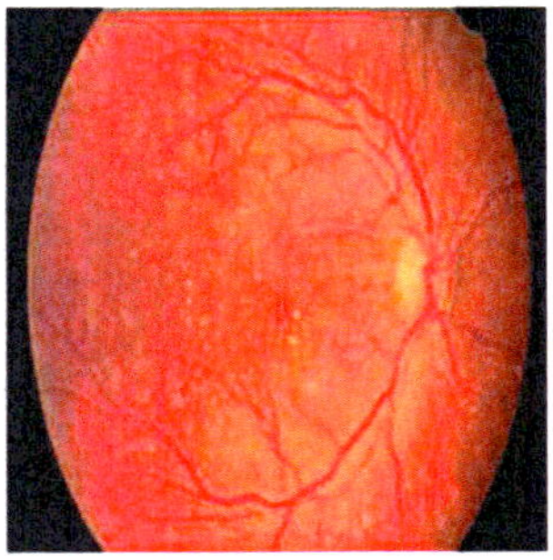

Fig. 8 DR affected eye

Fig. 9 Optic disk

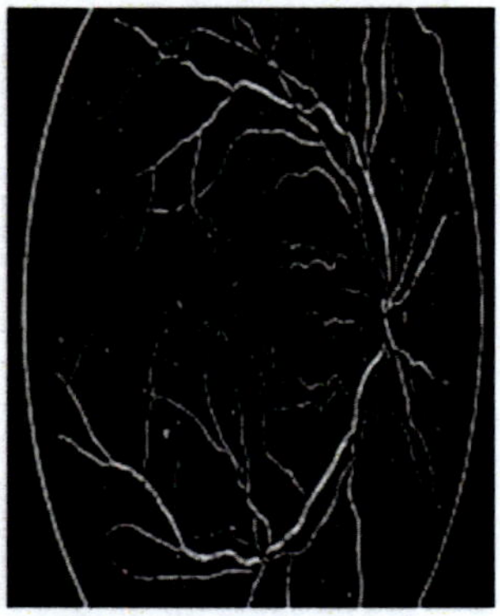

Fig. 10 Iris of eye

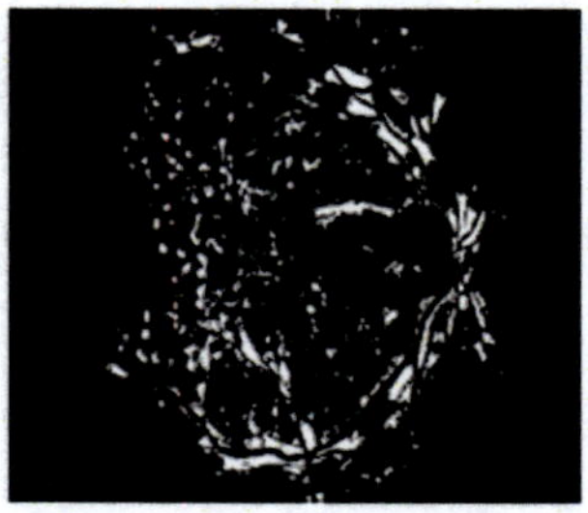

Fig. 11 Hemorrhages

and it assesses the earlier probabilities from the overall frequencies of the classes in preparing. In SVM classifier, direct part is utilized to plan the preparation information into the piece space [12, 13]. It is seen that the proposed BBO-based SVM technique has a higher characterization rate, since utilizing SVM classifier because of the way that an information point is seen as a μ-dimensional vector (a rundown of numbers), such focuses are isolated with a $(\mu - 1)$ dimensional hyperplane, i.e., direct classifier. There are numerous hyperplanes that may group the information. The best hyperplane is picked that speaks to the biggest partition, or edge, between the two classes which brings about higher characterization rate. The irregularity between the quantity of preparing tests per class in Bayes classifier brings about helpless selections of loads for straight choice limit that leads in misclassification. BBO-based SVM performed superior to the SVM and NB classification.

4 Performance Measures

In the literature, the specificity and sensitivity esteems are reliably detailed since they compare to the current clinical practice and have clear explanations in the clinical terms. Specificity on the solid populace and sensitivity esteem relies upon the sick populace (see Eq. 1). These qualities give the methods for examining what number of sound and what number of unhealthy patients are impeccably researched with a given technique. From the technique correlation viewpoint, the two qualities are not feasible since the two conveyances imbricate and the genuine dependability is perpetually an exchange off. In our assessment convention, we have chosen two evaluation standards: (1) the estimation is picture based and (2) is done separately to different diabetic retinopathy information (Sect. 2). The main guideline is approved by the way that it compares to the clinical practice where choices are 'tolerant based.' Spatial zone (pixelwise)-based evaluation can be practical in technique improvement, yet the trouble itself is that it depends on pictures. The subsequent standard is because of the exact certainty that most analysts focus just on one or a few information types and it is not compulsory for the down to earth technique to distinguish all the information.

For an appropriate correlation, the specificity and sensitivity esteems must be converged into a structure which can depict the conduct over various amalgamations of the qualities. Receiving operating curve (ROC) is a characteristic choice because of its notoriety and demonstrated inclination in homogeneous PC vision undertakings, for example, object class acknowledgment [7], clinical exploration [2], and face acknowledgment [10]. The ROC actualizes a graphical interpretation for 1-specificity (FPR) assurance and sensitivity (TPR). The ROC curve gives the way to the ideal investigation for the issue in order to locate the best technique boundaries for the undertaking or contrast exhibitions regardless with operating conditions. In our assessment, we adjusted the practices from [7], where every strategy is required to give a score to each test picture. A high score relates to a high likelihood that a finding is available in the picture. By controlling the gave scores, the ROC curve can be naturally produced.

The successive step is to compute the value of true positive (TP), false negative (FN), false positive (FP), and true negative (TN). TP depicts that the image is recognized as corresponding to the class (positive). FP is the indication that the image is to be spotted from the relevant class classification in addition that there is a fault in recognition. TN represents the image of a class member but not precisely the identified class (negative). FP stipulates that the image is identified as a class member but not be a member of a class. This study is then fractionated into three classes, as normal class (positive class), a class that is not normal (negative class), NPDR class (positive class), non-NPDR class (negative class), PDR class (positive class), and PDR (negative class) as shown in Fig. 12.

The performance of the method will be better when a ROC curve approaches closer to the top left corner which is shown in Fig. 13 [14].

At the point when a zone under ROC curve = 1, it mirrors that framework totally concurs with the ground truth divisions. Area under the ROC curve (AUC) gives the classifier's exhibition over the entire scope of cutoff focuses. From Table 1 and Fig. 13, obviously our technique performs well for all the databases. Here, the AUC esteem is above 0.985 which uncovers that the exhibition of classifier is phenomenal. Mathematical morphology helps in determining the value in a quite effective manner through mechanized fundus segmentation. Exudates from the eye with reduced noise are obtained using maxtree and tribute filtering results in exudate segmentation. This strategy improved past scientist's techniques [15, 16]. The Feature extraction utilizing exudate includes values, green channel homogeneity, the factual estimation of immersion pictures (mean, standard deviation, kurtosis, skewness), and entropy

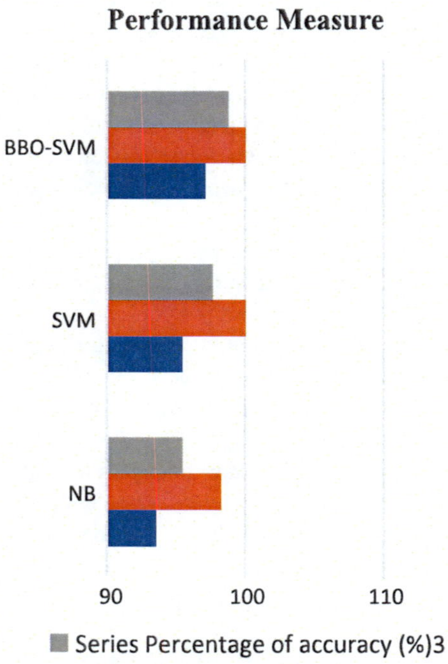

Fig. 12 Performance evaluation of algorithms

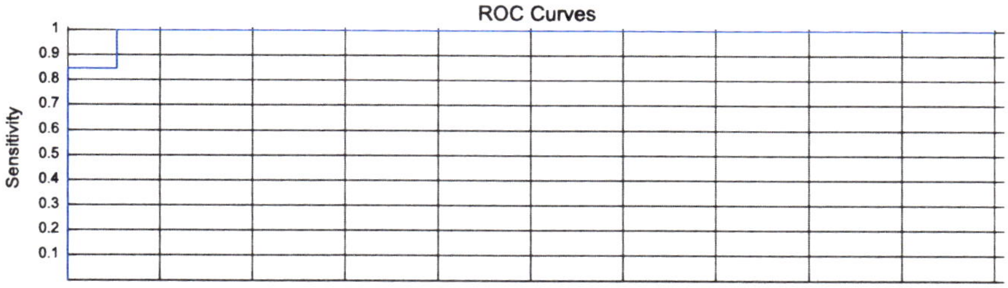

Fig. 13 ROC curve for BBO-SVM

Table 1 Performance evaluation of classification

Method	Sensitivity	Specificity	Accuracy
NB	93.55	98.23	95.45
SVM	95.45	100.00	97.62
BBO-SVM	97.11	100.00	98.76

green channel and can be utilized as a mention to order. Execution of BBO-based SVM was in brilliant class with a sensitivity estimation of 97.11%, specificity 100%, exactness of 98.76%, and AUC of 0.985. The supremacy of BBO-SVM is also revealed by its execution time. From the chart (Fig. 14), it is observed that the BBO-SVM takes less time when compared to SVM and NBC.

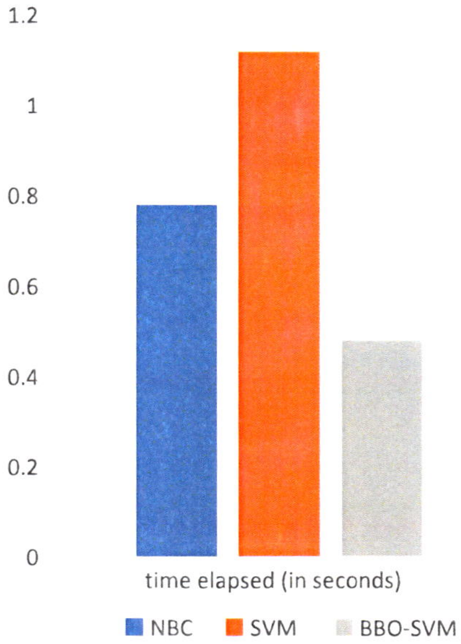

Fig. 14 Comparison of execution time for algorithms

References

1. Simon, D.: Biogeography-Based Optimization, vol. 12, no. 6, pp. 702–713. IEEE Computer Society Press, Washington, DC (2008)
2. Abell, J., Du, D.: A Framework for Multi-objective, Biogeography Based Optimization of Complex System Families, pp. 1–10. American Institute of Aeronautics and Astronautics, Reston (2010)
3. Dan, S.: Biogeography-based optimization. IEEE Trans. Evol. Comput. **12**(6), 702–713 (2008)
4. Du, D., Simon, D., Ergezer, M.: Biogeography-based optimization combined with evolutionary strategy and ımmigration refusal. In: IEEE International Conference on Systems, Man, and Cybernetics, pp. 1023–1028 (2009)
5. Ergezer, M., Simon, D., Du, D.: Population distributions in biogeography-based optimization algorithms with elitism. İn: The Proceedings of IEEE International Conference on System, Man and Cybernetics, San Antonio, U.S.A (2009)
6. Lowd, D., Domingos, P.: Naive Bayes models for probability estimation. In: Proceedings of the 22th International Conference on Machine Learning, pp. 529–536 (2005)
7. Lewis, D.D.: Naive Bayes at forty: the independence assumption in information retrieval. Lecture Notes in Computer Science, vol. 1398, pp. 4–15 (2005)
8. Park, D.C.: Centroid neural network for unsupervised competitive learning. IEEE Trans. Neural Netw. **11**(8), 520–528 (2000)
9. Sopharak, A., Nwe, K.T., Moe, Y.A.: Automatic exudate detection with a Naive Bayes classifier. In: International Conference on Embedded Systems and Intelligent Technology, Bangkok, Thailand (2008)
10. Rao, P.N., Devi, T.U., Kaladhar, D.: A probabilistic neural network approach for protein superfamily classification. J. Theor. Appl. Inf. Technol. **6**(1), 101–105 (2009)
11. Jayanthi, D., Devi, N., Swarna Parvathi, S.: Automatic diagnosis of retinal diseases from color retinal images. Int. J. Comput. Sci. Inf. Secur. **7**(1), 234–238 (2010)
12. Srivastava, D.K., Bhambhu, L.: Data classification using support vector machine. J. Theor. Appl. Inf. Technol. **12**(1) (2009)
13. Hsu, C.-W., Lin, C.-J.: A comparison of methods for multiclass support vector machines. IEEE Trans. Neural Netw. **13**(2) (2002)
14. Hanley, J.A., McNeil, B.J.: The meaning and use of the area under a receiver operating characteristic (ROC) curve. Diagn. Radiol. **143**, 29–36 (1982)
15. Priya, R., Aruna, P.: SVM and neural network based diagnosis of diabetic retinopathy. Int. J. Comput. Appl. **41**(1), 6–12 (2012)
16. Aquino, A., Gegundez-Arias, M.E., Marin, D.: Detecting the optic disc boundary in digital fundus images using morphological, edge detection, and feature extraction techniques. IEEE Trans. Med. Imaging 1860–1869 (2010)

An Efficient Energy Management of Hybrid Renewable Energy Sources Based Smart-Grid System Using an IEPC Technique

K. Bapayya Naidu, B. Rajani, A. Ramesh, and K. V. S. R. Murthy

Abstract In this paper, a grid-connected microgrid (MG) is proposed to find energy scheduling for optimal energy management. Here, the MG system has a photovoltaic system, wind turbine, battery storage (BS), as well as microturbine (MT). An ımproved emperor penguin colony (IEPC) technique can continuously track the necessary load demand of the MG system connected to the grid. Here, the huddling behavior of EPC is improved by crossover and mutation operator. The goal of the IEPC method is described by the participation of fuel cost, a variation of power per hour of the grid, cost of operation, and maintenance of the MG system connected with the grid. The convenience of RES, power demand, as well as state of charge of the storage element denotes limitations. The battery is employed from an energy source, stabilize and permit units of a renewable energy system to continue the operation in delivering steady as well as stable output power. Results of comparison demonstrate the superiority of the IEPC method as well as confirm its potential to solve the issue.

Keywords Optimal energy management · Forecasting errors · Renewable energy sources · Operation and maintenance cost · Batteries

K. Bapayya Naidu (✉) · K. V. S. R. Murthy
Department of Electrical and Electronics, Aditya Engineering College, Surampalem, Andhra Pradesh 533437, India
e-mail: kbapayanaaiidu@gmail.com

K. V. S. R. Murthy
e-mail: murthy.kvs@aec.edu.in

B. Rajani
Department of Electrical and Electronics, Aditya College of Engineering and Technology, Surampalem, Andhra Pradesh 533437, India
e-mail: drrajaniboddepalli2015@gmail.com

A. Ramesh
Department of Electrical and Electronics, Aditya College of Engineering, Surampalem, Andhra Pradesh 533437, India
e-mail: principal@acoe.edu.in

P. Karuppusamy et al. (eds.), *Sustainable Communication Networks and Application*, Lecture Notes on Data Engineering and Communications Technologies 55,
https://doi.org/10.1007/978-981-15-8677-4_53

1 Introduction

To meet the growing demand for renewable energy sources (RES) link wind and solar, the existing supply networks tend to achieve the goal of minimizing the greenhouse gas effects [1]. The random employment of RES control generates some complexity and proceedings as well as control of the distribution network. Micro-grid has been developed to minimize complexities with the existing electrical system [2]. An MG is thought to be a promising site for great penetration of renewable energy. The main merits of the MG system are to reduce the power loss, reliable power flow, reducing the energy cost to the consumer. The energy storage capacity of MG has a limited to generate. So, it is advantageous to combine multiple MGs into a multi-MG (MMG) system. Then every MG system may minimize their regional cost through energy trading as well as enlarges their viability through reserve sharing [3].

Energy management is an essential factor in MG. Intelligent energy management is critical to maintaining reliable power supply for dealing with unknowns in MG. In economics controlling the stored energy can minimize the operational costs. The set of initial sequences does not provide optimal control policy when they can adjust for different scenarios [4]. In normally, these modes, energy is transmitted depending on obtainable battery power as well as state of charge is saved if the output power exceeds the power requirement. In addition, charging and eliminating the effects will shorten the life of the storage system.

More optimization techniques are presented to develop optimal control techniques for battery energy; they are linear programming, dynamic programming, fuzzy logic as well as adaptive dynamic control. The current SOC charging and discharging limits were put forward to employ storage systems to send RES easily. To minimize the emission level and the operating cost, a microgrid was developed in the fuzzy logic system [5]. PM mechanism employing dynamic scheduling to 45 PV systems by storage with permit huge penetration of photovoltaic power into distribution networks are explained at the firefly algorithm, the useful life characteristics of a battery were implemented to get optimal performance of microgrid. In residential power, management was used to ADP. Adaptive dynamic Programming is evolved to ideally battery control in energy management.

2 Recent Research Works: A Brief Review

Several investigation operations have survived on literature based on optimal energy management on grid-connected microgrid by utilizing several methods. Here a few of them were reviewed.

To find energy programming in MG, Roy et al. [6] presented an intelligent method to EMS depend on RNN by the Ant-Lion Optimizer algorithm. The goal of the introduced technique was used to decrease the cost of electricity production through daily and real-time scheduling. Using RNN, demand response (DR) was assessed as well as extra indices were also considered. To determine generation, storage, as well as responsive load offerings, the ALO algorithm was evolved to solve economic dispatch problems. Liu et al. [7] developed a distributed energy management system for the MG community. The optimization depends on the energy management system schedules distributed energy resources as well as energy storage systems. In all repetitions, MG central controller regulates the scheduling of distributed energy resources and energy storage systems at the MG level. Optimization converges while the unbalanced power of entire buses was near to zero. The house's dynamic thermal model was included in HEMS for customer control of heating, ventilation as well as the air conditioning system.

Liu et al. [8] have developed a secure, distributed, transitive energy management method of multiple interconnected MGs. Based on the method, every MG was directed to MG distributed energy management system that interactions quantity and commercial price information by another MGs to conserve the privacy of information. While every MG performs like a price taker, S-DTEM dynamically optimizes their energy sales price as well as operation schedule to reduce their local cost through energy trade to another MGs/backbone. In the interim, that algorithm could reduce the cost of interconnected MGs. With the quadratic barrier functions, the finite-time convergence of S-DTEM could be analytically ensured to strong operational restrictions. A malicious MG-EMS can misbehave with forbidding optimal solution, and the misbehavior detection mechanism was suggested by finite-time convergence possessions. Yang et al. [9] have introduced a bilayer game theoretical framework to multi-MG (MMG) multi-MG interactive energy management (IEM). The upper layer of the framework has managed energy trade as well as consumption behavior of every MG. The cost model was designed based on economic factors as well as users' willingness. The lesser layer runs in greater frequency as well as adjusts the operations of MGs to reduce the upper and actual layers. By doing so, supply, as well as demand uncertainties and outage events, could be managed appropriately, and energy trade between MGs could be accomplished straightly without an intermediary. They have designed a decentralized interactive algorithm with a realize bilayer IEM framework utilizing the Nash equilibrium idea [10]. In addition, to improve system resilience, the emergency situation was covered.

Zeng et al. [11] have presented a completely distributed operational optimization of MEMS by great diffusion of renewable energy as well as demand response. The iterative algorithm of a better strategic response to get Nash equilibrium of game was presented. Finally, to verify efficiency and viability, the MG approach was simulated. The importance of the manuscript is as below. Initially, the schema, as well as the algorithm, is fully distributed. Second, the optimization process was open and dynamic. Third, consistency among individual rationality and general significance was ensured.

2.1 Background of the Research Work

Evaluation of current investigation operation displays, which energy management of hybrid renewable energy storage devices to MG. As energy management, dissimilar renewable devices becomes incorporated along to diesel power generator. At the energy management system, determining power generations is very hard. Energy source management is done through an energy supervisory system and in charge of the control energy sources of the renewable energy system as well as cost factors involved in the issue. Conventionally, many techniques, like fuzzy, neuro-fuzzy, optimization algorithms, are used MG energy management systems. By utilizing a fuzzy logic controller, it provides the best outcomes yet does not typify the single nature of fuzzy systems theory. On the contrary, PSO has shown to contain superior global search capabilities. Though, in the PSO algorithm, the velocity equation has stochastic variables; consequently, the better overall value varies indecisively. Consequently, the control methods of a renewable energy system are primarily intended to track power demand and regulate DC bus voltage of the connected MG system. So, to overwhelm these challenges, an integrated MG system is necessary to promise solutions. Very few papers depend on methods to solve these issues are introduced on bibliography; this disadvantage, as well as issues, have motivated this investigation work.

3 Mathematical Formulation of Energy Management

The mathematical execution of the energy management optimization problem as per the objective function is depicted in this section.

$$a(y)\colon \min\ f(x) \tag{1}$$

$$f(x) = \sum_{k}^{n} \omega_k(\theta_1(k) + \theta_2(k) + \theta_3(k) + \delta_k) \tag{2}$$

Here, in the energy storage system, the cost function charging and discharging mode is expressed as $\theta_1(k), \theta_2(k), \theta_3(k), \delta_k, \omega_k$. The total production cost is reduced for fulfilling the constraints of generation resources. The below equation was determined by the cost function which is given as follows,

$$\theta_k^{\alpha} = \sum_{i=1}^{n_\alpha} m_k^{i,\alpha}\left(\omega_k^{i,\alpha}\right) \tag{3}$$

$$\theta_k^{\beta} = \sum_{i=1}^{n_\beta} m_k^{i,\beta}\left(\omega_k^{i,\beta}\right) \tag{4}$$

$$\theta_k^{\lambda} = \sum_{i=1}^{n_\lambda} m_k^{i,\lambda}\left(\omega_k^{i,\lambda}\right) \tag{5}$$

By utilizing the Eqs. (3), (4) and (5) shows the dispatchable, non-dispatchable, resources, generated by the cost function of the energy load are determined. Here, $m_k^{i,\alpha}$, $m_k^{i,\beta}$, $m_k^{i,\lambda}$ indicates output power. The *i*th non-dispatchable, dispatchable resources are denoted as $\omega_k^{i,\alpha}\omega_k^{i,\beta}$, $\omega_k^{i,\lambda}$ and the number of non-dispatchable, dispatchable resources load is expressed as $n_\alpha, n_\beta, n_\lambda$. The cost which is devoured by ES is assessed in addition by using the accompanying equation,.

$$\theta_k^{\mu+} = \sum_{i=1}^{n_\mu} m_k^{i,\mu+}\left(\omega_k^{i,\mu+}, \gamma_k^{i,\mu}\right) \tag{6}$$

$$\theta_k^{\mu-} = \sum_{i=1}^{n} m_k^{i,\mu-}\omega_k^{i,\mu-}\left(1-\gamma_k^{\mu}\right) \tag{7}$$

$$\theta_k = m_k^{\tau}\omega_k^{\tau} \tag{8}$$

Here, when the system is encountered with the UP, the cost is expressed as ω_k^{τ} the power which is not supplied by the MG is expressed m_k^{τ}, γ_k^{μ} is indicated as the status of the ES operation mode. When $\gamma_k^{\mu} = 0$ then the ES is in discharging mode. When $\gamma_k^{\mu} = 1$ then, the ES is charging mode [11]. The below equations are evaluated for achieving the objective function. The power balance equation is depicted in the following

$$b:\left(\ell + \lambda^{-} + T\right) = A:\left(E + T + \lambda^{+}\right) \tag{9}$$

The dispatchable, non-dispatchable resources and the load are generated by an energy which is indicated in Eq. (9). For evaluating the objective function, the accompanying Eq. (10) is considered.

$$0 \le \sum_{i=1}^{n_\alpha} m_k^{i,\alpha} \le M_k^{m,\alpha} \tag{10}$$

Here, during the time period, k based on the non-dispatchable sources, the maximum power generation units is expressed as $m_k^{m,\alpha}$. In reference, the rest of the constraints are described.

4 Optimal Energy Management of Smart Grid with IEPC Technique

Photovoltaic, wind turbine, microturbine as well as storage system source is available on MG connected system. To consider power flows among energy sources as well as the grid, the proposed control system is utilized. To maintain grid power as for grid operators and satisfy accessible renewable energy power. The electrical power necessary to grid operators is provided from reference with input MG. Here, batteries are employed to balance as well as allow renewable energy system units to continue to operate in steady as well as stable output power to act as an energy source. In light of the concepts, wind/photovoltaic renewable energy resources are examined in non-dispatchable units, and MT denotes a dispatchable unit. The battery is considered from the energy storage device. The energy management system is solved to the MG system as well as the total generation cost function is analyzed as the use of the IEPC technique. The following section describes the detailed evaluation of the proposed technique.

4.1 Emperor Penguin Colonies Optimization

The biggest species of penguins are called Emperor Penguin. Between 110 and 130 cm is the height of a mature emperor penguin. This height is corresponding to a penguin, which is walk as well as extends its neck. IEPC technique represents the emperor penguins huddling behavior [12]. The important steps of Emperor Penguin Colonies optimization are to create a huddling boundary, around huddling, calculate the temperature, compute distance, get effectual mover. The huddles will hold it's dense for certain hours or certain days. The huddles moving slowly during this time. To provide the MG optimal management as well as acquiring MG, less generation cost IEPC algorithm is utilized. The inputs are assumed from a wind turbine, photovoltaic, microturbine as well as battery power. The main purpose of the IEPC algorithm denotes cost function reduction. In terms of objective, the process of optimal management is defined. The process of IEPC is determined as below,

Step 1: Initialization
First, the power of photovoltaic, wind turbine, microturbine as well as battery is used to initializing in an algorithm.

Step 2: Random Generation
It defines the initial population of penguin as well as matrix is represented as (11),

$$\Gamma \text{ penguins} = \begin{pmatrix} a_{1,1} & a_{1,2} & \cdots & a_{1,n} \\ a_{2,1} & a_{2,2} & \cdots & a_{2,d} \\ \vdots & \vdots & \vdots & \vdots \\ \vdots & \vdots & \vdots & \vdots \\ a_{n,1} & a_{n,2} & \cdots & a_{n,d} \end{pmatrix} \tag{11}$$

where $a_{m,n}$ represents the value of nth penguins mth variable, n represents the amount of penguins population size.

Step 3: Fitness Function
Calculate the fitness of every search agent.

Step 4: Huddling Process
Define the huddling boundary of the emperor penguin and calculating the temperature around the huddle.

Step 5: Updation
Updating every search agent position.

Step 6: Crossover and Mutation
By reorganizing the tunicate swarm location on methods for the consequent updating function utilizing crossover and mutation operator, the solution can be optimized. The updating equations are defined as per the following: Among the two individuals who create a new set of solutions, the crossover rate is achieved.

$$X_{\text{over}} = \frac{\delta}{\kappa} \tag{12}$$

where δ indicates the number of individuals crossover and κ indicates the length of individuals.

Step 6.2: Mutation Operator
The individuals are mutated randomly in light of the specific mutation rate at the process of mutation.

$$Y_{\text{mu}} = \frac{\pi}{\kappa} \tag{13}$$

where π represents mutation point as well as L denotes an individual's length.

Step 7: In a given search space, verify whether any search agent is out of boundary.

Step 8: Updating the optimal solution position which is previously obtained.

Step 9: When stop criteria are met, the algorithm would finish. Or else, move to step 5 (Fig. 1).

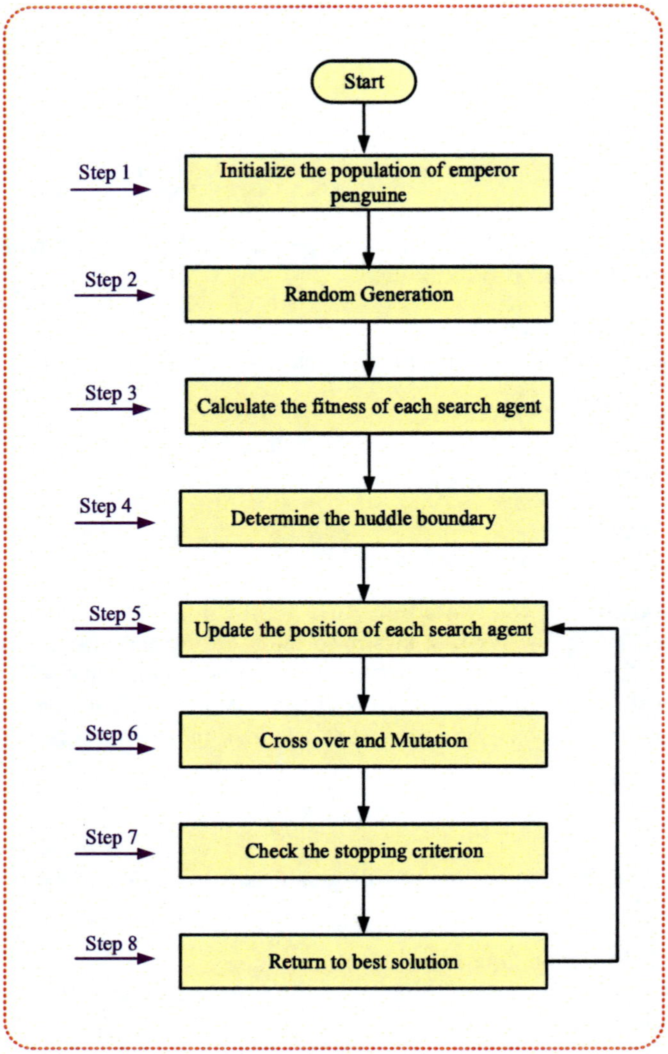

Fig. 1 The flow of the proposed IEPC technique

In the below section the result and discussions for the optimal energy management of SG along less cost with proposed technique are discussed, and the results and the performances of IEPC technique are comparing to other previous methods like GOA as well as ALO.

5 Result and Discussion

Results of simulation and discussion to OEM of SG along less cost with the proposed technique are discussed in this section. OEM means optimal energy management. The SG connected systems are photovoltaic, wind turbine, microturbine, ESS. Here, reducing operational costs and increasing the total generation of sources is performed by the proposed IEPC technique. The IEPC technique will enhance the capability of local search and reduced computational complex as well as reasonable randomness on generation. The performances are executed at the worksite of MATLAB/Simulink to prove the performance of the IEPC technique. The results and the performances of the IEPC technique are compared to other previous methods like GOA as well as ALO.

Figure 2 shows the analysis of demand power. It flows from 8 to 15 kW at the time period of 1–7 h. And it flows to the maximum demand of 17–25 kW in a time period of 10–15 h. Figure 3 depicts the power analysis of proposed with subplots (a) photovoltaic (B) wind turbine (c) microturbine (d) ESS. Here in subplot (a), the PV power varies from 0 to 6 kW at the time period of 7–15 h. In subplot (b) wind turbine flows from high power of 0–7.3 kW, and the in subplot (c) microturbine flows from high power of 0–12 kW, and in subplot (d) the ESS flows the maximum power of 0–3 kW. Figure 4 depicts proposed as well as existing technique analysis. When compared with the existing technique, the SOC of the IEPC technique is greater. Figure 5 depicts the cost comparison of proposed as well as existing technique. Here, from subplot (a) and (b) compared with the previous technique, the cost of the proposed technique is low. Figure 6 depicts the cost comparison of proposed as well as the previous technique. When compared with the existing technique, the cost of the IEPC technique is low.

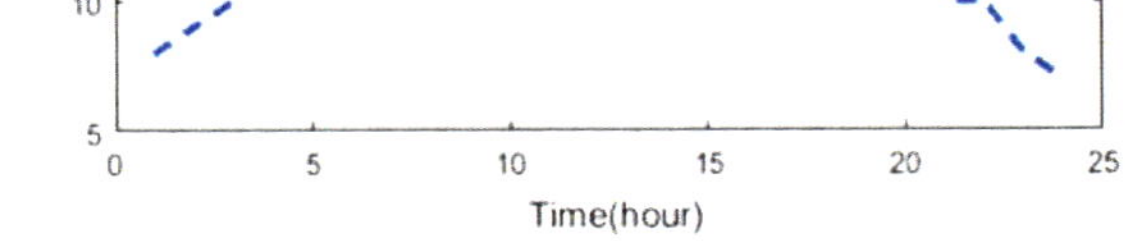

Fig. 2 Analysis of demand power

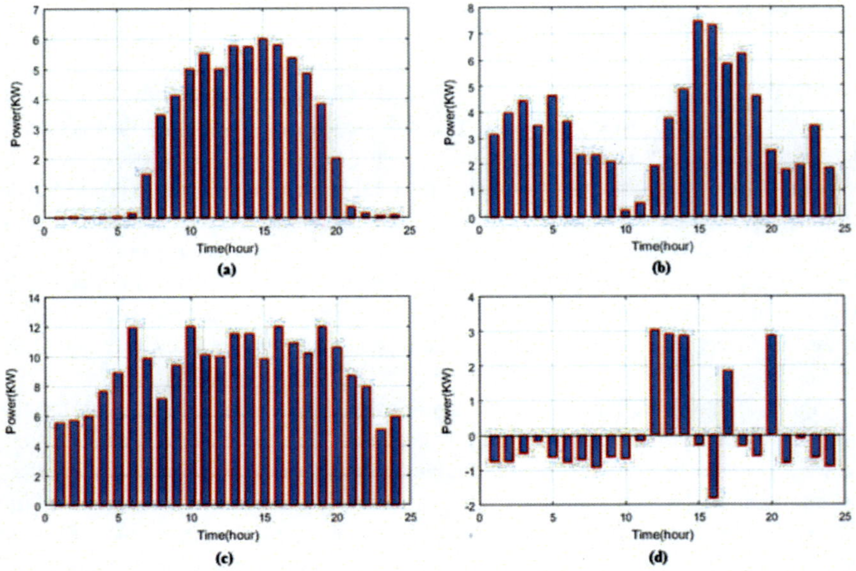

Fig. 3 Power analysis of proposed **a** photovoltaic, **b** wind turbine, **c** microturbine, **d** ESS power

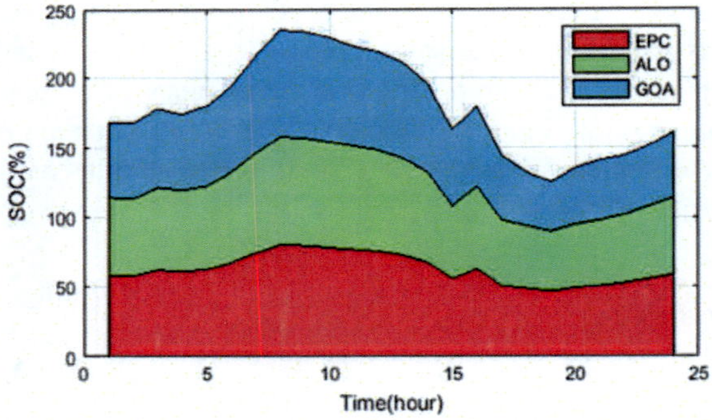

Fig. 4 SOC comparison of proposed and existing technique

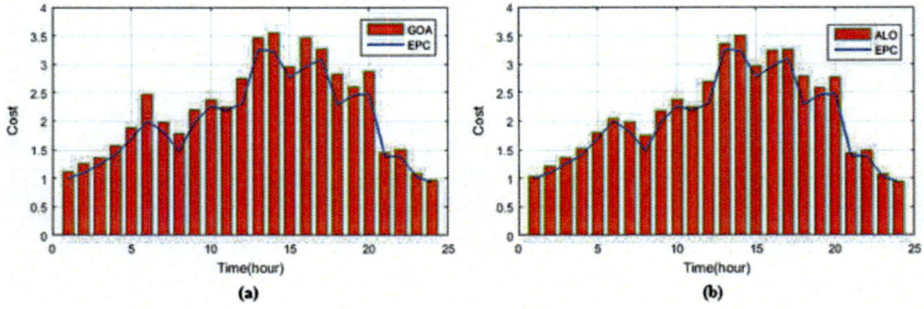

Fig. 5 Cost comparison of proposed as well as existing technique. **a** EPC-GOA. **b** EPC-ALO

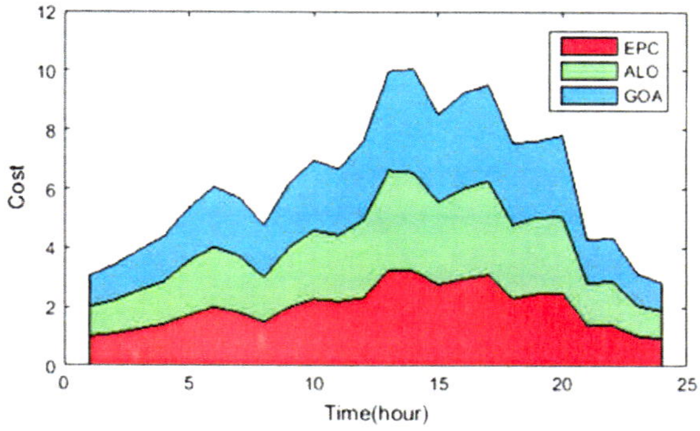

Fig. 6 Cost comparison of proposed as well as existing technique

6 Conclusion

This paper has introduced a process of a grid-connected photovoltaic, wind turbine, microturbine as well as ESS hybrid energy system by using the IEPC technique. The dissertation explores the system modeling as well as micro-grid distribution along with less effort by using the proposed hybrid technique. IEPC hybrid technique selects the micro-grid allocation as represented through load requirement along with minimal fuel cost, replacement cost as well as operating cost. IEPC technique is examined for different load requirement values as well as MG organization and the associated annual total costs are defined. The performance of IEPC technique is examined by using the comparative analysis of the existing methods such as GOA and ALO. The comparison result proves that the IEPC technique is more efficient than the other existing methods.

References

1. Hossain, M., Pota, H., Squartini, S., Abdou, A.: Modified PSO algorithm for real-time energy management in grid-connected microgrids. Renew. Energy **136**, 746–757 (2019)
2. Dong, G., Chen, Z.: Data-driven energy management in a home microgrid based on Bayesian optimal algorithm. IEEE Trans. Industr. Inf. **15**, 869–877 (2019)
3. Roslan, M., Hannan, M., Ker, P., Uddin, M.: Microgrid control methods toward achieving sustainable energy management. Appl. Energy **240**, 583–607 (2019)
4. Nasr, M., Nasr-Azadani, E., Rabiee, A., Hosseinian, S.: Risk-averse energy management system for isolated microgrids considering generation and demand uncertainties based on information gap decision theory. IET Renew. Power Gener. **13**, 940–951 (2019)
5. Yan, J., Menghwar, M., Asghar, E., Kumar Panjwani, M., Liu, Y.: Real-time energy management for a smart-community microgrid with battery swapping and renewables. Appl. Energy **238**, 180–194 (2019)
6. Roy, K., Mandal, K.K., Mandal, A.C.: Ant-lion optimizer algorithm and recurrent neural network for energy management of micro grid connected system. Energy **167**, 402–416 (2019)

7. Liu, G., Jiang, T., Ollis, T.B., Zhang, X., Tomsovic, K.: Distributed energy management for community microgrids considering network operational constraints and building thermal dynamics. Appl. Energy **239**, 83–95 (2019)
8. Liu, Y., Gooi, H.B., Li, Y., Xin, H., Ye, J.: A secure distributed transactive energy management scheme for multiple interconnected microgrids considering misbehaviors. IEEE Trans. Smart Grid **10**(6), 5975–5986 (2019)
9. Yang, X., He, H., Zhang, Y., Chen, Y., Weng, G.: Interactive energy management for enhancing power balances in multi-microgrids. IEEE Trans. Smart Grid **10**(6), 6055–6069 (2019)
10. Islam, S.N., Baig, Z., Zeadally, S.: Physical layer security for the smart grid: vulnerabilities, threats, and countermeasures. IEEE Trans. Ind. Inform. **15**(12), 6522–6530 (2019)
11. Zeng, J., Wang, Q., Liu, J., Chen, J., Chen, H.: A potential game approach to distributed operational optimization for micro grid energy management with renewable energy and demand response. IEEE Trans. Industr. Electron. **66**(6), 4479–4489 (2018)
12. Wang, H.: Sustainable development and management in consumer electronics using soft computation. J. Soft Comput. Paradigm (JSCP) **1**(01), 56 (2019)

Brain–Computer Interfaces and Neurolinguistics: A Short Review

Talal A. Aldhaheri, Sonali B. Kulkarni, and Pratibha R. Bhise

Abstract The brain–computer interface (BCI) domain has grown significantly and emerges as an important field of research and one of the vital fields for understanding and diagnosing the minds from a functional or perceptual viewpoint. The primary goal is to create communication channels between various systems and also with human brain to carry out specific tasks through brain signals without any movement or muscle effort. This paper contains various types of BCI, as it illustrates neural imaging techniques to gain brain signals, and comprehensive survey. All previous studies that were presented in the field of brain–computer interfaces in relation to neurolinguistics have been studied, by knowing that the previous studies on this field are still in their infancy and did not achieve the desired results. This review shows how cognitive neuroscience techniques are used to study language with the aim of showing its advantages to neurolinguistics.

Keywords BCI · EEG · Neurolinguistics · Brain signals · Second language · ERP · Language acquisition

1 Introduction

The development and growth of information and communication technology in this accelerating form make it the major target and the most sensitive research dimension for many researchers. BCI design is the science of controlling computers by using only the brain activity, without any peripheral and muscular movement.

T. A. Aldhaheri (✉) · S. B. Kulkarni · P. R. Bhise
Department of Computer Science and Information Technology, Dr. Babasaheb Ambedkar Marathwada University, Aurangabad, Maharashtra, India
e-mail: talalalthahri@gmail.com

S. B. Kulkarni
e-mail: sonalibkul@gmail.com

P. R. Bhise
e-mail: bhisepratibha@gmail.com

P. Karuppusamy et al. (eds.), *Sustainable Communication Networks and Application*, Lecture Notes on Data Engineering and Communications Technologies 55,
https://doi.org/10.1007/978-981-15-8677-4_54

A number of different monitoring techniques have been used to establish a brain–computer pathway. Many studies have been recently conducted to understand and clarify neurological linguistics and their impact on learning and acquiring foreign languages based on the recording of brain signals. Many of researchers have tested the behavioral performance of brain signal activities related to second language (L2) learning. Be that as it may, what are the activities or changes that occur in the human brain with progress in learning? How might the contrasts in brain's activities are distinguished or changes that clarify the achievements in learning?

The field of science concerning the development of models of the physiological mechanisms related to the language information processing by the human brain is now called as neurolinguistics. It serves to evaluate psychology and linguistic theories together, using aphasiology, brain imaging, electrophysiology, and computer modeling as tools [1]. Because of globalization and openness, many people learn a second and third language and speak it for work, study, travel, or for the purpose of intercultural communication. There are many studies related to bilingualism that show that familiarity with several languages gives the learner the advantages of cognitive in the long run, in addition to the great and varied benefits when compared to learning a single language (mother tongue) [2]. Generally learning and acquiring a first language is easier and more effective than learning and acquiring a second and third language, especially when the second languages are learned late in the life of the learner.

As demonstrated by the most well-known definition, any person who utilizes more than one language routinely can be seen him as bilingual [1]. That individual despite everything can be viewed as bilingual because of the day-by-day utilization of two dialects. At the point when individual lean toward one of the dialects over the other he is viewed as a lopsided or prevailing bilingual. By and large, the ideal language is the first language [3]. As over half of the total populace can be respected bilingual, with the incredible larger part being uneven, an expanded premium is seen in researching the unequal bilingual cerebrum, the portrayal of dialects in it and their communication. The impedance of the primary language or predominant language (L1) while playing out a plain assignment in the later gained second (much of the time more vulnerable) language (L2) was more than once appeared regarding the occasion-related potential (ERP $\pm$ EEG reaction time bolted to the objective improvement) with the components of intuitive interpretation [4, 5]. Functional magnetic resonance imaging (FMRI), electroencephalography (EEG) and functional near-infrared spectroscopy (FNIRS), and other are currently the most common a brain–computer interface non-invasive measurement modalities [6–8].

2 Brain–Computer Interfaces Systems (BCIs)

Brain–computer interfaces systems (BCIs) are a hardware- and software-based communication systems between a human and a computer that use covert brain activities to control devices. This communication pathway can be particularly beneficial

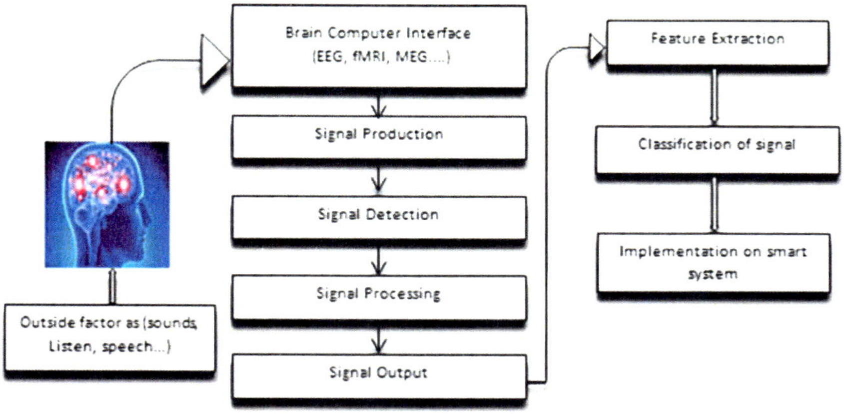

Fig. 1 Stages of the BCI cycle

to individuals with motor impairments as it does not require any motor movement. Therefore, one of the primary goals of BCI research is to provide a communication system for individuals who present as locked-in [9].

In a BCI, data flow through various computational stages, forming the BCI cycle (see Fig. 1). First, the brain activity is captured using one or more measurement modalities. The measurement modality is chosen based on the application of the BCI and also the nature of the activation task. In the next step, the preprocessing stage, the acquired data are prepared for analysis by various artifact and noise removal techniques. This stage is goal to enhance the noise-to-signal ratio to maximize the probability of correct brain activity detection.

After the signal has been acquired and processed to remove artifacts and noise, the discriminative information contained within is identified. Finding discriminative features from brain signals is extremely challenging, as task-irrelevant brain activities produce confounding signals that may overlap in both time and space with signals of interest [9]. In the classification step, also referred to as the machine learning step, the extracted features are used as inputs to a classification model that distinguishes among different mental states or activities. Then, the detected mental states are translated into related commands for controlling an output device, such as a neural prosthetics or wheelchair. The BCI cycle is closed when the user perceives the output. In this way, the user can judge the precision of the identified brain activity and adapt the mental activity accordingly.

2.1 Types of Brain–Computer Interfaces Systems

In this section, we will category brain–computer interfaces systems depending on their types as following:

1. **Invasive/non-invasive BCI systems**: These systems are generally and widely classified as non-invasive and invasive. In invasive kind of brain–computer interface systems, electrodes are embedded on the human's skull. It creates high-resolution signals, high signal-to-noise ratio, proportion yet purposes medical issues. There are different brain–computer interface techniques based on invasive to record the brain's activity, such as electrocorticography (ECoG). On the other hand, the non-invasive kind of brain–computer interface systems includes setting electrodes on scalp for recording of the mind electrical activity. The signal this type has low resolution with high signal-to-noise proportion. Be that as it may, the use of the non-invasive systems is easy and not expensive. Electroencephalography is most widely used non-invasive techniques. So, EEG-based brain–computer interface systems are the most used systems [9–11].
2. **Asynchronous/synchronous BCI systems**: BCI system can be either signal-based (synchronous), for example, at the point when mental activities are activated by outside stimulus or unsignaled (asynchronous), for example, at the point when the user chooses by persistent control that a psychological errand ought to happen and the resultant control signal be created. The primary case is that the EEG must be broke down just in predefined time windows by PC-driven. The last is user-driven and the EEG signals must be analyzed and classified persistently [9–11].
3. **Individual/universal BCI systems**: In the global (universal) brain–computer interface system, the electroencephalography data are gathered from different users to get the most accurate features and the best classification method for each one from users, but in the individual brain–computer interface systems, the electroencephalography data have gathered from one person knowing that no two persons are same. So every BCI system is quite different from the other, Therefore this kind of BCI systems is different with different users [9–11].
4. **Online/offline BCI systems**: The EEG signals are recorded in online BCI systems by using more electrodes, also BCI systems by utilizing more electrodes. These recorded EEG signals are put away and utilized later to build up the: brain–computer interface systems or for real: brain–computer interface researches, where the online systems are the real ongoing working frameworks which give input to client. This is unimaginable in the disconnected BCIs systems [9–11].
5. **Imagery/mental task BCI system**: Brain–computer interface systems for images mainly depend on a variety of different images that are required by the user to execute them brainily as in motion pictures such as the rotation of the geometry. The computer–brain interface system based on the mental task requires various mathematical tasks as in the task of visual counting, etc. [9–11].
6. **Endogenous/exogenous BCI systems**: The exogenous type of BCI system depends on the response of user's brain for the external stimulus and depends on the type and strength of the external stimulus, meaning that this brain depends on the interaction with certain tasks on external stimuli. While the endogenous BCI system depends on the functional and cognitive activity of the user's brain itself, he does not need any external stimuli when carrying out certain tasks [9–11].

Table 1 Techniques of structural brain imaging

Technique	Advantages	Disadvantages
X-ray scan	Easier and cheaper to use than other techniques	Invasive, few information, may cause skin cancer
Angiography	Detailed information and offering therapeutic as well as diagnostic options	Invasive, use X-rays and not readily available
Computed tomography (CT) scan	Non-invasive, widely available, rapid image acquisition and processing	Requires use to X-ray radiation, relatively lower spatial resolution than MRI
Ultrasound	Non-invasive, low cost, fast, used without radiation and not painful	Dependent on operator and patient
Magnetic resonance imaging (MRI)	Good spatial resolution, non-invasive, no use of any X-rays or radioactive materials and not painful	Limited temporal resolution, expensive, indirect measure of brain activity

3 Types of Brain Imaging Techniques

various acknowledged, safe imaging procedures being used today in research centers and specialized medical clinics all through the world which can be categorized as same as [12, 13]:

1. **Structural neuron imaging (anatomy)**: To visualize and analysis of anatomical properties of the brain structure including detection of brain damage and distortions, volume or thickness of a cortical area [12, 13] (Table 1).
2. **Functional (anatomy) neuron imaging**: There are basically two classifications, hemodynamic or physiological and electromagnetic strategy or electrophysiological imaging method [12, 13].

   (i) Electrophysiological or electromagnetic imaging techniques: It indicates the discovery and direct recording of the electrical activity of the brain cell [12, 13] (Table 2).

3. **Molecular imaging techniques**: Used to identify the biochemical activities of cells or atoms (molecules) in human body or animals [12, 13].

   (i) Physiological or hemodynamic imaging techniques: Used to recognize and gauge the changes in brain metabolism [12, 13] (Table 3).

4 Literature Review

In recent decades, the number increased of various researches and studies presented by many specialized scientists and researchers in the field of neurolinguistics and

Table 2 Electromagnetic or electrophysiological brain imaging techniques

Technique	Advantages	Disadvantage
Electroencephalography (EEG) Evoked related potentials (ERP)	High temporal resolution, silent, non-invasive, inexpensive, brain activity can be directly associated to a stimulus or event	No images, only brainwaves, low spatial resolution, and very complex in analysis of acquired data
Magnetoencephalography (MEG)	Non-invasive, without noises and speed temporal resolution	Expensive, poor availability
Electrocorticography (ECoG)	Good resolution, efficiently define conflict areas	Risky and highly invasive

Table 3 Physiological or hemodynamic brain imaging techniques

Technique	Advantages	Disadvantage
Positron emission tomography scan (PET)	Silent, more safer and effective, it detect Alzheimer's, epilepsy, and more	Invasive, high expensive, poor of temporal and spatial resolution
Single photon emission computed tomography (SPECT)	Less expensive, available, faster, and an improved image	Invasive, limited resolution
Functional magnetic resonance imaging (FMRI)	Non-invasive, shows excellent resolution of brain activity	Expensive, poor temporal resolution, loud environment

everything that falls under it in terms of the structural and functional brain structure to understand and acquire one or more languages and in terms of problems and diseases that affect the brain and negatively affect the understanding or pronunciation of languages and their association are neglected. There are many studies that attempt to understand the relationship between the structure and mechanism of the brain's work and between understanding and perceiving languages, including the following:

Akshara Soman et al., in 2019 [14], showed an experimental research to cognizance and understand the main differences that occurred in the representations of neural when the volunteers is presented and tested depending on signals of speech with a familiar and an unfamiliar language. Besides, this research is focused on language learning task and understand range of the develop of neural responses in the brain during this learning of language.

Hashim et al. [15] proposed a model to classify 'yes' and 'no' imagined words. Most of already experiments only focus on a binary classification problem while phoneme, vowel, and syllable or word format in not taken into their consideration. In the proposed work, the writer try to perform classification using 25 syllables classes from co-speech EEG signals. Co-speech means the recording of EEG signals when a volunteer is listening to speech audio.

Sereshkeh et al. [16], mentioned that different researches aimed for build speech visual or auditory by using feedback, selective attention test, and covert speech or the sensorimotor rhythms (SMR) reading. Despite the fact that these researches include component an auditory, usually in the form of auditory feedback, the detection or classification for specifically auditory imagery is not involved. Produce auditory imagery and motor imagery may be another complementary path and might deliver a different approach to achieve communication.

Gonzalez-Castaneda et al. [17] used suitable techniques of frequency transform like bump model or discrete wavelet transform combined with the algorithm of bag-of-words in order to distinguish the different between imagined words, such as (up, down, left, right, select). The EEG signal is treated as a signal of audio (sonification) text sentences consisting of words (textification). 27 subjects are inclusive in this work and the final report of classification was achieved 83.34% accuracy on average.

Herff et al. [18] referenced that brain-to-text system was the principal showing, which translated overt speech discourse goal deduced from broadly appropriated mind territories legitimately into content with words and telephones blunder rates going 25–60% and 50–80% for vocabulary running 10–100 word individually (all essentially above possibility). In spite of the fact that the performance of the brain-to-text system was modest relatively, however, it provides valuable standard to know how performance might develop when uploading vocabulary. Moreover, show examination exhibits that prephonatory activity is used to extend to decoding these above-chance results, according to the authors a raw analog for covert speech decoding intention.

A pair imagined word is chosen to classify individual words using ECoG by Martin et al. [19]. The overall of proposed work achieves 58% of binary classification accuracy. The suggested methods are guided by offline speech decoding research efforts to be used real-time experimental models in covert and overt speech production activities.

A smart learning tool based on the user's eye movement and brain waves is proposed by Jun-Su Kang, in 2015 [20]. By using and analyzing the futures of eye movement and brain waves, the proposed system is able to detect if a given word is known or not to a user while learning a foreign language. After that, the meaning of the given word is searched and is supported users with a vocabulary list of unknown words in real time.

Prat et al. [21] the grammar acquisition process is essentially related to the right functional areas of the brain is featured from these outcomes and are in the same way with recent researches concerning measurements the power of Resting-State-EEG before learning second language. The resting-state-qEEG predictors of second language efficiency was explored and indicated to the power of neural frequency bands of gamma and beta vibrations that bounded essentially in electrodes sites of brain's right hemisphere that was the strongest foreteller of second language learning rates. A role importance of the brain's right hemisphere and first language proficiency is correlated positively based on the concurrence between these results.

da Rocha et al. [22] presented study to collect EEG analysis tools in which help not only to study the concurrent parallel and massive suitable activity supporting human reasoning in details but also to differentiate the cognitive variety tasks one of them is language learning.

In 2016, Wong et al. [23]'s study that related to very young children who are able to take more than one language environment is occurred and noticed that the processing of their brains for the two languages is carried out in similar regions, specialized or linked to language processing operations.

Adescope et al. [24] mentioned that notwithstanding contrasts through language tasks (named as bilingual-signature) in cortical activity, scientists of neuro have affirmed that the monolingual brains are weaker compared to bilingual brain regarding with not only specific language but also more general functions. For instance, working memory improved and greater adaptability strategies in different situations are shown by bilinguals.

Presented a study related to some individuals who were better at learning tasks had bigger volumes of left less successful learners, is shown by Wong et al. in 2016 [23]. Besides, the auditory cortex findings in relation to linguistic HG and to phonetics compared to the pitch learning are convergent too with discoveries of bigger HG volumes in children with better frequency balance location limits, a limit that is known to associate phonological awareness with literacy skills.

Mårtensson et al. [25] examined hippocampal volumes and the cortical thickness of conscript interpreters during period three months of extreme languages studies. The final outcomes showed that uncovered in hippocampus volume and in cortical thickness of the left center frontal gyrus, second rate frontal gyrus, and superior temporal gyrus for translators regarding controls. The left superior temporal gyrus and right hippocampus were fundamentally increasingly flexible in interpreters getting higher capability in the unknown language or foreign.

Bouchard et al. [26–28] presented a study to show decoded discourse spectrograms during obvious discourse from around the STG likewise utilizing action in the high gamma band (70–170 Hz). Both clarified that decoded action can be used to resynthesize discourse with high relationship to the original discourse signal, with the past demonstrating that vowels can be decoded with higher exactness by utilizing anatomical data than should be possible utilizing ordinary strategies.

Four studies with different approach have taken and focused on level of bilingualism with group of children these studies tested what relations between bilingualism and executive function performance with the same children. The first study is longitudinal for children between 24–31 months and it is discovered that largest advantage is achieved by children who became more bilingual over children of monolingual implementing a group from activites of executive function at 31 months [29].

Study's number two was a longitudinal study also, for durationa one year period examed children which old them between 9 and 10 years and showed that level of executive control within the same children is increased by level of bilingualism [30].

Third study, sample of bilingual children who low socio and economic status, their old were 8–10 years are examined, performance on executive function tasks were predicted by degree of bilingualism with no prediction of contributing from any other variable [31].

The final study depended on what know with Immersion Education Programs (IEPs), tested on children who their ages ranged between seven and ten years. The results predicted by the degree of bilingual also show the effect range of performance in executive function tasks [32].

Temporo-parietal region varies in its power of beta and low-gamma frequency and effected by the rate learning of subsequent second language [33]. Accordingly, performance of the brain's right hemisphere might relate to the rate of learning second language. This result is complying with outcomes of concerning individual variations of grammar learning abilities that were collected by FMRI.

A recent study concerning has examined that the long-range temporal correlations (LRTC) and the spectral power that occur in alpha waves by small group of participants are investigated by Mahjoory et al. [34]. Accordingly, they noticed these oscillatory indexes with their corresponding neural origin assessments were correlated positively with memory range and performance it is when calculated from a switch-cost score from the test of attentional performance.

Dubois et al. [35, 36] the significant objective of this work is to decide whether a effective predictors of psychological contrasts between humans in healthy and clinical groups can be given by effective activity of coordinates brain's networks.

Campbell and Schacter [37] cognitive function is effected by aging and disease-related changes factors and might be appeared in functional or structural differences in networks of brain.

McNaughton and Smillie [38] in the operation of understanding the scope and size of actual brain network system cannot help in stable encode but also can assist to find similar contrasts in inspiration, influence, and insight, since quite a while ago contended by character scholars. In this way, this is a significant theme for preceded with examination.

Pérez et al. [39] showing the presence of neural synchrony between two individuals associated with a discussion has just been the initial step, affirmed Alejandro Pérez. 'There are numerous unanswered asks and obastacles left to move.' Pérez imagines that the useful capability of the examination is tremendous. 'Issues with correspondence happen each day.' They are wanting to take advantage of this disclosure of between mind synchronization with the objective of improving correspondence, he finished up.

Casula et al. [40] combined electroencephalogram (EEG) and transcranial magnetic stimulation (TMS) to decide the patterns of cortical rearrangement over the primary motor cortex (M1) and the posterior parietal cortex (PPC) of hemisphere of the brain both affected and unaffected. Transcranial magnetic stimulation which evoked other worldly force, independently for every frequency band, was arrived at the midpoint of among each channel to evaluate the global oscillatory action.

Iljina et al. [41] different researches have tried to remake speech output from brain signals in which opens up the probability of utilizing this procedure as a naturalistic mode of communication, as neural connect of imagined speech are turning into a focused on BCI challenge to be tended to.

Sudaryat et al. [42] presented study that targeted EEG recordings to help for reading different kinds of books such as academic science nonfiction book, general science nonfiction book, and science fiction book and to identify various attention cases. This mechanism can be used to detect the reading skills of students to understand the science. The recording outcomes showed that distinctive brain activities are appeared by using different kinds of books. The distinctions depended on the dominant part of the brain, amplitudes, and frequency patterns.

Dash et al. [43, 44] examined speaker-free neural discourse translating of five persistent expressions from magnetoencephalography (MEG) signals while eight subjects delivered speech covertly (imagination) or overtly (articulation). They have utilized both regulated and solo speaker adjustment procedures for executing a speaker autonomous model. Test results exhibited that the proposed adjustment-based speaker autonomous model has essentially improved deciphering execution. Undoubtedly, this is the primary show of the chance of speaker-free neural discourse deciphering (Table 4).

5 Discussion

In this study, we investigated various acknowledged, safe imaging techniques that were studied in research centers and specialized medical clinics all through the world which associated dynamics of neural activity, especially in the field of neurolinguistics. Previous studies related to knowing the functional and cognitive effects of the brain were identified when learning new languages.

6 Conclusion

Language is one of the most important means of communication between living organisms. The main pillar of brain–computer interface systems is the presence of a clear mechanism for communication between individuals and various artificial intelligence equipments in various fields, depending on the resulting brain signals. Therefore, understanding and acquiring the human language is one of the most important means of communication and interaction between human individuals and smart machines. With this short review, brain–computer interfaces were defined in terms of functional or cognitive terminology and their effectiveness in the field of neurolinguistics, its different types, and neuroimaging techniques were available to acquire brain signals that are used to analyze and understand the human language. This review was addressed in previous studies on the mechanism of the brain while learning or speaking any language, and research studies have been reviewed on the effects associated with learning and acquiring new languages.

Table 4 Summary of literature review

Author and year	Type of speech	Recording technique	Data acquisition device	Data collection	Classifier
Soman et al. (2019) [14]	Overt	EEG	By an BESS F-32 amplifier consists from 32 passive electrodes	17 Indian participants from three English, Japanese, and Hindi languages	Support vector machine (SVM) and linear kernel function
Hashim et al. (2018) [15]	Imagined	EEG	By Emotiv Epoc sensor device consists of 14-channels	Four volunteers have agreed to participate as subjects, all healthy males	k-NN algorithm
Sereshkeh et al. (2017) [16]	Imagined	EEG	BrainAmp consists of sixty-four active electrodes and DC amplifier	By 12 able-bodied fluently English-speaking	Multilayer perceptron neural networks (MLPNN)
Gonzalez-Castaneda et al. (2017) [17]	Imagined	EEG	High-resolution channels (14 electrodes) and the fast fourier transform (FFT)	Data were collected of the EEG signals from 27 individuals whose native language is Spanish	SVM Naive Bayes is a probabilistic classifier (NB)
Herff et al. (2017) [18]	Overt	ECoG	g.USB amplifier (Guger Technologies, Austria) it consists of 16 channel	Data were collected from a 42-year-old female patient	Complex linear model
Martin et al. (2016) [19]	Overt/imagined	ECoG	Via device its name as g.USBamp consists of 16 electrode	From five patients with epilepsy by implanted electrodes in their brain	Support vector machine (SVM) and Gaussian kernel

(continued)

Table 4 (continued)

Author and year	Type of speech	Recording technique	Data acquisition device	Data collection	Classifier
Kang et al. (2015) [20]	Imagined	EEG	Tobii 1750 eye-tracker to detects eye movement of the user, Brainno device (2 channels)	Data were collected of sixty-three participants in English word exam	Support vector machine (SVM)
Prat et al. (2016) [21]	Overt	qEEG	Wireless EPOC 16-channel headsets (Emotiv, Australia)	By sixteen participants speaking (French–English)	Mini-mental state examination (MMSE)
da Rocha et al. (2015) [22]	Overt	EEG (sLORETA)	Twenty electrodes were used based on the 10–20 international system	EEG's data were collected of forty-one students adults speaking (native Portuguese)	Principal component analysis (PCA)
Mårtensson et al. (2012) [25]	Overt	MRI	By scanner includes 32-channel (UFBI)	From thirty-one interpreters and controllers	SPSS 3D Gaussian kernel
Bouchard et al. [26–28]	Overt	ECoG	256-channel were putted on perisylvian cortex	Data were collected by four participants have epilepsy, they speaking (English)	L1-regularized linear regression
Barac et al. (2016) [32]	Overt	ECoG	64 active electrodes were used recording (BioSemi Active 2)	Get on data were from sixty-two children speaking English and some them another language also	Event-related potentials (ERP)
Dubois et al. (2018) [35, 36]	Overt	Resting-state fMRI	Via monitoring the spread of blood in brain	Data were collected of public data named HCP of 884 participants	Principal component analysis (PCA) and general linear model (GLM)

(continued)

Table 4 (continued)

Author and year	Type of speech	Recording technique	Data acquisition device	Data collection	Classifier
Pérez et al. (2017) [39]	Overt	EEG	Brain products GmbH includes 32 electrodes, also amplifier	EEG's data were collected of thirty participants speaking (Native Spanish)	Multiple linear regression model (MLRM)
Sudaryat et al. (2019) [42]	Overt	EEG	Four EEG channels were used for recording by a maximum impedance of 15Ω under 10–20 international system	Data were collected using open brain–computer interface and involved sixteen respondents speaking (English)	Power spectral density (PSD) analysis
Dash et al. (2018) [43, 44]	Overt	MEG	Elekta-Triux -Neuromag MEG includes 306 channels	By four participants recording five English phrases	Long short-term memory recurrent neural network (LSTM-RNN) and DNN
Yoshimura et al. (2016) [45]	Imagined	EEG/fMRI	Thirty-two electrodes were used based on the 10–20 international system	Data were collected of ten healthy human (by two Japanese vowels)	Sparse logistic regression (SLR)

References

1. Nakai, Y., Jeong, J.W., Brown, E.C., Rothermel, R., Kojima, K., Kambara, T., Shah, A., Mittal, S., Sood, S., Asano, E.: Three- and four-dimensional mapping of speech and language in patients with epilepsy. Brain **140**(5), 1351–1370 (2017). https://doi.org/10.1093/brain/awx051
2. Bialystok, E., Barac, R.: Emerging bilingualism: dissociating advantages for metalinguistic awareness and executive control. Cognition **122**, 67–73 (2012). https://doi.org/10.1016/j.cognition.2011.08.003
3. Abutalebi, J.: Neural aspects of second language representation and language control. Acta Physiol. (Oxf) **128**(3), 466–478 (2008)
4. Shi, S.J., Lu, B.L.: EEG signal classification during listening to native and foreign languages songs. In: 4th International IEEE/EMBS Conference on Neural Engineering, pp. 440–443 (2009)
5. Abutalebi, J., Green, D.: Bilingual language production: the neurocognition of language representation and control. J. Neurolinguistics **20**(3), 242–275 (2007)
6. Michel, C.M., Koenig, T.: EEG microstates as a tool for studying the temporal dynamics of whole-brain neuronal networks: a review. NeuroImage **180**, 577–593 (2018). https://doi.org/10.1016/j.neuroimage.2017.11.062
7. Milz, P., Pascual-Marqui, R.D., Achermann, P., Kochi, K., Faber, P.L.: The EEG microstate topography is predominantly determined by intracortical sources in the alpha band. NeuroImage **162**, 353–361 (2017). https://doi.org/10.1016/j.neuroimage.2017.08.058
8. Lerman-Sinkoff, D.B., Sui, J., Rachakonda, S., Kandala, S., Calhoun, V.D., Barch, D.M.: Multimodal neural correlates of cognitive control in the human connectome project. NeuroImage **163**, 41–54 (2017). https://doi.org/10.1016/j.neuroimage.2017.08.081
9. Nicolas-Alonso, L.F., Gomez-Gil, J.: Brain computer interfaces: a review. Sensors **12**(2), 1211–1279 (2012)
10. Wolpaw, J.R., et al.: Brain-computer interface technology: a review of the first international meeting. IEEE Trans. Rehabil. Eng. **8**(2), 164–173 (2000)
11. Kaur, M., Ahmed, P., Rafiq, Q.M.: Analyzing EEG based neurological phenomenon in BCI systems. Int. J. Comput. Appl. **57**(17), 40–49 (2012)
12. Zhuang, N., Zeng, Y., Tong, L., Zhang, C., Zhang, H., Yan, B.: Emotion recognition from EEG signals using mult idimensional ınformat ion in EMD domain. BioMed Res. Int. **2017**, 09 (2017)
13. Bhise, P.R., Kulkarni, S.B., Aldhaheri, T.A.: Brain computer ınterface based EEG for emotion recognition system: a systematic review. In: Proceedings of the Second International Conference on Innovative Mechanisms for Industry Applications (ICIMIA 2020), pp. 327–334 (2020). https://doi.org/10.1109/icimia48430.2020.9074921
14. Soman, A., Madhavan, C.R., Sarkar, K., Ganapathy, S.: An EEG study on the brain representations in language learning. Biomed. Phys. Eng. Express **5**(2), 025041 (2019)
15. Hashim, N., Ali, A., Mohd-Isa, W.N.: Word-based classification of imagined speech using EEG. İn: International Conference on Computational Science and Technology, pp. 195–204. Springer, Berlin (2017)
16. Sereshkeh, A.R., Trott, R., Bricout, A., Chau, T.: Eeg classification of covert speech using regularized neural networks. IEEE/ACM Trans. Audio Speech Lang. Process. **25**(12), 2292–2300 (2017)
17. González-Castañeda, E.F., Torres-García, A.A., Reyes-García, C.A., Villaseñor-Pineda, L.: Sonification and textification: proposing methods for classifying unspoken words from EEG signals. Biomed. Signal Process. Control **37**, 82–91 (2017)
18. Herff, C., Pesters, D.A., Heger, D., Brunner, P., Schalk, G., Schultz, T.: Towards continuous speech recognition for BCI, pp. 21–9 (2017)
19. Martin, S., Brunner, P., Iturrate, I., Millán, J.D.R., Schalk, G., Knight, R.T., Pasley, B.N.: Word pair classification during imagined speech using direct brain recordings. Sci. Rep. **6**, 25803 (2016)

20. Kang, J.S., Ojha, A., Lee, M.: Development of ıntelligent learning tool for ımproving foreign language skills based on EEG and eye tracker. In: The 3rd International Conference on Human. dl.acm.org (2015)
21. Prat, C., Yamasaki, B., Kluender, R., Stocco, A.: Resting state qEEG predicts rate of second language learning in adults. Brain Lang. **157–158**, 44–50 (2016)
22. da Rocha, A.F., Foz, F.B., Pereira, A.: Combining different tools for EEG analysis to study the distributed character of language processing. Comput. Intell. Neurosci. 865974 (2015). https://doi.org/10.1155/2015/865974
23. Wong, B., Yin, B., O'Brien, B.: Neurolinguistics: structure, function, and connectivity in the bilingual brain. BioMed Res. Int. (2016). https://doi.org/10.1155/2016/7069274
24. Adesope, O.O., Lavin, T., Thompson, T., Ungerleider, C.: A systematic review and meta-analysis of the cognitive correlates of bilingualism. Rev. Educ. Res. **80**(2), 207–245 (2010)
25. Mårtensson, J., Eriksson, J., Bodammer, N.C., Lindgren, M., Johansson, M., Nyberg, L., Lövdén, M.: Growth of language-related brain areas after foreign language learning. NeuroImage **63**, 240–244 (2012)
26. Bouchard, K.E., Conant, D.F., Anumanchipalli, G.K., Dichter, B., Chaisanguanthum, K.S., Johnson, K.: High-resolution, noninvasive imaging of upper vocal tract articulators compatible with human brain recordings. PLoS One **11**(3), e0151327 (2016)
27. Conant, D.F., Bouchard, K.E., Leonard, M.K., Chang, E.F.: Human sensorimotor cortex control of directly measured vocal tract movements during vowel production. J. Neurosci. **38**(12), 2955–2966 (2018)
28. Herff, C., Johnson, G., Diener, L., Shih, J., Krusienski, D., Schultz, T.: Towards direct speech synthesis from ECoG: a pilot study. In: 2016 38th Annual International Conference of the IEEE Engineering in Medicine and Biology Society (EMBC), pp. 1540–1543 (2016)
29. Crivello, C., Kuzyk, O., Rodrigues, M., Friend, M., Zesiger, P., Poulin- Dubois, D.: The effects of bilingual growth on toddlers executive function. J. Exp. Child Psychol. **141**, 121–132 (2016). https://doi.org/10.1016/j.jecp.2015.08.004
30. Riggs, N.R., Shin, H.S., Unger, J.B., Spruijt-Metz, D., Pentz, M.A.: Prospective associations betwseen bilingualism and executive function in Latino children: sustained effects while controlling for biculturalism. J. Immigr. Minor. Health **16**, 914–921 (2014). https://doi.org/10.1007/s10903-013-9838-0
31. Thomas-Sunesson, D., Hakuta, K., Bialystok, E.: Degree of bilingualism modifies executive control in Hispanic children in the US. Int. J. Bilingual Educ. Bilingualism (in press)
32. Barac, R., Moreno, S., Bialystok, E.: Behavioral and electrophysiological differences in executive control between monolingual and bilingual children. Child Dev. **87**, 1277–1290 (2016). https://doi.org/10.1111/cdev.12538
33. Kepinska, O., de Rover, M., Caspers, J., Schiller, N.O.: On neural correlates of individual differences in novel grammar learning: an fMRI study. Neuropsychologia **98**, 156–168 (2017). https://doi.org/10.1016/j.neuropsychologia.2016.06.014
34. Mahjoory, K., Cesnaite, E., Hohlefeld, F.U., Villringer, A., Nikulin, V.V.: Power and temporal dynamics of alpha oscillations at rest differentiate cognitive performance involving sustained and phasic cognitive control. NeuroImage **188**, 135–144 (2019). https://doi.org/10.1016/j.neuroimage.2018.12.001
35. Dubois, J., Adolphs, R.: Building a science of individual differences from fMRI. Trends Cogn. Sci. **20**, 425–443 (2016). https://doi.org/10.1016/j.tics.2016.03.014
36. Dubois, J., Galdi, P., Han, Y., Paul, L.K., Adolphs, R.: Resting-state functional brain connectivity best predicts the personality dimension of openness to experience. Pers. Neurosci. **1**(e6) (2018). https://doi.org/10.1017/pen
37. Campbell, K.L., Schacter, D.L.: Ageing and the resting state: is cognition obsolete? Lang. Cogn. Neurosci. **32**, 661–668 (2017). https://doi.org/10.1080/23273798.2016.1227858
38. McNaughton, N., Smillie, L.D.: Some metatheoretical principles for personality neuroscience. Pers. Neurosci. **1**, e11 (2018). https://doi.org/10.1017/pen.2018.9
39. Pérez, A., Carreiras, M., Andoni, J., Oabeitia, D.: Brain-to-brain entrainment: EEG inter-brain synchronization while speaking and listening. Sci. Rep. **7**(1) (2017)

40. Casula, E.P., Pellicciari, M.C., Ponzo, V.: Cerebellar theta burst stimulation modulates the neural activity of interconnected parietal and motor areas. Sci. Rep. **6**(36191) (2016). https://doi.org/10.1038/srep36191
41. Iljina, O., et al.: Neurolinguistic and machine-learning perspectives on direct speech BCIs for restoration of naturalistic communication. Brain Comput. Interfaces **4**, 186–199 (2017)
42. Sudaryat, Y., Nurhadi, J., Rahma, R.: Spectral topographic brain mapping in EEG recording for detecting reading attention in various science books. J. Turk. Sci. Educ. **16**(3), 440–450 (2019)
43. Dash, D., Ferrari, P., Malik, S., Wang, J.: Overt speech retrieval from neuromagnetic signals using wavelets and artificial neural networks. In: 2018 IEEE Global Conference on Signal and Information Processing (GlobalSIP), pp. 489–493 (2018)
44. Dash, D., Ferrari, P., Malik, S., Wang, J.: Automatic speech activity recognition from MEG signals using seq2seq learning. In: IEEE EMBS International conference on Neural Engineering Global Conference on Signal and Information Processing (GlobalSIP), pp. 340–343 (2018)
45. Yoshimura, N., Nishimoto, A., Belkacem, A.N., Shin, D., Kambara, H., Hanakawa, T., Koike, Y.: Decoding of covert vowel articulation using electroencephalography cortical currents. Front. Neurosci. **10**, 1–15 (2016)

An Efficacious Method for Face Recognition Using DCT and Neural Network

Mukesh Gupta and Deepika

Abstract Face images have close correlation and redundant information, by making the recognition system inefficient and increasing the complexity of the classifier. High information repetition and relationships to confront pictures result in wasteful aspects when such pictures are explicitly utilized for recognition. To reduce information redundancy we used DCT, which preserves only pertinent information of face, and complexity of classifier reduced as the length of feature vector gets reduced and normalization of a feature vector to increase the robustness of the system. Here in this work, a hybrid method of face recognition is presented in which a feature vector constructed using DCT provided to artificial neural networks. For classification, the multilayer perception (MLP) is used by an artificial neural network with a back-propagation algorithm. The recognition rate increased and the result was tested on a benchmark dataset.

Keywords Artificial neural network · Discrete cosine transform · Multilayer perceptron

1 Introduction

Nowadays, the uses of cameras are increasing, and the main reason for that is security. Due to global security concerns, the importance of utilizing biometrics is growing, and facial scans scored the highest compatibility among other characteristics. The last decade has shown dramatic progress in this area, emphasizing human–computer interaction, automated crowd surveillance, and biometric analysis. Face recognition

M. Gupta · Deepika (✉)
Department of Electronics and Communication Engineering, Government Engineering College, Ajmer, India
e-mail: deepikababani05@gmail.com

M. Gupta
e-mail: mg.gupta@gmail.com

P. Karuppusamy et al. (eds.), *Sustainable Communication Networks and Application*, Lecture Notes on Data Engineering and Communications Technologies 55,
https://doi.org/10.1007/978-981-15-8677-4_55

is the biometric identification of a person's face with a collection of faces. Face recognition scenarios can categorize into two ways [1], face verification (or authentication) and face identification (or recognition).

Face Verification: It is one to one process in which query face compared with template face, whose identity claimed. Verification rate versus false accept rate measured to evaluate the performance of the verification system, and an efficient verification system should balance both the rates.

Face Identification: It is one of many processes in which query face compared against all the template faces in the database and the identification of the query face completed by locating the image in the database with the highest similarity with that test image. The features of query image and the database compared and a similarity score found for each comparison. The similarity scores are numerically ranked, and the highest similarity score is first. If the similarity score is higher than the fixed threshold, an alarm is upraised and the system thinks that the query image belongs to the system's database. For evaluating the performance, a number of time system raises the alarm, and it accurately identifies an individual in the database, is called detection or identification rate. The number of times the system raises the alarm for an individual who is not in the database is known as a false detection rate.

In this paper, we address a face identification problem, a basic flow diagram of a face identification problem shown in Fig. 1. A face is three-dimensional objects that subject to varying pose, illumination, and expression, which identified using its two-dimension image; that is why face recognition has substantiated to be extremely difficult to imitate artificially. Performing face recognition, classification, and verification, the first question is how to represent a face so that a classifier can achieve accurate recognition without any computationally expensive. Mainly face recognition approach is classified into two categories: holistic and the feature-based [2, 3]. The holistic based approach used the complete face image using the pixel intensity, while in feature-based technique, facial features are used. In many feature extraction approaches, problems are introduced in localizing the features due to ill-posed, variation in illumination, scale, and orientation. Due to these difficulties, researchers try pixel images directly to neural networks. However, due to high information redundancy and robust correlation present in face images, these approaches give the unfortunate result. To overcome these problems preprocessing step is done to reduce the dimensionality. The selection of dimension reducing algorithm becomes the significant factor in calculating the system performance as there are many feature extraction techniques such as Principal Component Analysis (PCA), Linear Discriminant Analysis (LDA), Independent Component Analysis (ICA), Discrete Cosine Transform (DCT) [3, 4]. Then these facial features are classified using minimum

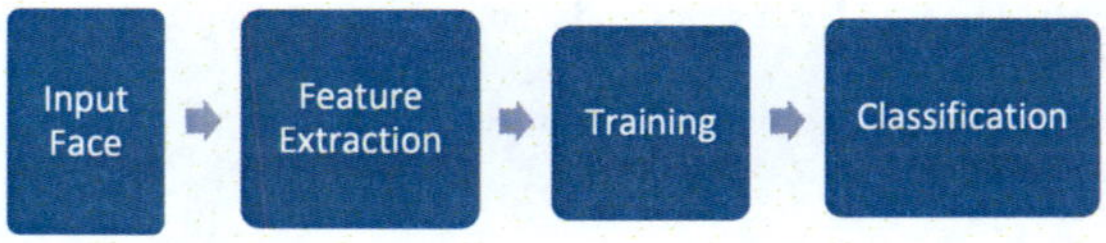

Fig. 1 Face recognition process

distance classifiers and by machine learning algorithms. Research interest in identifying people using their physiological and behavioral characteristics has increased, and face recognition has gained more attention than other characteristics.

The next section of the paper introduces the work that has so far done related to this. Then, our algorithm's methodology is presented. The authenticity contribution is through the application of DCT and the design of a neural network using supervised learning to learn the network and identify the query face. A detailed description of feature selection and network development is provided, followed by the developed methodology's testing result. The paper concludes with some limitations and future scope.

2 Related Work

Increasing the use of biometric systems and among different biometric indicator facial scans is sufficient to attribute. The problem related to face recognition is mature, and researchers consider these problems very challenging to solve. The field matured enough, which demand accurate and fast recognition. As face recognition problem is classified into three necessary steps: feature extraction, feature selection, and classification. Kirby and sirovich were the first to proposed Karhunen-Love-transform to successfully represent the facial images. DCT has a strong ability to data decorrelation and is very close to KLT but has a speedy realization compared to KLT. Initially, DCT was introduced by Ahmed and Rao [5] in the early seventies, and since then, DCT has become a renowned transformation to reduce redundancy and reconstruct the original image from the selected coefficients [2]. As small coefficients or high-frequency components related to noise can truncate, limited numbers of coefficients are enough to preserve most facial features. DCT transforms an image from spatial to a frequency domain, and image analysis in the frequency domain provides powerful means of modeling and removes noise. Another advantage of DCT is that it provides a reasonable settlement between computational complexity and information packing ability. The DCT is an efficient technique for coding face image and successfully applied in many face recognition applications [2, 5, 6]. DCT can be applied to the whole frontal face or on the local features like the nose, eyes,and lips since global and local features have different properties to provide complementary information. Both the features are sensitive to different variation factors such as illumination, orientation, etc. DCT applied to the entire image, but from global features, we select those features which enhance the information of most critical features and then be fed to classifiers so that classifiers can offer the best results. As we are selecting the most common facial features for face recognition, and then classifying the classes using ANN, called it a hybrid method. It offers a potentially enhanced feature vector. Now using this enhanced feature vector machine can possess the ability of generalization efficiently.

By paying attention to different human face recognition methodologies, scientists try to build an artificial face recognition system. The research interest in face recognition using machine learning has grown tremendously since the early 1990s. Joo et al. [7] proposed face recognition using DCT and RBF. They selected most discriminative features using Fisher's linear discriminant. Chen et al. [8] proposed DCT based face recognition to extract features, then a hierarchical radial bias function network is created using a predefined instruction set. Kothani et al. [9] described the structural learning for MLP. Rowley et al. [10] used a neural network for face a detection with detection rate between 77.9 and 90.3%. Chadha et al. [2] used DCT in two groups first applied on the entire face for global features and second for local feature, applied on the left eye, right eye, mouth, and nose. Ourada et al. [11] used neural networks to face recognition, they train their network using Gabor features with PCA and LDA for reducing dimension then classification using distance similarities or using decision boundary. Owajyan et al. [12] used ANN for recognizing different expressions of face images. The system successfully recognized the happy, angry, and neutral expression, and the recognition rate varies from 55 to 80%.

2.1 DCT

DCT transforms an image from spatial to a frequency domain and image analysis in the frequency domain provides powerful means of modeling and removes noise. DCT is used in facial recognition technologies due to its energy compaction property [5]. A function (shape) is described by a series of cosine instead of a series of positions. The DCT of a $N \times M$ image $f(i, j)$ is given by:

$$C(u, v) = \frac{2}{\sqrt{NM}}\alpha(u)\alpha(v)\sum_{i=0}^{N-1}\sum_{j=0}^{M-1} f(i, j)\cos\frac{(2i+1)u\pi}{2N}\cos\frac{(2j+1)v\pi}{2M} \quad (1)$$

where:

$$\alpha(u) = \begin{matrix} \frac{1}{\sqrt{N}}, & u = 0 \\ \sqrt{\frac{2}{N}}, & 1 \le u \le N-1 \end{matrix} \qquad \alpha(v) = \begin{matrix} \frac{1}{\sqrt{M}}, & v = 0 \\ \sqrt{\frac{2}{M}}, & 1 \le v \le M-1 \end{matrix}$$

When reconstructing the image using DCT coefficients, retaining only high magnitude or low-frequency components and rounding off the low magnitude or high-frequency components, still restore the real image using inverse transformation. The DCT transformation of an image is shown in Fig. 2 shows most energy is at the upper left corner. After calculating the DCT coefficients of the face image that extracts the features, zigzag scanning is done. DCT divides an image into three frequency bands: low-frequency band, which correlated with illumination condition, and high-frequency band contains noise and small variation [2]. The middle-frequency region represents the basic structure of the image, and these are useful for recognition. So in the DCT domain, the upper left corner depicts most of the

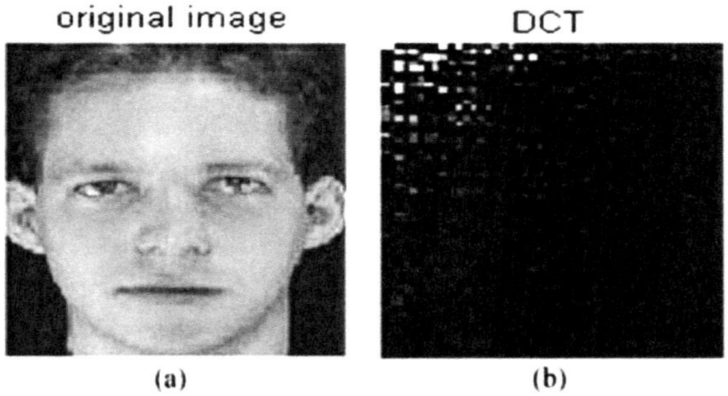

Fig. 2 **a** Face image and **b** its DCT image

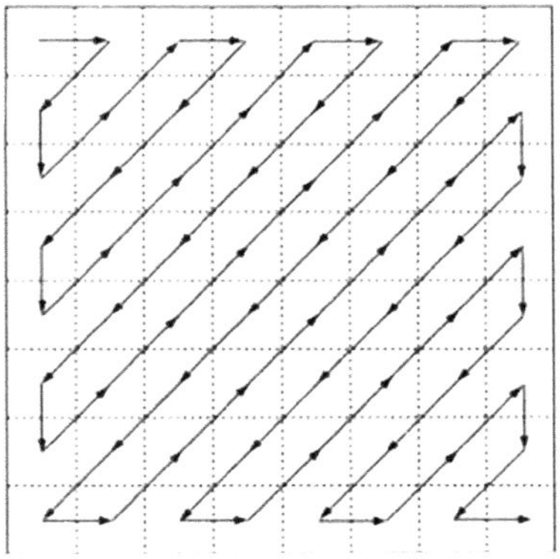

Fig. 3 Zigzag scanning

energy, and these frequency components reflect most of the image's information. According to the explication of DCT and distribution of its coefficients, DC coefficient $c(0, 0)$ represents the average luminance of the image. After removing this coefficient, variation due to illumination can be impressively reduced. Then low frequency features are selected, which changes high-dimensional facial image into low dimension space. More critical facial highlights of the face such as the position of eyes, outline of hair and face preserved. Compare to high-frequency features, these features are more stable and the human visual system is responsive to variation in such features (Fig. 3).

2.2 *ANN Using MLP with Back Propagation Algorithm*

ANN is an archetype of data classification, influenced by the biological nervous system, which consists of a significant number of highly coordinated processing

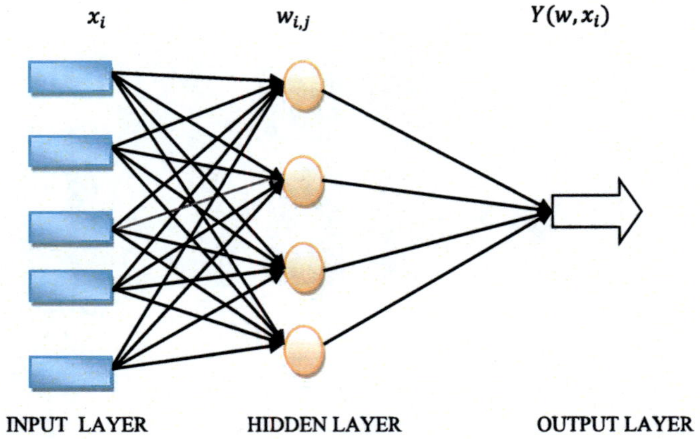

Fig. 4 ANN structure

element neurons functioning in tune to resolve particular problems, learning by example, as a human being. An ANN is constructed for particular tasks such as classification and pattern recognition using the learning process [9, 11, 12]. One such model of ANN is MLP, which consists of three or more layers as its name suggests, a feedforward model, which depicts the input data set to a set of congruous output [13].

An MLP is a fully integrated model in which each node in one layer is linked to every node of the next layer with a certain weight. MLP utilizes back propagation, which is a supervised learning approach to train the network. In this learning, the technique perceptron learned by changing the connection weight, and weight updated by an error calculated by comparing obtained output to the desired output. Since MLP is an entirely connected network, each node of the layer is ultimately linked with each node of the next layer through individual weights, as shown in Fig. 4. As MLP uses supervised learning, the perceptron is learned by changing the connection weight, and weight changed according to the amount of error generated when output is compared with the expected result [9]. For achieving the goal, we have to minimize our cost function that is the mean square error.

$$E = \frac{1}{N}\sum_{i=1}^{N}(d_i - Y(w, x_i))^2 \tag{2}$$

where d is desired output and Y is the output of the system and output predicted as:

$$Y = \sigma(\text{net}) \tag{3}$$

$$\text{net} = \sum_{i=1}^{N} w_i x_i \tag{4}$$

where N is the number of inputs w and x are the weight and input of ith node.

$$\frac{\delta E}{\delta w_i} = \left(\frac{\delta E}{\delta y}\right)\left(\frac{\delta y}{\delta \text{net}}\right)\left(\frac{\delta \text{net}}{\delta w_i}\right) \tag{5}$$

$\frac{\delta E}{\delta w_i}$ denotes the changes in error when weights are adjusted, $\frac{\delta E}{\delta y}$ denotes the changes in error when output is changed, $\frac{\delta y}{\delta \text{net}}$ denotes the changes in output when weighted sum are changed, $\frac{\delta \text{net}}{\delta w_i}$ denotes the changes in weighted sum when weights are changed. The local gradient indicates towards required change in synaptic weights and the updated weight will:

$$w_{i,j}(\text{new}) = w_{i,j}(\text{old}) + \Delta w_{i,j} \tag{6}$$

where:

$$\Delta w_{i,j} = -\eta \frac{\delta E}{\delta w_i}$$

The Back propagation algorithm yields estimation to the trajectory in weight space computed by gradient descent, in which learning parameter η is used. As η is smaller, changes in synaptic weights from one iteration to next will be smaller, and trajectory will be smoother in weight space but at a slow learning rate. Moreover, suppose the learning rate parameter is significant. In that case, it will speed up the learning rate but result in a more significant change in synaptic weight, which will make the network unstable. Modification is done by adding a momentum term as following to increase the learning rate without the danger of instability.

$$\Delta w_n = -\eta \frac{\delta E}{\delta w_n} + \alpha \Delta w_{n-1} \tag{7}$$

where α is momentum constant and n is the learning cycle of weight.

3 Proposed Method

In the face recognition system extracting feature vector, its classification should be adept so that system's overall efficiency will increase. The feature vector contains both global and local features. These prime steps are also listed, which includes extracting feature vector, training, and testing process:

- First DCT was processed on the whole input image. Then selected DCT were coefficients arranged into a linear stream by scanning them in a zigzag manner to make a feature vector.

- We initialized 40 classes for 40 different persons, i.e. one class for each person. There are 10 different images of each person in each class. So a total of 40 target vectors are assigned, each vector representing a different class that means if an image is from one of the 10 images of first-person, then the target vector is 1. If the image is from one of the 10 images of the second person, then the target vector is 2 and so forth.
- The calculated feature vector using DCT is passed to the neural net over the channels with assigned weights and bias.
- Using threshold value, i.e., activation function, activated neuron transmits data over the channel, and in the output layer neuron with the highest probability indicates the output.
- Then, the error is calculated by comparing obtained output with actual output and if the error is beyond limits, it back propagated to the network, and weights are adjusted using this information.
- The previous step is rerun until the stopping condition or the error produced by the system is within limits.
- Once the system is trained, then the system's performance is examined with images in the test data set.
- Then the output is recorded as true or false recognition, and then the recognition rate is determined by the formula:

$$r = \frac{\text{no.of correct recognised face}}{\text{total no.of test images}} \times 100$$

The prime steps of the proposed method are presented by the flow diagram in Fig. 5.

4 Experimental Setup and Results

The process was implemented in MATLAB. The ORL database images selected for training and testing by dividing the database images randomly in distinct ratio and performance is evaluated for respective ratio [14]. The efficiency of the ANN model was evaluated using MSE and recognition rate.

Three-layer ANN model used with one input, one hidden, and one output layer and only a small subset of DCT coefficients are used to form a feature vector. Then testing is done for a distinct pair of testing and training images. The neural network trained with different neurons in the hidden layer, i.e., 50, 70, 90 neurons, and hidden layer with 70 neurons, provides the best result, and in the output layer, 40 neurons taken representing 40 target vector one for each class (face id). The various network parameters are Activation Function: tansig, Training Algorithm: traingdm, Number of Epochs: 5000, Learning Rate: 0.2, Performance Function: mse.

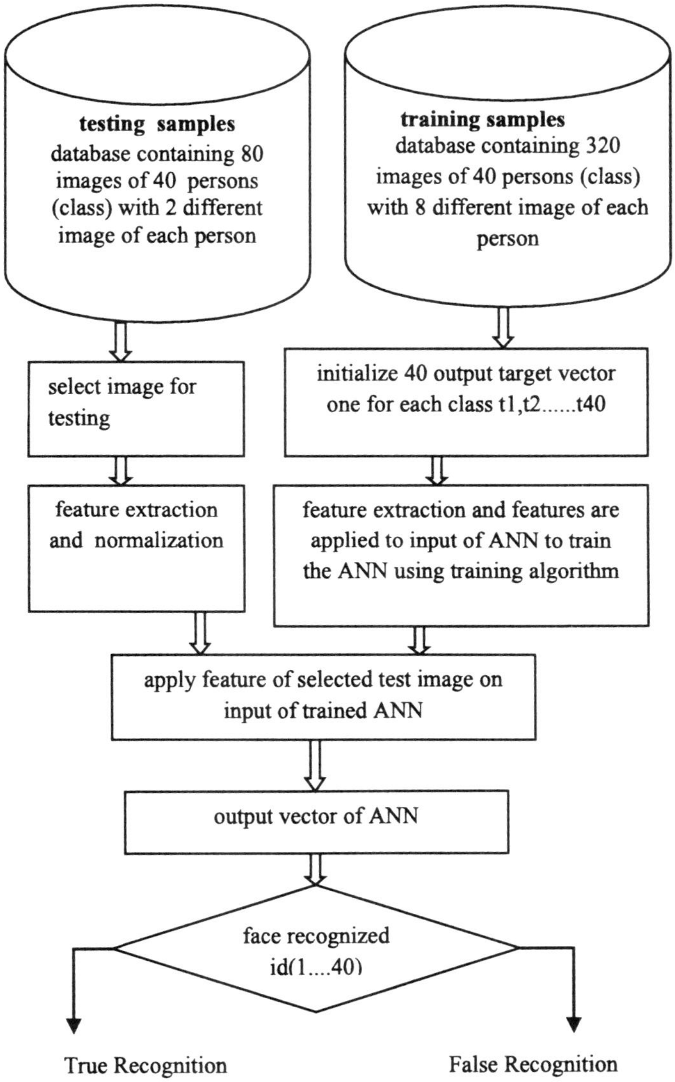

Fig. 5 Training and testing process of ANN

Our proposed method is processed to vindicate its performance on the randomly selected test and training data from the ORL database and the system's performance is plotted against the number of epochs, as shown in Fig. 6.

As we have to validate our system or evaluating the performance of the learning model, we use the test train split approach. In this approach, data randomly divided into test and training sets, then the model is first trained using images of the train data set, and then the test data set is processed for validation purposes. First, we split

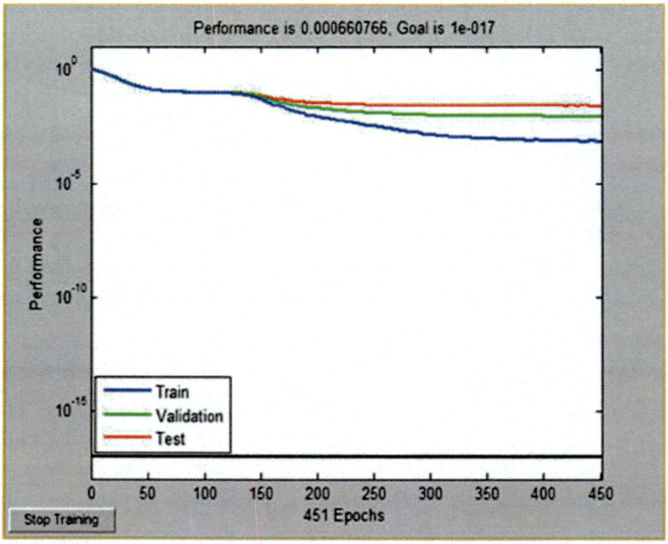

Fig. 6 Performance graph of ANN training process

data into 40:60, which means 160 images as training and 240 images for testing, so we took four images for each class (face id) as training and rested 6 out of 10 for testing. We split data into 50:50, which means 200 images for training and rest 200 for testing as five images from each class as training and rest five for testing. The system provides the best results for 80:20, which means 320 images for training and 80 images for testing or eight images from each class for training and the remaining two images of that class for testing, so there is no overlapping between test and training images. In 80:20 split average recognition rate is achieved 98.75%. As the number of training images increases, the training time will also increase, but the accuracy of the model increases with an increase in training images and for reliable is more considerate.

In this hybrid methodology, the efficiency of the system is validated with 40 classes, and as several classes increases, the recognition rate gradually decreases [8]. We can fix this problem by increasing the number of images of each class in the training data set; our proposed system has been validated for a different number of training images and the accuracy plot for different numbers of training samples of each class presented in Fig. 7. The plot provides the information that the recognition rate gradually increases or the error rate decreases with increasing the training samples per class. The receiver Operating Characteristics (ROC) curve plotted for different prediction cutoff and as shown in Fig. 8, it gives an area greater than 0.9, so we can say our FRS performs well in classifying facial images.

The result of the proposed hybrid method is compared with other popular methods, and our proposed method provides better results, as shown in Table 1.

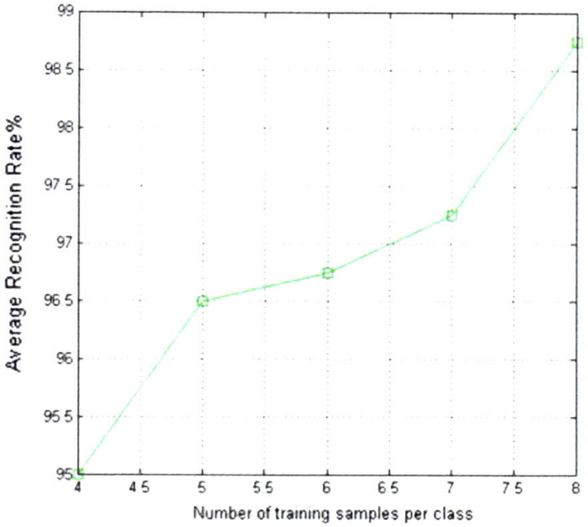

Fig.7 Performance with different number training sample

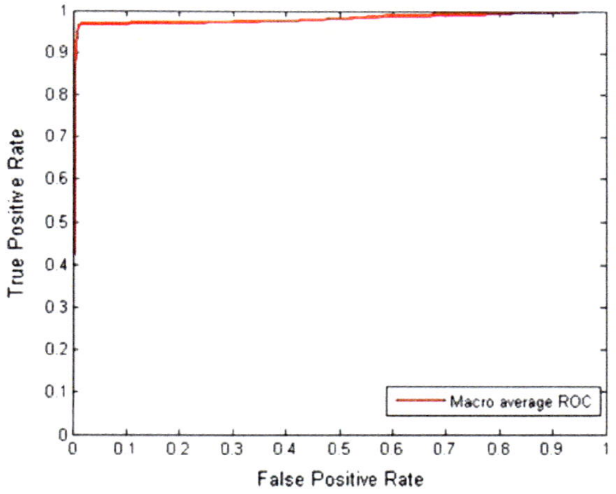

Fig. 8 ROC using macro averaging

5 Conclusion

Our system's recognition rate varies between 95 and 98.75% on varying the number of training samples per class. The feature vector calculated using DCT, which preserves only important features using zigzag scanning and decreases the computational requirements. Our proposed method processed in the ORL database provides better results. The proposed methodology can be processed on other databases to verify

Table 1 Comparison of face recognition on ORL database

S. No.	Methodology	Average recognition rate (%)
1	Global DCT + ANN (Proposed hybrid method)	98.75
2	Convolution NN [15]	96.2
3	NN [7]	94.64
4	DCT + RFNN [7]	97.68
5	Wavlet + PCA + ANN [16]	97.68
6	Weighted modular PCA [16]	87
7	DCT [17]	88.75

the efficiency of the system. This implies that the system is highly efficient and cost-effective and well suited for implementing real-time hardware.

References

1. Lu, X.: Image analysis for face recognition. https://www.cse.msu.edu/~lvxiaogu/publications/ImAna4FaceRcg_Lu.pdf (2003)
2. Chadha, A.R., Vaidhya, P.P., Mani Roja, M.: Face recognition using discrete cosine transform for global and local features. In: International Conference on Recent Advancements in Electrical Electronics and Control Engineering. IEEE, Sivakasi, India, 15–17 Dec 2011
3. Zaho, W., Chellappa, R., Phillips, P.J., Rosenfeld, A.: Face recognition: a literature survey. ACM Comput. Surv. **35**, 399–458 (2003)
4. Kumar, A., Rawat, K., Gupta, D.: An advanced approach of PCA for gender recognition. In: 2013 International Conference on Information Communication and Embedded Systems (ICICES), pp. 59–63. IEEE (2013)
5. Ahmed, N., Natrajan, T., Rao, K.R.: Discrete cosine transform. IEEE Trans. Comput. **C-23**, 90–93 (1974)
6. Azam, M., Almas Anjum, M.: Discrete cosine transformed based face recognition in hexagonal images. In: Proceedings of the 2nd International Conference on Computation and Automation Engineering (ICCAE) 2010, pp. 474–479. IEEE, Singapore, 26–28 Feb 2010
7. Joo, M., Chen, W., Wu, S.: High speed face recognition based on discrete cosine transform and RBF neural network. IEEE Trans. Neural Netw. **16**(3), 679–691 (2005)
8. Chen, Y., Yaou, Z.: Face recognition using DCT and hierarchical RBF model. In: Intelligent Data Engineering and Automated Learning (IDEAL 2006), vol. 4224, pp. 355–362. Springer, Berlin (2006)
9. Bashar, A.: Survey on evolving deep learning neural network architectures. J. Artif. Intell. **1**(02), 73–82 (2019)
10. Rowley, H.A., Baluja, S., Kanade, T.: Neural network based face detection. IEEE Trans. Pattern Recogn. Mach. Intell. **20**(1), 23–28 (1998)
11. Ouarda, W., Trichili, H., Alimi, A.M., Solaiman, B.: MLP neural network for face recognition based on Gabor features and dimensionality reduction techniques. In: Proceedings of International Conference on Multimedia Computing and System (ICMCS 2014), pp. 127–134. IEEE, Marrakech, Morocco, 14–16 Apr 2014
12. Kumar, D., Singh, R., Kumar, A., Sharma, N.: An adaptive method of PCA for minimization of classification error using Naive Bayes classifier. Procedia Comput. Sci. 9–15 (2015)

13. Achkar, R., Owayjan, M.: Control of an active magnetic bearing with multilayer perceptron using the torque method. In: Proceedings of the IEEE Tenth International Conference on Intelligent System Design and Applications, pp. 214–219. Cario, Egypt, Nov 2010
14. Vijayakumar, T.: Neural network analysis for tumor investigation and cancer prediction. J. Electron. **1**(02), 89–98 (2019)
15. Raj, J.S., Vijitha Ananthi, J.: Recurrent neural networks and nonlinear prediction in support vector machines. J. Soft Comput. Paradigm (JSCP) **1**(01), 33–40 (2019)
16. Mazaloom, M., Ayat, S.: Combinational method for face recognition: wavelet, PCA and ANN. In: Digital Image Computing: Techniques and Applications (DICTA 2008), pp. 90–95 (2008)
17. Kumar, A., Gupta, D., Rawat, K.: An advanced approach of face alignment for gender recognition using PCA. In: Babu, B., et al. (eds.) Proceedings of the Second International Conference on Soft Computing for Problem Solving (SocProS 2012), 28–30 Dec 2012. Advances in Intelligent Systems and Computing, vol. 236. Springer, New Delhi (2014)

Author Index

P. Karuppusamy et al. (eds.), *Sustainable Communication Networks and Application*, Lecture Notes on Data Engineering and Communications Technologies 55,
https://doi.org/10.1007/978-981-15-8677-4